International Conference on Polarised Neutrons for Condensed Matter Investigations (PNCMI 2016)

Journal of Physics: Conference Series Volume 862

Munich, Germany
4-7 July 2016

ISBN: 978-1-5108-4295-3
ISSN: 1742-6588

Printed from e-media with permission by:

Curran Associates, Inc.
57 Morehouse Lane
Red Hook, NY 12571

Some format issues inherent in the e-media version may also appear in this print version.

Printed by Curran Associates, Inc. (2017)

For permission requests, please contact the Institute of Physics
at the address below.

Institute of Physics
Dirac House, Temple Back
Bristol BS1 6BE UK

Phone: 44 1 17 929 7481
Fax: 44 1 17 920 0979

techtracking@iop.org

Additional copies of this publication are available from:

Curran Associates, Inc.
57 Morehouse Lane
Red Hook, NY 12571 USA
Phone: 845-758-0400
Fax: 845-758-2633
Email: curran@proceedings.com
Web: www.proceedings.com

TABLE OF CONTENTS

IOP Conf. Series: Journal of Physics: Conf. Series **862** (2017) 011001 doi:10.1088/1742-6596/862/1/011001

Proceedings of the 11th International Conference on Polarised Neutrons for Condensed Matter Investigation (PNCMI 2016)

Preface

The present volume of the *Journal of Physics: Conference Series* represents Proceedings of the 11th International Conference on Polarised Neutrons for Condensed Matter Investigation (PNCMI) held in Freising, Germany from July 4-7, 2016. The conference attended by more than 120 scientists from various academic, government, and industrial institutions in Europe, Asia and the Americas was organized by the Jülich Centre for Neutron Science of the Forschungszentrum Jülich.

The PNCMI-2016 continued the successful previous conferences in this series covering the latest condensed matter investigations using polarised neutrons and state-of-the-art methodologies, from effective polarization of neutron beams to wide-angle polarization analysis, as well as applications for novel instrumentation and experiments, with emphasis on prospects for new science and new instrument concepts.

The conference program included invited and contributed oral presentations and posters which demonstrated the activities using polarized neutrons all over the world and showed the deep interest in developing the topic. The presentations tackled all area of science including multiferroic and chirality, strongly correlated electron systems, superconductors, frustrated and disordered systems, magnetic nanomaterials, thin films and multilayers, soft matter and biology, imaging, as well as further developments in polarized neutron techniques and methods, including nuclear polarisation, Larmor techniques and depolarisation methods.

We would like to thank all speakers for their presentations and all attendees for their participation. We would also like to gratefully acknowledge the financial support by J-PARC and AIRBUS DS as Premium Sponsors and Swiss Neutronics, ISIS, LLB, PSI and Mirrotron as Standard Sponsors of this conference.

The next PNCMI will take place in Great Britain in 2018 and will be organized by ISIS.

Alexander Ioffe (Conference Chair)

Thomas Gutberlet (Conference Secretary)

Jülich Centre for Neutron Science of the Forschungszentrum Jülich

Peer review statement

All papers published in this volume of *Journal of Physics: Conference Series* have been peer reviewed through processes administered by the proceedings Editors. Reviews were conducted by expert referees to the professional and scientific standards expected of a proceedings journal published by IOP Publishing.

Discussion on data correction for Polarization Analysis with a ^{3}He spin filter analyzer

Earl Babcock*[1], Zahir Salhi[1], Emmanuel Kentzinger[2], Stefan Mattauch[1], Alexander Ioffe[1]

[1]Jülich Centre for Neutron Science (JCNS) at Heinz Maier-Leibnitz Zentrum (MLZ), Forschungszentrum Jlich GmbH 85747 Garching, Germany,
[2] Jülich Centre for Neutron Science JCNS and Peter Grünberg Institut PGI, JARA-FIT Forschungszentrum Jülich GmbH, 52425 Jlich, Germany

E-mail: * e.babcock@fz-juelich.de

Abstract. Fully polarized neutron reflectometry and grazing incidence small angle neutron scattering are effective methods to explore magnetic structures on the nm to μm length scales. This paper is an outline of how to fully correct for the polarization analysis (PA) inefficiencies of such an instrument and to determine the error contributions of the neutron polarizer and analyzer. This discussion considers the exact case of the polarization analysis instrumentation used on the MARIA neutron reflectometer at the MLZ or for a general polarized neutron scattering instrument using at least one ^{3}He neutron spin filter that has the capability for adiabatic fast passage nuclear magnetic resonance flipping of the ^{3}He polarization. This paper will work to build a conceptual understanding of how the inefficiencies of neutron polarization elements affect measured data in order to stress and encourage the application of PA corrections and to help perform successful measurements. Then, using data from a fully polarized neutron reflectometer test measurement we show how it is possible to recover signals on the order of, or even smaller than, the inefficiencies, or bleed-through, of the neutron polarization devices used.

1. Introduction

Neutron reflectometry and grazing incidence small angle neutron scattering (GISANS) with polarization analysis allow depth-resolved investigation of the magnetic structures in thin films and mesoscopic structures deposited on a surface. Neutron reflectometry gives access to in-plane averaged information and GISANS to correlations between in-plane variations, if those correlations appear on length scales smaller than the projection of the coherence volume of the beam on the sample surface. In the non-spin-flip (NSF) channels, one has access to the nuclear correlations and to the component of the fields parallel to the beam polarization vector and, in the spin-flip (SF) channels, one has access to the component of the field perpendicular to it. Therefore, by comparing the SF and NSF channels, one obtains vector information on the magnetic fields and their lateral correlations. Such studies are possible on the JCNS magnetism neutron reflectometer, MARIA [1] which uses a ^{3}He spin filter as the analyzer element.

In this paper we will follow the formalism presented by A. Wildes [2, 3] where the matrix method was used to describe the polarization (P), analysis (A), and flipper ($+F$) efficiency correction problem in polarized neutron scattering. Such corrections are implemented in software for polarized instruments at the Institute Laue Langevin and NIST for example but here we look at how such corrections affect measured data to obtain a conceptual understanding also discussing the error sources and propagation. Finally the discussion can be applied to analysis of data from our MARIA reflectometer [1] which is used as an example.

A typical neutron scattering instrument would have a polarizer, P followed by a polarization flipper F before the sample, then another flipper after the sample followed by a polarization analyzer A. This document will solve the equations for two sub-cases of those described in [2, 3], one where the neutron flippers have ideal efficiencies, and two where one flipper is of a known and arbitrary efficiency, but the other flipping efficiency is perfect. The first case is to aide the conceptual understanding and stress the importance of PA corrections, and the second is the actual case for our MARIA reflectometer [1] where the polarization analyzer utilizes adiabatic fast passage reversal of the ^{3}He nuclear polarization [4, 5, 6, 7] which makes A reversible negating the need for the second flipper. For this instance all the parameters of the PA and flipper ($+F$) efficiencies can be explicitly determined with neutron transmission measurements (see Appendix B for details), thus allowing full correction of the measured cross-sections up to statistical and systematic error levels. The resulting discussion can be used as a recipe for PAF data corrections on instruments such as MARIA, including those with possible ^{3}He polarization time-dependence.

In a full PA experiment, one would measure four scattering intensities, I^{00}, I^{01}, I^{10} and I^{11}, corresponding to the four possible combinations of the polarizer and analyser spin state. These intensities are then related to the two spin flip (Σ^{+-} and Σ^{-+}) and two non-spin flip (Σ^{++} and Σ^{--}) cross-sections with the relations derived in this work. From these four cross-sections the proportions of nuclear, magnetic and incoherent scattering can be obtained which can be applied for data from arbitrary scattering/detector geometries [8]. The main purpose of this discussion is both to guide in the practical data procurement and treatment in PA experiments, as well as to aide in a conceptual understanding of the relation between measured intensity data and the limits to the observable cross-section information contained within.

Beginning with the first case we derive the expressions for ideal neutron flipper inefficiencies as this greatly shortens the expressions. Even though not the practical case, the resulting expressions help one to understand how the often more intense non-spin flip cross-sections are being mixed with the much less intense spin-flip cross-sections in the measured intensities. This can be particularly useful to help understand the view on the live display during a measurement, thus giving an idea of when an observed spin-flip signal contains information on the sample cross-sections vs. simply being polarization leakage or "bleed through" of the bright non-spin flip states onto the spin flip states because of the PA inefficiencies. The resulting understanding can then be useful for intuitively performing effective neutron measurements. However as will be shown using a sample data set (see fig. 4), for an instrument such as MARIA even a low incident beam flipper inefficiency can lead to a spurious asymmetry in measured spin-flip intensities because this flipper is active in only two of the total four measured intensities. Therefore to correct the actual data, the discussion will then be extended to include the incident beam flipper inefficiency as well.

Finally, data obtained from a test measurement will be PAF corrected using the before described relations. This data will show visually the effects and robustness of the PAF inefficiency corrections by displaying the measured intensity plots next to the corrected cross-section plots with the resulting error contributions given. From this data it will be clear that in cases where there is no spin-flip scattering cross-sections, such as for saturated magnetic layers, the measured intensities are to first order simply the bleed through from the relative intensities of the non-spin-flip states times the sum of the polarization inefficiencies. As a result the amplitudes will go to zero in the corrected cross-section data, giving us confidence in data that does contain spin-flip scattering cross-sections even if their amplitudes lie below the PAF inefficiencies once the data is appropriately corrected.

IOP Conf. Series: Journal of Physics: Conf. Series **862** (2017) 012001 doi:10.1088/1742-6596/862/1/012001

2. Derivation of the PA correction equations

We use the polarization matrix method sumarized by A. Wildes in [2] to derive the PA corrections. The full matrices for the neutron polarization after scattering are as follows;

$$\begin{bmatrix} I^{00} \\ I^{01} \\ I^{10} \\ I^{11} \end{bmatrix} = \begin{bmatrix} (1-p_1) & 0 & p_1 & 0 \\ 0 & (1-p_1) & 0 & p_1 \\ p_1 & 0 & (1-p_1) & 0 \\ 0 & p_1 & 0 & (1-p_1) \end{bmatrix} \begin{bmatrix} (1-p_2) & p_2 & 0 & 0 \\ p_2 & (1-p_2) & 0 & 0 \\ 0 & 0 & (1-p_2) & p_2 \\ 0 & 0 & p_2 & (1-p_2) \end{bmatrix}$$
$$\times \begin{bmatrix} 1 & 0 & 0 & 0 \\ 0 & 1 & 0 & 0 \\ f_1 & 0 & (1-f_1) & 0 \\ 0 & f_1 & 0 & (1-f_1) \end{bmatrix} \begin{bmatrix} 1 & 0 & 0 & 0 \\ f_2 & (1-f_2) & 0 & 0 \\ 0 & 0 & 1 & 0 \\ 0 & 0 & f_2 & (1-f_2) \end{bmatrix} \begin{bmatrix} \Sigma^{++} \\ \Sigma^{+-} \\ \Sigma^{-+} \\ \Sigma^{--} \end{bmatrix} \tag{1}$$

In the experiment one measures the normalized intensities I^{00}, I^{01}, I^{10}, and I^{11} which are the convolution of the instrument's PA parameters and the scattering from the sample whereas Σ^{++}, Σ^{+-}, Σ^{-+} and Σ^{--} are the physical scattering contributions, or cross-sections themselves. The indices on I refer to the digital on (1) or off (0) state of the neutron flippers, i.e. a setting in the measurement script, whereas the indices on Σ refer to the polarization, or spin, of the neutrons before or after the sample being aligned parallel (+) or anti-parallel (-) to the magnetic holding field. The above matrix gives a set of linear equations that can be solved in terms of the intensities for Σ^{xy} if the flipper inefficiencies, f_1 for the incident beam, f_2 for the scattered beam, and the polarizing inefficiency $p_1 = (1-P)/2$ and the analyzer inefficiency $p_2 = (1-A)/2$ are known. Here P and A are the neutron polarization powers of the polarizer and analyzer respectively. We first solve the above relation for the case of ideal flippers (inefficiency=0) and arbitrary p_1 and p_2. Performing the algebra one obtains;

$$\Sigma^{++} = \frac{I^{00}(1+p_1p_2-p_1-p_2) + I^{01}(p_1p_2-p_2) + I^{10}(p_1p_2-p_1) + I^{11}p_1p_2}{(1-2p_1)(1-2p_2)} \tag{2}$$

$$\Sigma^{+-} = \frac{I^{01}(1+p_1p_2-p_1-p_2) + I^{00}(p_1p_2-p_2) + I^{11}(p_1p_2-p_1) + I^{10}p_1p_2}{(1-2p_1)(1-2p_2)} \tag{3}$$

$$\Sigma^{-+} = \frac{I^{10}(1+p_1p_2-p_1-p_2) + I^{11}(p_1p_2-p_2) + I^{00}(p_1p_2-p_1) + I^{01}p_1p_2}{(1-2p_1)(1-2p_2)} \tag{4}$$

$$\Sigma^{--} = \frac{I^{11}(1+p_1p_2-p_1-p_2) + I^{10}(p_1p_2-p_2) + I^{01}(p_1p_2-p_1) + I^{00}p_1p_2}{(1-2p_1)(1-2p_2)} \tag{5}$$

In these equations one can see the appearance of the incident polarization, $P = (1-2p_1)$, and the analyser efficiency $A = (1-2p_2)$, where the factor of two in these relations comes from the definition of a polarization (A and P) vs. the definition of the polarization inefficiencies used in eq. (1) which are simply the fraction of spins oriented the "wrong" direction [9]. Thus the data are normalized to the total instrumental analyzing power, and the factors multiplied by each measured PA state show how the mixed spins of the neutron beam contaminates the measured data. For a conceptual understanding of the importance of these corrections, let us assume $\Sigma^{++} \simeq \Sigma^{--}$, $\Sigma^{+-} \simeq \Sigma^{-+}$ and that the analyzing power and polarizing power are matched, i.e. $p_1 \simeq p_2$. Note, cross-section information in the analyzed data for the spin flip states diverges to 0 for a measured asymmetry of $I^{01}/I^{00} \simeq F_{\mathrm{SF}} \simeq ((I^{11}/I^{00})p_1 + p_2)/(1-p_1-p_2)$ where

IOP Conf. Series: Journal of Physics: Conf. Series **862** (2017) 012001 doi:10.1088/1742-6596/862/1/012001

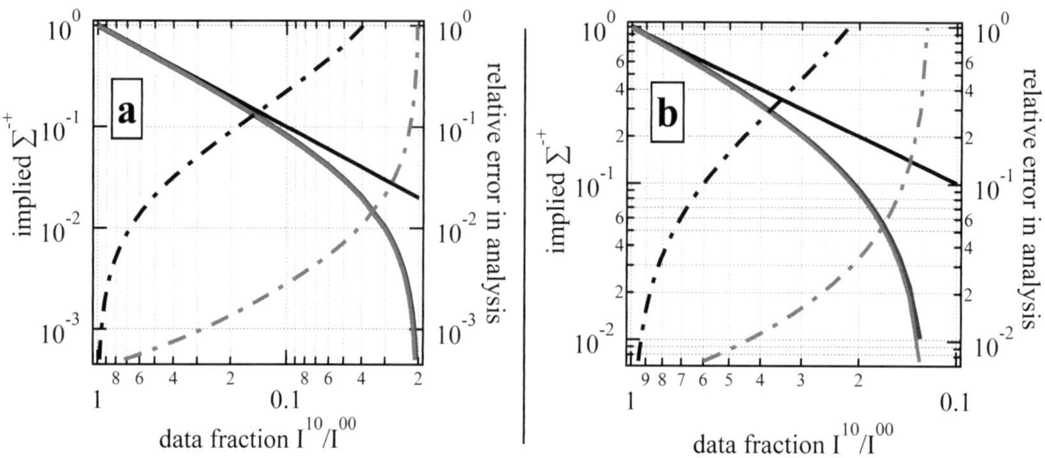

Figure 1. Influence of the the PA corrections on the observed signals for (a) balanced polarizer analyzer inefficiency of 98% or $p_1 = p_2 = 0.01$ and (b) for imbalanced PA with analyzing inefficiency of 80% or $p_2 = 0.1$ and a polarizer inefficiency of $p_1 = 0.01$. This shows that PA corrections become important for scattering at the level of the sum of the polarization inefficiencies. Left axis: the black line is the fraction of the measured signal I^{10}/I^{00}, the blue lines are the relative cross-section Σ^{-+} and the red line is the partially corrected signal where $^*\Sigma^{-+} = I^{10} - I^{00}p_2 - I^{11}p_1$. Right axis: the dot-dash lines are the respective relative error in assuming that the black (uncorrected intensities line) or red lines (first order corrected lines) represent the cross-section.

$F_{\rm SF}$ is the relative apparent spin flip scattering fraction of a measured feature for a given (α_i, α_f) pair compared to the non-spin-flip scattering where α_i is the neutron angle of incidence on the sample and α_f is the final angle after reflection from the sample. For practical cases this implies the error diverges for I^{01} or $I^{10} \leq p_1 + p_2$. Thus, for example if $p_1 = p_2 = 0.01$, a rather acceptable 98% polarization and analyzing power, then the spin flip cross-section data in the I^{01} or I^{10} channels is invisible for $I^{10} \simeq I^{01} \simeq 2\% \simeq p_1 + p_2$ if PA corrections are not made. As one can see in fig. 2 (a), for ratios of signal intensities of $0.1 \simeq I^{10}/I^{00}$ the mixing of the wrong spin states already leads to high error. One may be inclined to do a first order correction ignoring the crossed terms containing the p_1p_2 or p_x^2 terms, however in fig. 2 (a) one can see that using the approximation $^*\Sigma^{-+} \simeq I^{01} - I^{00}p_2 - I^{11}p_1$ still only allows recovery of signals on the order $F_{\rm SF} \simeq 2\%$ where the error diverges, and further this process actually over subtracts the spin mixing from the data. Fig. 2 (b) shows the same plots but for vastly imbalanced PA using a spin filter with $p_2 = 0.1$ and a polarizer with $p_1 = 0.01$ corresponding to a polarizer efficiency of 98% and an analyser efficiency of around 80%, nevertheless one arrives at the same conclusion. This shows conceptually that if one needs to observe spin flip signals overlapping strong non-spin-flip signals in Q space, i.e. occurring at the same (α_i, α_f), the corrections become critical at the level of $p_1 + p_2$ and therefore must be known to a good statistical accuracy. Conversely for intense spin-flip signals above this level, or for well resolved spin-flip and non-spin-flip signals in Q space, the first order correction when performed properly is readily sufficient to obtain up to 1% systematic accuracy in the processed data.

IOP Conf. Series: Journal of Physics: Conf. Series **862** (2017) 012001 doi:10.1088/1742-6596/862/1/012001

3. Full PA correction equations with an imperfect incident beam flipper, $+F$

The equations provided up to this point help one understand how PA inefficiencies affect measured intensity data, but they are not sufficient to correct measured data from an instrument with a non-perfect f_1 flipper inefficiency. Therefore solving eq. (1) assuming necessary corrections for p_1, p_2 and f_1, which is the actual case for MARIA, the above relations become;

$$\begin{aligned}\Sigma^{++} &= \frac{I^{00}((1+p_1p_2-p_1-p_2)(1-f_1)-f_1(p_1p_2-p_1))}{(1-2p_1)(1-2p_2)(1-f_1)} \\ &+ \frac{I^{01}((p_1p_2-p_2)(1-f_1)-p_1p_2f_1)+I^{10}(p_1p_2-p_1)+I^{11}p_1p_2}{(1-2p_1)(1-2p_2)(1-f_1)}\end{aligned} \tag{6}$$

$$\begin{aligned}\Sigma^{+-} &= \frac{I^{01}((1+p_1p_2-p_1-p_2)(1-f_1)-f_1(p_1p_2-p_1))}{(1-2p_1)(1-2p_2)(1-f_1)} \\ &+ \frac{I^{00}((p_1p_2-p_2)(1-f_1)-p_1p_2f_1)+I^{11}(p_1p_2-p_1)+I^{10}p_1p_2}{(1-2p_1)(1-2p_2)(1-f_1)}\end{aligned} \tag{7}$$

$$\begin{aligned}\Sigma^{-+} &= \frac{I^{10}(1+p_1p_2-p_1-p_2)}{(1-2p_1)(1-2p_2)(1-f_1)} \\ &+ \frac{I^{11}(p_1p_2-p_2)+I^{00}((p_1p_2-p_1)(1-f_1)-f_1(p_1p_2-p_1))+I^{01}(p_1p_2(1-f_1)+p_2f_1)}{(1-2p_1)(1-2p_2)(1-f_1)}\end{aligned} \tag{8}$$

$$\begin{aligned}\Sigma^{--} &= \frac{I^{11}(1+p_1p_2-p_1-p_2)}{(1-2p_1)(1-2p_2)(1-f_1)} \\ &+ \frac{I^{10}(p_1p_2-p_2)+I^{01}((p_1p_2-p_1)(1-f_1)-f_1(p_1p_2-p_1))+I^{00}(p_1p_2(1-f_1)+p_2f_1)}{(1-2p_1)(1-2p_2)(1-f_1)}.\end{aligned} \tag{9}$$

From analogy to the discussion where the flipper inefficiency was neglected, the result of this correction will again only become important when $f_1 \simeq F_{\mathrm{SF}}$. Since a typical RF gradient flipper as used on MARIA can have an inefficiency on the order of 1×10^{-3} the corrections should be included to obtain a dynamic range beyond this level.

4. Examples from PA corrections on polarized reflectometry data

We now apply the results of section 3 to correct the data from a test measurement of a $\mathrm{Fe}^{nat}\mathrm{Fe}^{58}$ superlatice grating. The grating produced in KU Leuven and on loan from the ILL was a superlatice of 10 repeats of 10 nm layers of natural abundance Fe followed by 10 nm of isotopically enriched Fe^{58} that was sputtered on a 2 cm by 2 cm Si substrate. This superlatice was then etched with photo-resist methods to print a grating with a period of 10 μm on it. This sample was then installed in the vertical sample magnet on MARIA where the detector arm scans the scattering in the horizontal plane with the grating lines also positioned vertically.

The MARIA instrument is a polarized neutron reflectormeter with a horizontal-plane scattering geometry that uses a $\Delta\lambda/\lambda = 10\%$ neutron velocity selector to choose the neutron wavelength [1]. The incident beam is polarized with a super-mirror and the incident beam flipper is a RF-gradient neutron flipper [6]. After the guide exit approximately 20 cm before the sample position, the neutron flight path was in air up to the position sensitive detector placed 200 cm after the sample position on the detector arm. For this experiment the analyzer was a 12 cm I.D. ^{3}He neutron spin filter cell called "Idafix" [10] that is 8 cm long with 1.16 bar neutron calibrated ^{3}He pressure and a 380 hour ^{3}He lifetime in ideal laboratory conditions. The cell was polarized off-line and placed about 60 cm after the sample inside an 80 cm long permanent

IOP Conf. Series: Journal of Physics: Conf. Series **862** (2017) 012001 doi:10.1088/1742-6596/862/1/012001

magnet μ-metal cavity producing a uniform magnetic field transverse to the neutron beam and sample magnet. This magnetic cavity had a rectangular solenoid RF-coil installed in it for the AFP flipping of the ^{3}He [6, 7].

For this test the sample was installed in a vertical electromagnet capable of up to 1.1 T with the 5 cm gap pole shoes we used. The detector in this position gives us an angular resolution of around 0.3 mRad. For this measurement we chose to make α_i steps of 0.9 mRad as a balance between resolution and the acquisition time required for reasonable statistics. Data was recorded for each α_i line of an $I^{\chi\psi}$ plot for 6 min, then the incident beam flipper was switched states to record the α_i line for the other flipper state where the superscripts χ and ϕ refer to the state (i.e. 1 or 0) of the incident beam polarization and analyzer respectively. After a full α_i scan for I^{00} and I^{10}, the ^{3}He AFP flipper was used to reverse the ^{3}He polarization and the detector arm was returned to the starting position and the α_i scan is repeated for the I^{01} and I^{11} plots. For this particular test experiment the ^{3}He spin filter was polarized offline in the JCNS spin-exchange optical pumping (SEOP) laboratory [11] and therefore had a time-dependent polarization. Since each pair of intensity plots took around 12 hours to acquire, the intensity data is corrected for the time dependent neutron transmittance of the ^{3}He spin filter and the PAF corrections also use its time dependent A to solve for the corrected cross-sections.

Both the neutron analyzing power and transmittance of the ^{3}He neutron spin filter cell are functions of the ^{3}He polarization P_{He}, and neutron wavelength λ (see eqs. (14) and (15) in Appendix B) and can be explicitly calibrated with transmittance measurements. Then using the determined ^{3}He cell parameters, the remaining P and F efficiencies can be obtained by measuring the instruments 2 flipping ratios $\mathcal{F}^{\text{on}}$ and $\mathcal{F}^{\text{off}}$ using the ^{3}He APF flipper to reverse the analyzer polarization with the incident beam flipper on or off respectively. Further possible time changing ^{3}He polarization, which normally follows an exponential decay, can be determined by repeating the transmittance calibration measurements periodically during the experiment. The result is all of the PAF efficiency parameters, and neutron transmittance can be explicitly characterized with simple calibration measurements on the direct neutron beam. A detailed description of the instrument calibration procedure used to perform the PAF corrections is given in Appendix B.

In order to correct the intensities to the time dependence, as follows from the full description in [3] the intensities $I^{\chi\psi}$ are normalized to the neutron transmittance. As stated before reflectometry data using a fixed neutron wavelength is obtained by rotating α_i stepwise recording one vertical α_i line of an $I^{\chi\psi}$ plot at a time. The measured intensities are thus transmittance normalized piecewise, using the deduced values for neutron transmittance T (from eq. 15 in Appendix B) for the actual time of the particular α_i line in the recoded intensity (i.e. $I^{\chi\psi}$) data. The resulting ^{3}He cell transmittance and sample geometry corrected intensity data are lastly normalized to the total reflected intensity below the critical edge. The resulting transmittance corrected and normalized intensity plots are shown in figs. 2 (a) and 3 (a).

Now that the data has been ^{3}He cell transmittance corrected over time and P, $A(t)$ and F have been calibrated, we apply the PAF corrections to the PA reflectometry measurement using the relations in section 3 to solve for the scattering cross-sections Σ^{xy}. As for the transmittance correction, we solve for the analyzing power A or equivalently p_2 at the time of a particular α_i line of an $I^{\chi\psi}$ plot which are then used in eqs. (6-9) to create the PAF (and ^{3}He time dependence) corrected cross-section plots. Thus there are different p_2 from the different times for each α_i line in the equations for a particular Σ^{xy}. This is because the polarization "beed-throughs" of each of the particular α_i lines in the four $I^{\chi\psi}$ plots need to be corrected for the inefficiencies at the time they were recorded in order to subtract/normalize the data the appropriate amount.

In the plots 2(a,b) and 3(a,b) we compare the corrected intensity data ($I^{\chi\psi}$) from the reflectometry measurement on the left (a) to the corrected spin-flip cross-section data ($\Sigma^{x,y}$) on the right (b). As stated before, the sample was an $Fe^{nat}Fe^{58}$ superlattice (10 repetitions of

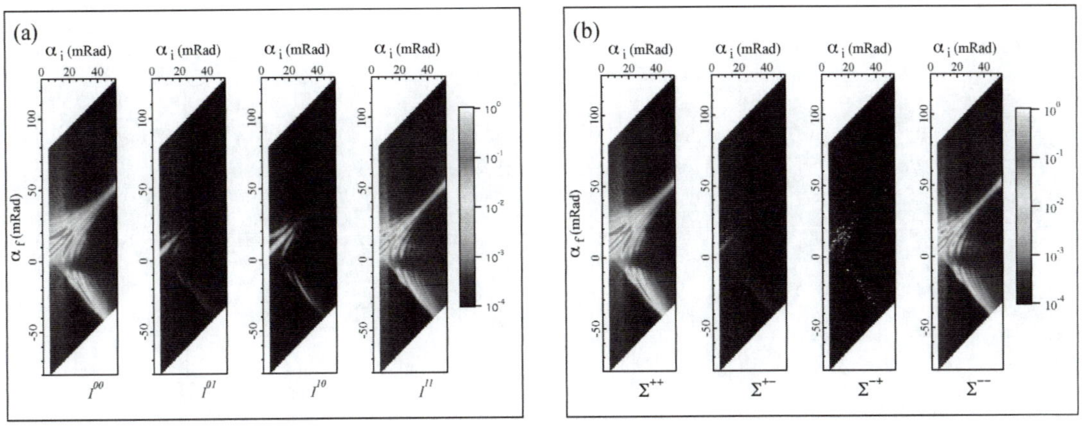

Figure 2. Polarized reflectivity and off-specular scattering data from the $Fe^{nat}Fe^{58}$ superlattice grating sample with a 0.7 T saturation field applied. The grating lines of the sample aligned parallel to the applied field. In the intensity plot to the left (a) a small spin-flip like signal can be seen and it also contains an apparent asymmetry in the spin-flip channels, graphs I^{01} and I^{10}. However, when one performs the full PAF corrections in the plots of the scattering cross-sections on the right (b), these signals disappear. The fact that the PAF corrections become important when $F_{\rm SF} \simeq p_1 + p_2 + f_1$ is shown visually in this data as the sum of the inefficiency terms was $\simeq 12 \times 10^{-3}$ for the I^{10} plot and $\simeq 8 \times 10^{-3}$ for the I^{01} plot which are the approximate amplitudes of the data in these two intensity plots for data on the specular line below the critical edge. Close inspection of the specular scattering i.e. $\alpha_i = \alpha_f$, shows a dynamic range of about $1\text{x}10^{-4}$ is achieved in the polarized mode.

10 nm of each Fe^{nat} and Fe^{58}) with a 10 μm pitch grating etched on it. The plots 3 show the magnetically unsaturated sample with a remanent sample magnetization, and the plots 2 show the sample at a high field of 0.7 T where the magnetization should be fully aligned for the high field measurement, but not necessarily for the low field remanent measurement. In fig. 2 before the PAF corrections one can clearly see the leakage of the imperfect PA, but after corrections any possible spin-flip cross-sections are reduced to zero, or below the 1×10^{-4} which is the S/N floor of this particular measurement. Next, in fig. 3 we see how much of the apparent spin-flip signal in the uncorrected data on the left is removed in the corrected data on the right, however a certain amount of spin-flip scattering remains.

To show that this data is indeed significant we consider the error contributions resulting from the combination of four measured intensities in each corrected cross-section plot. We perform a simple error analysis by taking the partial derivatives of the eqs. (2-5) with some simplifications (see Appendix A) and in terms of relative errors we obtain the usual result:

$$(\delta\Sigma^{++})^2 \simeq (\delta I^{00})^2 + (I^{00}\delta p_1)^2 + (I^{00}\delta p_2)^2 + (I^{01}\delta p_2)^2 + (I^{10}\delta p_1)^2 \tag{10}$$

$$(\delta\Sigma^{+-})^2 \simeq (\delta I^{01})^2 + (I^{01}\delta p_1)^2 + (I^{01}\delta p_2)^2 + (I^{00}\delta p_2)^2 + (I^{11}\delta p_1)^2 \tag{11}$$

$$(\delta\Sigma^{-+})^2 \simeq (\delta I^{10})^2 + (I^{10}\delta p_1)^2 + (I^{10}\delta p_2)^2 + (I^{11}\delta p_2)^2 + (I^{00}\delta p_1)^2 \tag{12}$$

$$(\delta\Sigma^{--})^2 \simeq (\delta I^{11})^2 + (I^{11}\delta p_1)^2 + (I^{11}\delta p_2)^2 + (I^{10}\delta p_2)^2 + (I^{01}\delta p_1)^2. \tag{13}$$

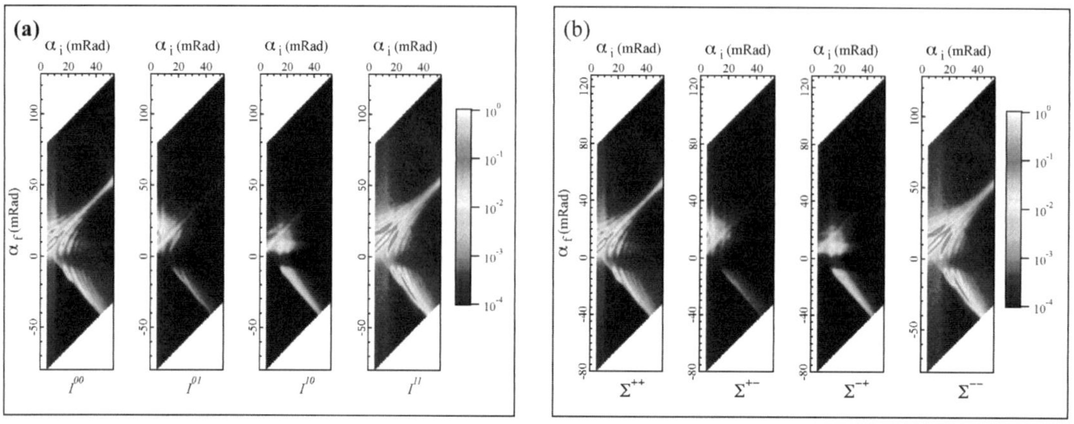

Figure 3. Polarized reflectivity and off-specular scattering data from the $Fe^{nat}Fe^{58}$ superlattice grating sample with remanent magnetization. The sample was magnetized by momentarily applying a strong field of 0.7 T anti-parallel to the neutron guide field and then switching it to a nominal 10 G (1 mT), thus leaving an anti-parallel remanent magnetization in the Fe of the grating. The grating lines, as in the saturated state, are parallel to the applied guide field. (a) The intensities are shown to the left, and (b) the PAF corrected data to the right. In the PAF corrected data one sees there is indeed remaining spin-flip scattering that is not a result of the instrument PAF inefficiencies. The main qualitative difference between the corrected cross-section data and the intensities data is that upon close inspection, the corrected spin-flip cross-sections along the specular, $\alpha_i = \alpha_f$, line are reduced as are the apparent ± 1 order diffraction lines compared to the uncorrected data. The resulting spin-flip signals are thus coming from the sample with good confidence and are likely caused by a slight component of the in-plane magnetization in the sample being perpendicular to the applied field.

Here one can see how the subtraction of the more intense non-spin flip states from the weaker spin-flip ones will increase the resulting errors bars of overlapping spin-flip data and since the flipper was a correction to the P and A corrections, it dropped out from the larger terms in the error budget. To show the effects visually, we used these formulas to calculate relative error bars for a the sample data set on the $Fe^{nat}Fe^{58}$ grating given in figs. 2 and 3 assuming a 1% accuracy of the p_1, p_2 and f_1 calibrations which are plotted in fig. 4 (a) and (b). The results indeed are acceptable as the error bars around features of interest where spin-flip scattering cross-sections are observed in fig. 4 (a) are clearly staying in the blue region which corresponds to 10% relative error or less for the data. Much of the remaining off-specular region are yellow corresponding to large (>0.5) error because of the very low scattering and subsequent lack of signal in these areas. The black areas (relative error >1) in fig. 4 (b) in the Σ^{+-} and Σ^{-+} around the specular line and where the diffraction pattern in the off-specular scattering was observed in the uncorrected data, as can be understood from eqs. 11 and 12 results from the subtraction of the much more intense I^{00} and I^{11} and its effect on the error budget, after subtraction of the more intense states the resulting corrected cross-sections in this area have amplitudes around the noise level of 1×10^{-4} but the error bars are equal to or slightly larger than this value. Finally a black vertical line of high error is visible at $\alpha_i \simeq 30$ mRad in the Σ^{+-} and Σ^{--} plots of 4 (b), this was likely caused by a temporary monitor or background instability, as due to the pairwise data acquisition these α_i lines were taken in the same 30 min block of time during the 2 day long measurement. Consequently we can now make a qualitative assessment of how the data interpretation changes once the PAF corrections are applied. Features of such data has been nicely discussed in [15, 16, 20] for example. Here we describe them in terms of the DWBA

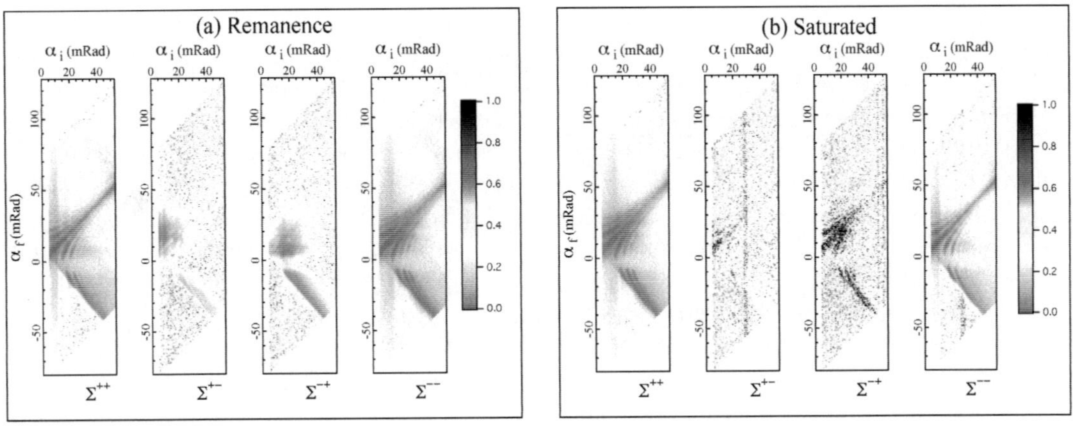

Figure 4. Plots of the relative error bars for PAF corrections for the remanent (a) and 0.7 T saturated (b) data calibrated to 1%. One can see in the areas of interest the relative error stays to approximately 10% or less, i.e the blue regions in the plots. In the (b) plots, the data around the specular line in the spin-flip channels has high relative error (black >100%) because the cross-sectionthat the amplitude was essentially 0 to within the S/N after the PAF corrections.

formalism [17] as this is the basis for modelling and fitting software at the MLZ [18].

The data in fig. 2 (a) of the non PAF corrected images at 0.7 T sample field indeed show reflectivity and off-specular scattering intensity for the sample at magnetic saturation, further there is an asymmetry between I^{10} and I^{01}. However, once one performs the full PAF correction, it is apparent that the asymmetry comes from the incident beam flipper which is only "on" for the I^{10} and I^{11} scans, and the remaining apparent spin flip signal was simply the bleed-through of the polarizer and analyzer due to their inefficiencies. As clearly seen in the corrected data in fig. 2 (b), since all magnetizations lay parallel to the field, the spin-flip scattering from the sample vanishes to the accuracy of the measurement. Along the line $\alpha_i = \alpha_f$ one obtains the specular reflectivity of the sample. Here, the critical angle of total reflection is bigger in the Σ^{++} channel than in the Σ^{--} channel because of the bigger average scattering length density. Away from this diagonal line, which is referred to as the off-specular scattering, the scattering wave vector has a component along the sample plane and up to three orders of Bragg scattering from the grating structure is observed. In the Σ^{++} channel, the scattering is on the average stronger than in the Σ^{--} channel. This can be explained if one considers the scattering amplitude in the framework of the distorted wave Born approximation (DWBA) [17]. In this formalism, the neutron-matter interaction potential is separated as a part obtained by in-plane averaging and a second part that describes the in-plane variations around this average. The first part leads to specular reflectivity and the second part to off-specular scattering. The scattering amplitude is no longer a Fourier transformation of the potential taken between the bra and ket that represent the wave function of the neutron after the sample (Born approximation), but a Fourier transform of the in-plane fluctuation potential taken between the bra and ket that represent the neutron wave function inside the laterally averaged potential. Those bras and kets are linear combinations of reflection and transmission amplitudes. As is evident from the specular intensities in the Σ^{++} and Σ^{--} channels, the reflection amplitudes for + neutrons are, on the average, more intense than in the − ones. This therefore explains why the off-specular scattering is more intense in the ++ channel and in the −− channel.

The fidelity of this high field 0.7 T data then gives us confidence in the data shown in fig. 3 (b). This figure shows the reflectivity and off-specular scattering from the sample at a field of 1 mT, after saturation in a strong anitiparallel magnetic field. For this remanent magnetization scan one can also see qualitative changes in the data. First, the intensity of the specular line in the spin-flip channels ($\alpha_i = \alpha_f$) is noticeably reduced, next what had appeared as weak diffraction from the grating lines around the specular line turned out to be PAF inefficiency bleed through. The corrected data (fig. 3 (b)) show two main differences with respect to the ones at saturation (fig. 2 (b)). The first one is that the Σ^{++} intensities are now weaker than the Σ^{--} intensities, a result of the fact that all magnetizations are now aligned almost anti-parallel to the incoming beam polarization. The second one is the presence of reflectivity and off-specular signal in the spin flip channels Σ^{+-} and Σ^{-+}. This is a sign that the laterally averaged magnetizations are not exactly anti-parallel to the external field, but have a small component perpendicular to it. One can also see that the off-specular signal is not symmetric with respect to the diagonal $\alpha_i = \alpha_f$. This can be explained, once again, in the framework of the DWBA [17]. In the Σ^{+-} channel, the neutron beam before interaction with the sample is polarized up and after interaction with the sample the neutrons are selected with spin projection down. As the reflection amplitudes of + neutrons are, on the average, smaller than in the one of - neutrons, one observes smaller intensities of off-specular scattering in the $\alpha_i > \alpha_f$ region than the ones in the $\alpha_i < \alpha_f$ region. The same type of explanation holds for the intensities in the Σ^{-+} channel. However, if one were to sum the Σ^{+-} and Σ^{+-} the resulting image would exhibit mirror-plane symmetry about the specular line.

The data corrections in this case now allow us to qualitatively see spin-flip effects approaching the 1×10^{-4} level even though the sum of the PAF inefficiency terms was around 1×10^{-2}. This data has reasonable quantitative confidence as the total relative error bars including the statistical uncertainties and the additional error from the PAF corrections are on the order of 10% or better for the areas of interest. If the error in the PAF inefficiency calibrations was on the order of 10% rather than 1%, then the relative error bars in the spin flip signals observed in the remanent data would be around 15% for the region near the specular line. Conversely, the statistical error bar just from the neutron counting rates for the I^{01} and I^{10} intensities in this area were about 3% themselves. Thus with 1% PAF inefficiency calibration the total error was around 3 times higher than the statistical error of the uncorrected data for this example and would be about 5 times higher with a less precise 10% PAF inefficiency calibration.

5. conclusion and acknowledgements

When one has a saturated magnetic structure with no spin-flip scattering such as in fig. 2, the data itself can help refine or verify the PAF inefficiencies used for the corrected cross-section data. However this may not be the case for all real experiments on more complex systems so one should have confidence in the instrument calibration data and PAF inefficiency correction procedure to ensure when potential spin-flip scattering cross-sections obtained after PAF corrections are indeed significant. As was shown in the data presented in figs. 2 and 3 one can indeed correct even time-dependent neutron PA data, in this case taken with a time-decaying ^{3}He spin filter if the PAF parameters can be calibrated to the appropriate level. For the data shown PAF corrected cross-sections were obtained to good confidence with a 1 to 1×10^{-4} dynamic range with statistically relevant error levels. Conversely the results of this paper should also make it clear the rule-of-thumb that non-PAF corrected polarized neutron data must be very carefully justified if the values of the relative cross-sections are on the order of the instrument's PAF inefficiencies unless such features are well separated from any intense non-spin-flip scattering features.

This paper explains the observed signal to noise including a discussion of error propagation and can serve as a recipe for data corrections in modern polarized neutron experiments, especially those employing a ^{3}He spin filter cell as the analyzer, and is meant to encourage a conceptual understanding of PAF corrections and their application to aid good measurement strategy and in analyzing and publishing data. MLZ is developing software for modelling and fitting layered structures with lateral correlations including the nuclear and magnetic structures using the DBWA [18]. We will work towards applying this software to the model multi-layered and patterned system data presented here, and presenting a discussion in future work.

We kindly acknowledge A. Wildes of the Institut Laue Langevin who provided the $Fe^{nat}Fe^{58}$ patterned multilayer sample and K. Temst whose group in KU-Leuven produced it.

Appendix A: error bar calculation for PAF corrected cross-sections

In this appendix we summarize the error analysis which was used to calculate the error bars shown in figs. 4. A polarized ^{3}He neutron spin filter and be fully calibrated (see following Appendix B) to nearly arbitrary accuracy, consequently is it useful to show an error analysis in order to understand how the statistical accuracy of the measurement, vs. the precision of the instrument calibration contributes to the error of the corrected neutron PA cross-section data. Using the first spin-flip cross-section Σ^{+-} (eq. (8)) as en example and performing a simple error analysis of the full formula by taking the partial derivatives one obtains:

$$
\begin{aligned}
(\delta\Sigma^{+-})^2 = {} & \left(\frac{\delta I^{01}((1+p_1p_2-p_1-p_2)(1-f_1)-f_1(p_1p_2-p_1))}{PAF'}\right)^2 + \left(\frac{\delta p_1 I^{01}(1-p_2)}{P^2AF'}\right)^2 + \left(\frac{\delta p_2 I^{01}(2f_1p_1-f_1-p_1+1)}{PA^2F'}\right)^2 \\
& + \left(\frac{\delta f_1 I^{01}(p_1-p_2p_1)}{PAF'^2}\right)^2 \\
& + \left(\frac{\delta I^{00}((p_1p_2-p_2)(1-f_1)-p_1p_2f_1)}{PAF'}\right)^2 + \left(\frac{\delta p_1 I^{00}(-p_2)}{P^2AF'}\right)^2 + \left(\frac{\delta p_2 I^{00}(-2f_1p_1+f_1+p_1-1)}{PA^2F'}\right)^2 + \left(\frac{\delta f_1 I^{00}(-p_1p_2)}{PAF'^2}\right)^2 \\
& + \left(\frac{\delta I^{11}(p_1p_2-p_1)}{PAF'}\right)^2 + \left(\frac{\delta p_1 I^{11}(p_2-1)}{P^2AF'}\right)^2 + \left(\frac{\delta p_2 I^{11}(-p1)}{PA^2F'}\right)^2 + \left(\frac{\delta f_1 I^{11}(p_1p_2-p_1)}{PAF'^2}\right)^2 \\
& + \left(\frac{\delta I^{10}(p_1p_2)}{PAF'}\right)^2 + \left(\frac{\delta p_1 I^{10}(-p_2)}{P^2AF'}\right)^2 + \left(\frac{\delta p_2 I^{10}(-p_1)}{PA^2F'}\right)^2 + \left(\frac{\delta f_1 I^{10}(-p_1p_2)}{PAF'^2}\right)^2
\end{aligned}
$$

for the relative error contribution $\delta\Sigma^{+-}$ of the Σ^{+-} cross-section. P and A are again the polarization and analyzer efficiencies related to the inefficiencies by $P = 1-2p_1$ and $A = 1-2p_2$ and the flipper efficiency is momentarily defined as $F' = 1-f_1$ to simplify the written expressions. The other terms for the Σ^{+-}, Σ^{-+}, and Σ^{--} can be obtained similarly. Using this eq. (14) to help us understand the dominate terms in the resulting error calculation, we discard all higher order terms containing products of p_1, p_2 and f_1 terms which will be small. Now we see that the flipper errors don't contribute to the eventual error bars in the first order approximation and by further inspection for a system with low polarization element inefficiencies p_1, p_2, and f_1 we can set the denominators to 1. This gives us the result of the compact set of eqs. (10-13) given previously in section 4 to estimate the error contributions.

Appendix B: instrument PAF efficiency calibration

One of the very powerful features of using a polarized ^{3}He spin filter with AFP spin flipping of the ^{3}He polarization employed, is the ability to fully characterize the neutron polarization parameters under the exact instrument configuration and conditions of the scattering experiment. Although the P^2T quality factor of a ^{3}He type analyzer may be lower than a large area super mirror array [19], their very deterministic nature which allows for simple calibration over all neutron wavelengths, their uniform and smooth analyzing power over large areas, and their ability to provide an arbitrarily high analyzing efficiency can outweigh the downsides of lower transmittance, required maintenance and possible time dependence for some instances. This method also avoids the complications of other neutron PA methods using two imperfect neutron flippers which require more involved instrument calibrations, a summary of which are included in the work of Williams [20] and also discussed in [3].

To quantify the PAF parameters we follow a method similar to that outlined in [21] where one simply measures the unpolarized neutron transmittance as a function of wavelength for the unpolarized and the polarized ^{3}He cell. In [21] Chupp et. al. used a "white" neutron beam and time of flight where used to determine the neutron wavelength but on many instruments such as MARIA where PA is installed, white-beam time of flight is not available to calibrate the ^{3}He spin filter cell. However MARIA [1] has a velocity selector that allows us to scan the neutron wavelength and determine discrete transmittance values and fit them as a function of neutron wavelength. The neutron analyzing (polarizing) power $A(\lambda)$ and transmittance $T(\lambda)$ of a ^{3}He cell are can be derived from the neutron polarization dependent 1/v cross-sections as in [22]. For a ^{3}He spin filter the neutron analyzing power $A(\lambda)$ and unpolarized neutron transmittance $T(\lambda)$ as functions of the neutron wavelength λ are

$$A(\lambda) = \tanh(P_{He}\Theta\lambda\sigma), \tag{14}$$

$$T(\lambda) = T_0\exp(-\Theta\lambda\sigma)\cosh(P_{He}\Theta\lambda\sigma) \tag{15}$$

where T_0 is the empty cell transmittance, $\sigma = 0.0732\,\mathrm{bar}^{-1}\mathrm{cm}^{-1}\text{Å}^{-1}$ is polarization dependent neutron absorption cross-section of ^{3}He, the ^{3}He polarization is P_{He} and Θ is the product of the spin filter cell's length in cm and ^{3}He pressure in bar at 25 °C. For the reblown GE180 cells used in the MARIA SEOP analyzer we have tested many cells and found them to have a relatively uniform 88% to 91% transmittance for 5 Å$< \lambda <$10 Å which is similar to that found by others [21]. Therefore when we fit the unpolarized cell data, we leave T_0 as wavelength dependent constant. Of course for highest accuracy one can also perform empty cell measurements to determine T_0 which would have to be characterized before preparation for SEOP cells which are permanently sealed after being filled with ^{3}He [23, 10], MEOP cells which are valved can be evacuated and essentially can be calibrated at any time [24]. Further in cases where the ^{3}He spin filter cell has single crystal silicon neutron windows [25] T_0 can be calculated from the Si neutron absorption cross-section and thickness.

To characterize the SEOP cells used on MARIA we measure the unpolarized neutron transmittance vs. wavelength typically with 4 to 5 steps from 5 Å to 10 Å for the cell when it is polarized, and completely depolarized. The polarized ^{3}He analyser on MARIA is installed on a translation stage so the transmittance measurements are performed comparing the cell in beam to the cell out of beam intensity measurements. Unpolarized cell transmittance data is fit to eq. (15) with P_{He} set to 0 (i.e. fit to an exponential) to determine the value of Θ. Then using this value of Θ the polarized cell data can be fit to eq. (15) to determine P_{He}. One can also simply take the ratio of the cell transmission when the ^{3}He polarized to its transmission with $P_{He} = 0$ and perform a $\cosh(P_{He}\Theta\lambda\sigma)$ fit to determine the product of P_{He} and Θ. Since SEOP polarized ^{3}He cells are permanently sealed, the unpolarized cell measurement to determine Θ must not be repeated but occasionally for routine checks, and strictly speaking if Θ and T_0 of a particular cell are sufficiently characterized only one neutron wavelength is required to determine P_{He}. Using P_{He} and Θ in eqs. (14) and (15) the analyzer's polarization efficiency and transmittance can be calculated for any neutron wavelength. For the cell used in the MARIA analyzer in this experiment, fits to unpolarized transmittance of the unpolarized cell over neutron wavelength gave us a value of $T_0 = 0.89$ and $\Theta = 9.27\,\mathrm{bar\,cm}$. The length l of this particular cell was about 8 cm, giving a cell pressure of 1.16 bar, however it is not required to know these numbers individually. Using the obtained Θ, the ^{3}He polarization from the cosh fit gave a starting value of $P_{He} = 71.5\%$ for the experiment described in this paper. The exponential fit of the unpolarized cell transmittance and the cosh fit of the polarized cell to unpolarized cell ratio over neutron wavelength typically give 1% errors in the determination of Θ and the product of P_{He} and Θ respectively.

Further in this particular measurement, the in-situ polarization of the ^{3}He analyzer was not yet available [11, 14], therefore the ^{3}He polarization was indeed a function of time. The time dependence of the decay of the ^{3}He polarization was determined by performing unpolarized neutron transmittance measurements of the direct neutron beam as a function on neutron wavelength at 12 hour intervals during the measurement (i.e. between each pair of I^{00}, I^{10} and I^{01}, I^{11} scans). During these calibration measurements the sample and incident beam polarizer are momentarily translated out of the neutron beam, and a transmittance scan for λ=5 Å-10 Å in five steps of the ^{3}He cell is recorded. An attenuator can be placed in the beam to prevent detector saturation if needed. The resulting transmittances are then fit to eq. (15) fixing the values of T_0 and Θ to those obtained from unpolarized cell measurements to give P_{He} at the different times. These P_{He} values are then fit to

$$P_{He}(t) = P^0_{He}exp(-t/t_1) \tag{16}$$

to give us t_1 which is the ^{3}He polarization decay time constant and where P^0_{He} is the initial ^{3}He polarization at time $t = 0$. For this experiment a t_1 of 156 ($\pm$4) hours was obtained which was then used to calculate T and A of the ^{3}He cell at anytime for the λ=10 Å neutron wavelength that we used for the reflectometry experiment. The initial and final analyzer efficiency were $A^i = 99.9988(1)\%$ or $p^i_2 = 6.1(6) \times 10^-5$ and $A^f = 99.971(1)\%$ or $p^f_2 = 1.5(1) \times 10^{-3}$ and the initial and final transmittance where $T = 6.4(1)\%$ and $T = 1.8(1)\%$ respectively. The relative transmissions are only important for comparing relative intensities taken at different times for the case of a time-changing P_{He}. Since the maximum time difference between α_i data used in the intensity corrections was approximately 12 hours, or less than 1/10 of a decay time constant, the effects of the correction is not as extreme as total change in transmittance over the whole experiment. A was very high and the corresponding T low in this measurement due to the ^{3}He cell which is optimized to also have a good value of A at the minimum MARIA wavelength of 4.5 Å. Additional corrections for the physical path length of the neutron through the ^{3}He cell[12] where small for data of interest in our case and not considered in the analysis, however an expanded discussion of this effect is also given below. Further the very high analyzing efficiency at 10 Å also allows us to determine the remaining P and F parameters accurately [13].

Once the ^{3}He polarization is determined to an appropriate accuracy, the incoming neutron polarization P and incident beam flipper efficiency F are determined using the polarized ^{3}He cell by performing two flipping ratio measurements of the polarized neutron beam. The flipping ratios are then related to the total neutron polarization $\mathcal{P}$ in the beam by

$$\mathcal{P} = \frac{\mathcal{F} - 1}{\mathcal{F} + 1} \tag{17}$$

where two flipping ratios are determined by using the ^{3}He AFP flipper to reverse the analyzer's polarization direction while measuring the transmitted neutron beam intensity for both ^{3}He polarization states with the incident beam flipper on and off. One obtains $\mathcal{F}^{\rm on}$ and $\mathcal{F}^{\rm off}$ which are the ratios of the intensities in the bright transmission state to the dark state with the incident beam flipper on or off respectively. This measurement then gives the total neutron polarization, or the product of all the neutron polarization efficiencies. For MARIA our incident polarization is a fixed double bounce reflection supper mirror polarizer which provides a stable neutron polarization parallel to the applied magnetic field on it. This polarization can then be flipped anti-parallel to the field with the incident beam RF gradient neutron flipper [6] which has inefficiency f_1. When the incident beam flipper is off it has no effect on the polarization efficiency so we measure PA, the product of the polarizer and analyzer polarizations. Conversely, when the incident beam flipper is active, or on, the small inefficiency of this flipper will lower the resulting flipping ratio and we obtain the product PAF. Therefore the ratio of $\mathcal{P}^{\rm on}/\mathcal{P}^{\rm off}$ gives the flipper efficiency F which is related to the flipper inefficiency by $f_1 = (1 - F)/2$.

Now, from the above eq. (14) we can solve for A to determine p_2 and then determine the value of P (i.e. p_1) independently. If the measurement is done using multiple neutron wavelengths we can perform this two flipping ratio analysis at the desired wavelengths used for a measurement because on MARIA, or any instrument using a supper mirror or crystal polarizer, the incoming neutron polarization and flipper efficiency can be a function of neutron wavelength and beam apertures or focusing. Consequently this method using a ^{3}He spin filter is particularly attractive for sensitive polarized neutron measurements because it is done for the specific instrument configuration of the actual sample measurement. On MARIA this is generally done by first aligning the sample, configuring the measurement, and then simply translating the sample out of the beam to quantify the instruments polarization efficiencies for the exact instrument and sample environment configuration using the direct beam at the time of the experiment.

From these flipping ratio measurements, the incident beam polarization P was 98.56(1)% or $p_1 = 7.2(1) \times 10^{-3}$ and the flipper efficiency $F = 1 - 2f_1$ was 99.20(2)% or $f_1 = 4.0(1) \times 10^{-3}$. These given errors follow from the 1% error in the determination of the ^{3}He cell parameters as discussed above and because the ^{3}He cell was "opaque" for the 10 Å neutron wavelength in this experiment providing an analyzing inefficiency an order of magnitude lower than the flipper and polarizer inefficiencies allowing for accurate determination of these parameters [13].

The cells used for analysis on MARIA are from GE180 tubing reblown into glass cylinders of suitable dimensions and thus have slightly rounded faces and corners. However the cells we used have a relatively low curvature in the neutron faces, and the area on the edges of the cylinder are masked with a neutron absorber. W C Chen et. al. [12] showed that the path length correction l/l_0 for such a cell is;

$$\frac{l}{l_0} \simeq 1 + \left[\frac{1}{2}\left(1 - \frac{l_0}{2R}\right) - \frac{D_{sc}^2}{Rl_0}\right] \tan^2(2\theta) - \frac{D_{cd}^2}{Rl_0} \frac{A_S}{2\pi(D_{sc} + D_{cd} + l_0)^2} \tag{18}$$

where l_0 is the assumed neutron path length through the middle of the cell, R is the radius of curvature of the ends of the cell, D_{sc} and D_{cd} are the sample to cell and cell to detector distances respectively, 2θ is the scattering angle, and A_s is the cross sectional area of the sample where a SANS transmission scattering geometry is assumed (i.e sample perpendicular to the incident beam). The cell used on MARIA for this experiment has $l_0 \simeq 8\,\text{cm}$, $R > 100\,\text{cm}$, $2\theta_{max} = 0.080\,\text{rad}$, $D_{sc} = 60\,\text{cm}$ and $D_{cd} = 132\,\text{cm}$. The sample was 2 cm x 2 cm, where for reflectometry it is placed at a small angle resulting in an over estimation of this sample size correction term for our geometry, however this term is very small in our case. MARIA uses an elliptically focusing guide to increase the neutron intensity on the sample [1] and the calibration measurement of A and T to determine the P_{He} and $\Theta = l_0[\text{He}]$ uses the same vertical focusing and is actually a cell-volume average so no further correction of A or T should be needed for the specular line which always passes through the center of the cell. Returning to the off specular scattering, using the path length correction from eq. (18) and eqs. (14), (15) for our cell, which has a path length ^{3}He density product of 9.27 bar cm with maximum and minimum polarization of 71.5% to 52.6% in this experiment, we can estimate the effects of the path length correction on our data. The maximum error in A using these values is below 1×10^{-4} and can be ignored. For T (eq. (15) used to correct the neutron transmittances) the resulting maximum correction could be up to 9%, however this is at the maximum α_f of ± 80 mRad from the specular scattering which passes through the center of the cell. The data of interest in this experiment had an $|\alpha_i - \alpha_f| < 25\,\text{mRad}$ so the possible path length dependent correction would be <1% for T in this range. Since as was described in section 4 the statistical error bars of our intensities were around 3% for the spin-flip scattering we have ignored the path length correction in our analysis. For some geometries and cell opacities (Θ) the path length corrections can become important and should be calculated and included when normalizing the intensities to the ^{3}He transmittance and correcting for the analyzing power. This can be done for each α_f line similar to the way the time dependence was corrected for the individual α_i lines.

References

[1] Heinz Maier-Leibnitz Zentrum, Forschungszentrum Jülich, Jülich Centre for Neutron Science 2015 *J of large-scale research facilities* **1** A8. [DOI: 10.17815/jlsf-1-29]

[2] Wildes A 1999 *Review of Scientific Instruments* **70** (11):4241-4245. [DOI: 10.1063/1.1150060]

[3] Wildes A R 2006 *Scientific Reviews: Neutron Polarization Analysis Corrections Made Easy, Neutron News* **17:2** 17-25 [DOI: 10.1080/10448630600668738]

[4] Abragam A 1961 *The Principles of Nuclear Magnetism* Oxford University Press, Oxford, England

[5] Jones G L, Dias F, Collett B, Chen W C, Gentile T R, Piccoli P M B, Miller M E, Schultz A J, Yan H, Tong X, Snow W M, Lee W T, Hoffmann C, Thomison J 2006 *Physica B: Condensed Matter* **385?386** 1131-1133 [DOI:10.1016/j.physb.2006.05.390]

[6] Babcock E, Petoukhov A, Chastagnier J, Jullien D, Leliévre-Berna E, Andersen K H, Georgii R, Masalovich S, Boag S, Frost C D, Parnell S R 2007 *Physica B* **397** 172-175. [DOI: 10.1016/j.physb.2007.02.093]

[7] McKetterick T J, Boag S, Stewart J R, Frost C D, Skoda W M A, Parnell S R, Babcock E 2011 *Physica B* **406** 2436-2438. [DOI: 10.1016/j.physb.2010.11.088]

[8] Ehlers G, Stewart J R, Deen P P, Andersen K H 2015 *European Physical J of Conf* **83** 03004. [DOI: 10.1051/epjconf/20158303004]

[9] Andersen K H, Cubitt R, Humblot H, Jullien D, Petoukhov A, Tasset F, Schanzer C, Shah V R, Wildes A R, 2006 *Physica B: Condensed Matter* **385-386** 1134-1137. [DOI: 10.1016/j.physb.2006.05.391]

[10] Salhi Z, Babcock E, Pistel P, Ioffe A 2014 *J of Physics Conference Series* **528(1)** 012015 [DOI: 10.1088/1742-6596/528/1/012015]

[11] Babcock E, Salhi Z, Theisselmann T, Starostin D, Schmeissner J, Feoktystov A, Mattauch S, Pistel P, Radulescu A, Ioffe A 2016 it J of Physics Conference Series **711** 012008 [DOI: 10.1088/1742-6596/711/1/012008]

[12] Chen W C, Erwin R, Watson S M 2013 *Physics Procedia* **42** 163-170. [DOI: 10.1016/j.phpro.2013.03.191]

[13] Klauser C, Chastagnier J, Jullien D, Petoukhov A, Soldner T 2012 *J of Phys: Conf Series* **340** 012011 [DOI: 10.1088/1742-6596/340/1/012011]

[14] Babcock E, Mattauch S, Ioffe A 2010 *Nuclear Instruments and Methods in Physics Research Section A Accelerators Spectrometers Detectors and Associated Equipment* **625** 43-46 [DOI: 10.1016/j.nima.2010.09.078]

[15] Theis-Bröhl K, Westphalen A, Zabel H, Rcker U, McCord J, Höink V, Schmalhorst J, Reiss G, Weis T, Engel D, Ehresmann A, Toperverg B P 2008 *New Journal of Physics* **10** 093021 [DOI:10.1088/1367-2630/10/9/093021]

[16] Zabel H, Theis-Bröhl K, Wolff M, Toperverg B P 2008 *IEEE Transactions on Magnetics* **44** no. 7, 1928-1934 [DOI:10.1109/TMAG.2008.924538]

[17] Toperverg B P 2015 *Physics of Metals and Metallography* **116** 1337-1375. [DOI: 10.1134/S0031918X15130025]

[18] Durniak C, Ganeva M, Pospelov G, Van Herck W, Wuttke J 2015, **BornAgain** *Software for simulating and fitting X-ray and neutron small-angle scattering at grazing incidence* http://www.bornagainproject.org

[19] Babcock E, Salhi Z, A Ioffe 2013 *J of the Physical Society of Japan* **82** (supp. A) SA030-10

[20] Williams

[21] Chupp T E, Coulter K P, Kandes M, Sharma M, Smith T B, Jones G, Chen W C, Gentile T R, Rich D R, Lauss B, Gericke M T, Gillis R C, Page S A, Bowman J D, Penttila S I, Wilburn W S, Dabaghyan M, Hersman F W, Mason M 2007 *Nuclear Instruments and Methods in Physics Research Section A: Accelerators, Spectrometers, Detectors and Associated Equipment* **574** 500-509. [DOI: 10.1016/j.nima.2007.02.091]

[22] Rich D R, Bowman J D, Crawford B E, Delheij P P J, Espy M A, Haseyama T, Jones G, Keith C D, Knudson J, Leuschner M B, Masaike A, Masuda Y, Matsuda Y,6, Penttila S I, Pomeroy V R, Smith D A, Snow W M, Szymanski J J, Stephenson S L, Thompson A K, Yuan V 2002 *Nuclear Instruments and Methods in Physics Research A* **481** 431-453. [DOI: 10.1016/S0168-9002(01)01331-6]

[23] Chen W C, Gentile T R, Fu C B, Watson S, Jones G L, McIver G L, Rich G L 2011 *J of Physics: Conference Series* **294** 012003 [DOI:10.1088/1742-6596/294/1/012003]

[24] Petoukhov A K, Andersen K H, Jullien D, Babcock E, Chastagnier J, Chung R, Humblot H, Lelivre-Berna E, Tasset F, Radu F, Wolff M, Zabel H 2006 *Physica B: Condensed Matter* **385?386** Part 2 1146-1148 doi:10.1016/j.physb.2006.05.394

[25] Andersen K H, Cubitt R, Humblot H, Jullien D, Petoukhov A, Tasset F, Schanzer C, Shah V R, Wildes A R 2006 *Physica B: Condensed Matter* **385-386** Part 2 1134-1137 [DOI: 10.1016/j.physb.2006.05.391]

IOP Conf. Series: Journal of Physics: Conf. Series **862** (2017) 012002 doi:10.1088/1742-6596/862/1/012002

Recent on-beam tests of wide angle neutron polarization analysis with a ^{3}He spin filter: Magic PASTIS on V20 at HZB

E Babcock[1*], Z Salhi[1], R Gainov[2,3], R Woracek[4], H Soltner[3], P Pistel[3], F Beule[5], K Bussmann[6], A Heynen[3], H Kämmerling[3], F Suxdorf[5], M Strobl[4], M Russina[2], J Voigt[6], A Ioffe[1]

[1]Jülich Centre for Neutron Science (JCNS) at Heinz Maier-Leibnitz Zentrum (MLZ), Forschungszentrum Jülich GmbH, 85747 Garching, Germany
[2]Helmholtz-Zentrum Berlin für Materialien und Energie GmbH (HZB), Hahn-Meitner-Platz 1, 14109 Berlin, Germany
[3]Zentralinstitut für Engineering und Technologie (ZEA-1), Forschungszentrum Jülich GmbH, 52425 Jülich, Germany
[4]European Spallation Source ERIC, S-221 00 Lund, Sweeden
[5]Jülich Centre for Neutron Science (JCNS) and Institute for Complex Systems (ICS) Forschungszentrum Jülich GmbH 52425 Jülich, Germany
[6] Jülich Centre for Neutron Science (JCNS) and Peter Grünberg Institut (PGI), JARA-FIT Forschungszentrum Jülich GmbH, 52425 Jlich, Germany

E-mail: `e.babcock@fz-juelich.de`

Abstract. A complete XYZ polarization analysis solution is under development for the new thermal time of flight spectrometer TOPAS [1], to be operated in the coming east neutron guide hall at the MLZ. Polarization Analysis Studies on a Thermal Inelastic Spectrometer, commonly called PASTIS [2], is based on polarized ^{3}He neutron spin filters and an XYZ field configuration for the sample environment and a polarization-preserving neutron guide field. The complete system was designed to provide adiabatic transport of the neutron polarization to the sample position while maintaining the homogeneity of the XYZ field. This system has now been tested on the polarized time-of-flight ESS test beam line V20 at HZB [3]. Down to the minimum wavelength of 1.6 Å on the instrument, the magnetic configuration worked ideally for neutron spin transport while giving full experimental freedom to change between the X, Y or Z field configuration. The ^{3}He cell used was polarized at the ^{3}He lab of the JCNS at the MLZ in Garching and transported to HZB in Berlin via car showing that such a transport is indeed feasible for such experiments. We present results of this test and the next steps forward.

1. Introduction

Beginning with the advent of ^{3}He polarization of practical magnitudes for neutron polarization and analysis, there have been proposed uses in wide-angle polarization analysis, especially for neutron spectrometry using high energy neutrons [4]. This idea developed into a uniform magnetic field insert housing a large-angle ^{3}He cell that can be placed around the sample environment of a typical wide-angle neutron spectrometer allowing polarization analysis over the full detector angle using a moderately sized ^{3}He cell and having minimal instrument modifications. One such application has been shown at NIST on the MACS spectrometer, however here only for "Z" polarization analysis, i.e. for the magnetic field perpendicular to the scattering vector. [5]. The full polarization analysis version of such as system has become known in the community by its acronym PASTIS, for Polarization AnlaysiS on a Thermal Inelastic Spectrometer [2]. Several projects towards such systems are in development at the ILL [6, 7] , ISIS [8], SNS[9] plus our upcoming spectrometer projects at the ESS, TREX and DREAM [10] as well as the system described here for use on TOPAS [1] and NEAT at HZB in the meantime [11].

The motivation for use of a ^{3}He spin filter in these is again largely due to the polarized ^{3}He "idealized" characteristics as a neutron analyzer which would not have alignment or sample size dependencies while having a deterministic neutron energy/wavelength dependence which can then be used on neutrons of arbitrarily high energy and high inelastic neutron energy transfer. Other neutron polarization analysis methods such as wide-angle neutron super mirrors have been applied for wide-angle diffraction and spectroscopy, but normally for cold-spectrum neutron instruments (i.e. approximately wavelengths greater than 2.4 Å or less than 15 meV energy) due to the limitations of of the high-energy neutron reflectivities of such devices [9, 12, 13]. For a thermal neutron spectrometer such as TOPAS [1] which will use incident neutron energies of up to 165 meV or 0.7 Å such devices would not provide the required performance leading us to work to further develop the magic PASTIS concept taking advantage of the characteristics of ^{3}He spin filter cells like the other groups cited earlier [5, 8, 9, 10].

This proceeding will summarize recent on-beam testing of the JCNS Magic PASTIS system in experiments on the V20 ESS test beamline at Helmholz Zentrum Berlin.

2. Magic PASTIS, from concept to prototype

Magic PASTIS was not only designed to yield a good ^{3}He lifetime or T_1 performance, but also to ensure fully adiabatic transport of the neutron polarization to the sample position. The PASTIS system plus neutron guide fields have been reported previously [1, 14]. In some ways this concept bears resemblance to the PASTIS 2.0 implemented at the Institut Laue Langevin [7], however the details of the magnetic systems are different, and while our system has small dead angles of around 3 degrees every 90 degrees in the horizontal plane, the vertical opening is higher than the ILL PASTIS 2.0, on the order of 40 degrees, and matched to the out-of plane vertical acceptance of our TOPAS detector systems. In laboratory measurements a large 22 cm outside diameter doughnut-shaped cell named "Homer" was used to experimentally optimize the field gradients. The cell which is filled to 0.5 bar ^{3}He pressure gave a ^{3}He lifetime in excess of 50 hours in each of the three field directions was reported in [14]. This value of T_1 obtained with a [^{3}He]=0.5 bar cell translates to a maximum magnetic field gradient of less than 1×10^{-3} cm^{-1} over the volume of the cell, and a magnetic plus dipole-dipole limited ^{3}He lifetime in excess of 100 hours for cells of over 1 bar pressure [14].

Similarly, the guide field system was constructed and commissioned first in laboratory conditions. The description of this guide field system was reported in [1] and it has recently been assembled. The results of the bench test in Jülich using a hall probe met the expectations from the calculations and also permitted us to test the newly developed control software for automization of the PASTIS power supplies. Finally a polarized ^{3}He cell was transported from Garching. The results of this test were twofold, 1: we verified that we can polarize a cell in Garching and transport it by car over large distances, 2: we did not observe adverse effects on the ^{3}He lifetime due to the addition of the guide fields via a measurement of the cell lifetime in Jülich using NMR free-induction decay measurements of the ^{3}He polarization over time. Directly after the positive verification of this test, the magic PASTIS coil system and guide fields plus power supplies and control electronics were shipped via courier transport to HZB in Berlin for the neutron tests on V20.

Figure 1. Photo of the magic PASTIS system and neutron X-Y-Z guide fields on V20. Right to left, the X-Y-Z guide fields and the incoming beam flight path, a sample in an aluminum cell in the center of PASTIS, the D=4.3 cm by L=15 cm ^{3}He cell used for analysis, and the four 40 cm neutron detectors mounted horizontally.

3. Magic PASTIS installed on V20

Using the experimental setup of V20 we have conducted the test of magic PASTIS in realistic conditions. The particular goals where to utilize the interface and test the practical handling of the system while obtaining the performance with polarized incident neutrons. Earlier tests of PASTIS systems, such as the tests on IN3 [2] showed that care must be taken with the guide fields to insure neutron polarization transport. Therefore this test was important to fully ensure and experimentally verify the neutron polarization transport beyond the expectations from calculations or field mapping.

The V20 instrument is a "test" beam line and is therefore very adaptable and has a very open geometry for installation of equipment [3]. The instrument essentially consists of a chopper system which can replicate the European Spalation Source (ESS) -proposed pulse structure [10, 15], an incident beam super-mirror bender polarizer, and a 6 m (can be extended to more than 10 m) open neutron flight path with adaptable detector options. The instrument's bender polarizer provided a good polarization over the range of our measurement. From data taken during this experiment with two different ^{3}He cells of the polarized and unpolarized neutron transmissions we determined the incident beam polarisation efficiency P_p to be above 97.5% for neutron wavelengths from 1.6 Å to 6.5 Å (see figs. 3,5).

Using a polarized ^{3}He cell, we also optimised an RF gradient (neutron AFP) flipper by simply maximizing the obtained neutron flipping ratio, i.e. the ratio of the flipper-off transmission to the flipper-on transmission as a function of neutron wavelength (see fig. 3). A photo of the RF-gradient flipper is shown in fig. 2. The gradient field was created with a stack of 4 strong 50 mm x 12 mm x 25 mm NbFeB magnets on either side of the double 1 mm thick iron pole plates top and bottom with 10 mm thick iron pillars on the other side to close the magnetic flux return path. This gives a smooth gradient field that varied from about 50 G on the incident side to 10 G on the exit of the 35 cm length. The RF field was created with a "Gachinga" style resonator coupled to a double-parallel 40 turn coil which had higher turn density in the middle of the 30 cm long coil. The resonator is simply a high current transistor driven LRC

Figure 2. Photo of the neutron RF gradient flipper. The RF coil in the middle is wound on a glass tube with silicon windows that can be filled with ^{4}He when desired. The NbFeB magnet stacks are to the left, and the thick iron pillars to the left. The arrangement produces a relatively smooth gradient field that varies from 50 G on the NbFeB magnet side to 10 G on the other.

oscillator that uses the coil for L and one tunes to the desired resonance by adjusting C. The coil and gradient field of this flipper was placed inside of the race-track style coils which are used to create a nominal 10 G neutron guide field over the instrument flight path for neutron polarization preservation. Operating the RF at a frequency of 88 kHz we obtained the maximum flipping ratios for 2.5 amps current. The resulting total neutron polarization was described by a $P_n = P_p \tanh(P_{He}\Theta\lambda)$ dependence very well where the flipping ratio $F(\lambda)$ is related to the beam polarization $P_n(\lambda)$ by $F(\lambda) = (1 + P_n(\lambda))/(1 - P_n(\lambda))$. Here P_p is incident beam polarization, λ the neutron wavelength, Θ the ^{3}He cells pressure-length-cross section product and P_{He} the cells ^{3}He polarization. Data for F in the three orthogonal directions is shown in fig. 3.

The ^{3}He polarization and density-length product can be determined independently from neutron polarization by measuring the unpolarized neutron transmission over neutron wavelength for the polarized and un-polarized ^{3}He cell in time of flight as described in [16]. Taking the ratio of polarized (T_p) to unpolarized (T_0) transmission gives $T_p(\lambda)/T_0(\lambda) = \cosh(P_{He}\Theta\lambda)$ which can be fit to determine P_{He} and Θ, such data is shown in sec. 4 fig. 5. The same measurement performed for the "Mary" cell used in the flipper optimization measurements gave us a value of ^{3}He polarization of 39% at the time of the end of the measurement and a pressure-length product of 17.5 bar cm.

These measurements also helped us verify the adibaticity of the neutron guide fields in the 3 orthogonal directions of PASTIS as each direction gave equivalent flipping ratios. Given the deterministic nature of the neutron polarization from polarized ^{3}He over wavelength, any deviation from the $P_p \tanh(P_{He}\Theta\lambda)$ dependence would imply a non-uniform incident beam polarization resulting from the incident beam polarizer, non-ideal flipping efficiency or polarization losses in the neutron transport. The latter case is ruled out because all three of the orthogonal directions of the PASTIS system and guide fields gave the same smooth $P_p \tanh(P_{He}\Theta\lambda)$ dependence and if there were a weak beam depolarization, it would neither be uniform over neutron wavelength nor is it likely that it would remain the same for the three very different field configurations of the PASTIS system. Also the flipper efficiency is assumed high as it is an adiabatic device and any deviation from adiabaticity would exhibit itself as oscillations in neutron polarization over neutron wavelength below a maximum value limited by the product of the incident beam polarization and ^{3}He analyzing power, especially at low wavelength. Incidentally, a slightly non-adiabatic neutron guide field would exhibit itself in the same way. Because we do no see such oscillations, and since this measurement gives us a

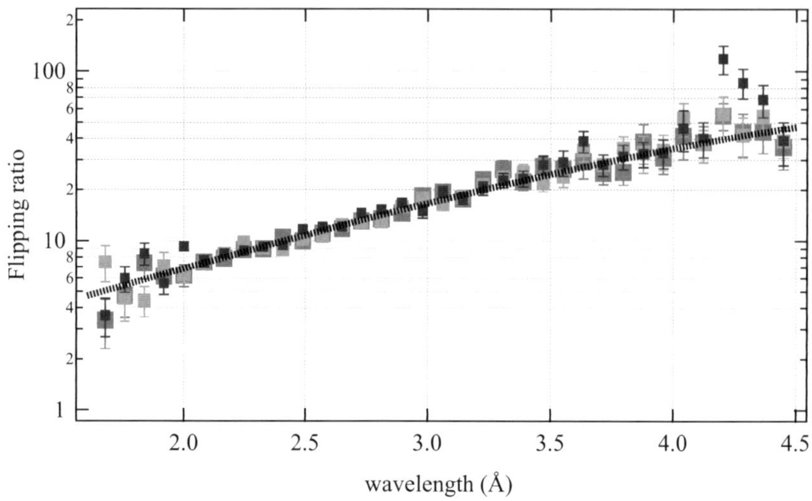

Figure 3. (Total neutron flipping ratio F_n which is related to the total neutron polarization P_n) by $F = (1+P_n)/(1-P_n)$ for each of the three orthogonal field orientations after optimization of the RF gradient flipper using the "Mary" cell which had 17.5 bar cm of ^{3}He and a polarization of 39% at the time of this measurement. Red is the vertical direction "Y", blue is the transverse direction "X" and green in the longitudinal direction "Z". This cell had a higher opacity (product of ^{3}He pressure and length) than the cell used for the overnight measurements, thus it had very low transmission above 4.5 Å for the dark state, i.e. incident beam polarization antiparallel to the ^{3}He analyzer polarization so the flipping ratios above this wavelength are noise/background limited. The dotted black line is a fit assuming a $P_n = P_p \tanh(P_{He}\Theta\lambda)$ dependence of the neutron polarization. This data allows us to conclude the incident beam polarization can be considered uniform with adiabatic transport in the X, Y and Z field directions over the range of neutron wavelengths we used .

P_p consistent with the overnight measurements in sec. 4 and gives P_{He} values consistent with the independent measurements of P_{He} from unpolarized neutron transmission, we conclude the polarization transport to be adiabatic and the flipper efficiency to be high within the errors of the measurement.

After the race-track guide fields, we placed our TOPAS X-Y-Z guide fields and the PASTIS system, as one can see in the photo of fig. 1. The guide fields plus PASTIS require a total of 11 independently set power supplies, and a 90° rotation of one of the permanent-magnet Halbach rings to obtain the 3 orthogonal field configurations. For this test the Halbach ring rotation was not automated, but simply rotated by hand between the two configurations required. The large number of power supplies did however require automation, to prevent errors as well as to minimise the time required to switch between field directions and thus minimize possible losses in ^{3}He polarization due to higher field gradients while switching. For this purpose we set the Delta-Elektronica power supplies via a profi-bus interface that was addressed over an ethernet port by a control computer running a PYTHON application and GUI. A screen shot of the GUI is shown in figure 4. This interface allowed us to set the default currents for all of the power supplies and enable and disable them in groups corresponding to the 3 orthogonal field directions. The application also insured that there is always one field ON in the PASTIS system, and enables the new set direction for the holding field before shutting off the prior field with a small time delay. After configuring the currents to the previously determined optimal values in

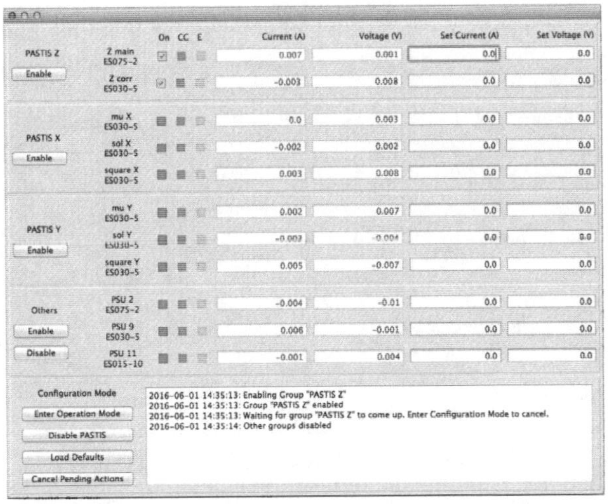

Figure 4. Screen shot from the GUI interface for control of PASTIS and the PASTIS guide field systems. This simple interface allows us to easily set the currents on the PASTIS power supplies and lock them to the given value. Once the current defaults for all the power supplies are set, this GUI allows one to simply "choose" the field direction with the press of a virtual button, which can be performed directly in this GUI or via a tcip command. The program prevents the inadvertent depolarisation of the ^{3}He cells during field switching by locking the set currents, and ensuring that the ^{3}He cell is always in an appropriate magnetic field during operation mode.

the GUI, one enters the operation mode and can simply give a command (mouse click) to choose the X, Y or Z field direction. This command can also be sent from any eventual instrument control computer and be integrated into measurement scripts.

4. Over-night measurement of cell parameters

After optimization of the installation and neutron flipper a "fresh" ^{3}He cell was transported from Garching and installed for measurements. A suitable c-shaped cell was not available for this measurement so we used a D=4.3 cm by L=15 cm cylindrical cell which has a 230 hour ^{3}He T_1 in laboratory conditions and 3.29 bar of ^{3}He. The cell cylindrical axis was placed perpendicular to the neutron beam off-center in the PASTIS coils behind a sample holder in the middle (see fig. 1). This cell is of comparable size and pressure to an eventual c-shaped cell optimized for a cold neutron spectrum, which would provide a neutron polarization of 94% at 2.4 Å with an initial ^{3}He polarization of 70% for example. This cell's 14.2 bar cm neutron path length in this configuration clearly doesn't compare to the approximately 30+ bar cm that will be required for the thermal neutron spectrum of TOPAS, production of suitable high performance cells for both the cold-neutron spectrum on NEAT and the thermal-neutron spectrum of TOPAS is the topic of continuing work [17].

The cell was removed from the polarizer at the MLZ in Garching at around 9 am and was eventually installed in PASTIS on V20 at around 7 pm. Using measurements of the flipping ratio and total transmission over time we were able to characterize the ^{3}He polarization of the cell. For each flipping ratio measurement, the PASTIS system was set to a different direction. Performing a global analysis of the transmission data taken from 7:30 pm until approximately noon the next day, shown in figure 5, we were able to determine the initial ^{3}He polarization of 50 % and on-beam lifetime of 109 hours. During this overnight measurement, the field direction was switched a total of 10 times, with measurements of the transmitted and the scatted beam

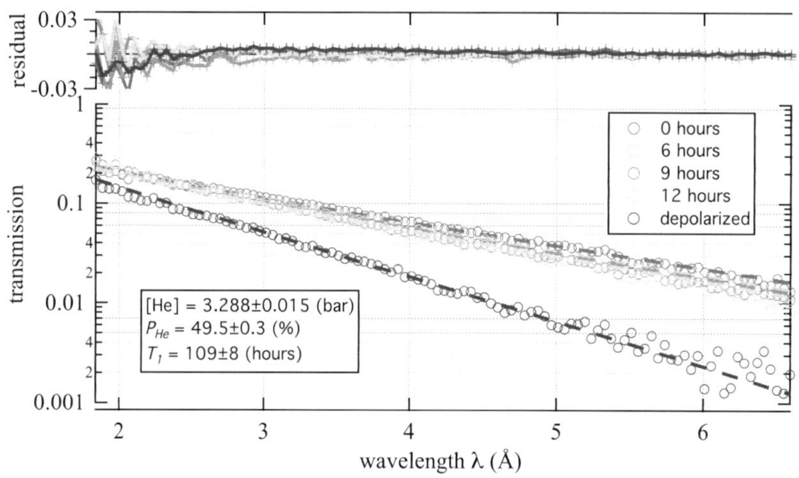

Figure 5. (Neutron transmission) Graph of the neutron transmission vs wavelength at different times and field orientations during our overnight measurement. A global analysis of this data gives a ^{3}He $T_1 = 109$ hours and an initial polarization of $P_{He} = 49.5\%$.

from a sample for the two neutron polarizations (flipper On and flipper Off) acquired for each field setting. The transmission measurements were typically recorded for 5 min each, and 1 hour each for the scattered beam. The starting polarization of 49.5 % is certainly not optimal but was sufficient for this test. Considering the transport via car and the 10 hour delay after polarization this value was a positive first result that we believe can be improved upon with further experience. The eventual regular use of this device on TOPAS will be in Garching where the cell polarization will take place nearby the instrument further helping to minimize any polarization losses during cell transportation. For the data acquisition, the position-sensitive-detector was not yet available on V20, thus we used an array of 4, 40 cm long ^{3}He detector tubes placed horizontal in order to achieve a wide effective Q-range. A beam stop was made of cadmium and was attached to the detector housing such that the transmitted beam could be blocked for the scattering measurements, and conversely extra cadmium masks where also attached to the detectors to block the scattered neutron beam while performing transmission measurements.

The sample used was a hydrated graphite powder in an aluminum holder. The expected signal was thus a combination of the diffuse non-spin flip, coherent scattering of the graphite powder plus a diffuse incoherent scattering from the hydrogen, with the usual 2/3 spin-flip scattering characteristic [18]. No difference in the spin-flip scattering is expected for the diffuse-incoherent only scattering with respect to field orientation, however the scattering experiment is performed in the 3 orthogonal directions in order to verify equivalence of the data and confirm that no artefacts come from the PASTIS or guide field installation. This sample was also measured after drying it in a 100° C oven for 2 hours, but for only one field direction. Polarized beam transmissions over time taken during these measurements are used to find the total beam polarization P_n at different times during the measurement, these data are shown in fig. 6.

While the beam intensity was sufficient for statistics on the transmitted beam, the scattered beam however, even after one hour of integration, provided only on the order of 10,000 counts for the entire spectrum from 1.6 Å to 6.5 Å . Even after binning the data in 0.25 Å steps, the relative errors remain on the order of 5-10% for the raw data, and since the transmission

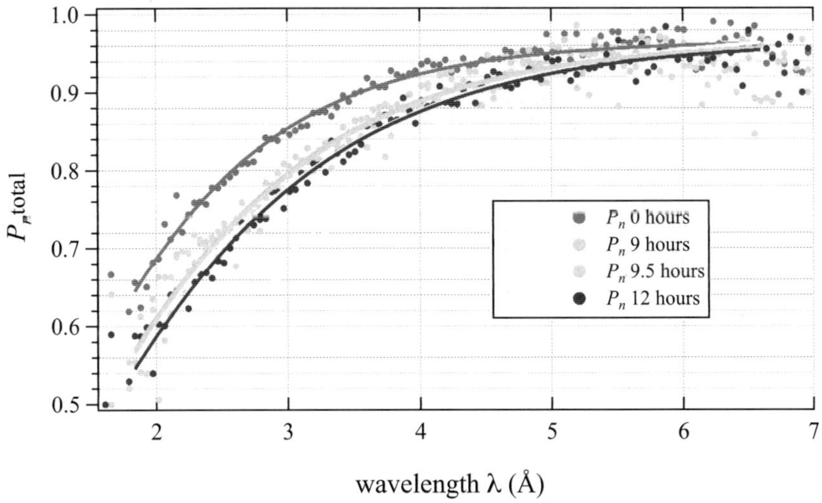

Figure 6. (total neutron polarization) Graph of the neutron polarization vs wavelength at different times and field orientations during our overnight measurement. The solid lines are fits of the data to $P_n = P_p \tanh(P_{He}\Theta\lambda)$.

of the empty cell used was anomalously low w.r.t. the sample measurements (a different but "identical" aluminum cell was used for the empty cell attempt) conclusions from the scattered-beam measurements are not readily obtainable. The data was however processed to separate the fraction of incoherent scattering from the coherent scattering as in [19, 20] and we did not see statistically significant differences in the scattered beam for the 3 different directions of the hydrated sample.

5. Conclusion and next steps

In realistic experimental conditions we have verified that the magic PASTIS field system and guide field configuration can perform to desired specifications in real world conditions. The ^{3}He lifetime met the desired benchmark of over 100 hours on beam while performing a number of X-Y-Z field directions changes that one might employ during a measurement. The neutron polarization transport was also verified to be adiabatic in agreement with the magnetic field calculations, up to the shortest available neutron wavelength of this instrument which was 1.6 Å. The control of the 11 power supplies used to control both the magic PASTIS system and the guide fields functioned as required to enable us to switch between field directions simply and quickly without losses of ^{3}He polarization.

Consequently further practical testing and proof-of-principle work is justified with actual TOF spectroscopy measurements, also with samples of scientific interests. In the near future we expect experiments to be conducted on NEAT at HZB [11] after the beam polarization is installed and commissioned and suitable long lifetime wide angle cells are available.

Acknowledgments

We acknowledge the many people who have contributed to the PASTIS and magic PASTIS throughout the years, notable colleagues who have been involved in discussions and prior work towards this method include A K Petoukov (ILL), J R Stewart (ISIS), K H Andersen (ESS), J Stride (Univ. of South Wales), T R Gentile, W C Chen and A Kirchhoff (NIST). We also

IOP Conf. Series: Journal of Physics: Conf. Series **862** (2017) 012002 doi:10.1088/1742-6596/862/1/012002

thank the technical support of the JCNS at MLZ in Garching and Jülich, the support of HZB, J. Plomp and A. Kusmin (TU Delft) who provided some equipment on V20 (fig. 2) and the students T. Theisselmann, J. Schmeissner, and D. Starostin who have assisted with the JCNS ^{3}He project over the years. Partial funding and support have come from the BMBF through the ESS design update project 05E10CJ1.

References

[1] Voigt J, Soltner H, Babcock E, Aldus R J, Salhi Z, Gainov R R and Brückel T 2015 *EPJ Web of Conferences* **83** 03016 [DOI:10.1051/epjconf/20158303016]

[2] Stewart J R, Andersen K H, Babcock E, Frost C D, Hiess A, Jullien D, Stride J A, Barthélémy J F, Marchal F, Murani A P, Mutka H, Schober H 2006 *Physica B* **385-386** 1142-1145 [DOI:10.1016/j.physb.2006.05.393]

[3] Woracek R, Hofmann T, Bulat M, Sales M, Habicht K, Andersen K, Strobl M 2016 *Nuclear Instruments and Methods in Physics Research Section A* [DOI:10.1016/j.nima.2016.09.034]

[4] Heil W, Andersen K H, Cywinski R, Humblot H, Ritter C, Roberts T W, and Stewart J R 2002 *Nuclear Instruments and Methods in Physics Research A* **485** 551-570 [DOI:10.1016/S0168-9002(01)00926-3]

[5] Chen W C, Gentile T R, Ye Q, Kirchhoff A, Watson S M, Rodriguez-Rivera J A, Qiu Y, Broholm C 2016 *J. of Physics Conference Series* **746(1)** 012016 [DOI:10.1088/1742-6596/746/1/012016]

[6] Rols S, *ILL Endurance Brochure* 2013 **PANTHER** p. 26 http://wwwnew.ill.fr/fileadmin/users files/documents/links/documentation/ILL Endurance Brochure 6_Mo.pdf

[7] Andersen K H, Jullien D, Petoukhov A K, Mouveau P, Bordenave F, Thomas F, Babcock E 2009 *Physica B* **404** 2652-2654 [DOI:10.1016/j.physb.2009.06.054]

[8] Beecham C J, Boag S, Frost C D, McKetterick T J, Stewart J R, Andersen K H, Bentley P M, Jullien D 2011 *Physica B* **406** 2429-2432 [DOI:10.1016/j.physb.2010.11.054]

[9] Winn B, Filges U, Garlea V O, Graves-Brook M, Hagen M, Jiang C, Kenzelmann M, Passell L, Shapiro S M, Tong X, 2015 *EPJ Web of Conferences* **83** 03017 [DOI:10.1051/epjconf/20158303017]

[10] Carlile C J, et al., Technical Design Report for the European Spallation Source *European Spallation Source* 2013 [DOI:10.13140/RG.2.1.2040.6483/1]

[11] Günther G, Russina M 2016 *Nuclear Instruments and Methods in Physics Research Section A* **828** 250-261 [DOI:10.1016/j.nima.2016.05.022]

[12] Stewart J R, Deen P P, Andersen K H, Schober H, Barthélémy J-F, Hillier J M, Murani A P, Hayesband T, Lindenau B 2009 *J. of Applied Crystallography* **42** Issue 1 69-84 [DOI:10.1107/S0021889808039162]

[13] Heinz Maier-Leibnitz Zentrum, DNS: Diffuse scattering neutron time-of-flight spectrometer 2015 *J. of large-scale research facilities* **1** A27 [DOI:10.17815/jlsrf-1-33]

[14] Salhi Z, Babcock E, Gainov R, Bussmann K, Kämmerling H, Pistel P, Russina M, Ioffe A 2016 *J. of Physics Conference Series* **711**(1) 012013 [DOI:10.1088/1742-6596/711/1/012013]

[15] Strobl M, Bulat M, Habicht K 2013 *Nuclear Instruments and Methods in Physics Research Section A* **705** 74-84 [DOI:10.1364/OE.23.000301]

[16] Chupp T E, Coulter K P, Kandes M, Sharma M, Smith T B, Jones G, Chen W C, Gentile T R, Rich D R, Lauss B, Gericke M T, Gillis R C, Page S A, Bowman J D, Penttila S I, Wilburn W S, Dabaghyan M, Hersman F W, Mason M 2007 *Nuclear Instruments and Methods in Physics Research Section A: Accelerators, Spectrometers, Detectors and Associated Equipment* **574** 500-509. [DOI: 10.1016/j.nima.2007.02.091]

[17] Salhi Z, Babcock E, Pistel P, Ioffe A 2014 2014 *J. of Physics Conference Series* **528(1)** 012015 [DOI:10.1088/1742-6596/528/1/012015]

[18] Moon R M, Riste T, Koehler W C, 1969 *Phys. Rev.* **181** 920 [DOI:10.1103/PhysRev.181.920]

[19] Gentile T R, Jonesa G L, Thompson A K, Barker J, Glinka C J, Hammouda B, Lynn J W 2000 *Journal of Applied Crystallography* **33**(33) 771-774 [DOI: 10.1107/S0021889800099817]

[20] Babcock E, Salhi Z, Appavou M-S, Feoktystov A, Pipich V, Radulescu A, Ossovyi V, Staringer S, Ioffe A 2013 *Physics Procedia* **42** 154-162 [DOI: 10.1016/j.phpro.2013.03.190]

IOP Conf. Series: Journal of Physics: Conf. Series **862** (2017) 012003 doi:10.1088/1742-6596/862/1/012003

Expected performance of time-gradient magnetic field SESANS diffractometer at pulsed reactor IBR-2

V Bodnarchuk[1], V Sadilov[1,2], S Manoshin[1], R V Erhan[1,3] and A Ioffe[4]

[1]Joint Institute for Nuclear Research, Frank Laboratory of Neutron Physics, Joliot Curie 6, 141980, Dubna, Moscow region, Russia

[2]Lomonosov Moscow State University, Faculty of Physics, Leninskiye Gory 1, 119991, Moscow, Russia

[3]Horia Hulubei National Institute for R&D in Physics and Nuclear Engineering, Department of Nuclear Physics, Reactorului 30, 077125 Bucharest - Magurele, Romania

[4] Jülich Centre for Neutron Science at Heinz Maier-Leibnitz Centre (JCNS@MLZ), Forschungszentrum Jülich GmbH, Lichtenbergstr. 1, 85747 Garching, Germany

bodnarch@nf.jinr.ru

Abstract. The application of the time-gradient magnetic field neutron spin-echo technique (TGF-NSE) aiming to extend the possibilities of the REFLEX reflectometer at the pulsed reactor IBR-2 (Dubna, Russia) towards a high and medium resolution small-angle scattering is considered. This technique requires the use of a linearly increasing magnetic field in the form of the sequence of saw-teeth pulses. Such a technique is well-suited to the pulsed structure of the neutron beam at the IBR-2 reactor. The wide range of neutron wavelengths employed in the time-of-flight (TOF) operation mode allows for the simultaneous coverage of a wide range of spin-echo lengths (corresponding to a wide Q-range in conventional SANS). Virtual experiments carried out using the VITESS simulation package demonstrate that such a setup at IBR-2 will allow for the coverage of the length scales from 100 Å to 6000 Å in a single instrument setting. The use of the TGF-NSE technique in the TOF mode leads to gaps in the measured SANS curves due to the jumps of magnetic field at the end of saw-teeth pulses. However, crucially, these do not harm the quality of the data. Moreover, the lost data can be retrieved by an additional measurement with a delayed start of the magnetic pulse sequence.

1. Introduction.

Recently, a new time-gradient magnetic field neutron spin echo technique (TGF-NSE) was proposed [1] and tested experimentally [2]. In this technique, the neutron speed is encoded by neutron spin (Larmor) precession in the magnetic field of spin rotators. Such magnetic field is perpendicular to the polarization vector of the neutron beam and increased linearly in time during the time-of-flight of neutrons through the instrument.

IOP Conf. Series: Journal of Physics: Conf. Series **862** (2017) 012003 doi:10.1088/1742-6596/862/1/012003

By inclining the spin rotators towards the beam axis one can implement the SESANS technique [3] to achieve a high sensitivity to the scattering angle at the sample. This allows the TGF-NSE setup can be used as a small-angle diffractometer or reflectometer [4]. Practically, the application of this TGF-SESANS technique requires the use of linearly increasing magnetic fields in the form of the sequence of saw-teeth pulses. A NSE signal occurs for neutrons passing through the spin rotators during the linear slope of the magnetic field. Consequently, this technique is very well suited to pulsed neutron beams. In this regard, the time-of-flight (TOF) technique a wide range of neutron wavelengths [5] which allows the simultaneous coverage of a wide range of spin-echo lengths (corresponding to a wide Q-range in conventional SANS [3]).

However, the use of the TGF-NSE technique in the TOF mode results in a particular feature that will be discussed in this article on the example of the TGF-SESANS instrument, which is considered for the realization at the pulsed reactor IBR-2 in Dubna, Russia.

2. TGF-SESANS technique at a pulsed beam of the IBR-2 reactor.

The idea of the TGF-NSE technique is to encode the neutron speed by the phase of the neutron spin (Larmor) precession that is acquired during the propagation across the area of magnetic field. Such field is linearly rising in time and perpendicular to the polarization vector of the neutron beam. The phase of the Larmor precession acquired during the neutron propagation across a pair of oppositely directed magnetic field areas (spin rotators) does not depend on the time of the entry of the neutron in the spin rotator, but only on the time required for the propagation in, and between, the two spin rotators forming the arm of the spin-echo spectrometer (figure 1). Similarly to all known NSE techniques [6, 7], the propagation of the neutrons across the second spin-echo arm, which is identical other, however has the opposite direction of magnetic fields, results in the restoration of the polarization of the non-monochromatic neutron beam (the spin-echo focusing).

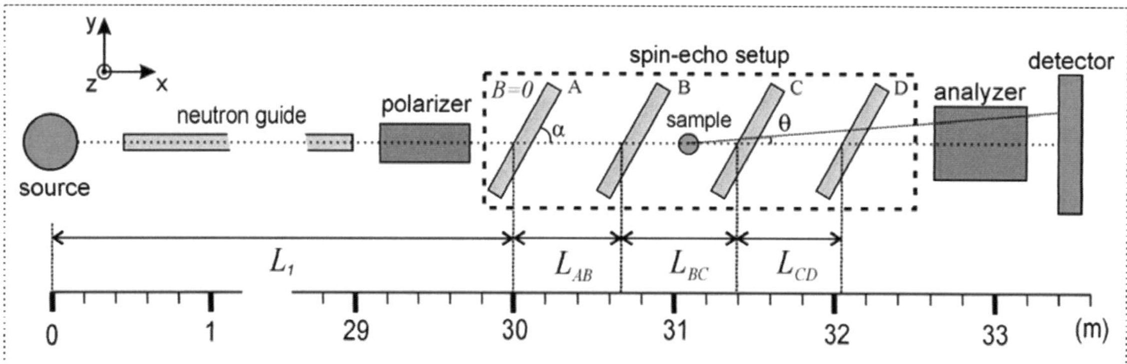

Figure 1. Layout of the SANS instrument using time-gradient magnetic fields (TGF-SESANS). Spin rotators A, B, C and D (of the thickness, a) inclined at an angle, α, to the beam axis are placed in the zero magnetic field volume created by the mu-metal shielding. θ is the angle of scattering on the sample placed between NSE arms. The length scale below shows the position of individual elements with respect to the pulsed neutron source (the reactor core).

The NSE signal occurs for neutrons passing through spin rotators during the linear slope of the magnetic field (figure 2); practically, a sequence of saw-teeth magnetic pulses is used. It should be noted that neutrons which pass the spin rotators during the jump of the magnetic field do not contribute to the NSE signal, as they will acquire a random Larmor precession phase that results in the depolarization of this part of the neutron beam.

The inclination of the spin rotators towards the beam axis of the TGF-NSE instrument yields features of SESANS [4]. In particular, a high sensitivity to the scattering angle θ at the sample placed between two arms. In this case the loss of polarization becomes a measure of the scattering angle θ.

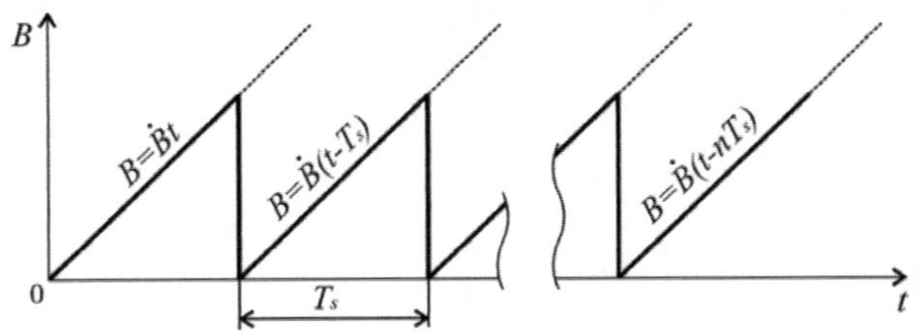

Figure 2. The time structure of the magnetic field in the spin rotators – the sequence of saw-teeth pulse. Only linearly rising parts of the pulses contribute constructively to the NSE signal

In this article we will discuss the use of this technique at neutron beamline 9 of the pulsed reactor IBR-2 in Dubna, where the polarized neutron reflectometer REFLEX [8] is currently operating. For this purpose REFLEX is considered to be additionally equipped with a TGF-SESANS setup formed by two pairs of identical spin rotators, which are tilted at a rather acute angle α (figure 1).

To evaluate the expected performance of TGF-SESANS at IBR-2 we assembled a virtual instrument using the VITESS simulation package [9] extended by dedicated modules to simulate the spin rotators that generate the linearly rising in time magnetic field [10, 11]. The distances between the spin rotators are 71 cm, the thickness of the spin rotators is $a = 2$ cm, the period of saw-teeth pulses is 5 ms, the amplitude of the pulse is 800 G and angular inclination relative to the instrument axis is 10°. The polarization of the neutron beam is assumed to be 100% over the entire wavelength range.

Evidently, neutrons with different wavelengths will experience either linearly rising magnetic field or field jumps. This magnetic field behavior is illustrated in figure 3, where the several possibilities of neutron propagation time over all four spin rotators are shown.

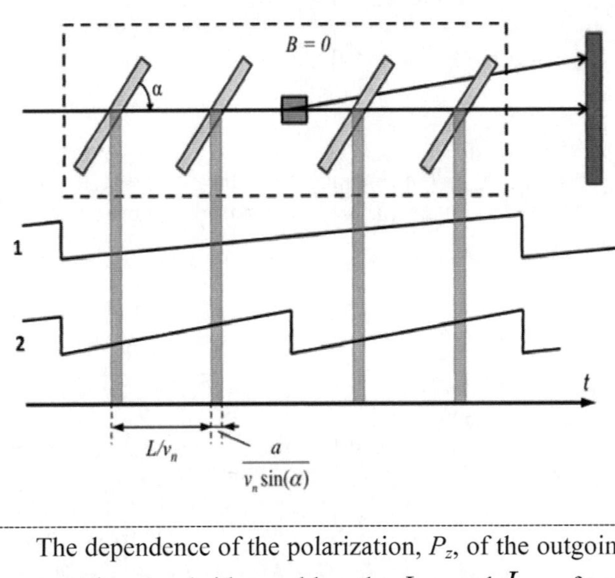

Figure 3. Time diagram illustrating the magnetic field behaviour experienced by neutrons with different wavelengths. For neutrons with shorter wavelengths only one saw-teeth pulse occurs during the neutron propagation time over all four spin rotators (denoted as 1 at the figure). For neutrons with longer wavelength one saw-teeth pulse occurs during the neutron propagation time over two spin rotators (denoted as 2).

The dependence of the polarization, P_z, of the outgoing neutron beam on the wavelength, λ, for the symmetric setup (with equal lengths L_{AB} and L_{CD} of two arms (see figure 1)) is presented in figure 4. Wavelength bands with $P_z = 1$ correspond to NSE focusing, i.e. to the normal operation of a NSE setup that takes place only for neutrons that experience the linearly rising magnetic fields. Wavelength bands with $P_z < 1$ in figure 4 (left) correspond to neutrons that during their propagation through the setup "experience" a jump of the magnetic field in at least one spin turner: in this case no regular NSE signal is observed.

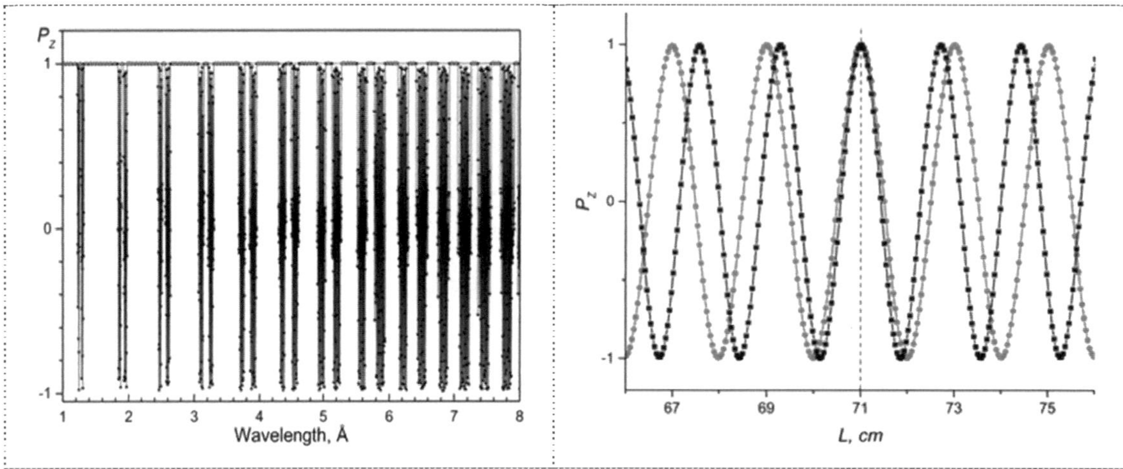

Figure 4. (left) Wavelength dependence of the polarization P_z of the outgoing neutron beam for the symmetric setup without sample. Red dots mark neutron wavelengths used for virtual experiments with samples (see Sect. 3). (right) NSE signals observed for $\lambda = (3.81 \pm 0.03)$ Å (full circles) and $\lambda = (4.02 \pm 0.03)$ Å (full squares) in wide and narrow wavelength bands, respectively.

The position of wavelength bands with $P_z = 1$ can be explained by means of the L-T diagram presented in figure 5. One can see that for $\lambda = 2.24$ Å the neutron propagation across the TGF–NSE setup (spin rotators from A to D) takes place during the rising time of a single *k*-th saw-teeth pulse. However, for $\lambda = 4.43$ Å the jump of the magnetic field takes place during the propagation between two spin-echo arms (between rotators B and C), so that neutrons experience two consequent pulses: *k* and (*k+1*).

The minimal wavelength of neutrons that experience the begin of *k*-th pulse is

$$\lambda_{min}^{wide} = \frac{h}{m_n}\frac{(k-1)T_s}{L_0} \tag{1}$$

where m_n is the neutron mass, h is the Planck constant, L_0 is the distance from the source to the spin rotator A and T_s is the duration of the saw-teeth magnetic field pulses. The maximal wavelength of neutrons that experience the end of *k*-th pulse is

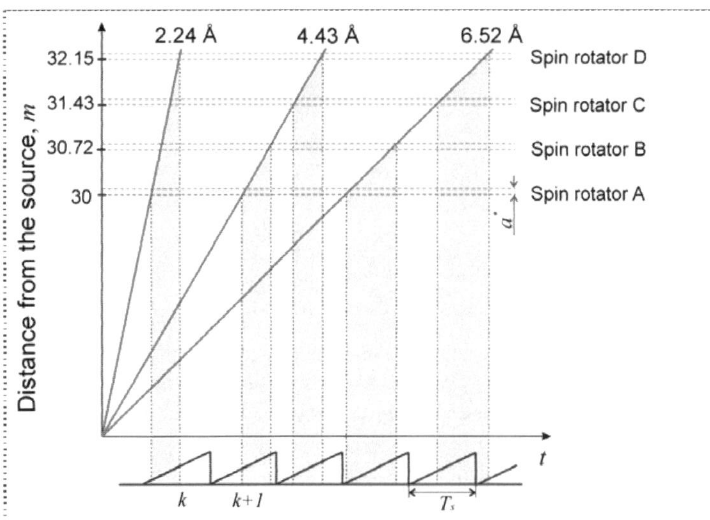

Figure 5. L-T diagram for the TGF-SESANS setup at the IBR-2. (Note that the vertical scale is distorted for the sake of simplicity.) At the bottom of the diagram the sequence of saw-teeth magnetic field pulses synchronized with the 0.24 ms wide reactor pulse (invisible on this scale) are shown. Horizontal lines indicate positions of spin rotators A-D (see figure 1), vertical dashed lines indicate the time when neutrons of different wavelengths cross spin rotators A-D.

$$\lambda_{max}^{wide} = \frac{h}{m_n} \frac{kT_s}{L_0 + L_{AB} + L_{BC} + L_{CD} + a^*} \tag{2}$$

where L_{AB}, L_{BC}, L_{CD} are distances between spin rotators of thickness a (see figure 1), and $a^* = a/sin(\alpha)$ is the effective thickness of the spin rotators.

The values calculated by Eqs. (1-4) (Table 1) are in excellent agreement with positions of the wide wavelength bands obtained by simulations (figure 4).

Table 1. The values of boundaries of wavelength bands calculated by Eqs. (1-4).

Wide bands		Narrow bands	
λ_{min} (Å)	λ_{max} (Å)	λ_{min} (Å)	λ_{max} (Å)
1.32	1.85	1.26	1.29
1.98	2.46	1.89	1.93
2.64	3.08	2.52	2.57
3.30	3.69	3.15	3.21
3.96	4.31	3.78	3.85
4.61	4.92	4.41	4.49
5.28	5.53	5.03	5.13
5.93	6.15	5.66	5.77
6.59	6.77	6.29	6.42

However, from figure 4 it can be seen that wide wavelength bands are separated by narrow ones. Positions of the latter correspond to the jumps of magnetic field between the NSE arms, so that the pairs of spin rotators A-B and C-D experience linear rising magnetic field, however from two adjacent pulses. The boundaries of these wavelength bands correspond to the end of $(k+1)$-th pulse and are given by

$$\lambda_{min}^{narrow} = \frac{h}{m_n} \frac{(k+1)T_s}{L_0 + L_{AB} + L_{BC}} \tag{3}$$

$$\lambda_{max}^{narrow} = \frac{h}{m_n} \frac{(k+1)T_s}{L_0 + L_{AB} + a^*} \tag{4}$$

These values are again in the excellent agreement with simulations (figure 4 (left)).

3. Virtual experiments.

As a sample for virtual experiments we use homogeneous monodisperse spherical particles with the diameter ranging from 100 Å to 6000 Å. Simulations were carried out only for neutrons within the wavelength bands providing $P_z = 1$ in the empty setup (marked by red dots at figure 4 (left))

The width of the wavelength bands corresponding to $P_z = 1$ is about 0.2 Å, which corresponds to time intervals of about 1.7 ms at the detector. These intervals can be easily selected using a standard multi-channel time analyzer with a channel width of 64 μs; the operation of such a setup is feasible at the IBR-2 reactor.

The neutron beam polarization at the exit of the instrument for a sample scattering at small angles ($\theta << 1$) is given by [12]:

$$P(\lambda) = \frac{\int_{\Omega} F cos(\Delta\Phi_{TGF}) d\Omega}{\int_{\Omega} F d\Omega}, \quad (5)$$

where the integration is carried out over the solid angle Ω subtended by the analyzer. F is the form-factor of the scattering particle that depends on the scattering angle, θ, neutron wavelength, λ, and the particle radius, R. The phase, $\Delta\Phi_{TGF}$, of the spin precession is given by [4]:

$$\Delta\Phi_{TGF} = 2\gamma B_{max} f_{saw} aL \frac{m_n^2}{h^2} \frac{cot(\alpha)}{sin(\alpha)} \lambda^2 \theta , \quad (6)$$

or, scattering angle, θ, can be expressed by momentum transfer $Q = 2\pi \cdot sin(\theta)/\lambda$ ($\theta << 1$),

$$\Delta\Phi_{TGF}(\theta) = QZ_{TGF} \quad (7)$$

where

$$Z_{TGF} = \frac{1}{2\pi} 2\gamma B_{max} f_{saw} aL \frac{m_n^2}{h^2} \frac{cot(\alpha)}{sin(\alpha)} \lambda^3 \quad (8)$$

is the spin-echo length obtained by the TGF-SESANS setup and is on a physical length scale which correlates to those of the sample being probed. Thus, providing direct information about the particle structure in real space without an intermediate Fourier transform and without the use of reciprocal space.

The results of virtual experiments shown in figure 6 are in a good agreement with the calculations made using Eq. 5. Small gaps ($P_z = 1$) in $P_z(\lambda)$ dependences are noted, however, they do not really harm the result. Moreover, these gaps can be easily closed by an additional measurement made with a slight offset of the saw-teeth sequence with respect to the neutron pulse, i.e. a delayed start of the pulse sequence.

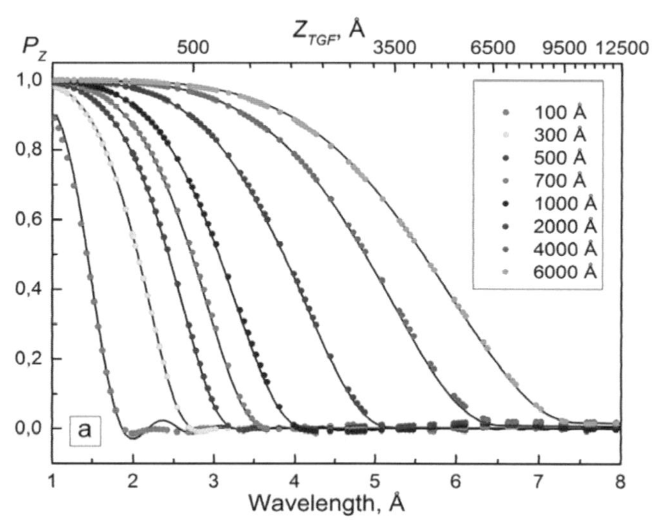

Figure 6. Results of virtual experiments with TGF-SESANS setup. Solid lines are the results of calculations using Eq. (9)

4. Conclusions

The application of the TGF-NSE technique aiming to extend the possibilities of the REFLEX reflectometer at the pulsed reactor IBR-2 (Dubna, Russia) towards a high and medium resolution small-angle scattering is considered. The use of linearly increasing magnetic fields in the form of saw-teeth pulses, is well suited to the pulsed structure of the neutron beam at the IBR-2 reactor.

The wide range of neutron wavelengths employed in the TOF operation mode allows for the simultaneous coverage of a wide range of spin-echo lengths (corresponding to a wide Q-range in conventional SANS, as the latter is proportional to the cube of the neutron wavelength, there is possibilities of SANS studies in a wide length scale). Virtual experiments carried out using VITESS simulation package demonstrate that such a setup at IBR-2 will allow for the coverage of the length scales from 100 Å to 6000 Å in a single instrument setting.

Finally, it should be noted that the use of the TGF-SESANS technique in the TOF mode leads to gaps in the measured SANS curves, caused by the jumps of the magnetic field at the end of the saw-teeth pulses. However, crucially, these do not harm the quality of the data. Moreover, the lost data can be retrieved by an additional measurement with a delayed start of the magnetic pulse sequence.

The financial support from RFBR (grant №14-22-01113-ofi-m) is acknowledged.

References

[1] Ioffe A, Nucl. Instr. and Meth. A **586** (2008) 31–35
[2] Ioffe A, Bodnarchuk V, Bussmann K, Müller R, Georgii R, Nucl. Instr. and Meth. A **586** (2008) 36–40
[3] Rekveldt M Th, Physica B **234–236** (1997) 1135-1137
[4] Ioffe A, Nucl. Instr. and Meth. A **634** (2011) S55-S58
[5] Dalgliesh R M, Langridge S, Plomp J, V.O. de Haan, A.A. van Well, Physica B **406** (2011) 2346–2349
[6] Mezei F, Z. Phyz. **255** (1972) 146
[7] Gahler R, Golub R, J. Phys. C**3-229** (1984) 45
[8] http://flnph.jinr.ru/en/facilities/ibr-2/instruments/reflex
[9] www.helmholtz-berlin.de/vitess
[10] Manoshin S, Ioffe A, Nucl. Instr. and Meth. A **586** (2008) 81–85
[11] Manoshin S.A., Belushkin A.V., Ioffe A., Physics of Particle and Nuclei, V **47**, pp. 667-680 (2016).
[12] Krouglov T, Kraan W H, Plomp J, Rekveldt M Th, Bouwman W G, J. Appl. Cryst. **36** (2003) 816

IOP Conf. Series: Journal of Physics: Conf. Series **862** (2017) 012004 doi:10.1088/1742-6596/862/1/012004

Spin-analyzed SANS for soft matter applications

W.C. Chen[1,2], J.G. Barker[1], R. Jones[1], K.L. Krycka[1], S.M. Watson[1], C. Gagnon[1,2], T. Perevozchivoka[1], P. Butler[1], T.R. Gentile[1]

[1] National Institute of Standards and Technology, Gaithersburg, Maryland 20899, USA
[2] University of Maryland, College Park, Maryland 20742, USA

E-mail: wcchen@nist.gov

Abstract.
The small angle neutron scattering (SANS) of nearly Q-independent nuclear spin-incoherent scattering from hydrogen present in most soft matter and biology samples may raise an issue in structure determination in certain soft matter applications. This is true at high wave vector transfer Q where coherent scattering is much weaker than the nearly Q-independent spin-incoherent scattering background. Polarization analysis is capable of separating coherent scattering from spin-incoherent scattering, hence potentially removing the nearly Q-independent background. Here we demonstrate SANS polarization analysis in conjunction with the time-of-flight technique for separation of coherent and nuclear spin-incoherent scattering for a sample of silver behenate back-filled with light water. We describe a complete procedure for SANS polarization analysis for separating coherent from incoherent scattering for soft matter samples that show inelastic scattering. Polarization efficiency correction and subsequent separation of the coherent and incoherent scattering have been done with and without a time-of-flight technique for direct comparisons. In addition, we have accounted for the effect of multiple scattering from light water to determine the contribution of nuclear spin-incoherent scattering in both the spin flip channel and non-spin flip channel when performing SANS polarization analysis. We discuss the possible gain in the signal-to-noise ratio for the measured coherent scattering signal using polarization analysis with the time-of-flight technique compared with routine unpolarized SANS measurements.
Key words: SANS polarization analysis, soft matter samples, spin-incoherent scattering, inelastic scattering, polarized ^{3}He, ^{3}He NSF

1. Introduction

Over the last three decades, small-angle neutron scattering (SANS) has been a powerful probe for determining the size, shape, distribution, inter- and intra-structures of particles or aggregates, such as colloidal particles, surfactant aggregates, and polymers, often existing in soft condensed matter and biology. For hydrogen-rich soft matter, hydrogen containing molecules yield a great deal of nuclear spin-incoherent scattering during SANS measurements. In order to determine the weak coherent scattering that carries the structural information of the sample, it is necessary to subtract out the nuclear spin-incoherent (NSI) background accurately, which is typically independent of the wave vector transfer Q . However, the subtraction can be challenging because accurate knowledge of the amount of this nearly Q-independent background is critical, particularly at high Q where the coherent scattering is much weaker than the background [1]. An over or under-estimation of the spin-incoherent background may change the slope of the scattering power law at high Q and thus lead to a different structural interpretation.

Polarization analysis is a powerful technique to separate nuclear coherent scattering from magnetic scattering or to separate coherent scattering from nuclear spin-incoherent scattering [2]. Polarization analysis in SANS has been developed at the National Institute of Standards and Technology using ^{3}He neutron spin filters (NSFs) [3, 4, 5, 6] and has recently been employed for hard condensed matter investigations such as magnetic nanoparticles [7], an exchanged-biased system [8], a ferroelectric material [9], and a giant magnetostrictor [10]. SANS polarization analysis has been demonstrated for separation of coherent and incoherent scattering in special soft matter samples [3, 11, 12, 13]. For soft matter samples, typically there is no magnetic scattering. For hydrogen, two-thirds of NSI single scattering occurs in the spin-flip (SF) channel and one-third occurs in the non-spin flip (NSF) channel, resulting in a two-to-one ratio of spin-flip scattering to non-spin flip scattering in the total spin-incoherent scattering if multiple scattering is negligible. However, multiple scattering in many samples reduces the two-to-one ratio to nearly one. The measured non-spin-flip signal contains both coherent scattering and $\frac{1}{3}$ to $\frac{1}{2}$ of the total NSI scattering. In addition there is a significant amount of inelastic scattering in light water [14, 15]. For a typical sample thickness, more than 50 % of the scattering is inelastic, as shown in Table 6 in Ref. [15]. This inelastic scattering appears at a much different wavelength compared to the elastic scattering. Therefore the polarizing efficiency and transmission for the inelastic part are different from the elastic part due to the strong wavelength-dependence of the ^{3}He analyzer. Moreover, the inelastic scattering mixes with the nuclear spin-incoherent scattering, making it impossible to be separated in the current SANS polarization analysis method [3, 4]. So the presence of the inelastic scattering makes separation of the coherent and incoherent scattering incorrect.

In this paper, we report the development of SANS polarization analysis in conjunction with the time-of-flight technique using a chopper to effectively remove the inelastic scattering background. The paper is organized as following. In Sec. 2 we present a brief description of the ^{3}He NSFs and their polarized neutronic properties. In Sec. 3, we describe the polarized beam setup. In Sec. 4, we describe a general scheme of how polarized SANS data can be reduced for soft matter applications. We discuss the polarized beam calibration and time-dependence performance in Sec. 5. We present how separation of coherent and incoherent scattering can be done in combination with determination of the effective spin flip efficiency due to a multiple scattering effect in Sec. 6 and we summarize the paper in Sec. 7.

2. Polarized ^{3}He spin filter as a spin analysis device

A ^{3}He NSF is a transmission-based neutron-polarizing device. It relies upon the strong neutron spin dependent absorption cross section for ^{3}He gas via the resonance reaction ^{3}He$(n,p)^3$H. The absorption cross section $\sigma_+ = \sigma_{\uparrow\uparrow} \approx 0$ when the neutron spin is parallel to the ^{3}He spin, while $\sigma_- = \sigma_{\downarrow\uparrow} = 10666$ b at 1.8 Å when the neutron spin is anti-parallel to the ^{3}He spin. The transmission through an unpolarized beam for two neutron spin states is given by

$$T^{\pm} = T_{\rm E}\exp\left[-\sigma(\lambda)nl\,(1 \mp P_{\rm He})\right] \tag{1}$$

where $T^{\pm}$ is the transmission with neutron spin parallel (+) or antiparallel (−) to the ^{3}He spin. $T_{\rm E}$ is the transmission of the glass windows of the ^{3}He cells that are fabricated from GE-180 glass [16]. $T_{\rm E}$ is typically 0.87. $P_{\rm He}$ is the ^{3}He polarization. $\sigma(\lambda)nl$ is the opacity (gas thickness) of the cell, linearly proportional to the ^{3}He gas density, the wavelength, and the length of the cell. The transmission for an unpolarized neutron beam passing through a polarized ^{3}He cell is then given by

$$T_{\rm n} = T_{\rm E}\exp\left(-\sigma(\lambda)nl\right)\cosh\left(\sigma(\lambda)nlP_{\rm He}\right) = T_0\cosh\left(\sigma(\lambda)nlP_{\rm He}\right) \tag{2}$$

where T_0 is the transmission for an unpolarized neutron beam passing through a unpolarized ^{3}He cell.

3. Experiment setup

The experiments were performed at the National Institute of Standards and Technology Center for Neutron Research on the NG-3 SANS beam-line as shown in Fig. 1a. The SANS instrument is equipped with an V-shape Fe/Si supermirror in a transmission geometry for polarizing the incident neutrons. A precession coil spin flipper is placed immediately after the polarizer with a flipping efficiency of ϵ. A ^{3}He analyzer with a capability of inverting ^{3}He polarization online was used to spin-analyze the scattered beam from the sample. The 2-dimensional position sensitive detector on NG-3 SANS has 128 by 128 array of pixels with a pixel resolution of 5 mm. An in-house-built chopper was located directly in front of the sample. The chopper-sample assembly was placed between a pair of longitudinal coils that were used to provide a sample field of 2.3 mT and define the neutron spin axis at the sample. The chopper had a weakly magnetic rotation shaft near the neutron beam that might depolarize the incident neutron spins (see Sec. 5). However we optimized the field of the longitudinal coils to minimize neutron depolarization from the magnetic shaft. Between the chopper and the neutron guide, we placed a $\frac{\pi}{2}$ spin rotation device that was able to adiabatically rotate the neutron spin from vertical to longitudinal direction along the beam path.

The sample was comprised of 0.5 mm-thick silver behenate powder and 1 mm-thick light water (H_2O). The silver behenate powder was loaded inside a 1.43 cm diameter cadmium aperture between single layers of scotch tape (we estimated to be 0.05 mm thick in total). Light water was loaded in a quartz cell with a window thickness of 2 mm and placed directly behind the powder. SANS measurements were performed at room temperature. The sample to detector distance was set to be 2.88 m to observe the first order diffraction peak of silver behenate at 0.1076 Å^{-1}. We chose a wavelength of 7.5 Å to enhance the separation between the inelastic and elastic time-of-flight (TOF) spectrum. To extend the momentum transfer range and be able to observe the first diffraction peak at 0.1076 Å^{-1} of the silver behenate sample, the ^{3}He analyzer cell was offset to the beam center by 3.5 cm. The chopper spun a four-slot (15 degree opening) disk and operated at 2200 rpm with a duty cycle of 16.7 %. Elastic scattering from a 1 mm thick glassy carbon sample was used to characterize the time-of-flight spectrum showing that inelastic data were stripped from data files by eliminating all events having a time shorter than 4 ms, corresponding to a wavelength of 5.8 Å. The elastic data were extracted out with all events having a time longer than 4 ms. So we had the option of either using or removing the inelastic data for polarization efficiency correction and data analysis.

During the experiment, the ^{3}He polarization was maintained in a compact, magnetically shielded solenoid [17] that is 25 cm in diameter and 35 cm long. The field of the solenoid was parallel to the sample field. For the experiment, ^{3}He gas was polarized offline in one of the NIST SEOP systems [18], then transported to the NG-3 SANS beam-line. The nuclear spin-polarized ^{3}He starts to decay with time once the lasers were turned off. The ^{3}He cells we chose have an intrinsic longitudinal relaxation time of 400 h. The ^{3}He polarization relaxation time was reduced from 400 h to 225 h due to an overall volume-average transverse field gradient of $|\vec{\nabla} B_{\perp}/B| \approx 5.4\times10^{-4}$ cm^{-1} from the stray field of the guide field, but is still long compared to the one-hour counting time. The ^{3}He cell was placed about 25 cm away from the sample to maximize the Q range for this study and under vacuum to reduce air scattering. The ^{3}He cell used for this test has a diameter of 10 cm, covering a maximum scattering angle of $2\theta = 11$ degrees. With a larger ^{3}He cell and different setup at 5 Å, we can cover a maximum scattering angle up to 18 degrees in the future.

The neutron transmissions on the main beam were measured with and without the ^{3}He cell in the beam to determine the ^{3}He polarization. SANS measurements were done with both the chopper and the ^{3}He analyzer in the beam. Polarized beam calibration was done by measuring the four spin-dependent transmissions on the main beam (see Sec. 5). For all measurements, the data were corrected for background and the time dependence of the ^{3}He polarization was corrected for polarized beam calibration.

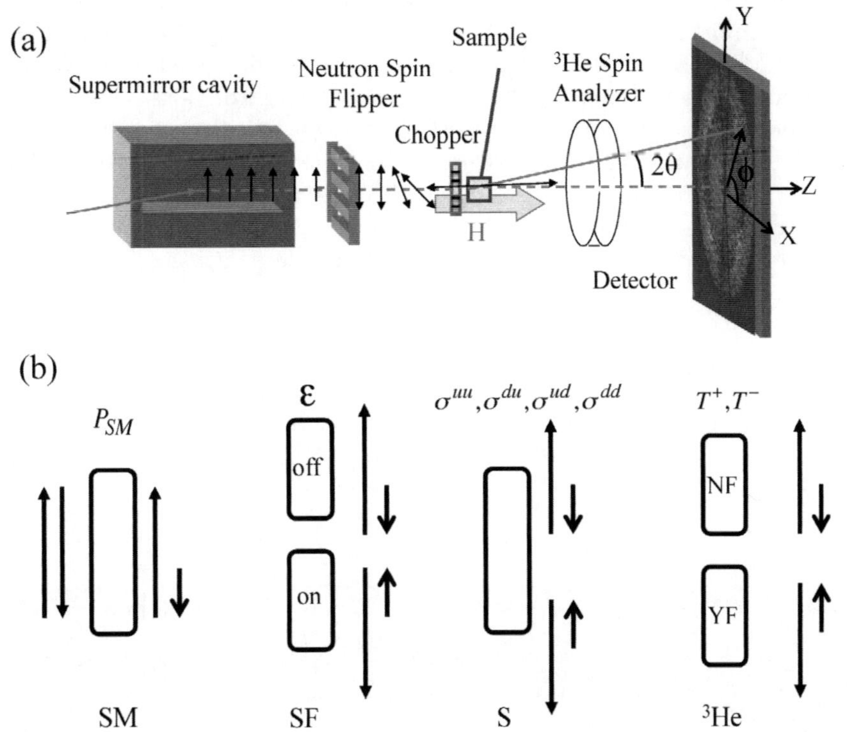

Figure 1. Color online. Schematic diagram of (a) polarized SANS setup (not to scale) with longitudinal polarization analysis and (b) possible neutron spin configurations after each polarizing element. The key instrument components shown in (a) from left to right are supermirror cavity polarizer, spin flipper, chopper and ^{3}He analyzer. An adiabatic $\frac{\pi}{2}$ spin rotation is necessary to rotate the neutron spin from the vertical direction to along the beam path between the spin flipper and chopper. The length of the arrows in (b) mimics the number of neutrons in either spin UP or DOWN. Although the neutron spins lie along the beam path in both the sample and the ^{3}He analyzer location, in the spin configuration illustration in (b), we draw them vertically for simplicity. The setup starts with a unpolarized beam. The supermirror polarizer (SM) has a polarizing efficiency of P_{SM}. The neutron spin flipper (SF) has a flipping efficiency of ϵ. The ^{3}He analyzer (^{3}He) transmits neutrons with polarization in the same direction as that produced by the SM polarizer when not flipped (NF) or in the antiparallel direction as that produced from the SM polarizer when flipped (YF). Neutron transmissions through a ^{3}He analyzer, T^+ or T^- are defined in the text. The four possible spin-dependent scattering cross sections from the sample are defined as σ^{uu},σ^{du}, σ^{ud}, and σ^{dd}.

4. data reduction

In this section, we describe how data reduction can be done for the spin-analyzed SANS data. The data reduction procedure has been described for hard matter samples [4]. For some magnetized samples, it is necessary to measure four spin dependent scattering cross sections. Here it is only necessary to measure two spin dependent scattering cross sections. Here we describe a simplified procedure for soft matter samples. In addition, inelastic scattering is accounted for in the polarization efficiency correction procedure. Fig. 1b shows all possible spin configurations after each polarizing device. By choice of energizing or de-energizing the neutron spin flipper and/or the ^{3}He adiabatic fast passage (AFP) nuclear magnetic resonance (NMR) based flipper, one can measure up to four spin-dependent scattering cross sections, I^{uu}, I^{du}, and I^{ud}, I^{dd}, where the first (second) index refers to the initial (final) neutron spin state and the u (d) symbol corresponds to the UP (DOWN) state of a neutron spin. The four microscopic differential spin-dependent scattering cross sections from the sample are defined as σ^{uu},σ^{du}, σ^{ud}, σ^{dd} accordingly. For soft matter samples without any magnetism, σ^{uu} is equal to σ^{dd} and σ^{ud} is equal to σ^{du}. The measurements were done with the ^{3}He polarization UP. The measured intensities $I^{uu}_{\rm SAM}$ and $I^{du}_{\rm SAM}$ contain small leakages from each imperfect polarizing device. By ray-tracing all possible neutron spin configurations after each polarizing device, $I^{uu}_{\rm SAM}$ and $I^{du}_{\rm SAM}$ can be expressed as

$$
\begin{aligned}
I^{uu}_{\rm SAM} &= \left(\frac{1+P_{\rm SM}}{2}T^{+}_{\rm SAM,uu} + \frac{1-P_{\rm SM}}{2}T^{-}_{\rm SAM,uu}\right)\left(\sigma^{uu}_{\rm SAM} + T_{\rm S}\sigma^{uu}_{\rm EMP}\right) \\
&+ \left(\frac{1+P_{\rm SM}}{2}T^{-}_{\rm SAM,uu} + \frac{1-P_{\rm SM}}{2}T^{+}_{\rm SAM,uu}\right)\left(\sigma^{du}_{\rm SAM} + T_{\rm S}\sigma^{du}_{\rm EMP}\right) + (1-T_{\rm S})\,BGD \qquad (3)
\end{aligned}
$$

$$
\begin{aligned}
I^{du}_{\rm SAM} &= \left[\frac{1-(2\epsilon-1)P_{\rm SM}}{2}T^{+}_{\rm SAM,du} + \frac{1+(2\epsilon-1)P_{\rm SM}}{2}T^{-}_{\rm SAM,du}\right]\left(\sigma^{uu}_{\rm SAM} + T_{\rm S}\sigma^{uu}_{\rm EMP}\right) \\
&+ \left[\frac{1-(2\epsilon-1)P_{\rm SM}}{2}T^{-}_{\rm SAM,du} + \frac{1+(2\epsilon-1)P_{\rm SM}}{2}T^{+}_{\rm SAM,du}\right]\left(\sigma^{du}_{\rm SAM} + T_{\rm S}\sigma^{du}_{\rm EMP}\right) + (1-T_{\rm S})\,BGD \quad (4)
\end{aligned}
$$

where $T_{\rm S}$ is the transmission of the sample. BGD is the measured beam blocked background of the instrument. $P_{\rm SM}$ is the polarizing efficiency of the supermirror polarizer. The subscripts in scattering cross sections either measured or spin-analyzed refer to either the sample run (SAM) or the empty run (EMP) similar to the routine unpolarized SANS measurements[19]. The subscripts in transmissions (T^+ or T^-) of the ^{3}He analyzer refer to spin configurations (uu or du).

The solution to Eqs. 3 and 4 is

$$
\begin{aligned}
(\sigma^{uu}_{\rm SAM} + T_{\rm S}\sigma^{uu}_{\rm EMP}) &= \frac{T^{11}_{\rm SAM}(I^{uu}_{\rm SAM} - BGD) - T^{01}_{\rm SAM}(I^{du}_{\rm SAM} - BGD)}{T^{00}_{\rm SAM}T^{11}_{\rm SAM} - T^{01}_{\rm SAM}T^{10}_{\rm SAM}} \\
(\sigma^{du}_{\rm SAM} + T_{\rm S}\sigma^{du}_{\rm EMP}) &= \frac{-T^{10}_{\rm SAM}(I^{uu}_{\rm SAM} - BGD) + T^{00}_{\rm SAM}(I^{du}_{\rm SAM} - BGD)}{T^{00}_{\rm SAM}T^{11}_{\rm SAM} - T^{01}_{\rm SAM}T^{10}_{\rm SAM}} \qquad (5)
\end{aligned}
$$

where $T^{00}_{\rm SAM}$, $T^{01}_{\rm SAM}$, $T^{10}_{\rm SAM}$, and $T^{11}_{\rm SAM}$ are given as

$$
\begin{aligned}
T^{00}_{\rm SAM} &= \frac{1}{2}\left((1+P_{\rm SM})T^{+}_{\rm SAM,uu} + (1-P_{\rm SM})T^{-}_{\rm SAM,uu}\right) \\
T^{01}_{\rm SAM} &= \frac{1}{2}\left((1+P_{\rm SM})T^{-}_{\rm SAM,uu} + (1-P_{\rm SM})T^{+}_{\rm SAM,uu}\right) \\
T^{10}_{\rm SAM} &= \frac{1}{2}\left[(1-(2\epsilon-1)P_{\rm SM})\,T^{+}_{\rm SAM,du} + (1+(2\epsilon-1)P_{\rm SM})\,T^{-}_{\rm SAM,du}\right] \\
T^{11}_{\rm SAM} &= \frac{1}{2}\left[(1-(2\epsilon-1)P_{\rm SM})\,T^{-}_{\rm SAM,du} + (1+(2\epsilon-1)P_{\rm SM})\,T^{+}_{\rm SAM,du}\right] \qquad (6)
\end{aligned}
$$

Similarly the solution to $\sigma^{uu}_{\rm EMP}$ and $\sigma^{du}_{\rm EMP}$ for the empty run is

IOP Conf. Series: Journal of Physics: Conf. Series **862** (2017) 012004 doi:10.1088/1742-6596/862/1/012004

$$\sigma_{\mathrm{EMP}}^{\mathrm{uu}} = \frac{T_{\mathrm{EMP}}^{11}(I_{\mathrm{EMP}}^{\mathrm{uu}} - BGD) - T_{\mathrm{EMP}}^{01}(I_{\mathrm{EMP}}^{\mathrm{du}} - BGD)}{T_{\mathrm{EMP}}^{00}T_{\mathrm{EMP}}^{11} - T_{\mathrm{EMP}}^{01}T_{\mathrm{EMP}}^{10}}$$
$$\sigma_{\mathrm{EMP}}^{\mathrm{du}} = \frac{-T_{\mathrm{EMP}}^{10}(I_{\mathrm{EMP}}^{\mathrm{uu}} - BGD) + T_{\mathrm{EMP}}^{00}(I_{\mathrm{EMP}}^{\mathrm{du}} - BGD)}{T_{\mathrm{EMP}}^{00}T_{\mathrm{EMP}}^{11} - T_{\mathrm{EMP}}^{01}T_{\mathrm{EMP}}^{10}} \quad (7)$$

Eqs. 5 and 7 describe an appropriate procedure in order to adequately determine σ^{uu} and σ^{du} from the sample. After polarization efficiency correction for both the sample and empty run, σ^{uu} and σ^{du} can be placed on an absolute scale using the incident beam flux measurement.

5. Polarized beam calibration

As the polarizing devices have imperfect efficiencies, proper corrections for small leakage of the undesired neutron spin for each polarizing element is necessary for data analysis. The aim of performing beam calibration is to determine the efficiencies of the supermirror polarizer and the neutron spin flipper. We measured four spin-dependent transmissions, $I_{\mathrm{Trans}}^{\mathrm{uu}}$, $I_{\mathrm{Trans}}^{\mathrm{du}}$, $I_{\mathrm{Trans}}^{\mathrm{ud}}$, $I_{\mathrm{Trans}}^{\mathrm{dd}}$. Similar to Eqs. 3 and 4, we can express the four measured transmitted intensities $I_{\mathrm{Trans}}^{\mathrm{uu}}$, $I_{\mathrm{Trans}}^{\mathrm{du}}$, $I_{\mathrm{Trans}}^{\mathrm{ud}}$, $I_{\mathrm{Trans}}^{\mathrm{dd}}$ in terms of the neutron spin leakage of each polarizing device. Then, the efficiencies of the supermirror and flipper can be solved as following:

$$P_{\mathrm{SM}} = \frac{I_{\mathrm{Trans}}^{\mathrm{uu}}\left(T_{\mathrm{Trans,ud}}^{+} + T_{\mathrm{Trans,ud}}^{-}\right) - I_{\mathrm{Trans}}^{\mathrm{ud}}\left(T_{\mathrm{Trans,uu}}^{+} + T_{\mathrm{Trans,uu}}^{-}\right)}{I_{\mathrm{Trans}}^{\mathrm{uu}}\left(T_{\mathrm{Trans,ud}}^{+} - T_{\mathrm{Trans,ud}}^{-}\right) + I_{\mathrm{Trans}}^{\mathrm{ud}}\left(T_{\mathrm{Trans,uu}}^{+} - T_{\mathrm{Trans,uu}}^{-}\right)} \quad (8)$$

$$(2\epsilon - 1) = \frac{I_{\mathrm{Trans}}^{\mathrm{dd}}\left(T_{\mathrm{Trans,du}}^{+} + T_{\mathrm{Trans,du}}^{-}\right) - I_{\mathrm{Trans}}^{\mathrm{du}}\left(T_{\mathrm{Trans,dd}}^{+} + T_{\mathrm{Trans,dd}}^{-}\right)}{P_{\mathrm{SM}}\left[I_{\mathrm{Trans}}^{\mathrm{dd}}\left(T_{\mathrm{Trans,du}}^{+} - T_{\mathrm{Trans,du}}^{-}\right) + I_{\mathrm{Trans}}^{\mathrm{du}}\left(T_{\mathrm{Trans,dd}}^{+} - T_{\mathrm{Trans,dd}}^{-}\right)\right]} \quad (9)$$

where T^{+} and T^{-} are defined above. For this experiment, the polarizing efficiency of the supermirror was determined to be 0.906 ± 0.008 at 7.5Å. This is lower than 0.94, a value that we have typically obtained without a sample and a chopper. The lower polarizing efficiency might be due to use of a larger aperture in front of the supermirror polarizer for higher incident beam intensity and a magnetic motor shaft of the chopper that was necessarily placed near the beam. The flipper efficiency ϵ was determined to be 0.979 ± 0.006. The ^{3}He analyzing efficiency was over 98.5 % through the entire test. The flipping ratio was measured to be 14 at the beginning of the experiment. Neutron spins after the sample were flipped by inverting the ^{3}He polarization of the analyzer using the AFP NMR technique. During each AFP inversion, the ^{3}He polarization loss was previously determined to be 0.003 % using both the NMR and neutron transmission method to monitor the ^{3}He polarization [20]. So we ignored this small loss when performing the polarized beam calibration.

The ^{3}He polarization decays exponentially with a time constant T_1 and $P_{\mathrm{He}}(t) = P_{\mathrm{He}}^{0} exp\left(-\frac{t}{T_1}\right)$ where P_{He}^{0} is the ^{3}He polarization at the beginning of the experiment. Accordingly, the polarized neutronic performance (analyzing efficiency and transmission of the ^{3}He analyzer, instrumental flipping ratio) changed with time, hence it is necessary to correct these time-dependent efficiencies and transmissions. To characterize T_1, we determined the ^{3}He polarizations at regular time intervals by measuring the neutron transmission of the polarized ^{3}He cell with unpolarized neutrons using Eq. 2. The opacity of the cell was determined to be 2.94 ± 0.02. During the experiment, T_1 was determined to be (224 ± 8) h.

6. results

We began with unpolarized measurements for the combined sample of silver behenate powder ($CH_3(CH_2)_{20}COOAg$) back filled with 1 mm-thick light water. Silver behenate powder has been used as a SANS wavelength calibration method where the first three diffraction peaks can be seen in the SANS instrument [21]. The silver behenate sample had a small NSI background from hydrogen along with the diffraction peaks [21]. Light water is a strong NSI scatterer. The combined silver behenate and water sample contained diffraction peaks as well as a significant nearly Q-independent NSI background. SANS polarization analysis was used to separate the coherent scattering from NSI scattering. The water sample was added to test our ability to remove a strong NSI signal from measured small angle scattering. For a direct comparison with the polarized beam experiment using a chopper, the unpolarized measurements were carried out using the same experimental setup as that used in the polarized measurements. The unpolarized data for the combined sample were corrected for background, detector efficiency, empty sample cell holder, and sample transmission using the SANS software [19]. The data were placed on an absolute scale using the direct beam flux measurement. The presence of the magnetically shielded solenoid and choice of 7.5 Å limited by the chopper allowed us to observe the first diffraction peak only as shown in Fig. 2. The data, radially averaged and on an absolute scale, are shown in Fig. 3. The first diffraction peak is located at 0.1076 $Å^{-1}$, as expected.

Polarized measurements with the chopper were performed for the non-spin-flip (UU) and spin-flip (DU) scattering with and without the chopper as discussed in the data reduction section. As depicted in Sec. 4, polarized data were corrected for background, detector efficiency and time-dependent transmission and polarization efficiency of the ^{3}He analyzer. Again, the data were also placed on an absolute scale using the direct beam flux measurement. Small-angle scattering from the ^{3}He glass container contributes an additional background and has been measured to be approximately Q-independent for the Q range interested here [17]. This was properly subtracted as shown in the reference [4]. This procedure was applied for the sample and empty

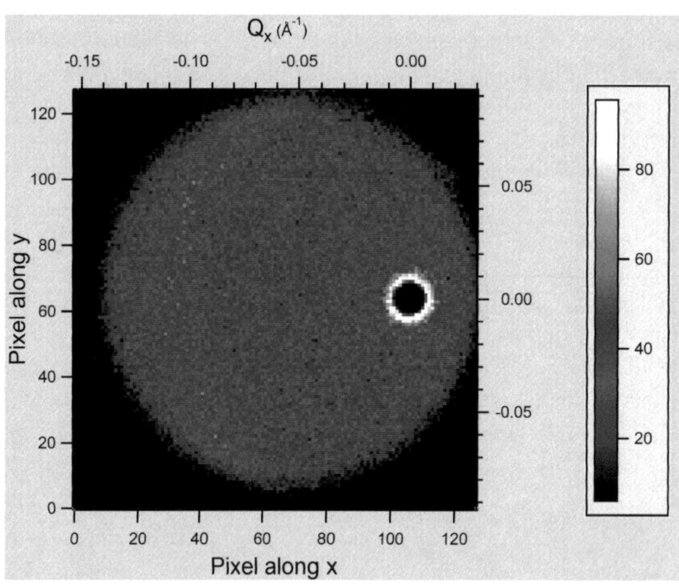

Figure 2. Color online. SANS image for the unpolarized measurements for silver Behenate powder back filled with light water. The large circle indicates the Q limit from the neutron shielding material attached to the mu-metal end cap hole of the magnetically shielded solenoid.

run separately. It has been reported that 1 mm-thick light water produced a significant inelastic scattering [14, 15]. With the chopper in the polarized beam configuration, polarization analysis can be done with options of either ignoring the inelastic data (elastic data only) or using both elastic and inelastic data. The latter is what has been demonstrated before [3] for a sample where inelastic scattering was ignored and multiple scattering was small. In both configurations shown in Fig. 3, σ^{du} is nearly Q-independent from part of spin-incoherent scattering and σ^{uu} shows a diffraction peak at 0.1076 $\mathrm{\AA}^{-1}$ as well as a nearly Q-independent spin-incoherent scattering background. The height and width of the diffraction peak with polarization analysis are the same as those from unpolarized measurements. However, both σ^{uu} and σ^{du} determined with only the elastic component yield a reduced spin-incoherent scattering component due to removal of the inelastic background.

For a hydrogen-containing soft matter sample using the SANS configuration shown in Fig. 1,

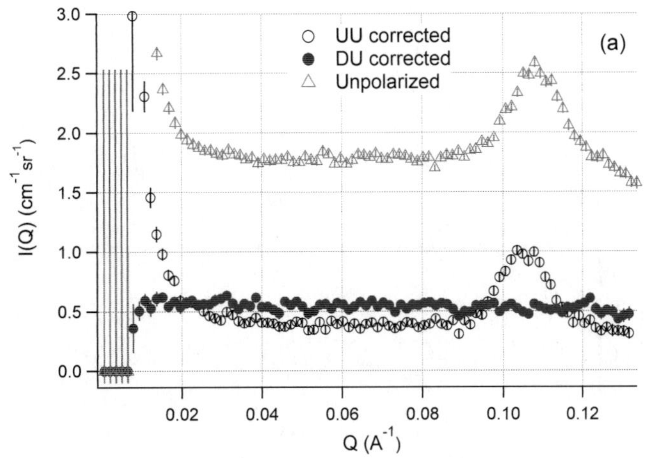

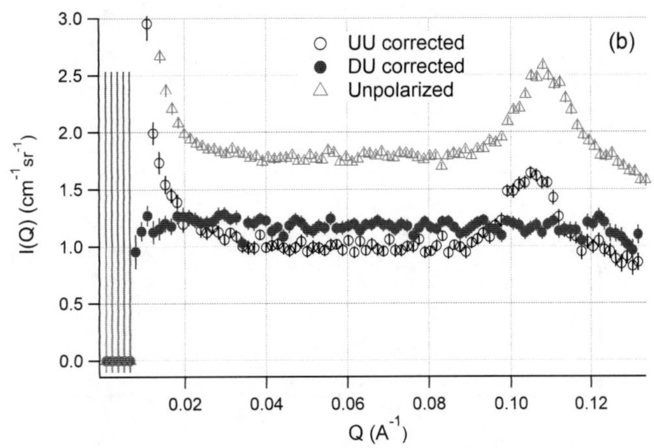

Figure 3. Color online. Polarization efficiency corrected, radially averaged data on an absolute scale ($\mathrm{cm}^{-1}\mathrm{sr}^{-1}$) for a sample of 0.5 mm-thick silver behenate back filled with 1 mm-thick light water for (a) the elastic data only and (b) both elastic and inelastic data. Non spin-flip data σ^{uu} are shown in black open circles; spin-flip data σ^{du} are shown in blue solid circles. Unpolarized data (red open triangles) are also shown for comparison. All uncertainties given in the paper represent one standard deviation.

IOP Conf. Series: Journal of Physics: Conf. Series **862** (2017) 012004 doi:10.1088/1742-6596/862/1/012004

two-thirds of nuclear spin-incoherent scattering (σ_{NSI}) shows in the spin-flip (SF) channel and one-third shows in the non-spin flip (NSF) channel. As stated previously, the measured non-spin-flip signal contains both coherent scattering (σ_{Coh}) and $\frac{1}{3}$ of spin-incoherent scattering. However, the ratios of $\frac{1}{3}$ and $\frac{2}{3}$ of the NSI scattering to the NSF and SF channels will not hold true if there is a significant fraction of multiple scattering events from the hydrogen-containing sample. Neutrons that scatter twice contribute $\frac{5}{9}$ and $\frac{4}{9}$ of the NSI scattering to the NSF and SF channel, respectively; Neutrons that scatter three (four) times contribute $\frac{13}{27}$ ($\frac{41}{81}$) and $\frac{14}{27}$ ($\frac{40}{81}$) of the NSI scattering to the NSF and SF channel, respectively and so on. The multiple scattering in such a sample effectively decreases the two-to-one ratio of the SF channel to NSF channel of the NSI scattering, approaching an one-to-one ratio. Assuming that the probability of the spin flip scattering in the total NSI scattering is p ($0.5 \leq p \leq \frac{2}{3}$), the measured non-spin flip scattering $\sigma^{\mathrm{uu}}_{\mathrm{SAM}}$ and spin-flip scattering $\sigma^{\mathrm{du}}_{\mathrm{SAM}}$ can be expressed as

$$\sigma^{\mathrm{uu}}_{\mathrm{SAM}} = \sigma_{\mathrm{Coh}} + (1-p)\sigma_{\mathrm{NSI}} \tag{10}$$

$$\sigma^{\mathrm{du}}_{\mathrm{SAM}} = p\sigma_{\mathrm{NSI}} \tag{11}$$

To estimate the effect of multiple scattering to the spin flip scattering for 1 mm-thick light water with a measured transmission of 0.5, we have done Monte-Carlo simulation where no inelastic scattering is assumed in the scattering kernel. The Monte-Carlo simulation has been

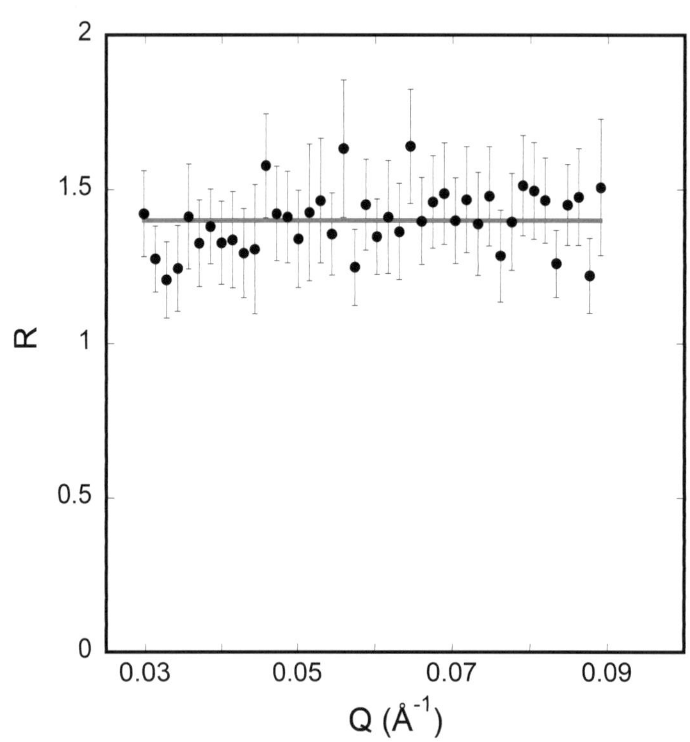

Figure 4. Color online. Ratio R ($R = \frac{\sigma^{\mathrm{du}}}{\sigma^{\mathrm{uu}}}$) in the Q range from 0.03 Å to 0.09 Å where coherent scattering is negligible. σ^{uu} and σ^{du} were polarization efficiency corrected with the elastic data. The ratio was fit to a constant, $\frac{p}{1-p}$. p is determined to be 0.583 ± 0.005, agreeing with the Monte Carlo simulation discussed in the text.

IOP Conf. Series: Journal of Physics: Conf. Series **862** (2017) 012004 doi:10.1088/1742-6596/862/1/012004

thoroughly described in the references [15, 22]. We found that the ratio $\frac{I_{n+1}}{I_n}$ of the intensity of successive orders of multiple scattered neutrons is 0.45. Here I_n is the intensity of neutrons that have been scattered n times. This agrees with the model in the reference [23]. From the Monte-Carlo simulation, we determined p to be 0.58. p can also be determined by fitting the ratio R of $\sigma^{\rm du}_{\rm SAM}$ to $\sigma^{\rm uu}_{\rm SAM}$ to a constant in the Q range from 0.03 Å^{-1} to 0.09 Å^{-1} where coherent scattering is negligible. The fitted constant is equal to $\frac{p}{1-p}$. Fig. 4 shows R versus Q. We found $R = 1.4$ and $p = 0.583 \pm 0.005$, which agrees well with the Monte-Carlo simulation. For samples that exhibit minimal multiple scattering, R would be close to 2 and p would be close to 0.667. Using Eqs. 10 and 11 and the determined value of p, the coherent scattering from silver behenate can be separated from the strong nuclear spin-incoherent scattering in the NSF channel. The results are shown in Fig. 5 for the elastic data (Fig. 5(a)) and for both the elastic and inelastic data (Fig. 5(b)). As a comparison, the unpolarized data are also shown. As expected, the nuclear spin-incoherent scattering background obtained from the elastic data is smaller than that from unpolarized measurements due to removal of the inelastic scattering.

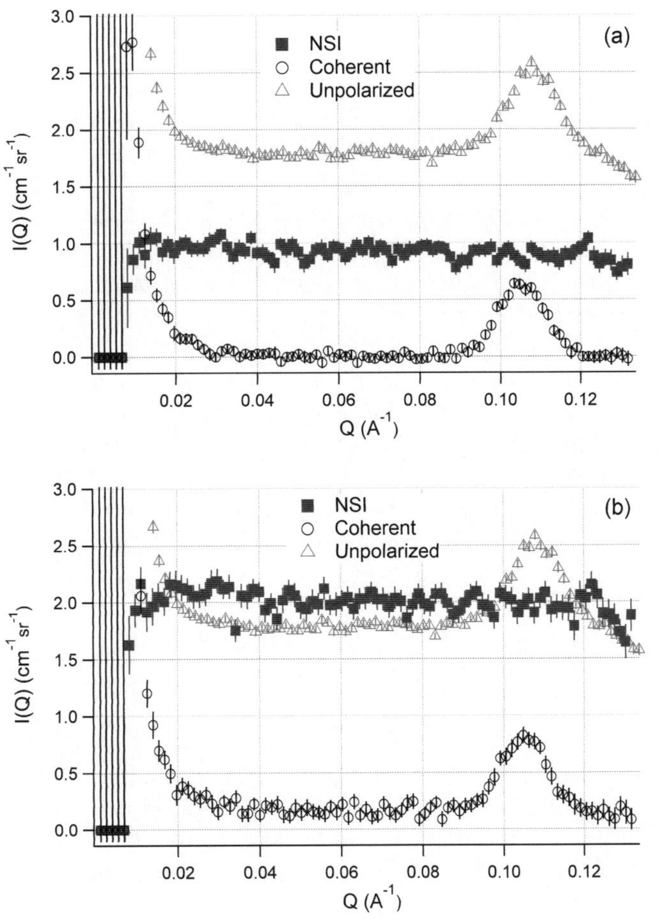

Figure 5. Color online. Separation of coherent scattering $\sigma_{\rm Coh}$ (black open circles) from nuclear spin-incoherent scattering $\sigma_{\rm NSI}$ (blue solid boxes) for a sample of 0.5 mm-thick silver behenate back filled with 1 mm-thick light water for (a) the elastic data only and (b) both elastic and inelastic data. Unpolarized data (red open triangles) are also shown for comparison.

However, the nuclear spin-incoherent scattering obtained using both elastic and inelastic data is higher than that from unpolarized measurements, leading to an unphysical result. Also σ_{Coh} has a residual nearly Q-independent background. This is because the inelastic scattering occurs at different wavelengths with a peak at 1.4 Å [15]. The polarizing efficiency and transmissions of the ^{3}He analyzer are strongly wavelength-dependent. We had an initial ^{3}He polarization of 0.81 during the test, which provided an analyzing efficiency of 0.983 and transmission of 0.496 for the desired spin state at 7.5 Å. This would be 0.417 for the analyzing efficiency and 0.784 for the transmission at 1.4 Å. Therefore, for hydrogen-containing soft matter samples that exhibit both inelastic scattering and nuclear spin-incoherent scattering, the general polarization analysis method [3] may not be adequate for separation of coherent scattering from NSI scattering without the time-of-flight technique to remove the inelastic scattering. If the sample exhibits multiple scattering, the effect of multiple scattering to polarized beam performance has to be taken into account.

To estimate whether there is any improvement in the determined signal-to-noise (S/N) ratio of coherent scattering using the method of polarization analysis combined with TOF, we assume that the uncertainty is due solely to Poisson counting statistics. As shown in Fig. 3(a), the nearly Q-independent background for polarization analysis with TOF is reduced by a factor of 4.5 compared with that from the normal unpolarized SANS measurement. During the experiment, the reduction in count rate was a factor of 6 from the chopper with a duty cycle of 16.7 %, and a factor of 5.5 from the supermirror polarizer and the ^{3}He analyzer. Hence the overall reduction in count rate from polarization analysis with TOF was a factor of 33. The factor of 4.5 improvement in the S/N ratio shown in Fig. 3(a) is cancelled by the factor of 33 reduction in count rate from polarization analysis with TOF. We have installed a superemirror polarizer with a higher polarization efficiency and transmission. The chopper could be operated at a duty cycle of 25 %. In the future, a ^{3}He polarization of 90 % may be achievable [24], resulting in a higher polarization efficiency and transmission for the ^{3}He analyzer. The overall reduction in count rate from polarization analysis with TOF would be a factor of 14. The improvement using polarization analysis with TOF would be still limited compared with a normal unpolarized SANS measurement. For a normal unpolarized SANS measurement for biological macromolecule samples, it is necessary to carefully prepare addition samples such as dilute solution to determine the Q-independent background [1]. However it does not require such a necessity using polarization analysis with TOF. In addition, the Q-independent background can not always be observed at large Q for certain hydrogen-rich samples. A Q-dependence of the measured SANS signal for the large molecular size of the alkines has been reported [25]. In this case, SANS polarization analysis combined with the time-of-flight technique can be used to separate coherent scattering from spin-incoherent scattering, hence offering a different approach to obtain the correct structural information without a need to determine the background accurately.

7. Conclusions

We have done polarization analysis with the time-of-flight technique using a chopper that was used to effectively remove the inelastic background for a sample of 0.5 mm-thick silver behenate back filled with 1 mm-thick light water. We have developed a data reduction procedure for SANS polarization analysis in conjunction with the time-of-flight technique. With such a setup, proper data reduction and the Monte Carlo simulation for multiple scattering estimation, we have unambiguously separated the first diffraction peak of silver behenate from the strong spin-incoherent background for a sample of silver behenate back filled with 1 mm-thick light water that is known to exhibit strong nuclear spin-incoherent scattering, inelastic scattering and multiple scattering. We show that the SANS polarization analysis method without an appropriate correction to inelastic scattering using the time-of-flight technique may not be adequate to separate coherent scattering from spin-incoherent scattering for samples that show inelastic scattering. Polarization analysis combined with the time-of-flight technique reduced Q-independent background from hydrogen by a factor of 4.5 compared with a normal unpolarized SANS measurement for the sample of 0.5 mm-thick silver behenate back filled with 1 mm-thick light water. The factor of 4.5 improvement in the signal-to-noise ratio is cancelled by the factor of 33 reduction in count rate from polarization analysis with TOF. However, for samples where the Q-independent background can not observed at large Q, SANS polarization analysis in conjunction with the time-of-flight technique offers a different approach to determine the structural information from coherent scattering.

Acknowledgments

The authors thank J. Anderson, J. Fuller, and A. Kirchhoff of the NIST Optical Shop for assistance with cell fabrication. We benefitted in using the Monte Carlo simulation code written by Gordon Jones. The work utilized facilities supported in part by the National Science Foundation under Agreement No. DMR-1508249.

References

[1] Rubinson, K. A., stanley, C., & Krueger, S., (2008) *J. Appl. Cryst.*, **41**, 456-465.
[2] Moon, R., Riste, T., Koehler, W., (1969) *Phys. Rev.*, **181**, 920931.
[3] Gentile, T. R., Jones, G. L., Thompson, A. K., Barker, J., Glinka, C. J., Hammouda, B., and Lynn, J. W. (2000) *J. Appl. Cryst.*, **33**, 771-774.
[4] Krycka, K. L., Booth, R. A., Borchers, J. A., Chen, W. C., Conlon, C., Gentile, T. R., Hogg, C., Ijiri, Y., Laver, M., Maranville, B. B., Majetich, S. A., Rhyne, J., and Watson, S. M., (2009) *Physica B*, **404**, 2561-2564.
[5] Chen, W. C., Erwin, R., McIver, J. W. III, Watson, S., Fu, C. B., Gentile, T. R., Borchers, J. A., Lynn, J. W., and Jones, G. L. (2009) *Physica B*, **404**, 2663-2666.
[6] Krycka, K. L., Chen, W. C., Borchers, J. A., Maranville, B. B., Watson, S. M., (2012) *J. Appl. Cryst.*, **45(3)**, 546-553.
[7] Krycka, K. L., Booth, R. A., Hogg, C. R., Ijiri, Y., Borchers, J. A., Chen, W.C., Watson, S.M., Laver, M., Gentile, T.R., Dedon, L. R., Harris, S., Rhyne, J. J., and Majetich, S.A., (2010) *Phys. Rev. Lett.*, **104**, 207203; Krycka, K. L., Borchers, J. A., Ijiri, Y., Hasz, K., Rhyne, J. J., Booth, R. A., and Majetich, S.A., (2014) *Phys. Rev. Lett.*, **113**, 147203.
[8] Dufour, C., Fitzsimmons, M. R., Borchers, J. A., Laver, M., Krycka, K. L., Dumesnil, K., Watson, S. M., Chen, W. C., Won, J., Singh, S., (2011) *Phys. Rev. B*, **84**, 064420.
[9] Ramazanoglu, M., Laver, M., Ratcliff, W., Watson, S. M., Chen, W. C., Jackson, J., Kothapalli, J., Seongsu, L, Cheong, S. -W., and Kiryukhin, V., (2011) *Phys. Rev. Lett.*, **107**, 207206 (2011).
[10] Laver, M., Mudivarthi, C., Cullen, J. R., Flatau, A. B., Chen, W. C., Watson, S. M., and Wuttig, M., (2010) *Phys. Rev. Lett.*, **105**, 027202.
[11] Gaspar, A. M., Busch, S., Appavou, M. -S., Haeussler, W., Georgii, R., Su, Y., and Doster, W., (2010) *Biochim. Biophys. Acta*, **1804**, 76-82.
[12] Ioffe, A., Babcock, E., Pipich, V., Radulescu, A., (2011) *J. Phys.: Conf. Series*, **294**, 012013.
[13] Babcock, E., Salhi, Z., Appavou, M-S., Feoktystov, A., Pipich, V., Radulescu, A., Ossovyi, V., staringer, S., and Ioffe, A.,(2013) *Phys. Procedia*, **42**, 154-162.
[14] Ghosh, R. E., Rennie, A. R., (1999) *J. Appl. Cryst.*, **32**, 1157-1163.
[15] Barker, J. G., Mildner, D. F. R., (2015) *J. Appl. Cryst.*, **48**, 1055-1071 and references therein.
[16] GE Lighting Component Sales, Bldg. 315D, 1975 Noble Rd., Cleveland, OH 44117. Certain trade names and company products are mentioned in the text or identified in an illustration in order to adequately specify the experimental procedure and equipment used. In no case does such identification imply recommendation or endorsement by the National Institute of Standards and Technology, nor does it imply that the products are necessarily the best available for the purpose.
[17] Chen, W. C., Gentile, T. R., O'Donovan, K. V., Borchers, J. A., and Majkrzak, C. F., (2004) *Rev. Sci. Instrum.*, **75**, 3256-3263.
[18] Chen, W. C., Gentile, T. R., Erwin, R., Watson, S.M., Ye, Q., Krycka, K.L., and Marvanville, B.B., (2014) *J. Phys.: Conf. Ser.* **528** 012014
[19] Kline, S. R. (2006) *J. Appl. Cryst.*, **39(6)**, 895.

[20] Optimization of the ^{3}He spin inversion by the AFP NMR method is not important for this test, so will not be discussed in details here and will be published elsewhere.
[21] Gilles, R., Keiderling, U., Wiedermann, A., (1998) *J. Appl. Cryst.*, **31**, 957-959.
[22] Barker, J. G., unpublished.
[23] Mayers, J., & Cywinski, R., (1985) *Nucl. Instr. and Meth.*, **241**, 519531.
[24] We have recently achieved a ^{3}He polarization of 89 %.
[25] Arleth, L., Pedersen, J. S., (2000) *J. Appl. Cryst.*, **33**, 650-652

IOP Conf. Series: Journal of Physics: Conf. Series **862** (2017) 012005 doi:10.1088/1742-6596/862/1/012005

PASTIS2 and CROCODILE: XYZ-wide angle polarisation analysis for thermal neutrons

Mechthild Enderle, David Jullien, Alexander Petoukhov, Pascal Mouveau, Ken Andersen[1], Pierre Courtois

Institut Laue-Langevin, CS 20156, 38042 Grenoble Cedex 9, France

E-mail: enderle@ill.fr

Abstract. We present a wide-angle device for inelastic neutron scattering with XYZ-polarisation analysis (PASTIS2). PASTIS2 employs a banana-shaped Si-walled ^{3}He-filter for the polarisation analysis and allows pillar-free neutron scattering for horizontal scattering angles 0-100°. The guide field direction at the sample can be chosen vertical or with 45° incremental steps in the horizontal scattering plane. When PASTIS2 is implemented on a polarised neutron beam, the incident neutron spin can be flipped with an easy-to-optimise broad-band adiabatic resonant flipper (CROCODILE) independent of the guide field direction at the sample position. We have tested the performance of this new device on the polarised thermal triple-axis spectrometer IN20 at the Institut Laue-Langevin, equipped with Heusler monochromator and the FlatCone multi-analyser, and discuss its potential for future instruments.

1. Introduction

In spite of intense efforts at all major neutron centers, inelastic thermal neutron scattering with wide-angle polarisation analysis is still outperformed by the point-by-point measurement of triple-axis instruments like IN20. Although wide-angle ^{3}He-spin filter cells have been developed, the achievement of a magnetic field system with sufficient field homogeneity for more than one field direction has been a major challenge. The few wide-angle devices for thermal neutrons that are in operation world-wide (including those with supermirror polarisers) are restricted to a single guide-field direction at the sample, or introduce shadowing pillars in the angular range of the analyser/detector, and/or present such a low polarising efficiency that a triple-axis instrument is probably still more efficient even when it comes to measuring entire sections of the four-dimensional energy-wave vector space. In this paper, we describe a ^{3}He-filter device, PASTIS2, which when combined with the neutron spin flipper CROCODILE presents a major progress towards "parallel" polarisation analysis in the thermal neutron range.

PASTIS2 is the successor of PASTIS1 [1], a first-generation device for wide-angle XYZ-polarization analysis which housed a banana-shaped quartz-glass walled ^{3}He-filter. In PASTIS1, eight vertical pillars intersected the horizontal scattering plane and shadowed a substantial part of the angular range, the quartz glass of the cell introduced considerable background. PASTIS2 provides 135° of free access in the horizontal scattering plane, it allows to choose either a vertical magnetic guide field at the sample or field directions in 45° increments in the horizontal plane, important for polarisation analysis of single crystals [2, 3]. We show that changes in the guide field direction can

[1] present address: ESS ERIC, Box 176, 22100 Lund, Sweden

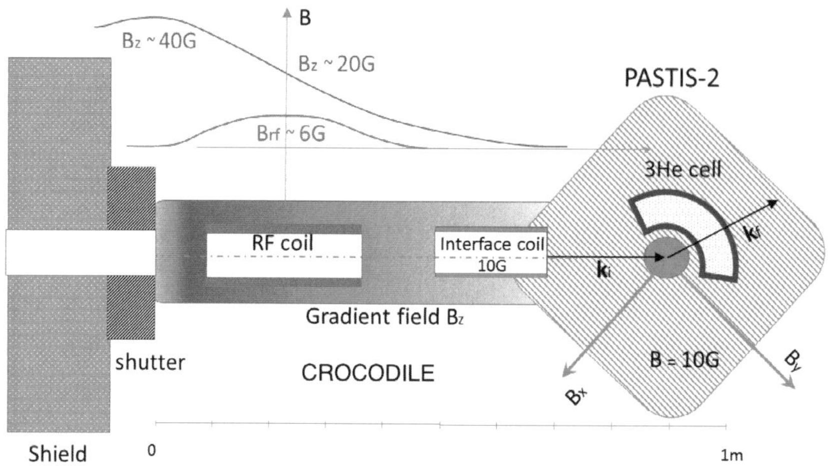

Figure 1. General layout: Monochromator protection and shutter, CROCODILE, and PASTIS2.

be made with negligible polarisation decay of the ^{3}He-filter. The adiabatic fast-passage (AFP) neutron spin flipper CROCODILE permits to flip the incident neutron polarisation independent of the guide field direction at the sample, and flips energy-independent within a broad band of incident neutron energies with an efficiency f above 0.99. We show that it is possible to correct for the finite polarisation and transmission of the ^{3}He-filter and their time-dependences in an inelastic experiment, resulting in a final residual "leakage" of only 2.5% of a phonon into a spin-flip channel, which corresponds to an effective flipping ratio of 40. The Cs-coated Si-walls of the ^{3}He-filter cell have a transmission of over 0.986, and do not introduce any measurable background. The cell covers the full angular range (75°) of the multi-analyser FlatCone. This means that IN20 with FlatCone, PASTIS2, and CROCODILE permits full polarisation analysis over a 75°-range of scattering angles simultaneously. While IN20 with FlatCone provides one energy transfer per measurement, PASTIS2 and CROCODILE work for a broad energy band, and could be associated with broad-band incident or final energies as e.g. on time-of-flight spectrometers. There, PASTIS2 and CROCODILE would allow full polarisation analysis for the full range of scattering angles *and* a broad band of energy transfers simultaneously.

2. Overall Experimental Setup

Equipped with a large horizontally focused Heusler monochromator and analyser, IN20 provides a polarised flux of about 10^8n/cm^{2}s (at k_i =4.5 Å^{-1}). IN20 in its single analyser mode allows to measure energy transfers between 1 meV and 100 meV, provides an energy resolution of 0.9 meV and a flipping ratio of typically 27 for elastic scattering at 2.662Å^{-1}. These specifications allow polarised studies of even weak signals like diffuse inelastic spin-1/2 magnetic continua from 50 mm^3 single crystals as well as the so-called spherical polarisation analysis in diffraction from e.g. 0.01 mm^3 thin-film manganites. For the tests reported here, we used the (polarising) Heusler(111) monochromator, CROCODILE and PASTIS2 (figure 1), and the multi-analyser unit FlatCone [4]. FlatCone covers an angular range of 75° of the scattered neutrons, its 31 vertically bent Si(111) analysers select scattered neutrons with final wave vector k_f = 3Å^{-1} independent of their spin state, and scatter them into the vertically displaced detectors. PASTIS2 (figure 2) is a coil device with non-saturated μ-metal cores [5] providing magnetic guide fields in the

Figure 2. Left: PASTIS2 with 135° free horizontal access. Right: ^{3}He-filter cell, made from Cs-coated single-crystalline Si (cylindric parts) and pyrex (top and bottom).

sample region and houses a banana-shaped Si-walled wide-angle ^{3}He-filter. The filter cell fully covers the 75° horizontal scattering angle of FlatCone. PASTIS2 has 135° free access in the horizontal scattering plane, and allows different field directions at the sample. Throughout this article, we denote with $H||x, y$ horizontal field directions at $-135°$ and $-45°$ to the incident wave vector k_i, and with z the vertical field direction. As described below, PASTIS2 has been modified considerably with respect to its initial version, PASTIS2.0 [5]. The incident neutron polarisation can now be flipped with the broad-band adiabatic fast-passage (AFP) flipper CROCODILE described in more detail below. We performed experiments on the attenuated direct beam, as well as on a Ge single crystal of about 2 cm^3 size, mounted in the $(hh\ell)$-scattering plane inside an Orange cryostat that was kept at 300 K.

3. PASTIS2 design changes

For operation with a horizontal guide field at the sample, the initial PASTIS2.0 [5] lacked sufficient field homogeneity. The experimentally determined RMS gradient ($> 2 \cdot 10^{-3}$cm^{-1}) did not meet the expectations from the simulations ($1 \cdot 10^{-3}$cm^{-1}) [5], and limited the polarisation decay time T_1^{mag} to about 50 hours (h), with an overall T_1 of 20 h in situ in early experiments. The original PASTIS2.0 was improved by replacing the wire-wound x- and y-coils with new coils wound from Cu-tape around the μ-metal cores. The residual field inhomogeneity was measured with a 3D field-mapping robot, and the measured field map was employed in subsequent simulations to design adapted correction coils. A total of 22 correction coils were integrated and optimised via parallel resistances, so that each PASTIS2 field direction (x, y, z) is finally fed by only one power supply. This allows to use standard power supplies to feed PASTIS2 (otherwise high-stability power supplies would have to be employed). With this optimisation, the relative field homogeneity in the ^{3}He-banana cell region was improved by more than a factor 2 for horizontal field directions, from $> 2 \cdot 10^{-3}$ cm^{-1} to $0.85 \cdot 10^{-3}$ cm^{-1}. The polarisation decay time due to field inhomogeneities, T_1^{mag}, measured with a banana-shaped Cs-coated Pyrex glass cell (T_1^{wall} =650 h, 2 bar ^{3}He) in ideal laboratory conditions, nearly quadrupled, from 100 h to 380 h. This translates to an overall $T_1 \approx 100$ h in a Si-cell with a typical $T_1^{wall} = 200$ h and 2 bar ^{3}He as used for experiments. The dedicated ^{3}He-filter cells, briefly described in [5], have an inner (outer) radius of 6 cm (12 cm), and a height of 5 cm, figure 2. They provide a vertical take-off $\pm 9.5°$. The vertical walls are cut from single-crystalline Si, top and bottom are Pyrex. Pyrex and Si-parts are glued, and the walls are Cs-coated, permitting $T_1^{wall} \approx 200$ h. The background scattering from this type of cell is negligible, in fact, an evacuated cell scatters less neutrons than the corresponding volume of air.

4. CROCODILE - near-adiabatic neutron spin transport and flipping

For horizontal field directions, PASTIS2 presents a sign change of the magnetic field on the incident neutron path that complicates the adiabatic neutron spin transport. Moreover, neutron spin flipping with IN20's standard Mezei-flipper in the incident beam is influenced by stray fields from PASTIS2. The guide field on IN20's monochromator is vertical in order to permit horizontal focusing, and this vertical guide field is not suited for a field flipper. The neutron spin flipping and adiabatic transport difficulties are overcome with the entrance flipper CROCODILE, sketched in figure 3. CROCODILE incorporates a vertical guide field (in prolongation of the vertical guide field of IN20's monochromator protection) that decreases in strength towards the sample position and provides a horizontal guide field at the position where the PASTIS2-field changes sign. The longitudinal gradient G_0 of the vertical guide field $B_0(s) = B_0 + G_0 s$ along the neutron path s permits the integration of an adiabatic fast-passage (AFP) neutron flipper [6, 7]. In an AFP flipper, a transverse rotating RF magnetic field is applied. The constant RF frequency ω_0 corresponds to the Larmor frequency of the static field in the central region of the field gradient $\omega_0 = |\gamma| B_0(s_c)$, the amplitude $B_1(s)$ is maximum at the resonance position s_c, but falls off to near-zero towards the beginning and the end of the static field's gradient region [6]. In a frame that rotates around the static field with ω_0 and travels with the neutron velocity v, the neutron spin experiences a field $\boldsymbol{B}_{eff}$ with slowly changing direction,

$$\boldsymbol{B}_{eff}(s) = \left(B_0(s) + \frac{\omega_0}{\gamma}\right)\hat{\boldsymbol{z}} + \boldsymbol{B}_1(s), \tag{1}$$

from $+\hat{\boldsymbol{z}}$ to $-\hat{\boldsymbol{z}}$ for $B_0 > 0, G_0 < 0$, since $B_1 \approx 0$ at the beginning and the end of the gradient region. In the rotating neutron frame the neutron spin follows this effective field adiabatically, and hence in the laboratory system, the spin is flipped after the passage through the flipper.

Apart from the adiabatic condition, the AFP-flipper is independent of the neutron velocity, and hence broad-band. Stray fields will simply shift the "resonance position" slightly, the AFP-flipper is therefore less sensitive to stray fields than a Mezei flipper.

The condition for adiabatic passage,

$$v|G_0| << |\gamma| B_1^2, \tag{2}$$

is much easier to meet for cold [7] and ultracold neutrons [8] than for thermal neutrons, in particular in presence of geometric constraints for the length of the gradient region. Nevertheless, flipping efficiencies $f > 0.97$ have been reported for thermal neutrons with $\lambda > 0.9$Å with an AFP flipper which had $\gamma B_1^2/vG_0$ of the order of 2 [9]. In CROCODILE the rotating RF-field is replaced by a linearly oscillating one, realised with a solenoid parallel to the neutron beam. This linearly oscillating RF-field can be understood as a superposition of two oppositely rotating RF-fields of half the amplitude, one of which rotates in the sense of the Larmor precession of the neutron spin, the other one has negligible effect on the neutron spin. In CROCODILE, the static magnetic field gradient is about 33 Gm^{-1}, the (linear) RF amplitude at the resonance position is 6 G (fig.1), leading to an equivalent rotating amplitude of 3 G. The value of $\gamma B_1^2/vG_0$ ranges between 2.7 and 1.9 (at k_i = 3 Å^{-1} and 4.3 Å^{-1}, respectively). We demonstrate here, that in spite of the operation in only near-adiabaticity, it is possible to achieve efficient broad-band flipping of thermal neutrons.

5. Data analysis methods

The unpolarised transmission $\tilde{T}$ of the ^{3}He-cell, and its analysing power A are given by

$$\tilde{T} \;=\; T_0 \exp(-\theta\lambda)\; \cosh(\theta\lambda\; P_{3He}) \tag{3}$$

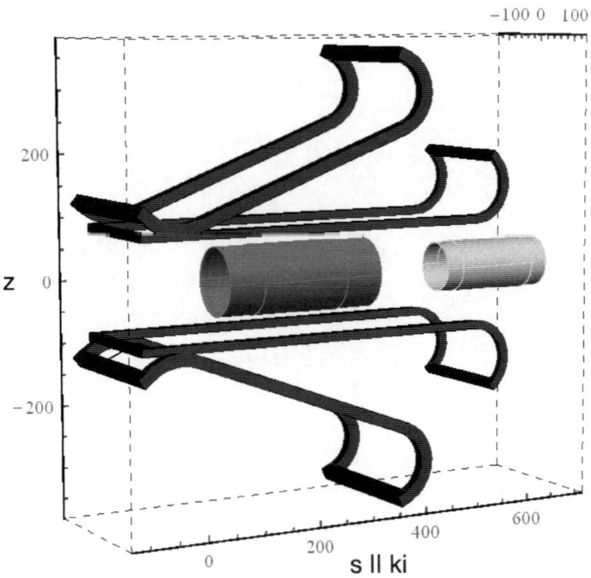

Figure 3. Scheme of CROCODILE. Blue: coils for the static vertical gradient field of the AFP-flipper and adiabatic transport; red: solenoid for the RF-field oscillating with the Larmor frequency; green: solenoid providing a supplementary static guide field parallel to k_i.

$$A = \tanh(\theta\lambda\, P_{3He}), \tag{4}$$

where T_0 is the transmission of the evacuated cell, P_{3He} is the polarization of the ^{3}He nuclei, $\lambda = \frac{2\pi}{k_f}$ the wave length of the scattered neutrons, and θ the ^{3}He-cell opacity. At temperature $T = 295$ K, θ is given by

$$\theta = 0.0728 \frac{\text{Å}^{-1}}{\text{bar cm}} \cdot p_{3He} \cdot \ell, \tag{5}$$

with the ^{3}He-pressure in bar, and the length ℓ of the cell in cm. With an incident polarised beam, prepared by a polariser of efficiency P_i, a flipper with efficiency f, followed by a purely nuclear scatterer with cross section σ^{++} (absent or subtracted nuclear spin-incoherent scattering), and analysis with a ^{3}He-filter, the non-spin flip and spin-flip intensities are, irrespective of the field direction, given by

$$I_{NSF} = T_0\sigma^{++}\exp(-\theta\lambda)\left[\cosh(\theta\lambda P_{3He}) + P_i \sinh(\theta\lambda P_{3He})\right] \tag{6}$$
$$I_{SF} = T_0\sigma^{++}\exp(-\theta\lambda)\left[\cosh(\theta\lambda P_{3He}) - f P_i \sinh(\theta\lambda P_{3He})\right]. \tag{7}$$

This holds likewise for measurements with the attenuated direct beam, Ge Bragg peaks, or phonon measurements. All measured elastic intensities were corrected for the non-negligible FlatCone detector deadtime of $\tau = 3.97\ 10^{-6}$ s.

For the data correction of the inelastic measurements, we need to know P_i, θ, f as well as $P_{3He}(0)$ and the decay time T_1 for the time-dependent ^{3}He-polarisation $P_{3He}(t) = P_{3He}(0)e^{-t/T_1}$. We have determined P_i independently at $k_i = 3$ Å^{-1} on the direct beam, using the "magic box", a ^{3}He-cell with large and known ^{3}He-polarisation [10], and an integrated AFP-flipper of efficiency 1 [11]. The value of θ was calculated from the ^{3}He-pressure and cell geometry, eq. (5).

IOP Conf. Series: Journal of Physics: Conf. Series **862** (2017) 012005 doi:10.1088/1742-6596/862/1/012005

The flipping efficiency of CROCODILE, which includes the polarisation transport in the beam section between shutter and sample, was determined with the CROCODILE as only flipper, and a ^{3}He-cell for the polarisation analysis, and is described below. The standard method with a second flipper would have been impractical.

In order to disentangle flipper efficiency and time-dependent polarisation and transmission we measured $I_{NSF}(t)$ and $I_{SF}(t)$ on a Ge(111) Bragg peak, and additionally the intensity with the ^{3}He-filter removed, σ^{++}. The empty-filter transmission T_0 of the Si-walled banana-cell was measured separately, $T_0 > 0.986$, and agreed well with the calculated value, $T_0 = 0.992$. The two equations for $I_{NSF}(t)$ and $I_{SF}(t)$ (6,7) contain only two unknowns, P_{3He} and f. We used I_{NSF} to determine $\exp(\theta\lambda P_{3He})$, and the measured flipping ratio R to determine f

$$R = \frac{I_{NSF}}{I_{SF}} = \frac{1 + P_i \tanh(\theta\lambda P_{3He})}{1 - fP_i \tanh(\theta\lambda P_{3He})} \quad (8)$$

$$f = -\frac{1}{R} + \frac{1 - \frac{1}{R}}{P_i \tanh(\theta\lambda P_{3He})}. \quad (9)$$

A second series of measurements was performed on the attenuated direct beam, with a highly polarised cylindrical ^{3}He-filter instead of the Si-banana, for better precision. The beam attenuation was adapted to the low transmission of the ^{3}He-cell, the intensity with removed cell was too close to saturating the detector and was not used for the analysis. Since R does not depend on $T_0\sigma^{++}$, and $f, P_{3He}(0)$, and T_1 do not depend on time, it is nevertheless possible to extract $f, P_{3He}(0)$, and T_1 from a series of flipping ratio measurements at different times, even without knowing σ^{++}.

The inelastic measurements were interlaced in regular intervals (about once per hour) by measurements of the Ge(111) Bragg peak, in order to monitor the decay of the polarisation and the transmission.

6. Results

We measured the in-situ polarisation decay in an initial test without CROCODILE, using the attenuated direct beam, a simple solenoid for the adiabatic transport between vertical guide field in the incident beam section and PASTIS2, and a Mezei-flipper, optimized for each field direction and incident energy. Applying the PASTIS2-field horizontally at 135° to k_i, $H||x$, we find $T_1 = 51$ h. In a second step we varied the PASTIS2-field direction as fast as possible, rotating between x, y- and z-directions. After 100 rotations we obtain an overall $T_1 = 58$ h, indicating that the rotation itself plays a negligible role for the polarisation decay. Note that employing the CROCODILE which interferes less with the ^{3}He-filter cell, we find $T_1 = 70(2)$ h for $H||x$, and $T_1 = 113$ h for $H||z$, to be compared to 20 h with the initial PASTIS2.0.

In order to test the influence of the PASTIS2 stray field on the flipper performance of CROCODILE, we measured the flipping ratio on the direct beam for magnetic field directions every 45° in the horizontal plane as well as for vertical guide field. We employed a ^{3}He-cell with high initial polarisation for optimum sensitivity in the polarisation analysis. The raw polarisations spread from 0.8954(15) to 0.8977(15). Corrected for the ^{3}He-polarisation decay with time, the polarisations of all directions are identical within 0.4%. This shows that stray fields from PASTIS2 have negligible influence on the flipping efficiency of CROCODILE. For an experiment with full polarisation analysis this means that systematic errors are suppressed to less than one percent of the unpolarised scattering cross section.

Employing the separation method described in the previous section we find the flipper efficiency of CROCODILE is $f_{exp} > 0.999$ for vertical guide field, $f_{exp} = 0.998 \pm 0.003$ for horizontal guide field direction at 135° to the incident beam at $k_i = 3$ Å^{-1}, in very good agreement with the calculated performance of CROCODILE, $f_{calc} > 0.985$ for a beam section of 40x40mm^2 and $k_i \in [1.57, 4.5]$Å^{-1}.

IOP Conf. Series: Journal of Physics: Conf. Series **862** (2017) 012005 doi:10.1088/1742-6596/862/1/012005

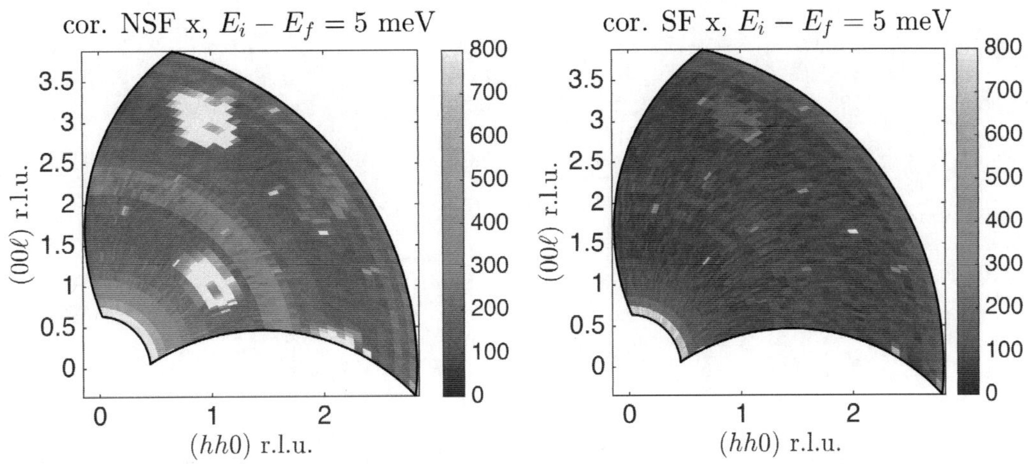

Figure 4. Intensity-color maps of the $(hh\ell)$-plane of Ge at 5 meV energy transfer after correction for the time-dependent polarisation and transmission. The truncated colorscale refers to Counts/5 min, the phonon intensities reach 10000 Counts/5 min. Small residual Ge-Bragg peaks arise from $2k_i$ and inelastic scattering in the Si-analysers.

By measuring an acoustic phonon emerging from Ge(111), with k_i ranging from 3.38 Å^{-1} to 4.0 Å^{-1} and corresponding incident energies between 23.6 meV and 33.2 meV, we tried to test the broad-band functioning (polarisation transport and flipping) of CROCODILE. For these measurements with finite energy transfer, f and the incident polarisation were not separated. We corrected the data assuming $f = 1$ and an incident polarisation $P_i = 0.9136$ independent of k_i. The resulting effective flipping ratios are about 18-20 for k_i up to 4.0 Å^{-1}, indicating that CROCODILE indeed works broad-band.

In order to test the feedthrough of a phonon into the spin-flip channel we measured on the Ge single crystal a sector of the $(hh\ell)$-plane at an energy transfer of 5 meV, with $k_f = 3$ Å^{-1}. This sector includes the acoustic phonons near (111) and (113). We interlaced the inelastic measurements with flipping ratio measurements on the Bragg peak (111) in order to determine the ^{3}He-cell polarisation $P_{3He} = P_{3He}(0)\exp(-t/T_1)$ between start and end-time, $P_{3He}(0) = 0.776(3)$ and $T_1 = 70(2)$ h, as described under the data analysis methods. All data were taken with a horizontal guide field $H||x$ at the sample. The raw flipping ratio decayed from 12 to 8 during the measurements. Figure 4 displays the data after correction for polarisation decay and transmission changes with time. The non-spin flip channel (NSF) displays the acoustic phonons and weaker aluminium powder rings from the cryostat, the spin-flip channel (SF) is essentially empty. The correction assumed $P_i = 0.9136$ as measured for $k_i = 3$ Å^{-1}, and the experimentally determined $f = 0.998$, see above. $\theta\lambda$ and T_0 play only a role for the overall intensity scale. After corrections, only approximately 2.5% of the phonon intensity leap into the SF-channel, corresponding to an effective flipping ratio of 40. Figure 5 displays the intensities integrated over the entire acoustic phonon ring (integrating over 0.9 Å^{-1} thick slices in the (00ℓ) direction). Here, raw NSF and raw SF label the raw data, while cor. NSF and cor. SF denote the corrected intensities. A dataset taken on the same sample, by simple removal of the cell and changing the monochromator to Si(111), shows that the signal in fully polarised mode is reduced by only an average factor of 12 compared to fully unpolarised mode. This factor contains the difference in monochromator reflectivity as well as the factor 2 that remains even for a perfect polariser/analyser and a hypothetical ^{3}He-filter transmission of 1.

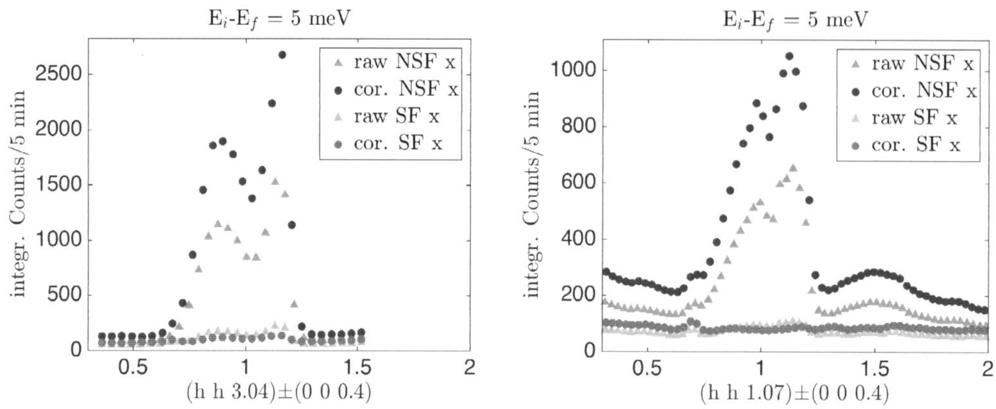

Figure 5. Acoustic Ge-phonons near (113) (left) and near (111) (right), integrated over 0.8 r.l.u. (0.9Å^{-1}) in the (00ℓ)-direction. Raw NSF and SF denote the raw data, cor. NSF and SF are the corresponding data, corrected for the time-dependent polarisation and transmission.

7. Conclusion

IN20 with FlatCone, the improved PASTIS2, and CROCODILE permits full polarisation analysis for a 75°-range of scattering angles simultaneously, with in situ decay times T_1 of the order of 100 h, for variable field directions in the horizontal plane as well as vertical field. We have established a correction method that suppresses polarisation leakage to the level of 2.5%, equivalent to an effective flipping ratio of 40. We have demonstrated that the AFP-flipper CROCODILE works with $f > 0.99$ in a broad range of thermal energies independently of the field direction in PASTIS2, in spite of being operated only near-adiabatically. Due to its insensitivity to stray-fields, such an AFP-flipper, integrated into the monochromator protection, would be extremely useful for polarised experiments in large horizontal or vertical magnetic fields in IN20's single-detector mode. More generally, both PASTIS2 and CROCODILE are broad-band, and hence could be associated with broad-band incident or final energies as on e.g. time-of-flight spectrometers.

References

[1] Stewart R, Andersen K H, Babcock E, Frost C D, Hiess A, Jullien D, Stride J A, Barthelemy J F, Marchal F, Murani A P, Mutka H, and Schober H, 2005 *Physica* B **385-386** 1142
[2] Ehlers G, Stewart J R, Wildes A R, Deen P P, and Andersen K H 2013 *Rev. Sci. Inst.* **84** 093901
[3] Schweika W 2010 *J. Phys. Conf. Ser.* **211** 012026
[4] Kempa M, Janousova B, Saroun J, Flores P, Boehm M, Demmel F and Kulda J 2006 *Physica* B **385-386** 1080
[5] Andersen K H, Jullien D, Petoukhov A K, Mouveau P, Bordenave F, Thomas F and Babcock E 2009 *Physica* B **404** 2652
[6] Abragam A, 1961 *The Principles of Nuclear Magnetism*, Clarendon, Oxford
[7] Bazhenov A N, Lobashev V M, Pirozhkov A N, and Slusar V N 1993 *Nucl. Inst. Meth.* A **332** 534
[8] Holley A T, Broussard L J, Davis J L, Hickerson K, Ito T M, Liu C Y, Lyles J T M, Makela M, Mammei R R, Mendenhall M P, Morris C L, Mortensen R, Pattie R W, Rios R, Saunders A, and Young A R 2012 *Rev. Sci. Inst.* **83** 073505
[9] Babcock E, Petoukhov A K, Chastagnier J, Jullien D, Lelièvre-Berna E, Andersen K H, Georgii R, Masalovich S, Boag S, Frost C D, and Parnell S R 2007 *Physica* B **397** 172
[10] Petoukhov A K, Guillard V, Andersen K H, Bourgeat-Lami E, Chung R, Humblot H, Jullien D, Lelièvre-Berna E, Soldner T, Tasset F, and Thomas M 2006 *Nucl. Inst. Meth.* A **560** 480
[11] Petoukhov A K, Andersen K H, Jullien D, Babcock E, Chastagnier J, Chung R, Humblot H, Lelièvre-Berna E, Tasset F, Radu F, Wolff M, and Zabel H 2006 *Physica* B **211** 012026

Magnetic structure of the noncentrosymmetric heavy-fermion superconductor $CePt_3Si$

B Fåk[1], M Enderle[1] and G Lapertot[2]
[1]Institut Laue-Langevin, CS 20156, F-38042 Grenoble Cedex 9, France
[2]Université Grenoble-Alpes and Commissariat à l'Energie Atomique, INAC-Pheliqs, F-38000 Grenoble, France
E-mail: fak@ill.fr

Abstract. We have studied the magnetic structure of the noncentrosymmetric heavy-fermion superconductor $CePt_3Si$ using elastic polarized neutron scattering with field-projected (longitudinal) XYZ polarization analysis. We find that the magnetic moment has components both in the tetragonal basal plane and along the tetragonal four-fold axis. Such a magnetic structure is not symmetry-allowed if the magnetic phase transition is second order, and may imply that $CePt_3Si$ has lower symmetry in the superconducting state than originally thought.

The tetragonal heavy-fermion compound $CePt_3Si$ has received strong attention in recent years because it shows unconventional superconductivity despite its noncentrosymmetric crystal structure [1, 2, 3]. This lack of inversion symmetry leads to an antisymmetric spin-orbit coupling of Rashba type, which mixes spin-singlet and spin-triplet Cooper-pairing channels [4]. The superconducting transition occurs at $T_c = 0.75$ K, well inside the magnetically ordered state that sets in at $T_N = 2.2$ K. This suggests that magnetism may play an important role in the pairing mechanism, further highlighted by the fact that the Cooper pairs are formed by heavy quasiparticles [1].

The magnetic excitations in $CePt_3Si$ are well understood. The $J = 5/2$ multiplet of the $4f^1$ Ce^{3+} ion is split by the tetragonal point symmetry into three doublets. The doublet ground state is $0.46|\pm 5/2\rangle + 0.89|\mp 3/2\rangle$, as shown by a study combining polarized inelastic neutron scattering with soft x-ray absorption spectroscopy [5]. The main exchange interactions have been determined from the dispersion of the somewhat damped spin-wave excitations in the magnetically ordered phase by unpolarized inelastic neutron scattering (INS) [6]. Above T_N, Kondo-type spin fluctuations with an anisotropic wave-vector dependence are observed by INS [6]. As a consequence of the antisymmetric spin-orbit coupling, the spin susceptibility is anomalous with an orthorhombic anisotropy, as shown by polarized INS [7]. No resonance or any change in the magnetic excitation spectrum is reported in the superconducting state, despite intense search [6, 8].

The magnetic structure of $CePt_3Si$ is characterized by a propagation vector $\mathbf{k} = (0, 0, 1/2)$, but the moment direction is not known. Neutron scattering measurements show magnetic Bragg peak intensity for wavevector transfers $\mathbf{Q} \parallel \mathbf{k}$, which implies that the ordered moment is not parallel to the tetragonal c axis. It was therefore assumed that the magnetic structure consisted of sheets of collinear moments aligned along an unknown axis in the a–b plane, with the sheets stacked antiferromagnetically along the c axis [9]. Based on this assumption, the magnitude of the ordered moment was determined to 0.16±0.01 μ_B [6, 9], substantially reduced from the 1.72 μ_B that can be calculated from the ground-state doublet [5]. The hypothesis of the moments lying in the tetragonal basal plane received further support from representation analysis, which showed that for a second-order magnetic phase transition with $\mathbf{k} = (0, 0, 1/2)$ in the $CePt_3Si$ space group $P4mm$ (No. 99), the moments are either in the basal plane or along the c axis [6]. Very surprisingly, the intensities of the magnetic Bragg peaks were found to increase substantially on the application of a magnetic field **B** along the [001] axis [10]. On the assumption of moments aligned in the basal plane, this suggested that the moment magnitude would double for a field of only 6.6 T along the c axis.

Because of the small moment, all magnetic structure determinations in $CePt_3Si$ are based on elastic neutron scattering measurements on triple-axis spectrometers (TAS), a technique that is not very precise in determining moment directions, since it is difficult to normalize the intensities measured on different Bragg peaks. We have therefore studied the magnetic structure using elastic polarized neutron scattering with field-projected (longitudinal) XYZ polarization analysis on the IN20 triple-axis spectrometer. The use of polarized neutrons allows to eliminate the Q-dependent instrumental normalization factors that make unpolarized TAS measurements problematic. We used a standard set-up with Heusler monochromator and analyzer and a Helmholtz coil at the sample position. To minimize contributions of unpolarized second-order neutrons from the monochromator, we used an incoming neutron wavevector of $k_i = 2.57$ Å^{-1} and two pyrolytic graphite filters. The sample was the same image-furnace grown single crystal of mass 6 g used in many of the previous measurements [5, 6, 7], mounted with the a and c axes in the scattering plane (the lattice parameter are $a = 4.072$ and $c = 5.442$ Å). The flipping ratio on nuclear Bragg peaks exceeded 23, which makes polarization corrections unnecessary.

For the determination of the magnetic structure, we measured thirteen magnetic Bragg peaks, namely $(0, 0, 1/2)$, $(0, 0, 3/2)$, $(0, 0, 5/2)$, $(1, 0, \pm 1/2)$, $(1, 0, \pm 3/2)$, $(1, 0, \pm 5/2)$, $(2, 0, \pm 1/2)$, and $(2, 0, \pm 3/2)$ with seven different polarized neutron scattering cross-sections $\alpha\beta$: xx, $x\bar{x}$, $\bar{x}x$, yy, $y\bar{y}$, zz, and $z\bar{z}$, where α and β are the incoming and scattered neutron polarization, respectively, the directions x, y, z correspond to the neutron polarization being along **Q**, perpendicular to **Q** in the scattering plane, and perpendicular to the scattering plane, respectively, and $\bar{x}$ means antiparallel to x.

Since neutron scattering observes only the component of the magnetic moment **M** that is perpendicular to **Q**, here called $\mathbf{M}_\perp$, it is straightforward to show that the measured cross-sections $I_{\alpha\beta}$ are related to the y and z projected moment components $M_{\perp y}$ and $M_{\perp z}$ via

$$I_{x\bar{x}} - I_{y\bar{y}} = |M_{\perp,y}|^2 - 2\mathrm{Im}\{M^\star_{\perp,y} M_{\perp,z}\} \quad (1)$$

$$I_{yy} - I_{xx} = |M_{\perp,y}|^2 \quad (2)$$

$$I_{x\bar{x}} - I_{z\bar{z}} = |M_{\perp,z}|^2 - 2\mathrm{Im}\{M^\star_{\perp,y} M_{\perp,z}\} \quad (3)$$

$$I_{zz} - I_{xx} = |M_{\perp,z}|^2 \quad (4)$$

$$I_{x\bar{x}} - I_{\bar{x}x} = -4\mathrm{Im}\{M^\star_{\perp,y} M_{\perp,z}\}. \quad (5)$$

The intensities $I_{\alpha\beta}$ were normalized to the beam monitor, and the background obtained from measurements with the sample rotated away from the Bragg position by 2° or in the paramagnetic phase at $T = 7$ K was subtracted. The chiral term $\mathrm{Im}\{M^\star_{\perp,y} M_{\perp,z}\}$ obtained from (5) is zero within the precision of the present measurements. This is expected, since the propagation vector of $CePt_3Si$ does not lead to any chirality.

In the actual geometry of the experiment, with θ denoting the angle between **Q** and the tetragonal basal plane, the magnetic moment components along the crystallographic axes, M_a, M_b, and M_c, are related to $M_{\perp,y}$ and $M_{\perp,z}$ via

$$|M_{\perp z}|^2 = |M_b|^2 \quad (6)$$

IOP Conf. Series: Journal of Physics: Conf. Series **862** (2017) 012006 doi:10.1088/1742-6596/862/1/012006

and

$$|M_{\perp y}|^2 = |M_a|^2 \sin^2\theta + |M_c|^2 \cos^2\theta - 2\mathrm{Re}\{M_a^\star M_c\} \sin\theta\cos\theta = \\ = |M_a|^2 \sin^2\theta + |M_c|^2 \cos^2\theta - 2M_a M_c \sin\theta\cos\theta, \quad (7)$$

where in the last equality we have assumed real (Fourier components of the) magnetic moments, which is justified in the case of a $\mathbf{k} = (0, 0, 1/2)$ propagation vector. By measuring $|M_{\perp y}|^2$ for several inequivalent values of $\mathbf{Q}$, and hence several values of θ, both M_a and M_c can be extracted from (7).

We introduce the notation $|M_{\perp y}|^2_{\pm l}$ for measurements performed at $\mathbf{Q} = (h, 0, \pm l/2)$ with $h > 0$. The mixed $M_a M_c$ term can be estimated from the difference

$$|M_{\perp y}|^2_{+l} - |M_{\perp y}|^2_{-l} = -2M_a M_c \sin(2\theta). \quad (8)$$

Within the precision of the measurements, $M_a M_c = 0$ (see table 1). This mixed term can also be eliminated from (7), by taking the average of $|M_{\perp y}|^2_{+l}$ and $|M_{\perp y}|^2_{-l}$.

The ratio of the in-plane components of the magnetic moment are obtained for reflections with $h = 0$ ($\theta = \pi/2$),

$$\alpha^2 \equiv \left.\frac{|M_{\perp y}|^2}{|M_{\perp z}|^2}\right|_{h=0} = \frac{|M_a|^2}{|M_b|^2}. \quad (9)$$

Averaging over the measured magnetic Bragg reflections with $h = 0$, we find $\bar{\alpha}^2 = 0.65 \pm 0.05$ for both spin-flip and non-spin-flip cross-sections, and hence $\bar{\alpha} = |M_a|/|M_b| = 0.81 \pm 0.03$ (see table 1). This implies that the population of in-plane S domains is unequal, and also that the in-plane component of the moment is not along a [110]-type axis. In other words, the basal-plane projected moment is either in a general direction with four in-plane S domains or along a [100]-type axis with two unequally populated in-plane S domains. A single-domain sample would be needed in order to determine a general in-plane direction of the magnetic moment.

The out-of-plane moment ratio is obtained from (6) and (7) after averaging over $|M_{\perp y}|^2_{+l}$ and $|M_{\perp y}|^2_{-l}$ for $h \neq 0$ Bragg peaks, as discussed,

$$\frac{|M_c|^2}{|M_b|^2} = \frac{1}{\cos^2\theta} \left.\frac{|M_{\perp y}|^2}{|M_{\perp z}|^2}\right|_{h\neq 0} - \bar{\alpha}^2 \tan^2\theta, \quad (10)$$

using the value of $\bar{\alpha}$ determined above. Note that by normalizing the data to $|M_b|^2 = |M_{\perp z}|^2$ for each $\mathbf{Q}$ value, absorption corrections as well as Lorentz and magnetic form factors are factored out. This is one of the advantages of the polarized neutron scattering technique. We find that

Table 1. Moment ratios of in-plane $|M_a/M_b|^2$, out-of-plane $|M_c/M_b|^2$, and mixed $M_a M_c/|M_b|^2$ contributions from spin-flip and non-spin-flip cross-sections, and their average. Statistical errors (one standard deviation) are given in parenthesis and observed standard deviations are given as $\pm$ values.

Moment ratio	Spin-Flip	Non-Spin-Flip	Average
$\lvert M_a/M_b\rvert^2$	0.68(4)±0.03	0.62(4)±0.05	0.65(3)±0.05
$\lvert M_c/M_b\rvert^2$	3.2(2)±1.2	3.2(2)±0.8	3.2(1)±1.0
$M_a M_c/\lvert M_b\rvert^2$	0.08±0.12	0.06±0.11	0.07±0.11

$|M_c/M_{ab}|^2 = 3.2 \pm 1.0$ (or $|M_c/M_{ab}| = 1.8 \pm 0.3$, see table 1) for both spin-flip and non-spin-flip cross-sections. This implies that the moment component along the tetragonal c axis is *substantial*, in contrast to assumptions hitherto.

The observation that the ordered moment direction is neither in the basal plane nor along the tetragonal axis has two important implications. One is that the field-induced intensity increase of magnetic Bragg peaks observed by Kaneko *et al.* [10] cannot be directly related to the size of the magnetic moment, since a change in the domain population or a reorientation of the magnetic moment in applied field would also change the Bragg peak intensity. For a magnetic field in the [001] direction, a domain repopulation is not likely since all S domains have the same energy. However, the Zeeman energy due to the magnetic field will compete with the anisotropy that produces the initial tilt angle of the staggered moment direction, and the total energy of the system will be lowered if the staggered magnetic moment components turn further towards the basal plane. Such a tilting will increase the intensity of the (0,0,1/2) Bragg peak, as observed experimentally [10].

The second implication is that the magnetic order established at T_N leads to a lower symmetry than expected for a second order phase transition. This means that the phase transition is either (weakly) first order or that there are two phase transitions close to each other near T_N. Both scenarios lead to a lowering of the symmetry in the magnetically ordered phase, and hence in the superconducting phase. Since the symmetry of the gap function and of the spin-orbit coupling depend on the crystal symmetry, a change in the latter may require a reexamination of existing theoretical models for the superconductivity in $CePt_3Si$.

References

[1] Bauer E, Hilscher G, Michor H, Paul Ch, Scheidt E W, Gribanov A, Seropegin Yu, Noël H, Sigrist M and Rogl P 2004 Heavy fermion superconductivity and magnetic order in noncentrosymmetric $CePt_3Si$ *Phys. Rev. Lett.* **92** 027003

[2] Bauer E, Kaldarar H, Prokofiev A, Royanian E, Amato A, Sereni J, Brämer-Escamilla W and Bonalde I 2007 Heavy fermion superconductivity and antiferromagnetic ordering in $CePt_3Si$ without inversion symmetry *J. Phys. Soc. Jpn.* **76** 051009

[3] Fujimoto S 2007 Electron correlation and pairing states in superconductors without inversion symmetry *J. Phys. Soc. Jpn.* **76** 051008

[4] Gor'kov L P and Rashba E I 2001 Superconducting 2D system with lifted spin degeneracy: mixed singlet-triplet state *Phys. Rev. Lett.* **87** 037004

[5] Willers T, Fåk B, Hollmann N, Körner P O, Hu Z, Tanaka A, Schmitz D, Enderle M, Lapertot G, Tjeng L H and Severing A 2009 Crystal-field ground state of the non-centrosymmetric superconductor $CePt_3Si$: a combined polarized soft X-ray absorption and polarized neutron study *Phys. Rev.* B **80** 115106

[6] Fåk B, Raymond S, Braithwaite D, Lapertot G and Mignot J-M 2008 Low-energy magnetic response of the noncentrosymmetric heavy-fermion superconductor $CePt_3Si$ studied via inelastic neutron scattering *Phys. Rev.* B **78** 184518

[7] Fåk B, Adroja D T, Enderle M, Böhm M, Lapertot G and Mineev V P 2014 Anomalous spin response in the non-centrosymmetric metal $CePt_3Si$ *J. Phys. Soc. Jpn.* **83** 063703

[8] Inosov D S, Bourges P, Ivanov A, Prokofiev A, Bauer E and Keimer B 2011 Dispersion and damping of zone-boundary magnons in the noncentrosymmetric superconductor $CePt_3Si$ *J. Phys.: Condens. Matter* **23** 455704

[9] Metoki N, Kaneko K, Matsuda T D, Galatanu A, Takeuchi T, Hashimoto S, Ueda T, Settai R, Ōnuki Y and Bernhoeft N 2004 Magnetic structure and crystal field excitation in heavy fermion superconductor $CePt_3Si$ *J. Phys.: Condens. Matter* **16** L207

[10] Kaneko K, Stockert O, Fåk B, Raymond S, Skoulatos M, Takeuchi T and Ōnuki Y 2014 Magnetic-field-enhanced antiferromagnetism in the noncentrosymmetric heavy-fermion superconductor $CePt_3Si$ *Phys. Rev.* B **89** 241105(R)

IOP Conf. Series: Journal of Physics: Conf. Series **862** (2017) 012007 doi:10.1088/1742-6596/862/1/012007

Efficient polarization analysis for focusing neutron instruments

Jochen Stahn[1], **Artur Glavic**[1]

[1] Laboratory for Neutron Scattering and Imaging, Paul Scherrer Institut, Villigen PSI, Switzerland

E-mail: `artur.glavic@psi.ch`

Abstract. Polarized neutrons are a powerful probe to investigate magnetism in condensed matter on length scales from single atomic distances to micrometers. With the ongoing advancement of neutron optics, that allow to transport beams with increased divergence, the demands on neutron polarizes and analyzers have grown as well. The situation becomes especially challenging for new instruments at pulsed sources, where a large wavelength band needs to be polarized to make efficient use of the time structure of the beam.

Here we present a polarization analysis concept for highly focused neutron beams that is based on transmission supermirrors that are bend in the shape of equiangular spirals. The method allows polarizations above 95% and good transmission, without negative impact on other beam characteristics. An example of a compact polarizing device already tested on the AMOR reflectometer is presented as well as the concept for the next generation implementation of the technique that will be installed on the Estia instrument being build for the European Spallation Source.

1. Introduction

Neutron radiation has several properties desired in the investigation of many condensed matter systems. In addition to a suitable wavelength on the Å length scale, high sensitivity to lighter elements and energies useful for spectroscopic studies, the main advantage the neutrons pose over x-ray photons is the sensitivity to magnetic induction inherent to the magnetic moment of the particle. The neutron is a spin 1/2 particle and thus has only two allowed values for the z-component of its magnetic moment, aligned parallel or anti-parallel to an external magnetic field. One speaks of spin-up/-down polarization, denoted by $|+\rangle/|-\rangle$.

Although interaction with magnetic moments in the investigated sample alone can reveal information about the magnetic structure, e.g. magnetic unit cell, symmetry and local moment from powder diffraction, the determination of the neutron polarization before and after the scattering process yields additional knowledge about the direction of magnetization[1, 2, 3]. There are basically two principles for the interaction of polarized neutrons with magnetic moments in the sample. Only the projection $\vec{M}_\perp$ of the magnetization vector $\vec{M}$ onto the plane perpendicular to the scattering vector $\vec{Q}$ is detected. Secondly, the relative orientation of this projection to the neutron polarization vector $\vec{P}$ determines the strength of the interaction and change of polarization state:

$$\text{non spin-flip scattering:} \quad \vec{M}_\perp || \vec{P} \tag{1}$$

$$\text{spin-flip scattering:} \quad \vec{M}_\perp \perp \vec{P} \tag{2}$$

In the first case the scattering potential is increased in the parallel and decreased in the anti-parallel case.

Neutron beams can essentially be polarized using one of three methods:

- Magnetic Bragg-scattering in special Heusler alloy crystals, where the nuclear and magnetic cross-sections balance out in the structure factor to only allow one spin state to get reflected[4, 5, 6, 7]. The practical advantage of this technique is the simple implementation at instruments that already use crystal monochromators without the need for extra space or complicated arrangements. Polarization efficiency, however, is rather limited and the technique is not suitable for polychromatic beams.
- The spin dependent absorption cross-section of ^{3}He can be used to polarize neutrons by generating a partially nuclear spin polarized gas volume where the neutron beam is transmitted through [8, 9, 10]. The much larger attenuation of the undesired spin-state results in a polarized neutron beam. Main advantages of this method are the possibility to specifically optimize the polarization efficiency to the desired application and the relatively low dependence of the device properties on the neutron trajectory. Especially the latter can be used for polarization analysis with large angular coverage. A main limitation of ^{3}He polarizers is the sensitivity to magnetic field perturbation that make it difficult to use them near high-field magnets. Secondly, the wavelength dependence of the absorption process leads to significant losses in transmission for longer wavelength, especially relevant for instruments with large wavelength bands. The transmission over the bandwidth of a time of flight instrument can vary by a factor of 5 or larger.
- Reflection of an artificial multilayer or supermirror with different reflectivities for spin-up and spin-down neutrons[11, 12, 13]. These devices can use varying mirror geometries adapted to the specific application, but the physical principle is the same. While limited by the small reflection angle needed for each mirror, these solid state devices can reach very high neutron polarizations with only minimal losses to the intensity of the desired neutron polarization state. In environments with constrained space the necessary system size and/or the high magnetic field needed to saturate the magnetic coating can lead to interference with other instrument components.

Here we present a system based on the latter method. The fundamental properties of reflecting polarization mirrors are:

- Reflectivity R is a function of the momentum transfer normal to the reflecting surface $q_z = \frac{4\pi}{\lambda} \sin\alpha$ with the neutron wavelength λ and the incident angle α.
- The effective upper critical q_z for the two spin states are $q_c^{|+\rangle} := m\, q_c^{\mathrm{Ni}}$ and $q_c^{|-\rangle}$, where the critical q_z of natural Ni $q_c^{\mathrm{Ni}} \approx 0.0022\,\text{Å}^{-1}$ is used as a reference (m=1).
- The polarization efficiency of the polarizer for the reflected beam is

$$P_R = \frac{R^{|+\rangle} - R^{|-\rangle}}{R^{|+\rangle} + R^{|-\rangle}} \tag{3}$$

$$\approx 1 - \frac{R^{|+\rangle}}{R^{|-\rangle}} \quad \text{for } q_c^{|-\rangle} < q_z < q_c^{|+\rangle}. \tag{4}$$

Here $R^{|-\rangle} \ll R^{|+\rangle}$ was used. Typical polarizers based on Fe/Si or FeCoV/Ti:N supermirrors (SM) reach $P = 92\% \ldots 99\%$ with a single reflection.

IOP Conf. Series: Journal of Physics: Conf. Series **862** (2017) 012007 doi:10.1088/1742-6596/862/1/012007

- For the transmitted beam the situation is different: The transmission $T^{|-\rangle} = 1 - R^{|-\rangle}$ is close to 100% for $q_z > q_c^{|-\rangle}$, but $T^{|+\rangle} = 1 - R^{|+\rangle}$ can reach up to 50% for high m. After a single transmission the polarizing efficiency

$$P_T = \frac{T^{|-\rangle} - T^{|+\rangle}}{T^{|+\rangle} + T^{|-\rangle}} \tag{5}$$

thus is in the range 30% to 90%.
For a good polarization, several SM coatings must be used in sequence. Often the supporting substrate (a Si waver) is coated on both sides, and sometimes a stack of substrates is used.

The advantage of using the transmitted beam is, that its trajectory is hardly affected by the polarizer. In addition such devices can be used to suppress long wavelength contamination from frame overlap in pulsed applications, as the transmission for both spin states below the critical edge of the substrate is 0. This manuscript deals essentially with a special geometry of a SM used in transmission that allows polarization of neutron beams with large divergence. We make use of the small focus size in modern, highly focused neutron instruments that provide the necessary boundary condition, namely a limited local angular distribution of neutron trajectories.

2. Basic principle

For a given angular distribution $\Delta\alpha$ and wavelength spread $[\lambda_{\mathrm{min}}, \lambda_{\mathrm{max}}]$ the necessary m of the coating can be calculated with

$$\begin{aligned} q_c^{|-\rangle} &< \frac{4\pi}{\lambda_{\mathrm{max}}} \sin\alpha_{\mathrm{min}} \\ \alpha_{\mathrm{min}} &> \arcsin\frac{q^{|-\rangle}\lambda_{\mathrm{max}}}{4\pi} \end{aligned} \tag{6}$$

and

$$m\, q_c^{\mathrm{Ni}} \geq \frac{4\pi}{\lambda_{\mathrm{min}}} \sin\left(\alpha_{\mathrm{min}} + \Delta\alpha\right) \quad , \tag{7}$$

This limits the application of SM polarizers to cases where either the α- or the λ-range are moderate or where a lot of space is available (due to the small angles of incidence α on the polarizer). The latter is the case if a (part) of the neutron beam guide is coated with a polarizing SM or if a polarizing cavity[14] is used.

A transmission filter used for polarizing a beam or for frame-overlap suppression (i.e. used as a low-pass for λ) should ideally intersect the beam at the same optimized angle $\hat{\alpha}$ everywhere within the beam. For a parallel beam this is fulfilled by a flat surface, inclined relative to the beam by $\hat{\alpha}$. Although a highly beam divergence ($\Delta\theta$) seems to be incompatible with this condition on the first glance, the knowledge of a point of origin/convergence is enough to generate a curved surface that intersects with the beam under the same angle over the whole divergence range. The corresponding surface has the shape of a logarithmic spiral, also called *equiangular* spiral.

3. Features of an equiangular spiral

The general parametric representation of an equiangular spiral is

$$\begin{aligned} x &= a\,\mathrm{e}^{b\theta} \cos\theta \\ y &= a\,\mathrm{e}^{b\theta} \sin\theta \end{aligned}$$

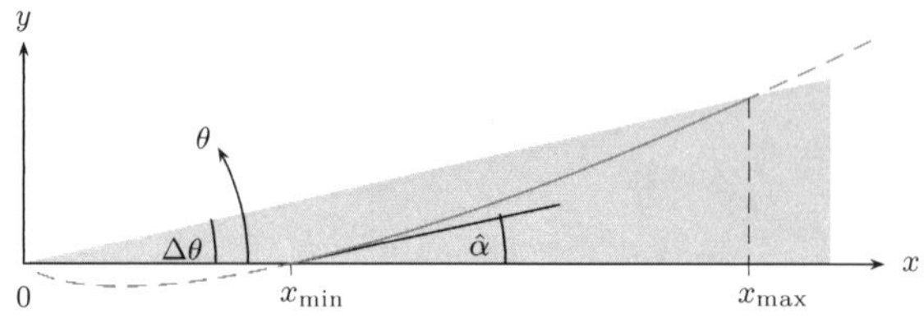

Figure 1. Geometry of the equiangular spiral polarizer in its scattering plane. The dashed line is the spiral, the solid part the actually realized transmission reflector covering the beam (gray area) emerging from the spiral pole.

where a is a scaling factor and $b = 1/\tan\hat{\alpha}$ relates to the intersection angle $\hat{\alpha}$ of the spiral with an arbitrary trajectory $y = x\,\tan\theta$ through the pole at $x = y = 0$.

A feature of the equiangular spiral is that a central stretching, i.e. changing a is equivalent to a rotation $\theta \to \theta + \theta'$ around the pole.

For a small $\hat{\alpha}$ and a rather low divergence $\Delta\theta \ll 10°$ the spiral can be approximated by the function

$$f(x) \;=\; \hat{\alpha}\,x\,\ln\left[\frac{x}{x_{min}}\right] \tag{8}$$

where x_{min} is a scaling factor given by the intersection of $f(x)$ with the horizon $f(x_{min}) = 0$.

4. Single lamella device

The simplest realization of a polarization- or wavelength filter using the presented concept is a single face with spiral bending in one direction (illustrated in figure 1) and no bending in the perpendicular direction. Such a device can in principle cover arbitrary divergences $\Delta\theta$, where the length of the device growth exponentially with $\Delta\theta$.

The optimization of the device, i.e. the chioce of $\hat{\alpha}$ and m, has to take into account the final width of the source $\pm\Delta y$. This leads to angular errors at the polarizer of $\Delta\alpha = \arcsin\Delta y/x$ largest for the point closest to the pole x_{min}. Thus eqns. (6) and (7) are replaced by

$$\hat{\alpha} \;>\; \arcsin\frac{q_c^{|-\rangle}\,\lambda_{\rm min}}{4\pi} + \arcsin\frac{\Delta y}{x_{min}} \tag{9}$$

and

$$m \;>\; \frac{4\pi}{q_c^{\rm Ni}\,\lambda_{\rm min}}\left(\sin\hat{\alpha} + \frac{\Delta y}{x_{min}}\right) \tag{10}$$

4.1. Demonstrator device

A polarizer based on this principle was designed and build for the Selene demonstrator at the AMOR reflectometer of PSI[15]. The device was designed at the Laboratory for Neutron Scattering and manufactured by SwissNeutronics. The rather limited space of 40 cm from source to guide entrance and the relatively large source with a height of $2\Delta y = 1\,$mm necessitate a split assembly, i.e. the device has a V-shape comparable to a polarizing V-cavity[14].[1]

Figure 2 shows the first design sketch and the final device. The technical parameters of the latter are $\lambda \in [4, 16]\,$Å, $x_{min} = 170\,$mm and $x_{max} = 395\,$mm (point furthest from the pole). The

[1] Despite the similar shape, this is no cavity. The polarization is based on exactly one interaction of the beam with the polarizing reflector, there are no outer (reflecting) walls.

IOP Conf. Series: Journal of Physics: Conf. Series **862** (2017) 012007 doi:10.1088/1742-6596/862/1/012007

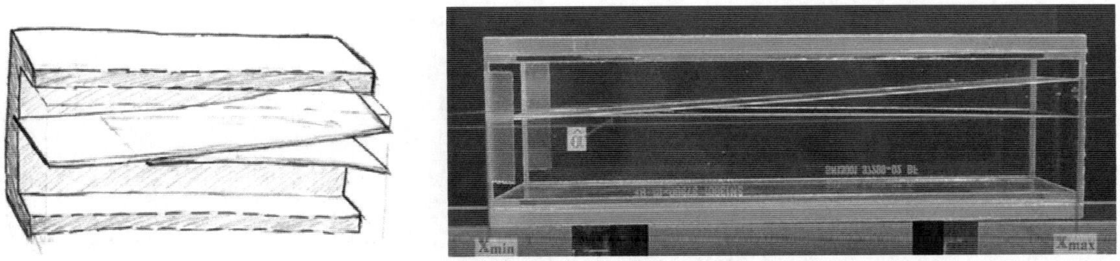

Figure 2. Left: First sketch for a frame-overlap filter and polarizer for the Selene prototype set-up, based on transmission/reflection through/at spiral-curved surfaces. Right: Final device, produced by SwissNeutronics. The glass box is 215 mm long and has an inner height and width of $50 \times 50\,\mathrm{mm}^2$. The virtual source is to be 170 mm before the entrance (to the left).

divergence to be covered is $\Delta\theta = 1.8°$. The angle of incidence is $\hat{\alpha} = 1.5°$ and the polarizing coating is a $m = 4.2$ FeCoV / Ti:N supermirror, covered with Ni to simultaneously act as a low-pass for $\lambda < 16.8\,\text{Å}$, i.e. $q_{\min} = q_c^{\mathrm{Ni}}$ in this case.

This polarizer was characterized using the Selene guide to focus the divergent beam behind the source at the pole position, a 1 mm wide slit behind the neutron guide, to an analyzer. The latter was a Si wafer coated with a polarizing $m = 6$ SM used in reflecting geometry. The magnetization field was 2000 Oe to ensure a high efficiency. The spin state of the neutron beam could be switched using an RF flipper located between polarizer and analyzer. In the following we make the conservative (for the calculated performance of the polarizer) assumption that the flipper is 100% efficient.

The efficiency of the polarizer P_p is a function of θ and λ, while the efficiency of the analyzer P_a depends on the angle of incidence on its surface, ω, and on λ. The relative alignment of the analyzer ω_a determines the relation $\omega = \theta + \omega_a$. This can be used to estimate a lower boundary for P_p by performing several measurements with varying ω_a. The measured spin asymmetry $P_{\omega_a}(\lambda,\theta)$ for each pixel is a product of $P_p(\lambda,\theta)$ and $P_a(\lambda,\omega_a+\theta)$. Using $P_a \leq 1$ this gives $P_p(\lambda,\theta) \geq P_{\omega_a}(\lambda,\theta)\,\forall\omega_a$. Figure 3 shows an intensity map for the lower boundary for P_p obtained from measurements with 4 different ω_a positions. The gradual decay of P_a with q and its rapid drop at $q_c^{|+\rangle}$ leads to the visible intensity steps in the upper left corner of the map.

The overall polarization is surprisingly good for only two polarizing transmission interfaces. Therefore the device is now used regularly for user experiments at AMOR. The darker line for $\theta \approx 0.22°$ corresponds to trajectories traveling through an area of strongly reduced transmission of the Selene guide, which leads to worse signal to noise ratio for the spin-flip channel and thus reduction of the measured polarization. The brighter region at $\theta \approx 0°$ to $0.3°$ corresponds to trajectories intersecting both substrates, i.e. 3 or 4 transmissions through SM interfaces. The lower performance for $\lambda \approx 4\,\text{Å}$ and $\theta > 0.6°$ or $\theta < -0.6°$, respectively, means that the angle of incidence at the highest θ angles is larger than expected, which is the result of a too large distance of the polarizer from the source.

A better alignment (tilt and distance to the source) is expected to deliver

$$\begin{aligned} P_p &\geq 93\% \text{ for } \lambda > 4\,\text{Å} \\ &\geq 96\% \text{ for } \lambda > 7\,\text{Å} \end{aligned}$$

over the full $\Delta\theta = 1.8°$, and

$$P_p \geq 97\% \text{ for } \lambda > 4\,\text{Å}$$

over the range at the device front, where the two mirrors overlap, $\theta = 0.15°$. The latter value can also be seen as a lower boundary for the performance of a device with two lamella in sequence.

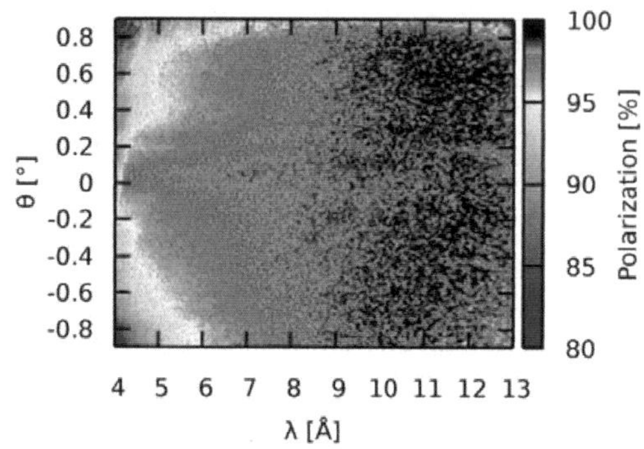

Figure 3. Measured polarization efficiency of the demonstrator device, analyzed by a $m = 6$ FeCoV/Ti:N supermirror on Si. The polarization was flipped with a RF spin-flipper. The band in the middle with improved polarization corresponds to transmission through both reflectors. Sharp lines visible in the top-left corner originate from the analyzer, while the small reduction around $\theta \approx 0.15°$ is an artifact from the Selene guide transmission.

The limitations of a single-lamella transmission polarizer are

- $\Delta y/x_{min}$ must be small, typically below $\arcsin 0.35° = 12$
- The coating is restricted to $m < 7$, where P_T drops considerably towards this limit, while the costs grow. This restricts $\hat{\alpha}$
- The wavelength band to be polarized is restricted due to the too high transmission for $|-\rangle$ neutrons at large q_z and total reflection from the substrate for low q_z.
- For small $\hat{\alpha}$ and large λ a parallel offset of the beam is caused by refraction. The offset is in the order of a few ten microns and thus only relevant for very good focusing systems.

5. Multi lamellae devices planned for Estia

Estia will be a focusing reflectometer at the European Spallation Source (ESS) that is being build in Lund, Sweden. The instrument concept is based on the Selene neutron guide, but in contrast to the focusing option available at AMOR, the elliptical mirrors will form the complete beamline neutron transport system. Increasing the size of the focusing neutron optics lead to an increased space available for the polarizing optics.

The goal is to achieve state of the art neutron polarization analysis while keeping the focusing capability and large intensity towards which the instrument is optimized. Therefore, larger devices with more than one lamella will be constructed for the beamline. Two separate mirrors for the polarizer will be placed close to the focus in the middle of the Selene guide and a double lamella device on the detector arm as an analyzer.

5.1. polarizer

Figure 4 shows a conceptual sketch of the Estia polarization concept. Two equiangular spirals intersect the beam before and after the focal point in the middle between the ellipses of the Selene neutron guide, where a large free space is available to place these components. A large polarization (>99%) and decent transmission[2] (>40%) can be expected from this geometry, as 4 SM interfaces (with m=5.0 Fe/Si) intersect with the neutron path.

The change in intersection direction inherent to the symmetry of the mirror has two additional advantages:

[2] The transmission of an unpolarized beam is >40% as >80% of spin-down neutrons are transmitted.

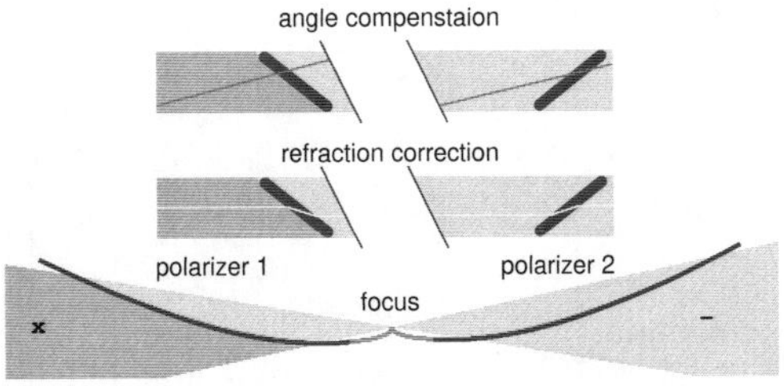

Figure 4. Polarizer concept for the Estia reflectometer. Two double sided supermirrors will be place before and after the focus between the two Selene guides. The symmetry of the mirrors has the property that the reflection direction on the second mirror is opposite to that of the first intersection. This results in a compensation of polarization efficiency for trajectories that don't go through the focal point as well as a correction of the wavelength dependent beam offset produced by the refraction within the silicon wafer.

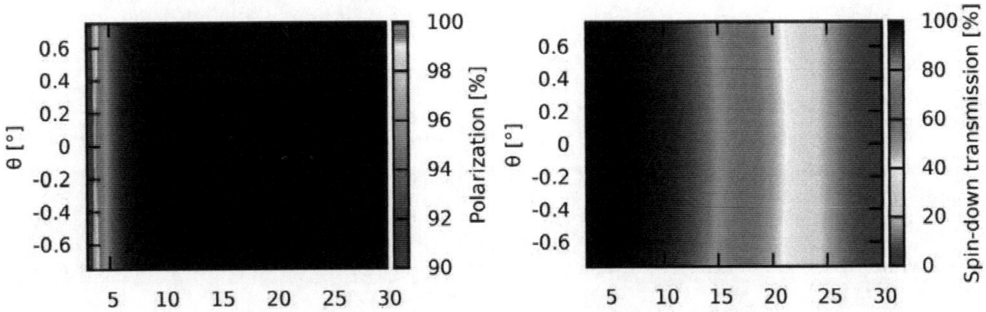

Figure 5. Calculated polarization efficiency (left) and transmission (right) of Estia double polarizer. The primary instrument wavelength band is between 4 Å and 11 Å

- Trajectories that do not pass through the focal point due to the finite size of the source will hit the second mirror with the inverse angular offset $\delta\alpha$ present at the fist mirror. Therefore a too large angle of incident that reduces polarization efficiency of one mirror leads to improved efficiency in the second mirror and vice versa.
- The small linear offset due to refraction in the first silicon wafer will mostly be reversed in the second mirror as the angle of incidence has the same magnitude but opposite sign.

With $\hat{\alpha} = 1.65°$ and m=5 the polarization system can cover the whole accessible wavelength range from 4 Å to 25 Å and, at the same time, allow frame overlap suppression above 30 Å by using the silicon total reflection. Coatings will be made of Fe/Si in the case of Estia, as the large neutron flux would lead to high activation of the component when using Co containing supermirrors. The two devices will be independent, allowing the user to choose between the use of one or two polarizing devices if higher intensity is desired.

5.2. analyzer

Analysis on Estia can not follow the same approach as the polarization, as there is not focal point behind the sample. To still be able to reach good polarization efficiency two subsequent transmissions need to be performed. When the analyzer spirals are curved in the direction perpendicular to the reflection plane and the geometry (scaling factor a) is correctly chosen, it is possible to not only measure the transmitted beam with the position sensitive detector but

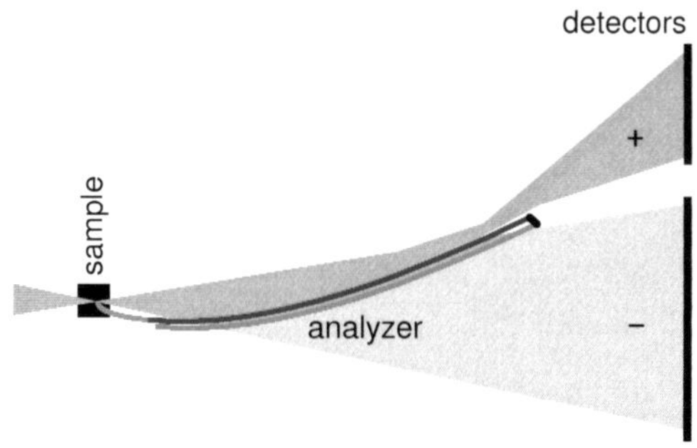

Figure 6. Analyzer concept for the Estia reflectometer. The first mirror is single sided coated as a reflection polarizer while the second is double sided and its reflection is absorbed. Capturing both, transmitted and reflected beam allows simultaneous measurement of each spin channel with sufficient polarization efficiency.

also the neutrons reflected from the first analyzer mirror.

For this reason the Estia analyzer, which will be placed together with the detector on a movable table, will consist out of two supermirror lamellae (fig. 6). The first will be a m=4 reflection polarizer with just one SM coated side to produce a well polarized spin-up state. After this a second, double sided SM with slightly larger $\hat{\alpha}$ and m-value will be placed, together with an absorber at the end of the device to capture neutrons only reflected on the second lamella. As a result the transmitted neutrons will have passed three reflective interfaces before reaching the detector, leading to a relatively good polarization.

A great gain of detecting both spin-states at the same time can be achieved when analyzing reflectivity from an unpolarized neutron beam, allowing polarized measurements with only slightly reduced intensity (85%) compared to unpolarized measurements. The advantage for polarization analysis is often limited, as the spin-flip signal is much weaker than the non spin-flip reflectivity and dominates the measurement time. On the other hand, if two polarized beams, spin-up and spin-down, hit the sample surface at the same time, this method can be used for complete polarization analysis in time resolved measurements. A later upgrade to Estia, installing two separate Selene neutron guides which share a common focus at the sample position, will make this simultaneous 4-state measurement possible.

6. conclusion

We have presented a novel concept to use curved polarizing supermirrors in transmission geometry to cover large beam divergences for instruments with beam focusing. A demonstration device for the focusing option of the AMOR reflectometer was build and tested on the beamline, showing the expected polarization efficiencies. Following the same general concept, the Estia reflectometer at ESS will be build using two spiral shaped polarizers and a double lamella analyzer. With this setup the instrument will achieve a high polarization of 99% over a broad wavelength band with only minor impact on instrument intensity and allow simultaneous measurement of spin-up and spin-down reflectivity on the analyzer side. After full extension of the instrument with two separate Selene beam paths, the ability to measure all 4 spin-states at once will allow novel time resolved polarization analysis studies.

As all other polarization devices, the equiangular spiral has its limitations imposed by the SM reflectivity that restricts the range of neutron wavelength and maximum focus size to which the system can be applied. With the ongoing advances in pulsed neutron sources and neutron optics, more instruments will have improved focusing capabilities to be able to measure tiny samples.

In this scenario, current ^{3}He systems are unable to deliver sufficient polarization efficiencies or suffer from significant attenuation when applied to beams with typical wavelength bands used at pulsed sources. Other solid states devices as e.g. cavities, mirror fans or bender systems significantly distort the neutron trajectories and thus the focusing ability of the instrument (or detector resolution). The presented devices are optimized for the focusing geometry and therefore offer significant advantages over the established polarization components.

References

[1] Halpern O and Johnson M H 1939 *Phys. Rev.* **55** 898–
[2] Blume M 1963 *Phys. Rev.* **130** 1670–
[3] Lovesey S W 1986 *Theory of Neutron Scattering from Condensed Matter: Volume 2: Polarization Effects and Magnetic Scattering. The International Series of Monographs on Physics* (Oxford University Press, New York)
[4] Nathans R, Shull C, Shirane G and Andresen A 1959 *Journal of Physics and Chemistry of Solids* **10** 138–146 ISSN 0022-3697
[5] Kulda J, Courtois P, Saroun J, Thomas M, Enderle M and Flores P 2001 *Neutron Optics* **4509** 13–20
[6] Takeda M, Nakamura M, Shimojo Y and Kakural K 2008 *Nuclear Instruments & Methods In Physics Research Section A-accelerators Spectrometers Detectors and Associated Equipment* **586** 229–238
[7] Groitl F, Keller T, Quintero-Castro D L and Habicht K 2015 *Review of Scientific Instruments* **86** 025110
[8] Heil W, Dreyer J, Hofmann D, Humblot H, Lelievre-Berna E and Tasset F 1999 *Physica B-condensed Matter* **267** 328–335
[9] Chen W C, Armstrong G, Chen Y, Collett B, Erwin R, Gentile T R, Jones G L, Lynn J W, McKenney S and Steinberg J E 2007 *Physica B-condensed Matter* **397** 168–171
[10] Parnell S R, Washington A L, Li K, Yan H, Stonaha P, Li F, Wang T, Walsh A, Chen W C, Parnell A J, Fairclough J P A, Baxter D V, Snow W M and Pynn R 2015 *Review of Scientific Instruments* **86** 023902
[11] Mesei F 1976 *Communications On Physics* **1** 81–85
[12] Stahn J and Clemens D 2002 A remanent transmission neutron polarizer
[13] Padiyath J, Stahn J, Horisberger M, Ay M and Boni P 2006 *Applied Physics Letters* **89** 253111
[14] Krist T, Lartigue C and Mesei F 1992 *Physica B* **180** 1005–1006
[15] Stahn J and Glavic A 2016 *Nuclear Instruments and Methods in Physics Research Section A: Accelerators, Spectrometers, Detectors and Associated Equipment* **821** 44–54 ISSN 0168-9002

IOP Conf. Series: Journal of Physics: Conf. Series **862** (2017) 012008 doi:10.1088/1742-6596/862/1/012008

Magnetic field imaging of a model electric motor using polarized pulsed neutrons at J-PARC/MLF

K Hiroi[1], T Shinohara[1], H Hayashida[2], J D Parker[2], K Oikawa[1], M Harada[1], Y Su[1] and T Kai[1]

[1]J-PARC Center, Japan Atomic Energy Agency, 2-4 Shirakata-shirane, Tokai, Ibaraki 319-1195, Japan
[2]Comprehensive Research Organization for Science and Society 162-1 Shirakata, Tokai, Naka, Ibaraki 319-1106, Japan

E-mail address : kosuke.hiroi@j-parc.jp

Abstract. We have been developing a magnetic field imaging method using polarized pulsed neutrons for quantitative evaluation of magnetic fields in industrial equipment. To reduce the power loss of these products, it is important to quantitatively evaluate the distribution of not only strength but also direction of the internal field. In this study, we attempted to visualize the directional distribution of a magnetic field in a model electric motor. Because of the limited wavelength range that we could use in the present analysis, the polarization distribution image of the model motor was confirmed to reflect the directional distribution of the magnetic field outside the motor. It was also clarified that usage of shorter wavelength neutrons would enable us to visualize the field inside the motor. Then polarized neutron imaging was found to be a potentially powerful method for observing the directional distribution of the driving field of a rotating motor.

1. Introduction

Electric motors, which are widely used in many electric machines to produce mechanical force from electric power, consume large amounts of the world's electricity. Thus, improving the energy efficiency of electric motors is very important for the reduction of power loss. Evaluating the magnetic field in a motor under driving state and comparing it with the design value are necessary for the development of motors with high energy efficiency. However, there is no appropriate technique to observe the field distribution in a practical motor, and hence, the actual field distribution in a motor under driving state has not been evaluated before now. Neutron imaging [1] is a promising technique to solve such problems because it has been used for the non-destructive visualization of internal structures of bulk objects. Recently this technique was applied to a novel magnetic field imaging method using neutron spin analysis [2-3]. In our previous work, we successfully demonstrated quantitative magnetic field imaging using a polarized pulsed neutron beam [4].

Here, we briefly explain the principal of our magnetic field imaging method. Neutrons experience Larmor precession when they travel in a field whose direction is not parallel to their spin vector, and its precession angle φ can be expressed as

$$\varphi = \omega_L t = \frac{\gamma}{v} \int_{\text{path}} \boldsymbol{B} \cdot ds = \frac{\gamma m \lambda}{h} \int_{\text{path}} \boldsymbol{B} \cdot ds \quad , \tag{1-1}$$

where B is a given magnetic field, ω_L is an angular velocity of the precession proportional to B, t is the time a neutron travels through the field, γ is the gyromagnetic ratio of a neutron, h is Plank's constant, and m, v, and λ are the mass, velocity, and wavelength of the neutron, respectively. According to this equation, φ is proportional to the neutron wavelength and strength of the field integrated along the neutron flight path. Then, by measuring the wavelength dependence of the neutron polarization, which is a projection of the neutron spin vector onto a quantized axis of neutron polarization, we can observe the precession of the neutron spin as an oscillatory behavior in the wavelength dependent polarization. Assuming that the magnetic field along the neutron beam trajectory is uniform, the expression of such oscillations was described by Rekveldt [5] as

$$P = D \cdot P_0 = \begin{bmatrix} 1-(1-\cos\varphi)\cdot(1-n_x^2) & (1-\cos\varphi)n_x n_y - n_z\sin\varphi & (1-\cos\varphi)n_z n_x - n_y\sin\varphi \\ (1-\cos\varphi)n_x n_y + n_z\sin\varphi & 1-(1-\cos\varphi)\cdot(1-n_y^2) & (1-\cos\varphi)n_y n_z - n_x\sin\varphi \\ (1-\cos\varphi)n_z n_x - n_y\sin\varphi & (1-\cos\varphi)n_y n_z - n_x\sin\varphi & 1-(1-\cos\varphi)\cdot(1-n_z^2) \end{bmatrix} \cdot P_0, \quad (1\text{-}2)$$

where $\boldsymbol{P_0}$ and $\boldsymbol{P}$ are the polarization vectors of the incident and transmitted neutron beam, and n_i is a directional cosine of each component of the field. By choosing the direction of $\boldsymbol{P_0}$ and $\boldsymbol{P}$ appropriately, each component of depolarization matrix **D** can be measured, and then the integrated field strength and direction can be determined using these relations. Furthermore, the analysis of the wavelength dependent polarization with spatial resolution enables us to obtain a quantitative field image of a bulk object and the surrounding space. This method is thought to be suitable for observing the field distribution in moving products like rotating motors, because we can perform the measurement nondestructively with a large field of view up to several centimeters. However, there are some anticipated difficulties in applying this method to the measurement of actual products, e.g., the field cannot be assumed as homogeneous along the beam trajectory neither in direction nor in strength, the strength of the field integral is too strong to be analyzed, etc. In this study, we performed polarized neutron imaging of a model electric motor using a pulsed neutron beam and investigated the relationship between observed polarization images and the field of the motor using simulations.

2. Experimental procedures

Magnetic field imaging experiments using a polarized pulsed neutron beam were performed at beamline BL10 "NOBORU" [6] in the Materials and Life Science Experimental Facility (MLF) of J-PARC. Figure 1 shows a schematic illustration of our experimental setup. In this system, the z direction was taken along the path direction of the neutron beam, and we obtained polarization distribution images in the xy plane. The pulsed neutrons were polarized and analyzed using a polarizer and analyzer that consisted of stacked bend magnetic super-mirrors [7]. The quantized axis of neutron polarization for the polarizer and analyzer was chosen as the y (vertical) direction. We set collimators behind both the polarizer and analyzer to remove neutrons reflected from the super-mirrors. To obtain the neutron polarization, P, we flipped the neutron spin polarity of the incident beam using an adiabatic fast passage (AFP) spin flipper [8] that was placed behind the polarizer and measured the neutron transmission intensity with the spin flipper switched off (I_{off}) and on (I_{on}). The neutron polarization was then obtained by calculating $P = (I_{\text{off}} - I_{\text{on}})/(I_{\text{off}} + I_{\text{on}})$ [9]. Additionally, spin rotators, composed of two pairs of orthogonally arranged spin-precession coils which are used to control the polarization direction of the incident and transmitted beam, were located together with the object in the magnetic shield chamber to reduce the effect of any environmental magnetic fields [10]. A damping current is applied to the precession coils in inverse proportion to time and synchronized with the generation of a neutron pulse to give a constant spin rotation for a wide wavelength range of neutrons. Since the rotation angle of the neutron spin is proportional to the amplitude of the field applied by the coils, it is possible to adjust the rotation angle of the polarization vector arbitrarily by controlling the applied electric currents. The μPIC-based Neutron Imaging Detector (μNID) [11] was used as a two-dimensional detector. Distance between the neutron source and the object was 14.18 m, and the distance between the source and the detector was 14.87 m.

Figure 2(a) shows a photograph of a small model motor that we used as an object. This motor is a brushless-type that consists of a central rotor, which is a quadrupole magnet of samarium cobalt (SmCo), and a stator with driving coils around the rotor. The gap distance between the rotor and the stator is about 2 mm. In this experiment, we are especially interested in the quantification of the field in this gap because it is an important characteristic for determining the performance of this type of motor. As a pilot study to visualize the magnetic field in the moving motor, we performed an imaging experiment with a model motor in a static state and attempted to evaluate its magnetic field. The integrated field strength in the gap was roughly estimated to be several T·mm from the magnetization of the rotor magnet and width of the motor. The expected period of polarization oscillation that will appear in the analysis of the field in the gap was shorter than 0.035 Å from equation (1-1). Analyzing such short-period oscillation seems to be difficult due to the restriction of our polarization analysis equipment, which has a minimum bin width of wavelength $\delta\lambda$ longer than 0.01 Å originating from the intrinsic pulse width of our neutron source. Therefore, quantification of the field strength in the gap by analyzing the frequency of polarization oscillation was difficult. On the other hand, focusing on a diagonal term of equation (1-2), the polarization degree P of neutrons that pass through a strong magnetic field can be expressed as follows,

$$P/P_0 = 1-\left(1-<\cos\varphi>\right)\cdot\sin^2\theta = \cos^2\theta \quad (2\text{-}1)$$

where θ is the angle between the magnetic field and the polarization direction of the neutron. While the oscillation term, which depends on the Larmor precession angle φ, is smeared out due to the coarse $\delta\lambda$, information about field direction, which corresponds with θ, is still observable in the polarization distribution images. Accordingly, by analyzing the change in polarization depending on the polarization direction using spin rotators, the field direction can be evaluated independent of the strength of the field. In addition, this analysis method is also applicable to continuous sources because the energy-resolved analysis is not necessary.

An imaging experiment was performed to confirm the validly of this method in analyzing the motor field. The polarization vector of the incident beam, which is along the y direction, is rotated into the desired direction by the first pair of precession coils located in front of the object. After transmission through the object, the polarized neutrons enter the second pair of precession coils behind the object where the initial rotation is exactly reversed. In this way, we controlled the neutron spin orientation of the incident and transmitted beam and measure the depolarization component along any arbitrary direction.

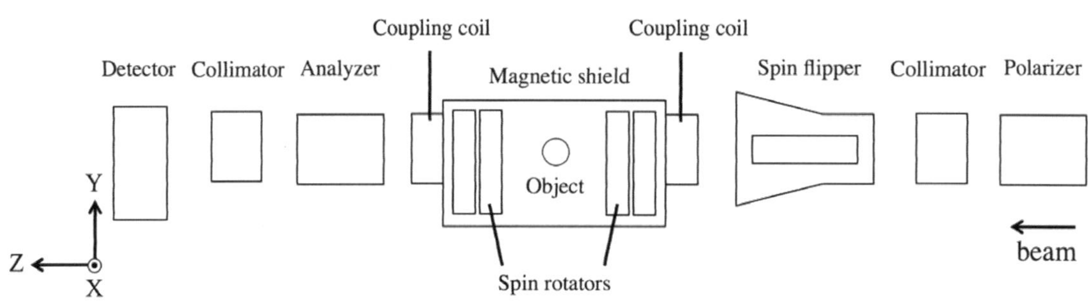

Figure 1 A schematic illustration of the experimental setup at BL10

3. Results and Discussion

3.1. Evaluation of the field direction in the motor using polarized neutron imaging

We positioned the model motor with the rotation axis parallel to the z direction in order to observe the field in the gap. In this arrangement, the field in the gap was mainly directed in the xy plane, and then we measured polarization distribution images while changing the direction of P_0 in the xy plane. Figure 2(d) shows polarization distribution images of the model motor in a static state obtained for various polarization directions. As can be seen from this figure, the polarization distribution in the gap gradually changes as the polarization direction is turned, rotating 90 degrees when polarization direction was turned from the y to the x direction. This result indicates that the directional information of the field in the gap was preserved in the observed polarization images. Figure 2(b) presents the wavelength dependence of P and P_0 at the gap position for the polarization direction along the y axis. The oscillatory behavior that is indicative of Larmor precession was not observed in the wavelength dependence of P, which was expected from the above discussion. On the other hand, the value of P decreased compared to that of P_0. This could be explained by the equation (2-1): the reduction of the polarization degree occurred due to the disagreement between the direction of field in the motor and the polarization direction of P_0. We analyzed this polarization change by fitting with the equation (2-1) revised as

$$P/P_0 \propto \cos^2(\theta' + \alpha), \tag{3-1}$$

where θ' is the angle of the polarization direction from the y axis and α corresponds to the angle of the motor field from y axis. It is worthwhile mentioning that we can extract the information of field direction in the xy plane from this analysis even if the motor field has a z component, because it does not affect the phase component of the sinusoidal form in the θ' dependence of P/P_0 but affects the amplitude of the waveform according to the equation (1-2). Figure 2(c) shows the θ'dependence of P/P_0 at the gap position of the motor. We measured several polarization images changing the polarization direction from the y (0°) to the x direction (90°). The number of directional steps was five: 0°, 30°, 45°, 60°, 90°, which was adequate to fit the experimental results to equation (3-1). In this figure, the averaged value of P/P_0 over a wavelength range from 4 to 8 Å was used because P/P_0 was almost independent from the neutron wavelength. P/P_0 plotted against θ' clearly exhibited a sinusoidal shape and could be well fitted by the equation (3-1). We performed this analysis position by position and obtained the two-dimensional distribution image of the field direction of the motor by plotting α at each position (Figure 3). From this image, it was found that the field in the left and right side of the rotor was aligned along the y direction and the top and bottom side was along the x direction.

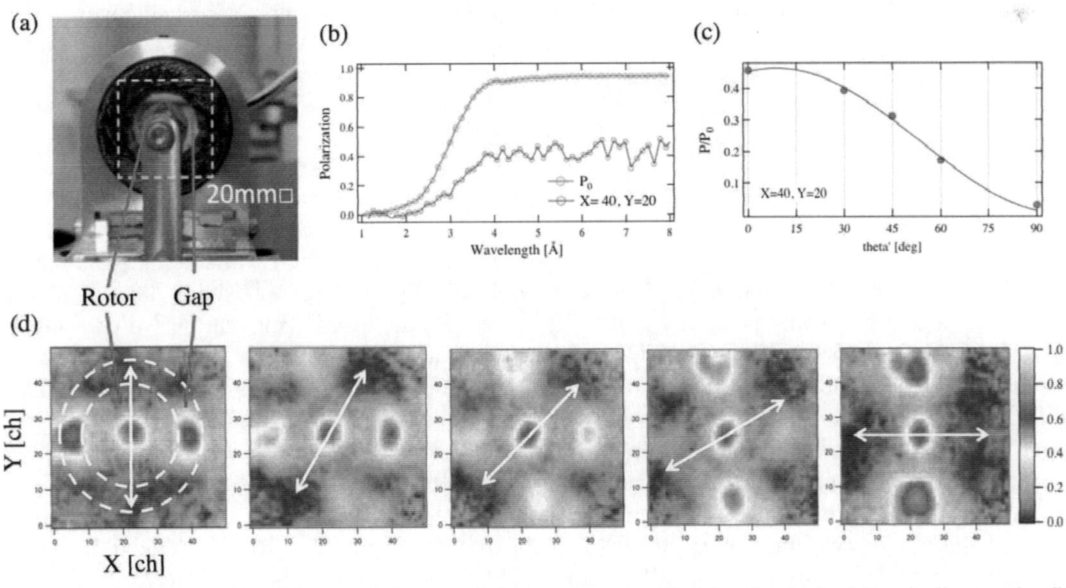

Figure 2 (a) Photograph of the model motor. The area surrounded by the dashed line indicates the field of view for the polarization images. (b) Wavelength dependence of polarization at the gap position of (X, Y) = (40 [ch], 20[ch]) in the polarization image and that of the incident beam P_0 (the polarization axis is in the *y* direction). (c) θ' dependence of P/P_0 at the same position as (b). The solid line indicates the result of fitting using equation (3-1). (d) Polarization distribution images obtained with various incident polarization directions (1 [ch]= 0.4 [mm]).

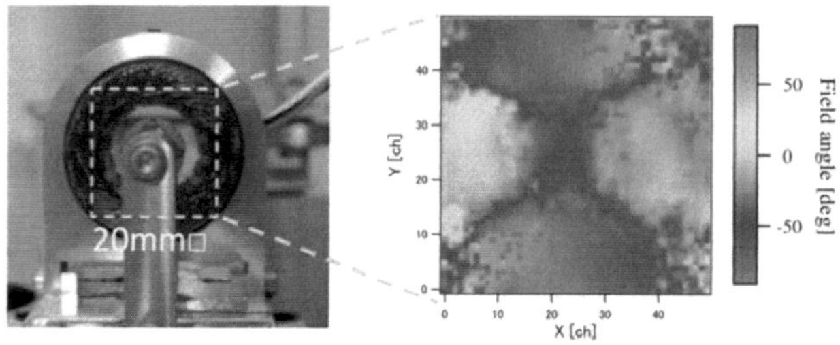

Figure 3 Distribution image of the field direction around the rotor obtained by the analysis of the depolarization as a function of incident polarization direction.

3.2 Field simulation of the model motor and comparison with the experimental result

To confirm the validity of this experimental result, we performed a simulation study of the magnetic field inside and outside the model motor and compared the result with the field distribution obtained from the experiment. Magnetic fields inside and outside the model motor were calculated using the commercial simulation software JMAG [12]. In this simulation, we calculated the *x*, *y* and *z* components of the internal field and the leaked field outside the model motor. Then, we determined the field direction in the *xy* plane by calculating $\tan^{-1}(B_x/B_y)$ where B_x and B_y are the *x* and *y* components of the field. Figure 4(a-c) show the distributions of field direction at different positions along the neutron beam trajectory. The distribution of the leaked field outside the motor (Figure 4(c)) was clearly different from that inside the motor (Figure 4(a)). Comparing these figures with the experimental result shown in Figure 3, the distribution for the leaked field (Figure 4(c)) appears to best reproduce the experimental results. The reason why the experimentally obtained distribution image resembles that outside the motor, not inside the motor, can be explained by the adiabatic transition of neutrons in the motor. We calculated the adiabaticity parameter *E* along the beam path, which was defined as

$$E = \left|\frac{\omega_L}{\omega_B}\right| = |\gamma B| \Big/ \left|\frac{d\theta_B}{dz} \cdot v\right| , \qquad (3\text{-}2)$$

where ω_B was the rotation speed of the magnetic field in the transmission path and $d\theta/dz$ was the change of field direction along the path. If the change of field direction is much slower than the Larmor precession ($\omega_L >> \omega_B$), the component of the polarization vector parallel to the field is conserved. Such an adiabatic rotation occurs when $E > 10$ [13]. Figure 4(d) shows the calculated adiabticity parameters in some positions at the gap of the motor. In this calculation, we considered neutrons with the wavelength of 4 Å. *E* exhibits its maximum at the center of the motor whose value is larger than 500. The value of *E* decreases to about 100 at the edge of the motor and becomes smaller than 10 at the position 8 ~ 9 mm away from the edge. According to these results, we surmised the motion of neutron spin in the motor as follows: (i) Polarized neutrons enter the upstream leaked field region where E >10, maintaining its polarization vector along the controlled direction by the spin rotator since the magnetic field in the region where $E < 10$ is very small. (ii) Then neutron spins experience Larmor precession around the field in the region where $E > 10$. The direction of the component of the polarization vector parallel to the field follows the direction of the field by adiabatic rotation. (iii) After passing through the motor field, neutrons travel through the region where $E < 10$ again, and then the component of the polarization vector selected by the spin rotator is projected to the polarization axis of the analyzer. The directional information of the component of the polarization vector parallel to the field is a key information in estimating field direction. However, this information is changed by the adiabatic rotation in the beam path, and then information of the directional distribution in the motor is masked by that of the leaked field. In this way, the distribution image of the field direction obtained by this analysis is thought to become that of the leaked field where $E \sim 10$. In order to check the correctness of this supposition, we simulated the polarization distribution image by calculating the Larmor precession of the neutron spin inside and outside the motor using the magnetic field simulation results.

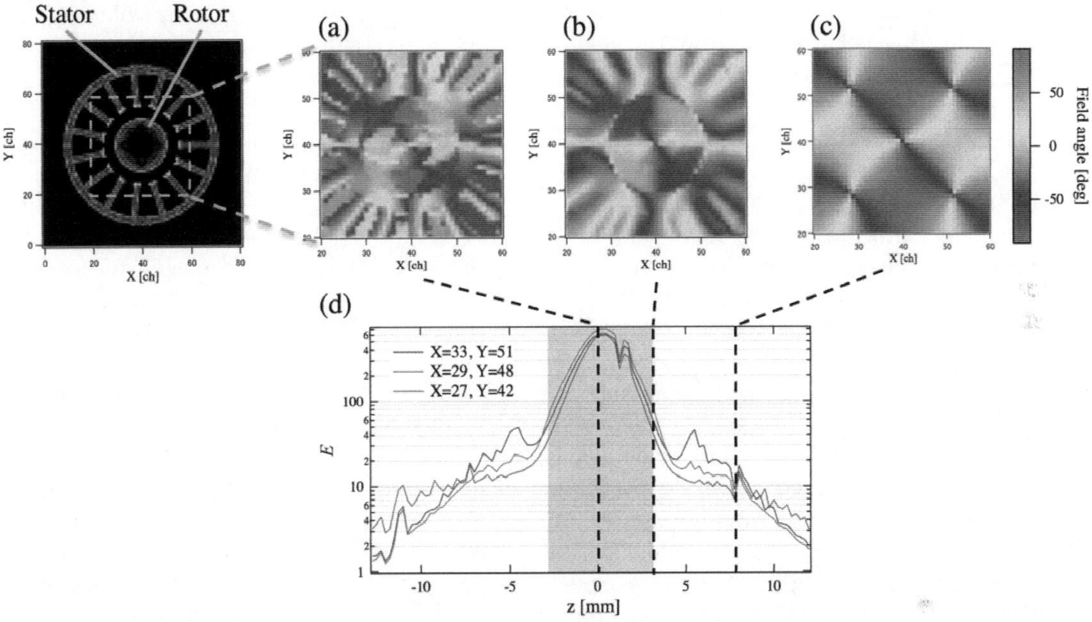

Figure 4 Distribution images of the field direction (a) at the center of the motor, (b) at the edge of the motor, (c) outside the motor (edge +8.5mm) (d) Logarithmic-scale plot of adiabaticity parameters along z direction in some positions of the gap. The grey zone indicates the motor position and the dashed lines indicate the position of (a), (b) and (c).

3.3 Simulation of the polarization distribution images

We calculated the Larmor precession of a neutron spin in the motor based on the magnetic field simulation result and simulated the polarization distribution image of the model motor. The motion of the neutron spin in a given magnetic field can be expressed as

$$\frac{d\boldsymbol{\sigma}_n}{dt} = \gamma(\boldsymbol{\sigma}_n \times \boldsymbol{B}) , \tag{3-3}$$

where $\boldsymbol{\sigma}_{\mathrm{n}}$ is a unit vector parallel to the neutron spin. We calculated a change of neutron spin vector along the beam path by numerically solving the above equation, and the neutron polarization P was obtained by projecting the final state of $\boldsymbol{\sigma}_{\mathrm{n}}$ onto the polarization vector of the analyzer $\boldsymbol{A}$ by calculating

$$P = \frac{\sigma_{nx}A_x + \sigma_{ny}A_y + \sigma_{nz}A_z}{|\boldsymbol{\sigma}_n||\boldsymbol{A}|} \tag{3-4}$$

and averaging over the same wavelength rage as used in the experiment (i.e., from 4 Å to 8Å). By performing these calculations position by position, we simulated the polarization distribution images using the results of the motor field in Section 3.2. We took into account the approximate beam divergence for the experimental condition at BL10 and assumed that polarized neutrons were only precessed by the magnetic field of the motor. Figure 5(a) shows the polarization images calculated from the simulation results for several polarization directions (corresponding to those measured during the experiment). The analysis of section 3.1 was performed on these simulated polarization images, and the distribution image of the field direction obtained by this calculation is shown in Figure 5(b). The simulated polarization distribution images and corresponding directional distribution image exhibit a similar tendency to that obtained from the experimental data. These results justify the correctness of our experimental results and the supposition about the reason why the obtained image of field direction was similar to that outside the motor.

On the other hand, comparing the distribution image of the field direction obtained by this analysis (Figure 5(b)) and that predicted by the field simulation (Figure 4(c)), it is seen that the two results do not perfectly coincide with each other. The disagreement originates not only from image blurring due to the beam divergence but mainly from the incorrect assumption of a uniform field distribution along the z direction in the conversion from polarization distribution images to the directional distribution image using the equation (3-1). This interpretation means that a uniform field along the z direction should be exactly quantified. To check the validity of this method, we performed another simulation where we removed the leaked field component from the result of the field simulation of Section 3.2 and performed the polarization simulation without beam divergence. Since the field distribution in the motor is almost uniform along the beam path direction, the directional distribution of the field should be precisely visualized in this condition. Figure 6 shows the polarization images for several polarization directions and the distribution image of the field direction obtained by the analysis. The directional distribution obtained by this simulation agrees well with that in the motor (Figure 4(a)). This result shows that in certain cases of restricted geometries (e.g. in-plane) and a uniform field along the flight path, our analysis method can quantify the orientation distribution of the field vector with respect to two dimensions. While such a suitable target is limited, this method is useful to quantify the field direction no matter how strong its magnitude is.

In the case of a field direction whose distribution changes along the beam path, the influence from such field variation should be carefully considered, and the precise determination of such a field is difficult using the present analysis method. One idea for discussing the field distribution of the motor is to use the polarization distribution images themselves. Although it was difficult to straightforwardly quantify the field direction of the motor, we confirmed that the polarization distribution image could be predicted from a prior knowledge of the field distribution. Using polarization distribution images, we can directly compare the experimental results with predictions of simulation without any assumptions and discuss where actual field distribution deviates from the designed one. This measurement will be helpful to inspect the distribution of real magnetic fields of industrial products. It remains to be investigated, however, whether the differences of model simulations and measurements would originate in variations of reality with respect to the idealized model or from certain issues of the measurement technique.

In our analysis method, the main reason why we can only obtain the directional information outside the motor comes from the limitation in wavelength range of the available polarized neutron beam, which was dictated by the polarizer performance. Since the adiabaticity parameter E decreases inversely proportional to the velocity of the neutron, utilization of polarized neutrons with shorter wavelengths $\lambda < 1$ Å will enable us to visualize inner field information of the motor. Moreover, it will become possible to evaluate the strength of the field in the motor with such short-wavelength polarized neutrons, because analysis of shorter-period oscillation in wavelength dependent polarization becomes possible by means of a narrower $\delta\lambda$ in such a wavelength range. Finally, we mention about the possible application of this technique. An attractive subject is to observe an actual driving motor. The polarization distribution measurements will be a help to identify the critical problem of an energy loss of the motor, which cannot be found only from the design work based on the numerical simulations. So as to utilize this technique, it is necessary to combine with the AC field observation method, which we have already established in our previous work [14]. Then this technique enables us to visualize the time-dependent change of the polarization distribution, and we can investigate the movement of the field of the rotating motor. Thus, our magnetic field imaging method

IOP Conf. Series: Journal of Physics: Conf. Series **862** (2017) 012008 doi:10.1088/1742-6596/862/1/012008

using polarized pulsed neutrons is regarded to contribute to the development of high efficiency motors as one of the useful tools for the characterization of driving motors.

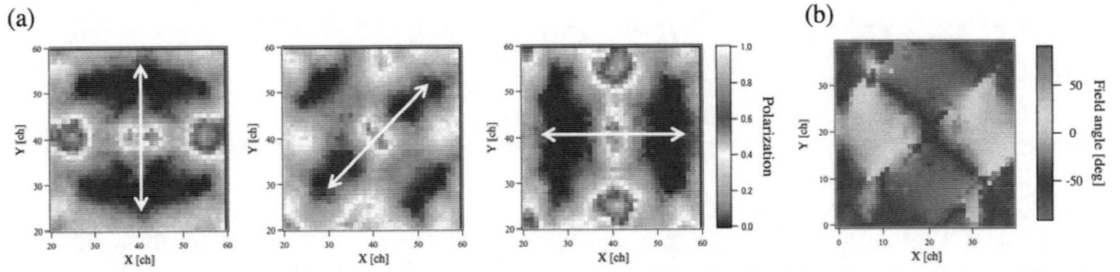

Figure 5 (a) Polarization distribution images of the motor field obtained by the simulation of Section 3-2. The white arrows indicate polarization direction. (b) Distribution image of the field direction obtained by the analysis.

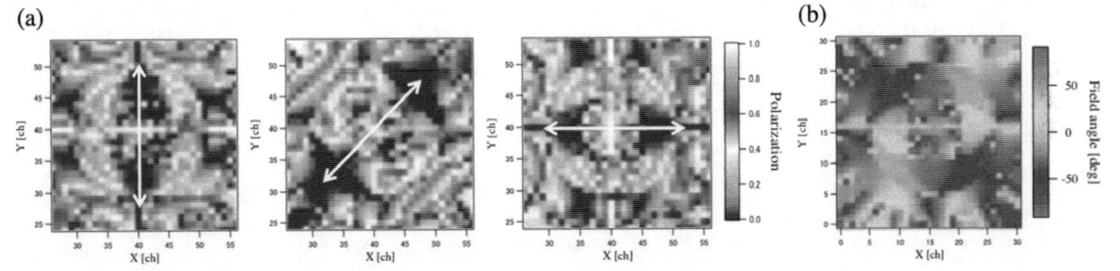

Figure 6 (a) Polarization distribution images of the motor without leaked field. The white arrows indicate polarization direction. (b) Distribution image of the field direction obtained by the analysis.

4. Conclusion

We attempted to quantify the directional distribution of a magnetic field in a model electric motor using polarized pulsed neutron imaging. We analyzed the change in the polarization distribution image depending on the incident polarization direction and obtained the distribution image of the field direction. Using simulation, the field distribution image was confirmed to be similar to that of the leaked field from the motor, and the information of the field inside the motor could not be extracted due to the limitation of the present analysis method. However, the polarization distribution image is considered to be useful to investigate the field distribution of a motor because the polarization distribution image directly reflects the directional distribution of the field outside the motor and we can directly compare experimental results with simulations. In addition, utilization of shorter-wavelength polarized neutrons in imaging experiments will enable to investigate inner field distribution of the motor.

Acknowledgement

We would like to thank Mr. Takao Imagawa of Hitachi, Ltd., for providing the model electric motor. This work was supported by the Photon and Quantum Basic Research Coordinated Development Program from the Ministry of Education, Culture, Sports, Science and Technology, Japan.

References

[1] Kardjilov N, Manke I, Hilger A, Strobl M, and Banhart J 2011 *Mater. Today.* **14** 6 248.
[2] Kardjilov N, Manke I, Strobl M, Hilger A, Treimer W, Meissner M, Krist T and Banhart J 2008 *Nat. Phys.* **4** 399
[3] Strobl M, Kardjilov N, Hilger A, Jericha E, Badurek G, Manke I 2009 *Physica B* **404**, 2611
[4] Shinohara T, Sakai K, Ohi M, Kai T, Harada M, Oikawa K, Maekawa F, Suzuki J, Oku T, Takata

S, Aizawa K, Arai M and Kiyanagi Y 2011 *Nucl. Instr. and Meth. A* **651** 121
[5] Rekveldt M Th 1973 *Z. Phys.* **259**, 391
[6] Maekawa F, Oikawa K, Harada M, Kai T, Meigo S, Kasugai Y, Ooi M, Sakai K, Teshigawara M, Hasegawa S, Ikeda Y and Watanabe N 2009 *Nucl. Instr. and Meth. A* **600** 335
[7] Krist T, Kennedy S J, Hicks T J and Mezei F 1998 *Physica B* **241–243** 82
[8] Weinfurter H and Badurek G 1989 *Nucl. Instr. and Meth. A* **275** 233
[9] Mitsuda S Yoshizawa H and Endoh Y 1992 *Phys. Rev. B* **45** 17 9788
[10] Wada N, Shinohara T, Sato H, Hasemi H, Kamiyama H and Kiyanagi Y 2015 *Physics Procedia* **69** 427
[11] Parker J D, Harada M, Hattori K, Iwaki S, Kabuki S, Kishimoto Y, Kubo H, Kurosawa S, Matsuoka Y, Miuchi K, Mizumoto T, Nishimura H, Oku T, Sawano T, Shinohara T, Suzuki J.-I, Takada A, Tanimori T, Ueno K, Ikeno M, Tanaka M and Uchida T 2014 *J. Phys.: Conf. Ser.* **502** 012048
[12] www.jmag-international.com
[13] Stewart R, Lecture notes of 12th Oxford School on Neutron Scattering
[14] Hiroi K, Shinohara T, Hayashida H, Wada N, Oikawa K, Harada M, Kai T and Arai M 2015 *JPS Conf. Proc.* **8**, 035001

Instrument developments and recent scientific highlights at the J-NSE

Oxana Ivanova[1], **Stefano Pasini**[2], **Michael Monkenbusch**[2], **Olaf Holderer**[1]

[1] Jülich Centre for Neutron Science (JCNS) at Heinz Maier-Leibnitz Zentrum (MLZ), Forschungszentrum Jülich GmbH, Garching, Germany

[2] Jülich Centre for Neutron Science JCNS and Institute for Complex Systems ICS, Forschungszentrum Jülich GmbH, Jülich, Germany

o.holderer@fz-juelich.de

Abstract. The J-NSE neutron spin echo spectrometer faces now 10 years of successful user operation at the FRM II research reactor at the Heinz Maier-Leibnitz Zentrum (MLZ). We present scientific highlights and instrumental developments of the last decade, for example the development of grazing incidence neutron spin echo spectroscopy (GINSES) at the J-NSE and investigations of the dynamics at solid-liquid interfaces with this new option. Polymers in confinement have been a prominent topic, as well as the internal dynamics of proteins. The scientific questions also triggered instrumental developments such as a new polarizer and a new neutron guide concept. Finally, the future of the J-NSE will be addressed with a short presentation of the current upgrade program with superconducting main coils with reduced intrinsic field integral inhomogeneity.

1. Introduction

Neutron spin echo spectroscopy (NSE) provides the highest energy resolution of all neutron scattering techniques. It has been first proposed by Ferenc (Feri) Mezei in 1972 [1]. The first instrument of this kind was the IN11 at ILL, put into operation in 1977 [2]. Neutron beam monochromaticity and resolution are decoupled in NSE spectroscopy by encoding velocity changes during the scattering process in a change of polarization of a polarized neutron beam. NSE spectroscopy measures the normalized intermediate scattering function S(q,t)/S(q,t=0), which is on the one side the Fourier transform from energy to time of the scattering function S(q,ω) measured in e.g. Time-Of-Flight spectroscopy or Three-Axis-Spectroscopy, and on the other hand the Fourier transform in space of the van-Hove correlation function G(r,t), which can be extracted e.g. from particle trajectories in computer simulations. NSE accesses length scales in the range of some nm and Fourier times of 0.1-100's of ns, translating in an energy resolution in the sub-μeV region. Thermal fluctuations in soft matter systems, such as membrane fluctuations in microemulsions or vesicles, internal domain motion of proteins or polymer motion in a melt are prominent examples, where the length- and energy-scale of the systems are exactly in the working region of NSE spectroscopy [3-7]. In this contribution we show the scientific field of use of the J-NSE spectrometer and the modifications and upgrades of the last decade and give a short perspective on the major main coil upgrade planned for the near future.

2. Instrument upgrades

IOP Conf. Series: Journal of Physics: Conf. Series **862** (2017) 012009 doi:10.1088/1742-6596/862/1/012009

The J-NSE spectrometer has been transferred to the FRM II reactor at MLZ in 2006 after 10 years of operation at the DIDO reactor in Jülich [8-9], involving an upgrade of the electronics and with the advantages of a neutron guide end position. During the 10 years of operation at the FRM II the spectrometer has experienced some smaller instrumental improvements. The most prominent ones were the efforts for better correction coils for an improved field integral homogeneity. Each main solenoid is equipped with 3 correction coils which provide radially symmetric corrections to the field integral and are essential for achieving the required homogeneity.
The currently installed design are "Pythagoras"-coils first proposed at the ILL. Instead of the originally installed spiral coils the new design consists of 2 rectangular coils, each providing a quadratic correction in one direction, x^2 or y^2, providing a $x^2 + y^2 = r^2$-correction when mounted with 90° with respect to each other.

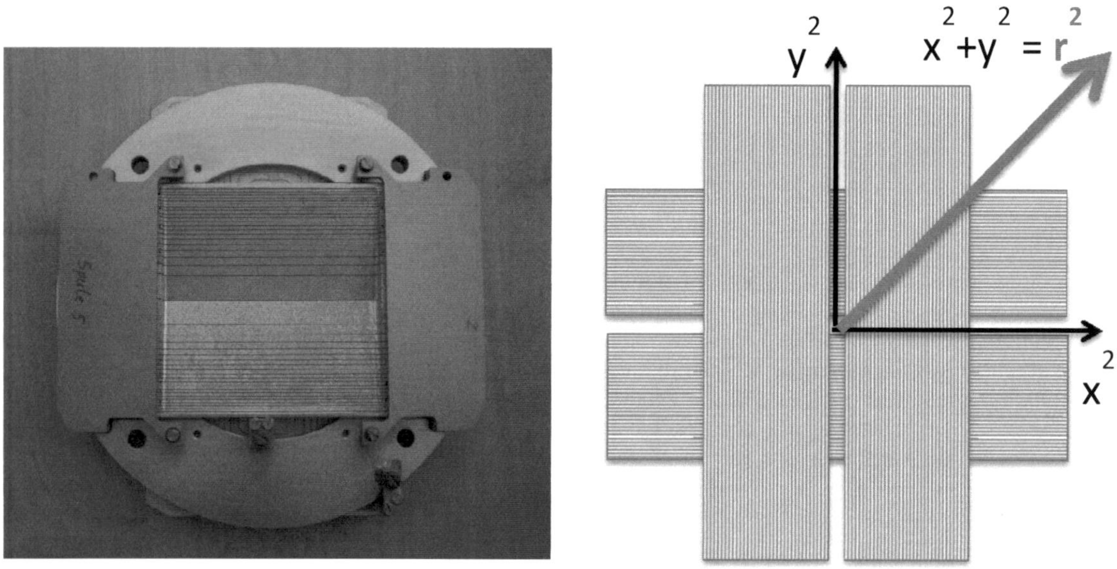

Fig. 1: Pythagoras correction coils at the J-NSE (left). The horizontal coils is visible, a second one perpendicular to it is on the back side. The schematic drawing (right) shows the effect of the two perpendicular coils, a x^2 and y^2 correction leading to a radially symmetric r^2 correction of the magnetic fields.

This correction coil design has the disadvantage compared to the spiral Fresnel coils that only quadratic corrections can be realized, higher order corrections nonetheless still play a role, especially for the largest correction coil.
The advantage on the other hand is the better mechanical precision due to the easier cutting scheme, the small amount of material in the beam, increasing the transmission significantly, and at the same time the good heat removal, since the path of each current zone to the side frame is much shorter as in the case of the spiral Fresnel coils.

The polarization of the neutron beam has been done in two steps, for short wavelengths (λ<8Å) in a bent part of the neutron guide, for longer wavelengths by a single reflection in a second polarizer at the instrument. The change between the two operation regimes involved a rotation of the whole spectrometer by 4°. The polarization concept has been modified by a new polarizer with significantly better performance, which polarizes the whole wavelength band from 4.5-17 Å used at the J-NSE. The rotation of the instrument became obsolete, simplifying the operation significantly with a slightly better polarization as before.

The renewal of the bent part of the neutron guide has been started, since first signs of degradation became visible. It is now replaced by a non-polarizing guide, leaving the whole polarization work to the new single polarizer in front of the main solenoids.

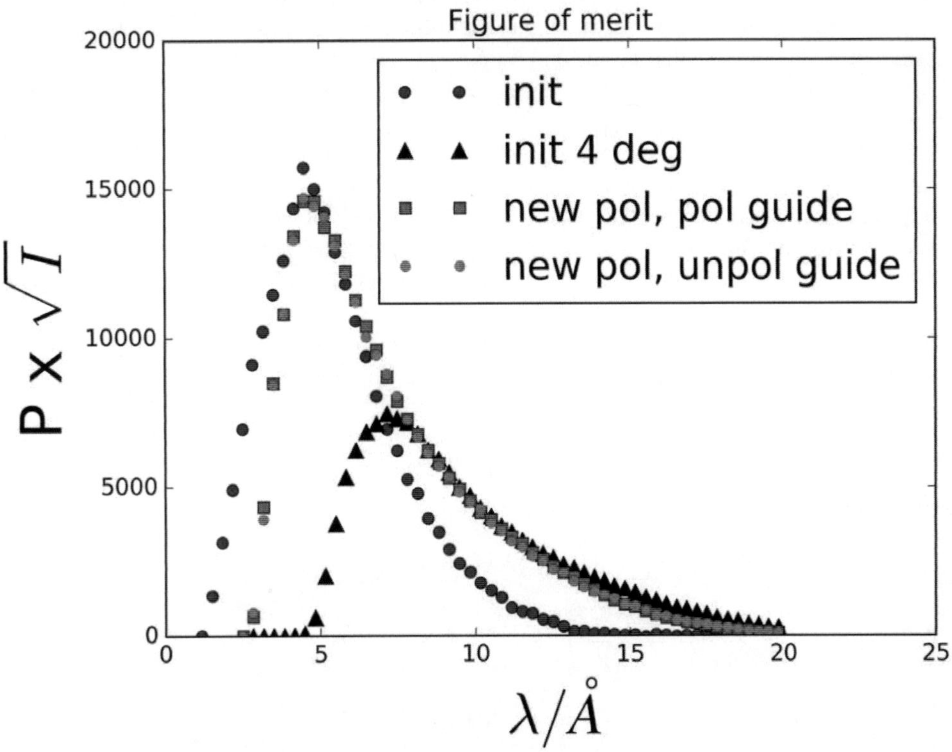

Fig. 2: Figure of merit (polarization x intensity$^{1/2}$) from McSTAS simulations for the new instrument configuration with the second polarizer only compared to the initial configuration.

3. Scientific usage of J-NSE

The scientific topics of the last decade came mainly from the domains of soft matter and biology. The development of the grazing incidence NSE technique (GINSES), where the near-interface dynamics is probed with an evanescent wave opened a new field of NSE experiments, where the influence of a rigid wall on surfactant membrane fluctuations [3-4] or microgels adsorbed at the interface [5] have been studied. Currently, efforts are made for increasing the applicability of this technique by different measures of intensity improvements. Polymer dynamics, one of the primary research fields for NSE, evolved towards complex systems where e.g. confinement effects play a crucial role [6]. The internal fluctuations of sub-domains in proteins has been studied with the J-NSE for example in immunoglobulin, a Y-shaped human antibody protein [7]. NSE allows to study the proteins in a physiologically relevant temperature range in solution.

4. Major upgrade project

Recent developments at the ILL and the proposed NSE instrument for the ESS [10] aimed for a configuration of the main precession coils with lower field-integral intrinsic inhomogeneity. This has been achieved through an optimization of the magnetic field shape and a consequent minimization of

IOP Conf. Series: Journal of Physics: Conf. Series **862** (2017) 012009 doi:10.1088/1742-6596/862/1/012009

the amount of required additional corrections. The 2.5-times improved homogeneity makes higher Fourier times possible than those previously accessible, by equal wavelengths. Instead of a single main precession coil the optimized configuration consists of a set of five pairs of superconducting coils with different diameter that are fringe field compensated in a similar way as the solenoids of the SNS-NSE spectrometer. The new design is based on an extension of the Zeyen and Rem [11] approach and it uses a combination of analytical calculations and numerical simulations to shape the magnetic field to an optimal form that minimizes the first order terms in an expansion of the intrinsic integral field inhomogeneity.

Assuming in the simulations linearly increasing radial current density at the position of the correction elements (e.g. Fresnel coils), the new layout allows reaching a residual inhomogeneity of 1 ppm with only two (radial) correction coils (instead of three like in the NSE's at MLZ and SNS). An optimization of the currents and of the positions of these elements shows that higher homogeneity of the field integral translates into weaker corrections and hence into weaker current in the outset (and most significant) large correction element.

Therefore the conclusion is that the final resolution of the spin echo spectrometer will be higher - even if corrections coils are used – because of the reduced intrinsic inhomogeneity of the coil.

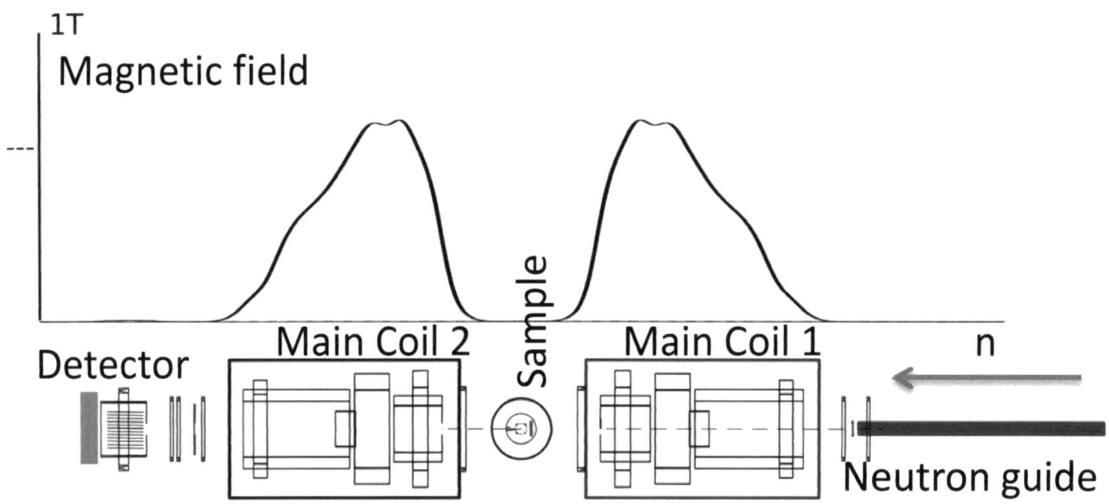

Fig. 3: Top: Magnetic field strength along the flight path of the neutrons (from right to left). Bottom: corresponding coil configuration for the intrinsically more homogeneous field configuration, with the neutrons coming from the right. The sample position is in the middle in the region of low field strength, the detector is at the left side.

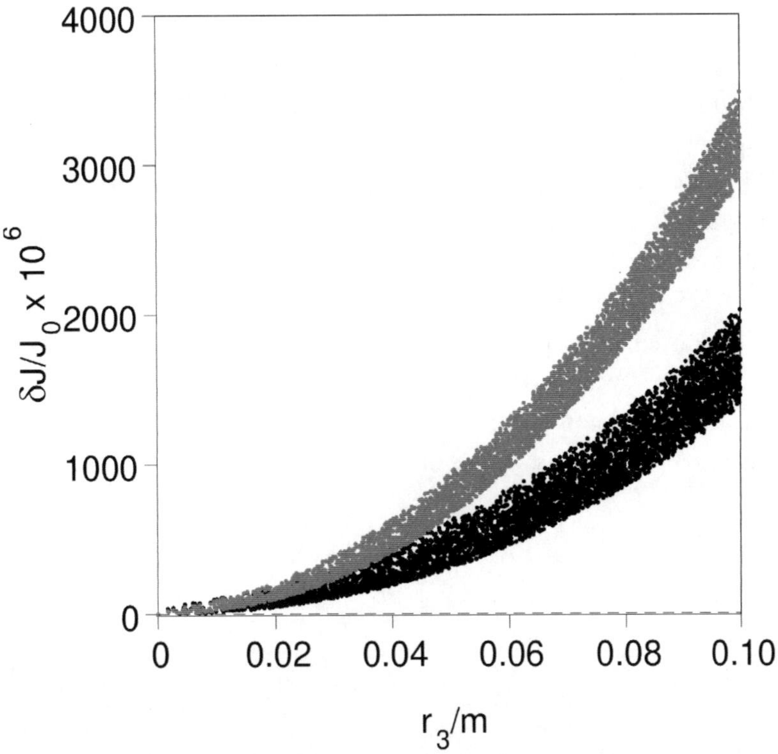

Fig. 4: Deviation of the field integral of a neutron path, $\delta J/J_0$ as a function of distance r_3 from the central axis at the flipper position, for the current configuration (red, top point cloud) and the configuration with optimized field shape (black, bottom point cloud). The deviation should be as small as possible also for off-axis flight paths ($r_3 \neq 0$).

5. Conclusion – NSE – Quo Vadis?

The spacial and temporal domain of nanometers and 1-1000 nanoseconds will be also the future domain of spin echo spectroscopy. The technique will be an indispensable tool with the capabilities provided at the modern facilities which cannot be replaced by other means. The tendency will be to observe smaller effects and changes in line shape accessible with the new capabilities of the upgraded NSE instruments. As an example, proteins in solution, i.e. in its natural environment show in addition to normal diffusion also the contribution of internal protein subdomain motions. Small effects have to be measured in this case up to several 100's nanoseconds, in order to properly determine the internal contribution to the dynamics (see e.g. [7] for the internal dynamics of Immunoglobulines). Dynamics at an interface, e.g. of polymers [5] or membranes [4] may also be a field where NSE can contribute a lot if high precision and sufficiently long Fourier times are provided. The domain of soft matter, with polymer melts, complex liquids and composite materials tends to go into the direction of more complexity of the systems (confinements, multi-component systems). The dynamics, e.g. the "nanorheology" in these samples, will remain an important issue. Also some structural changes in multicomponent soft matter systems may only be visible by its effects on the dynamics of the system. There NSE will provide unique information on these systems, which can be combined with other structural information, e.g. from small angle scattering, to provide a complete view of the sample. The currently upgraded J-NSE spectrometer with its new main coils and the increased dynamic range is therefore well prepared for the future scientific challenges in the field of slow dynamics.

IOP Conf. Series: Journal of Physics: Conf. Series **862** (2017) 012009 doi:10.1088/1742-6596/862/1/012009

6. References

[1] Mezei, F., Z. Phys. 255, 146-160 (1972)
[2] Mezei F. (ed.), Lecture Notes in Phyics 128, Springer Verlag Berlin, 1979
[3] Frielinghaus H. et al, Phys. Rev. E, **85**, 041408 (2012)
[4] Lipfert F. et al., Phys. Rev. E, **85**, 042303 (2014)
[5] Gawlitza K. et al, Macromolecules **48**, 5807 (2015)
[6] Krutyeva M. et al, Phys. Rev. Lett. **110**, 108303 (2013)
[7] Stingaciu L. R. et al, Sci. Rep. **6**, 22148 (2016)
[8] Monkenbusch M. et al, Nucl. Instr. Methods Phys. Res. A **399**, 301 (1997)
[9] Holderer O. et al, Meas. Sci Technol. **19**, 034022 (2008)
[10] Pasini S. et al, Meas. Sci Technol. **26**, 035501 (2015)
[11] Zeyen C.M.E., and Rem, P.C., Meas. Sci Technol. **7,** 782–791 (1996)

Spin-echo small-angle neutron scattering study of the structure organization of the chromatin in biological cell

E G Iashina[1,2],W G BOUWMAN[3], C P DUIF[3], M V FILATOV[2] and S V Grigoriev[1,2]

[1] Petersburg Nuclear Physics Institute NRC KI, Gatchina, 188300 St-Petersburg, Russia
[2] Saint-Petersburg State University, 198504 Saint Petersburg, Russia
[2] Delft University of Technology, Mekelweg 15, 2629 JB Delft, The Netherlands

E-mail: yashina@lns.pnpi.spb.ru

Abstract. Spin-echo small-angle scattering (SESANS) technique is a method to measure the structure of materials from nano- to micrmeter length scales. This method could be important for studying the packaging of DNA in the eukaryotic cell. We measured the SESANS function from chicken erythrocyte nuclei which is well fitted by the exponential function $G(z) = \exp(-z/\xi)$, where ξ is the correlation length of a nucleus (in experimental data $\xi = 3,3$ μm). The exponential decay of $G(z)$ corresponds to the logarithmic pair correlation function $\gamma(r) = \ln(\xi/r)$. As the sensitivity of the SESANS signal depends on the neutron wavelength, we propose the SESANS setup with the changeable wavelength in the range from 2 to 12 Å. Such option allows one to study in great detail the internal structure of the biological cell in the length scale from $10^{-2}\mu$m to 10 μm.

1. Introduction

It is well known that the cell is the basic structural and functional unit of all known living organisms. The cell nucleus acts like the control center of the cell. In this nucleus, a complex of macromolecules can be found which consists of DNA and proteins is called the chromatin. All the information about the biological organism is written in the double-helix DNA sequence.

The large-scale chromatin organization in interphase nuclei is fundamentally different from the small-scale chromatin structure [1-6]. There are several levels of chromatin organization. At the simplest level, chromatin is a double-helical structure of DNA. After that DNA is complexed with histones to form nucleosomes. The nucleosomes fold up to produce a 30-nm fiber. One of the most exciting questions that scientists are trying to answer is how DNA is packed inside the nucleus of a cell. Although DNA macromolecules extend up to a few meters in unraveled condition, they are packed really tight in the nucleus. The most mysterious feature of DNA in the nuclei is the way it is packed so thatany part can be accessed, read, transcripted and translated at any time in the process of replication.

Nowadays the multiscale fractal model of DNA arrangement inside the nuclei [3, 4] is the most promising to explain this mystery. Experimental proof of this theory is complicated firstly

[2] Present address: Department of Physics, University of Bristol, Tyndalls Park Road, Bristol BS8 1TS, UK.

because of the necessity to investigate nuclei under conditions as close as possible to those inside the organism and secondly by the simultaneous coexistence of DNA together with proteins inside the nuclei. Neutron scattering seems to be the most suitable instrument for such investigation due to such advantages of neutrons as the high penetration depth, the non-damaging and the possibility to use the contrast variation method. The last is very important as it allows one to illuminate a separate structure of DNA and proteins in the nuclei.

Previously, the arrangement of chromatin in intact chicken erythrocyte nuclei was investigated by small (and ultra-small) angle neutron scattering [4, 3]. The experimental results demonstrated the power-law character of the neutron scattering intensity versus the scattering vector $I \sim Q^{-D}$ with parameter $D \approx 2.5$ in the Q-range from $1.5 \cdot 10^{-2}$ to $4 \cdot 10^{-1}$ nm^{-1} and $D \approx 3$ in the Q-range from $3 \cdot 10^{-3}$ to $1.5 \cdot 10^{-2}$ nm^{-1}. It means that at lenght scales between 15 nm and 1 μm the nucleus exhibites properties of bi-fractal structure. On the length scales between 15 and 400 nm chromatin exhibites properties of a mass fractal with a fractal dimension of approximately 2.5.

In recent years, the data obtained via novel experimental techniques to study the genome structure, in particular Hi-C methods [7, 8], seem to provide data supporting the topological fractal globule approach. The fractal globule is a compact polymer state that emerges during polymer condensation as a result of topological constraints which prevent one region of the chain from passing across another one [8, 9]. The majority of scientists suggest that the chromatin structure is a fractal globule.

To answer the question how DNA is packed inside the interphase nucleus at the micron scale, the new method spin-echo SANS (or SESANS) was used. The technique allows one to measure a full pair correlation function in samples over distances from 20 nm to 20 μm. The ability to measure the correlation at the micron scale is incredibly important in biological cell studies. In the present study we report the result of the spin-echo SANS measurements from chicken erythrocyte nuclei. Analyzing these results we propose to optimize the parameters of the SESANS setup in order to obtain the full information on the pair correlation function of the chromatin inside a biological cell.

2. Principles of spin-echo SANS

The method is based on the Larmor precession of polarized neutrons transmitted across two successive precession devices before and after the sample, which encodes the scattering angle into a precession angle [10].

Spin-echo SANS technique is a method to measure structures of materials in real space. These structures can be determined over three orders of magnitude in length scale from 20 nm to 20 μm. The measured SESANS signal G(z) depends on the spin-echo (SE) length z, which is equal along the beam axis coordinate in real space

$$z = C\lambda^2 BL \cot(\theta_0)/2\pi, \tag{1}$$

where C is a constant, λ is the neutron wavelength, θ_0 is the angle of inclination of the precession devices front and end faces towards beam axis, B is the magnetic field in the precession device, and L its length [10].

As shown by Kruglov [11] the SESANS function G(z) is the projection of the spatial pair correlation function $\gamma(r)$ along the beam axis

$$G(z) = \frac{2}{\xi}\int_z^\infty \frac{\gamma(r)r}{\sqrt{r^2 - z^2}}dr, \tag{2}$$

where ξ is the correlation length. The inverse transformation is found as

$$\gamma(r) = -\frac{\xi}{\pi}\int_r^\infty \frac{G'(z)}{\sqrt{z^2 - r^2}}dz. \tag{3}$$

IOP Conf. Series: Journal of Physics: Conf. Series **862** (2017) 012010 doi:10.1088/1742-6596/862/1/012010

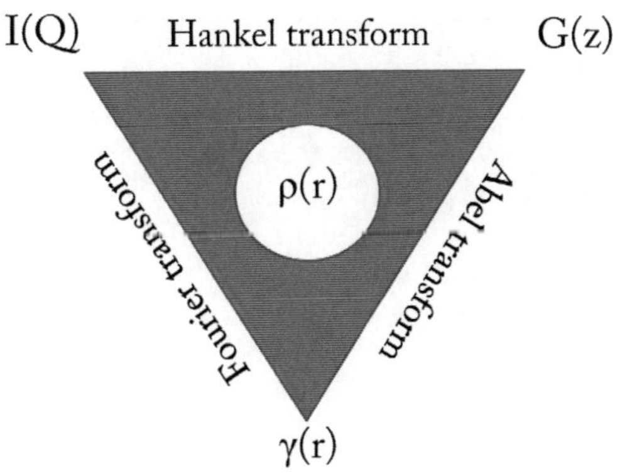

Figure 1. The relationship between the correlation function $\gamma(r)$, the scattering cross section I(Q) and the SESANS correlation function G(z) [11].

In addition, the SESANS correlation function $G(z)$ in terms of the differential scattering cross section $I(Q)$ of a sample is

$$G(z) = \frac{1}{k_0^2\sigma}\int_{R^2}\cos(zQ_z)I(Q)dQ_ydQ_z, \tag{4}$$

where k_0 is a neutron wave number, σ is the total scattering cross section and $Q = (0, Q_y, Q_z)$ is the scattering vector. The so-called Fourier – Abel – Hankel cycle relates the three functions $I(Q)$, $G(z)$ and $\gamma(r)$ to each other. The relationships are illustrated schematically in Fig.1 for isotropic distributions, where the figure shows the inverse related Abel – Hankel – Fourier transformation [12]. Nevertheless, in SANS measurements only the scattering intensity is measured (and not the phase of the scattered neutron wave), so critical information is lost which prevents the full retrieval of the original density (the 'phase problem'). The same situation occurs in the SESANS method. It should be noted that both methods give information about the spatial pair correlation function.

The polarization P/P_0 in a SESANS experiment is measured as a function of the total scattering cross section σ, the thickness of the sample l and the spin-echo length z

$$P(z)/P_0 = \exp(l\sigma[G(z) - 1]). \tag{5}$$

3. Study of the interphase chicken erythrocyte nuclei

The sample of isolated chicken erythrocyte nuclei was investigated at the SESANS-instrument at the HOR reactor of the Delft University of Technology, The Netherlands. To maximally increase the neutron scattering intensity we used the densest solution possible for these samples. To avoid the problem of agglomeration of nuclei, the detergent Triton X-100 was added to the solution, then clusters of order of a micron and more were mechanically destroyed by passing the solution through a syringe.

The SESANS instrument uses a constant wavelength $\lambda = 2$Å. The measurements were carried out in solution of a $D_2O > 95$ % to reach the maximum contrast for the scattering. Three different thicknesses of the sample (2, 4, 10 mm) were used to obtain the SESANS signal from the chromatin in the nucleus. The noticeable function had been measured for the sample of 10

IOP Conf. Series: Journal of Physics: Conf. Series **862** (2017) 012010 doi:10.1088/1742-6596/862/1/012010

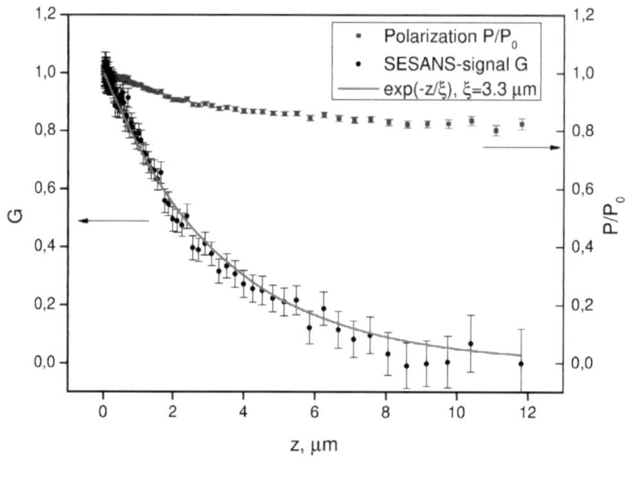

Figure 2. The polarization P/P_0 and the SESANS function G in dependence on the SE length z from the isolated chicken erythrocyte nuclei. The line is an exponential decay with the best fitted parameter $\xi = 3.3 \pm 0.1$ μm.

mm only. Fig.2 presents the polarization P/P_0 as a function of the SE length z. As one can see the polarisation changes from $P/P_0 = 1.00 \pm 0.01$ at small z to $P/P_0 = 0.85 \pm 0.01$ at $z = 8$ μm. The constant level of polarisation P/P_0 at large z determines the value of the product σl in Eq. (5). Thus the SESANS function of $G(z)$ can be definitely determined from the measured function $P/P_0(z)$ using Eq.(5). The SESANS function $G(z)$ is also shown in Fig.2. We were able to fit this function using the simple exponential decay $G(z) = \exp(-z/\xi)$, where $\xi = 3.3 \pm 0.07$ μm is the correlation length of a nucleus.

The pair correlation function can be obtained using the Abel transform Eq.(3)

$$\gamma(r) = \frac{1}{\pi}\int_r^\infty \frac{\exp(-z/\xi)}{\sqrt{z^2 - r^2}}dz = \frac{1}{\pi}K_0(r/\xi), \qquad (6)$$

where $K_0(x)$ is the Macdonald function of the 0th order. In its turn, the Hankel transform of exponential decay of $G(z)$ gives for the neutron cross section $I(Q) = A \cdot (1 + (Q\xi)^2)^{-3/2}$, which was observed in [3] for the large scale chromatin organization. However the upper border with fractal structure had not been found with SANS experiment because of an instrument resolution. Thus, the complementarity of SANS and SESANS can be realized via the suitable spacial ranges for SANS from 1 to $5 \cdot 10^2$ nm and that for SESANS from $5 \cdot 10^1$ to $2 \cdot 10^4$ nm. As we are interested in the correlation inside rather than outside the sample, we can restrict ourself by a small value of r/ξ. With this focus to the internal structure of the nucleus, we can simplify Eq.(6) when $r/\xi << 1$. We obtained the simplified expression of the pair correlation function using Eq.(6) and it's asymptotic form gives

$$\gamma(r) = \ln(\xi/r). \qquad (7)$$

Thus the exponential law of the SESANS function corresponds to the logarithmic pair correlation function. The correlation function increases with an additive rather than multiplicative constant, upon reducing the ruler length by a fixed rescaling factor. This leads to a logarithmic law instead

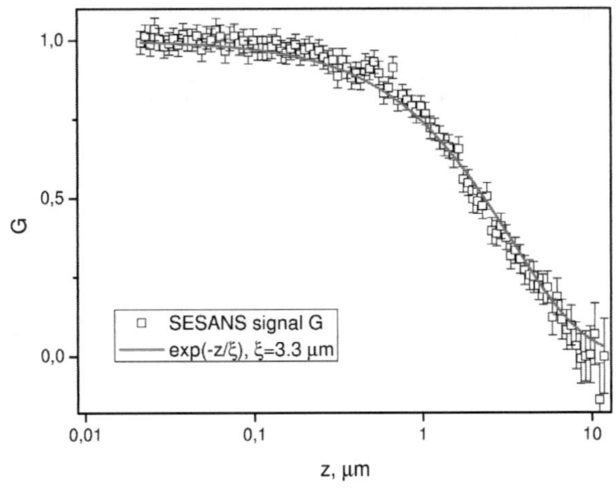

Figure 3. The lin-log plot of the SESANS function G as a function of the SE length from isolated chicken erythrocyte nuclei.

of the usual power law for the fractals. The logarithmic correlation function correspond to a hierarchically branched structure. A good sample of such structures are trees. Such a structure provides high accessibility of enzymes to specific gene site.

4. Criteria for SESANS measurements of a biological cell

The possibilities to extract the SESANS function $G(z)$ appear to be limited for the biological cell due to weak neutron scattering attributed to the nuclei of the cell. Fig.2 demonstrates the maximum of the possible signal determined by the maximum value of σl with a thickness of $l = 10$ mm and maximal σ provided by the maximum contrasting of the D_2O. However, as it is shown in Fig.3, the obtained meaningful data set for the SESANS function $G(z)$ is limited by the range of z from 1 to 12 μm, in spite of the fact that polarization data P/P_0 were measured within 3 decades of the SE length z from 0.02 to 12 μm. This limitation is caused by the experimental errors of the measurements of the polarisation P/P_0, which can be at the best measured with a accuracy of 1 %. This value is related not with the statistical errors of the neutron measurements but with the stability of many devices used to manipulate the orientation of the neutron spin in the SESANS setup. This limitation can be overcome if one is able to vary not only the thickness of the sample l but also the total neutron cross section via the neutron wavelength. The cross section for the small-angle neutron scattering depends of the wavelength as λ^2. Therefore the change of the wavelength from $\lambda = 2$ Å to $\lambda = 12$ Å will change the sensitivity of the SESANS method by 36 times. In the other words, as it is impossible to increase the thickness of the specimen to infinity, one should change the other parameter, σ, which can be easily done via variation of the neutron wavelength λ.

Often the SESANS method is applied to measure the size of the particle (10^{-1}-10 μm). However, the key information about the fractal properties of the particle is hidden at the small values of z. As was argued above for the biological cell the small values of z are inaccessible. Therefore we formulated the criteria necessary to perform a complete study of such biological objects using the SESANS method:

- one must have the possibility to change the neutron wavelength;

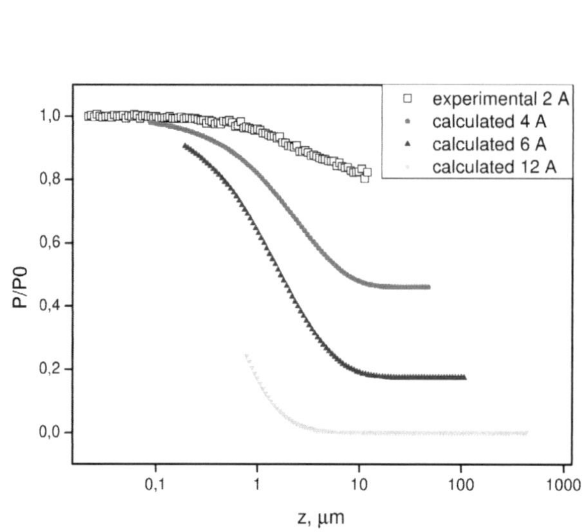

Figure 4. The lin-log plot of the SESANS correlation function from chicken erythrocyte nuclei at the different wavelengths: $\lambda = 2$ Å (experiment), $\lambda = 4, 6, 12$ Å (calculations) taken the same parameters of the setup such of B, inclination angle θ_0 and the lenth of the SE arm L in Eq.1.

- one must use different thickness;
- one must use the contrast-variation technique.

Here we introduce the sensitivity of SESANS is an important parameter which is independent on the polarization efficiency or statistics. The sensitivity is the difference between P_0 and P_{inf} normalized to P_0. The bigger is this number (which runs from 0 ($P_0 = P_{inf}$)) to 1 ($P_{inf} = 0$), the higher is sensitivity of the measurements of G(z) for those z at which $P/P_0 > 0$. Of course, we do not want to set $P_{inf} = 0$, as we must know its true level, but not the one which is restricted by the measurement conditions. To demonstrate the sensitivity of the SESANS signal on the neutron wavelength, the polarisation P/P_0 was calculated for the three different wavelengths $\lambda = 4, 6, 12$ Å, using Eq.(5) and the SESANS function from chicken erythrocyte nuclei ($G(z) = \exp(z/\xi)$).

Fig.4 presents the calculated values of the polarization $P/P_0(z)$ ($\lambda = 4, 6, 12$) Å and the experimental polarization P/P_0 for $\lambda = 2$ Å. The experimental polarization P/P_0 ($\lambda = 2$ Å) approaches a constant ($P/P_0 \cong 1$) for $z < 1$ μm. The sensitivity of the calculated polarization increases with increasing wavelength λ in the same z-range. However, the polarization P/P_0 with $\lambda = 12$ Å can be well measured for relatively small z only, since it tends to zero at z larger than 1 μm. In this case the value of σl can not be determined. It is clear from this consideration that the optimum measurements can be obtained with $\lambda = 6$Å. In general, in order to really cover the 3 decades (from $10^{-2}\mu$m to 10 μm) in measurements of the SESANS function $G(z)$, one must perform such measurements with at least 3 different values of the neutron wavelength. Such protocol of SESANS measurements is analogous to the protocol of the SANS measurements at 3 different sample-detector positions (2, 10, 18 m) aiming to provide a wide Q-range from 10^{-3} to 1 nm^{-1}.

5. Conclusions

In summary: the use of the beam with $\lambda = 6$ Å leads to an increase of the sensitivity of the SESANS method and opens the range of z [10^{-2} - 10] μm for the study of biological cells.

A SESANS setup with a changeable wavelength in the range from 2 to 12 Å will be highly appreciated by the community of life sciences.

6. References

[1] D. Gerlich and J. Ellenberg, Curr. Opin. Cell Biol. 15, 664 (2003).
[2] A. Belmont, Curr. Opin. Cell Biol. 15, 304 (2003).
[3] D.V. Lebedev, M.V. Filatov, A.I. Kuklin, A.Kh. Islamov, E. Kentzinger, R. Pantina, B.P. Toperverg, V.V. Isaev-Ivanov, FEBS Letters, 579, 14651468, (2005)
[4] A. V. Ilatovskiy, D. V. Lebedev, M. V. Filatov, M. G. Petukhov and V. V. Isaev-Ivanov, Journal of Physics: Conference Series 351, 012007 (2012)
[5] K. Metze, Expert Rev. Mol. Diagn. 13(7), 719735 (2013)
[6] T. Misteli, Science 291, 843 (2001).
[7] E. Lieberman-Aiden et al., Science 326, 289 (2009).
[8] L.A. Mirny, Cromosome Res. 19, 37 (2011).
[9] M. V. Tamm, L. I. Nazarov, A. A. Gavrilov, and A. V. Chertovich, PRL, 114, 178102 (2015)
[10] M.Th. Rekveldt, Nucl.Instr.Meth. B 114, 366, (1996)
[11] Krouglov T., Bouwman W.G., Plomp J., Rekveldt M.Th., Vroege G.J., Petukhov A.V., Thies-Weesie D.M.E. J.Appl.Cryst., 36 1417, (2003)
[12] R. Andersson, L.F. van Heijkamp, I.M. de Schepper, W.G. Bouwman, J. Appl. Cryst. 41 (2008), 868-885.

Development of a polarized ^{3}He neutron spin filter for POLANO at J-PARC

T Ino[1,2,3], M Ohkawara[4], K Ohoyama[5], T Yokoo[1,2,3], S Itoh[1,2,3], Y Nambu[4], M Fujita[4], H Kira[6], H Hayashida[6], K Hiroi[7,2], K Sakai[7,2], T Oku[7,2], and K Kakurai[6]

[1] Institute of Materials Structure Science, KEK, Tsukuba, Ibaraki 305-0801, Japan
[2] J-PARC Center, Tokai, Ibaraki 319-1195, Japan
[3] SOKENDAI - The Graduate University for Advanced Studies, Tsukuba, Ibaraki 305-0801, Japan
[4] Institute for Materials Research, Tohoku University, Sendai 980-8577, Japan
[5] Graduate School of Science and Engineering, Ibaraki University, Hitachi, Ibaraki 316-8511, Japan
[6] Comprehensive Research Organization for Science and Society, Tokai, Ibaraki 319-1106, Japan
[7] Japan Atomic Energy Agency, Tokai, Ibaraki 319-1184, Japan

takashi.ino@kek.jp

Abstract. We have developed a polarized ^{3}He neutron spin filter (NSF) for a new polarized neutron spectrometer, POLANO, at the Japan Proton Accelerator Research Complex (J-PARC). POLANO aims to utilize high energy neutrons polarized by a ^{3}He NSF and spin analyzed by an array of magnetic supermirrors for inelastic neutron scattering. The ^{3}He gas is continuously polarized *in-situ* by spin-exchange optical pumping to provide a highly and stably polarized neutron beam. The POLANO ^{3}He NSF is designed to polarize neutrons with energies as high as 200 meV and fit in a restricted space. It is equipped with adiabatic fast passage NMR that enables one to flip the ^{3}He spins, and consequently, the neutron spins.

1. Introduction

POLANO is a polarization analysis inelastic neutron spectrometer that has been newly constructed at the Materials and Life Science Experimental Facility (MLF) of the Japan Proton Accelerator Research Complex (J-PARC) for the investigation of the spin dynamics of materials [1]. POLANO is the first spectrometer dedicated to polarized neutron experiments at MLF with wide momentum-energy coverage for the investigation of dynamical properties of materials. It employs a ^{3}He neutron spin filter (NSF) for the initial beam polarization with incident neutron energies as high as 200 meV and an array of magnetic supermirrors for the spin analysis of the scattered neutrons. A converging neutron beam is focused at the sample position with a beam size of 20 mm by 20 mm, which is 50 mm by 50 mm at the ^{3}He NSF position. The basic instrumental parameters are summarized in Table 1.

The ^{3}He NSF of POLANO has to satisfy certain criteria that include a specific range of neutron energies, continuous operation for weeks, ability to fit in a restricted space, spin flipping, and the J-PARC safety standards. Such a ^{3}He NSF had been designed as described in a previous paper [2], and we have realized one that will be installed in the spectrometer. The ^{3}He NSF can fit in the restricted space available, i.e. 60 cm along the beamline and 60 cm wide, between the Fermi chopper and sample vacuum chamber as illustrated in figure 1. The ^{3}He gas is polarized *in-situ* by spin-exchange optical pumping (SEOP) to maintain a stable ^{3}He polarization for weeks. It is equipped with an adiabatic fast passage (AFP) NMR that is capable of flipping the ^{3}He spins instantaneously [3]. Each component of the POLANO ^{3}He NSF is described in the next section.

Table 1 Basic instrumental parameters of POLANO.

L_1 (moderator-sample distance)	17.5 m
L_2 (sample-detector distance)	2.5 m
L_3 (Fermi chopper-sample distance)	1.85 m
Energy resolution	$\Delta E/E_i$ 3 ~ 5 % at elastic position
Momentum resolution	$\Delta Q/k_i$ 1 ~ 2 %
Sample size	20 mm × 20 mm
Detector coverage	- 30 ° to 130 ° (horizontal) - 10 ° to 10 ° (vertical)

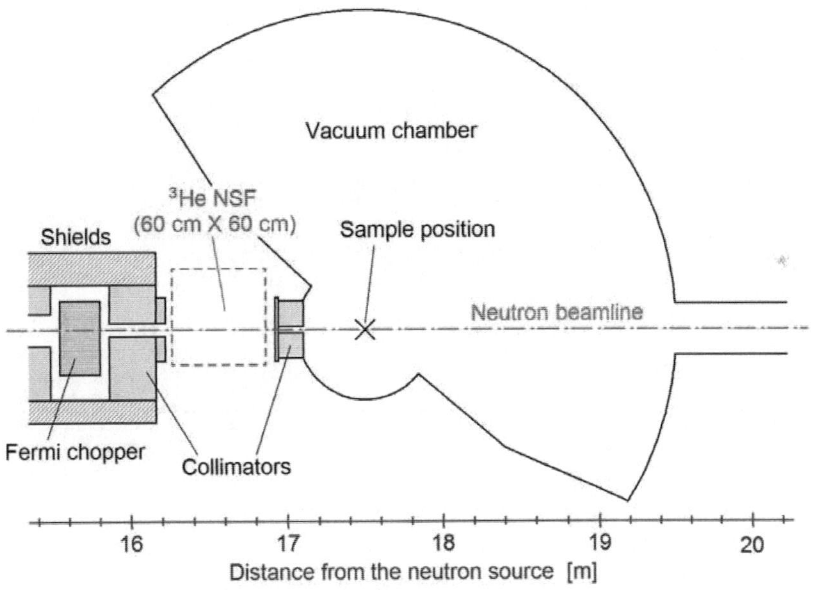

Figure 1 Schematic view of the POLANO spectrometer around the sample position (neutron beam propagation from left to right). The ^{3}He NSF is situated in the space shown by a dashed square between the Fermi chopper and vacuum chamber.

2. ^{3}He NSF of POLANO

In POLANO, the neutron beam profile at the position of the ^{3}He NSF is a square with a side length of ~5 cm, and neutrons with energies up to 200 meV need to be polarized. However, at the first stage of the POLANO project, the aim was to concentrate primarily on a safe and stable operation of the ^{3}He NSF. Hence, we restricted the coverages of the beam profile and neutron energies to a diameter of ~5 cm, collimated by sintered B_4C at the entrance of the ^{3}He NSF, and 100 meV, respectively.

Figure 2 illustrates a schematic view of the ^{3}He NSF of POLANO. A ^{3}He cell is situated inside an oven at the center of a magnetically shielded solenoid. A laser with a wavelength of the Rb D_1 resonance is guided to the ^{3}He cell using mirrors for the optical pumping of Rb. These components are described in the following subsections.

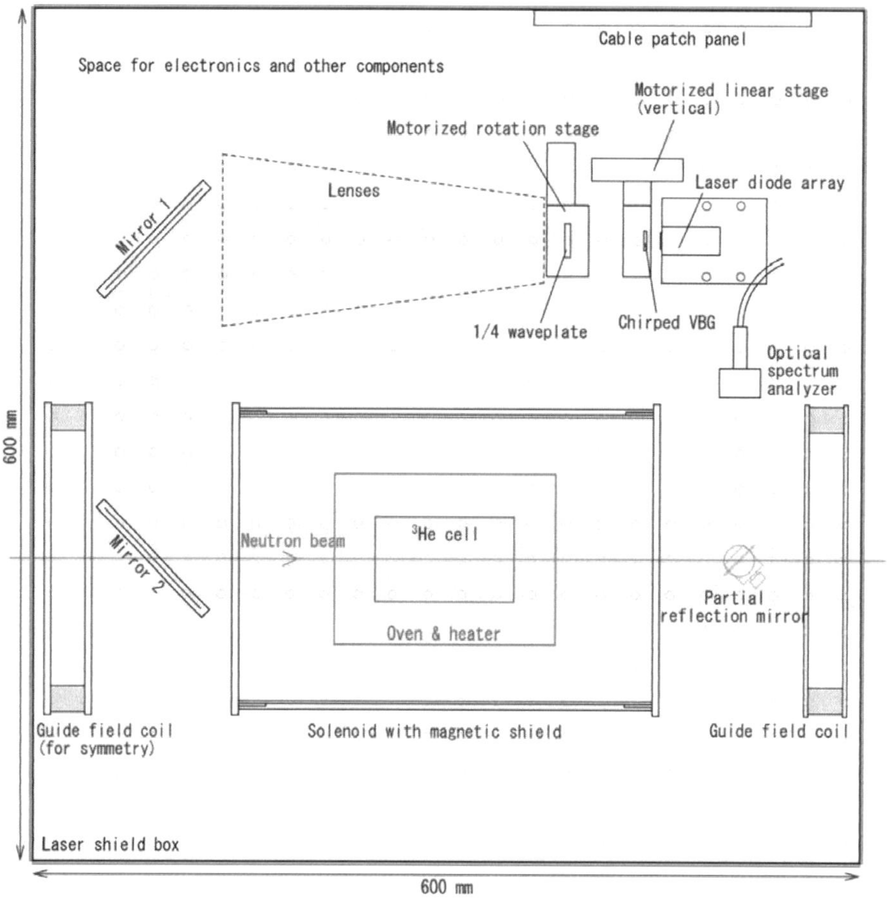

Figure 2 Schematic view of the ^{3}He NSF of POLANO (for details, see text).

2.1. ^{3}He cells

The neutron polarization P_n and transmission T_n for an unpolarized neutron beam passing through polarized ^{3}He gas are given by

$$P_n = \tanh(P_{He}\sigma\rho d), \quad (1)$$

$$T_n = e^{-\sigma\rho d}\cosh(P_{He}\sigma\rho d), \quad (2)$$

where P_{He} is the ^{3}He polarization, σ is the spin-averaged neutron absorption cross section that depends on the neutron energy, ρ is the number density of ^{3}He atoms, and d is the thickness of the ^{3}He gas. In polarized inelastic neutron scattering measurements, a high polarization such as $P_n \geq 0.90$ or 0.95 is often desired. Figure 3 shows the variation of the effective ^{3}He gas thickness ρd as a function of the neutron energy necessary to achieve a high polarization of between $P_n = 0.90$ and 0.95 for a

IOP Conf. Series: Journal of Physics: Conf. Series **862** (2017) 012011 doi:10.1088/1742-6596/862/1/012011

given value of $P_{He} = 0.75$. For the time being, POLANO utilizes polarized neutrons with energies ranging from 10 to 100 meV that can be realized using polarized ^{3}He gas with effective ^{3}He thicknesses ranging from 10 to 35 atm·cm as seen in figure 3. We are preparing suitable ^{3}He cells for these neutron energies.

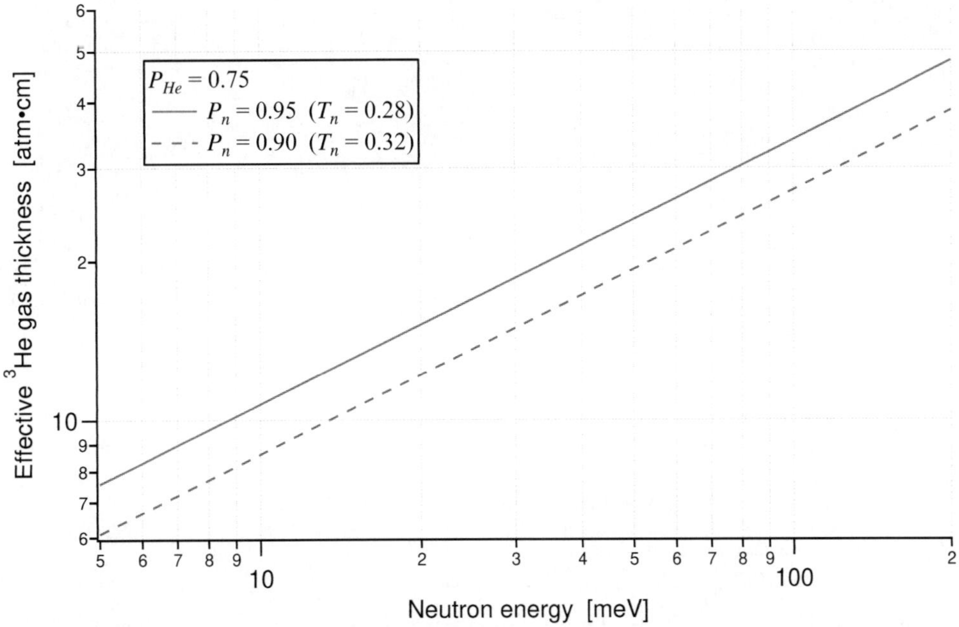

Figure 3 Variation of the effective ^{3}He gas thickness ρd as a function of the neutron energy in order to achieve a high polarization of $P_n = 0.90$ or 0.95.

Alkali metal hybrid SEOP [4] that uses a cell containing pressurized ^{3}He gas, ~0.1 atm of N_2 gas, and small quantities of Rb and K, is widely used to enhance the optical pumping efficiency. During the optical pumping of Rb in hybrid SEOP, a ^{3}He cell is heated to 200–250 °C, thereby increasing the gas pressure inside the cell from 3 atm at ambient temperature to approximately 5 atm at the elevated temperature. Finite element method calculations and pressure tests of glass cells are being performed to develop pressure safe ^{3}He cells. Two of our ^{3}He cells are presented in Fig. 4.

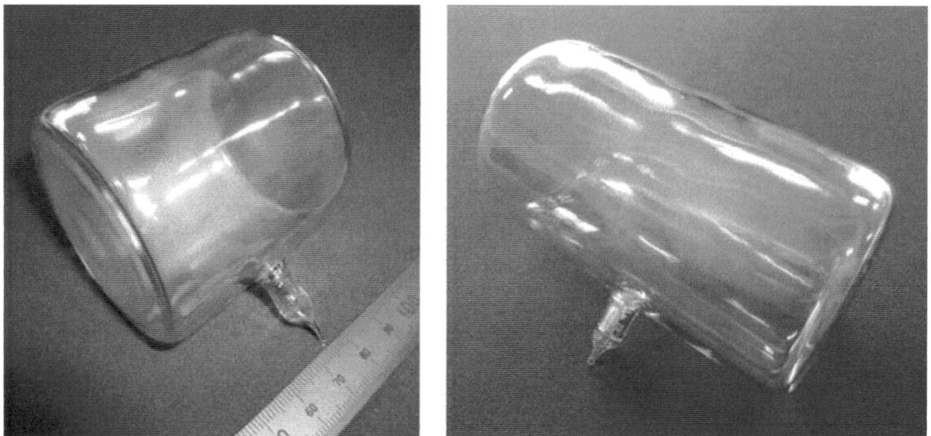

Figure 4 Typical ^{3}He cells. Outer diameters of both cells are 60 mm. Outer lengths are 60 mm (left) and 100 mm (right). Wall thicknesses are approximately 2 mm and ^{3}He gas pressures are ~2.5 atm at ambient temperature for both cells.

2.2. Static magnetic field

Magnetic field homogeneity is one of the important requirements for ^{3}He NSF. ^{3}He spin relaxation caused by a magnetic field gradient follows the equation

$$\Gamma_{mag} = D\frac{\left|\vec{\nabla}B_x\right|^2 + \left|\vec{\nabla}B_y\right|^2}{B_0^2}$$

where D is the diffusion constant of ^{3}He atoms, and B_0 is the strength of the primary magnetic field. B_x and B_y denote transverse magnetic fields [5]. A magnetically shielded solenoid with compensation coils was designed including neutron guide coils that fit in the available space. There are two requirements that the magnetic field needs to satisfy: (1) the field gradient $\sqrt{\left|\vec{\nabla}B_x\right|^2 + \left|\vec{\nabla}B_y\right|^2} / B_0 << 1\times10^{-3}$ cm^{-1} at the ^{3}He cell position so as to reduce the ^{3}He spin relaxation to an acceptable level, and (2) the neutron guide field strength > ~2 mT for $|Z| < 350$ mm from the solenoid center in order to suppress the depolarization of the neutron spins where the magnetic field direction changes by 90 degrees along a neutron path of 20 cm after the guide coil [6]. The magnetic coils are schematically shown in figure 5. Calculation results of the field gradient around the ^{3}He cell position and the guide field strength are shown in figure 6.

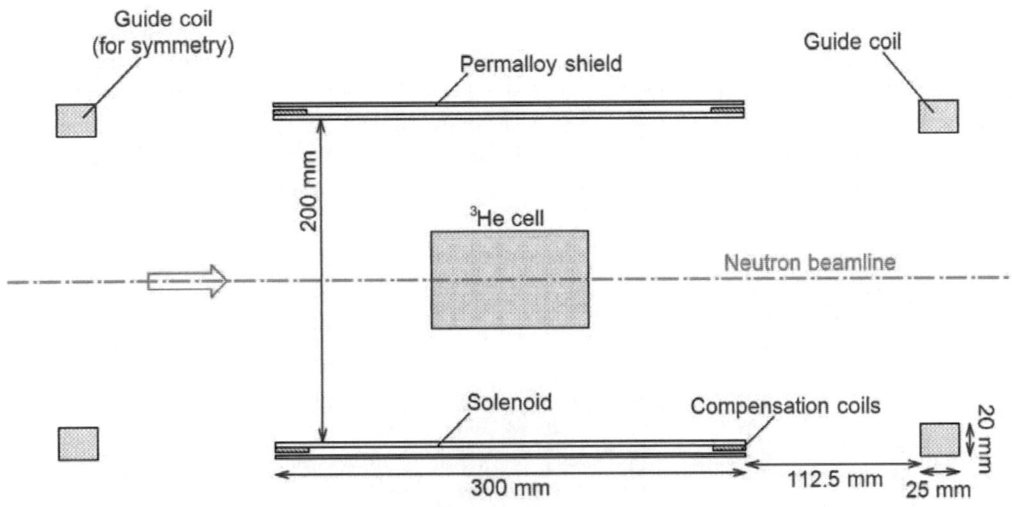

Figure 5 Cross-sectional view of the magnetically shielded solenoid with compensation coils and neutron guide coils. The ^{3}He cell shown is cylindrical with a diameter and length of 60 mm and 100 mm, respectively. The coil currents are 559 A-turn, 41 A-turn, and 462 A-turn for the solenoid, each compensation coil, and each guide coil, respectively. The permalloy shield is 0.6 mm thick. The left guide coil ensures the symmetry of the magnetic system.

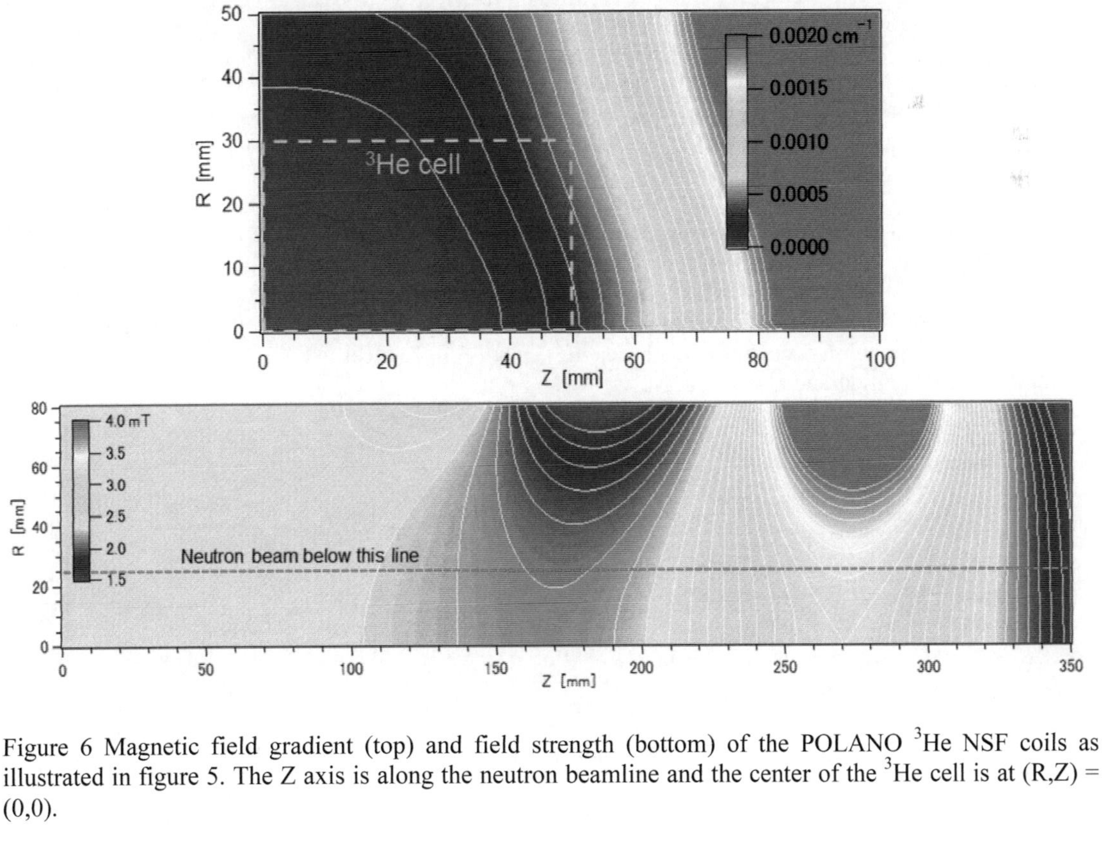

Figure 6 Magnetic field gradient (top) and field strength (bottom) of the POLANO ^{3}He NSF coils as illustrated in figure 5. The Z axis is along the neutron beamline and the center of the ^{3}He cell is at (R,Z) = (0,0).

2.3. Oven, heater, and AFP

In the POLANO ^{3}He NSF, a ^{3}He cell is heated to 200–250 °C in an oven made of aluminum that is wrapped with a flexible polyimide heater. The polyimide heater is specially designed to minimize magnetic field production due to the heater current. It consists of two layers of nickel-chromium foil that have identical strip patterns, separated by a polyimide film. The heater currents passing through the strips of the two layers are opposite in direction to each other so that the magnetic fields produced by the heater currents are canceled. Additional polyimide films are attached to both surfaces of the heating element for electrical insulation [7]. Figure 7 shows a photograph of a cylindrical aluminum oven wrapped with polyimide heater.

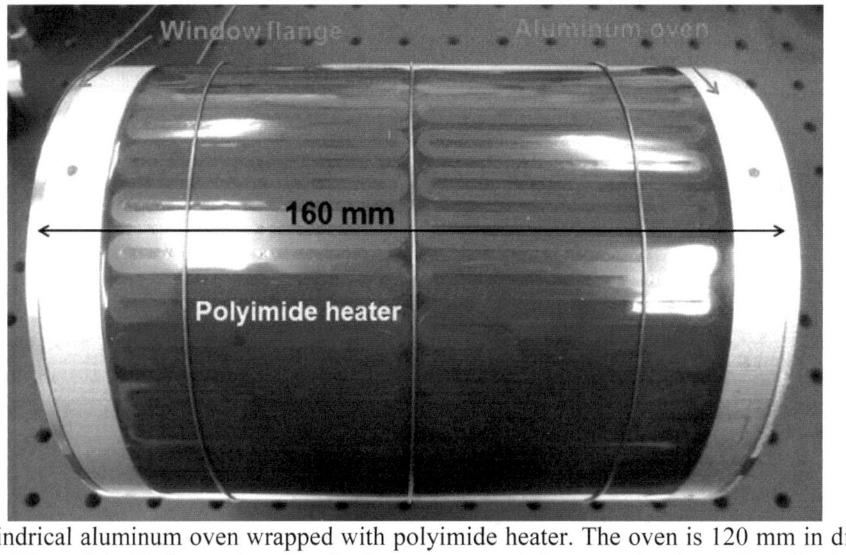

Figure 7 Cylindrical aluminum oven wrapped with polyimide heater. The oven is 120 mm in diameter and 160 mm long. The polyimide heater is secured using nonmagnetic brass or titanium wires. During operation, polyimide foam is attached to the polyimide heater for thermal insulation (the foam has been removed before taking the photograph).

Inside the oven, a cosine-theta coil is arranged to produce a uniform alternating magnetic field for AFP NMR [8]. The frequency of the coil current is swept and the ^{3}He spins are adiabatically flipped in a static magnetic field when the frequency crosses the Larmor resonance value. The sweeping speed used in the POLANO ^{3}He NSF is 100 kHz/s. The depolarization of the ^{3}He spin by one AFP spin flip was measured to be 0.0005, which, though not negligible, is acceptably small for practical use. Figure 8 shows the cosine-theta coil inside the cylindrical aluminum oven.

IOP Conf. Series: Journal of Physics: Conf. Series **862** (2017) 012011 doi:10.1088/1742-6596/862/1/012011

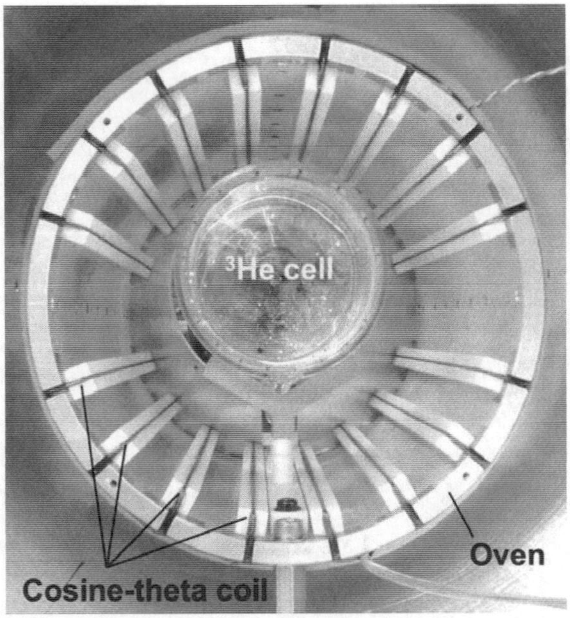

Figure 8 Photograph of the inside of a cylindrical aluminum oven. A cosine-theta coil is arranged to produce a uniform alternating magnetic field for AFP NMR. An oven window flange has been removed before taking the photograph.

2.4. *Laser and optical system*

A high power laser diode array (LDA) with a chirped volume Bragg grating (VBG) in the POLANO ^{3}He NSF emits a laser light with a power of ~70 W and a wavelength width < 0.2 nm (full width at half maximum) at the Rb D_1 frequency [9]. The center wavelength can be tuned remotely by moving the translation position of the chirped VBG. A quarter waveplate that converts linear polarization into circular polarization is placed after the chirped VBG. Its rotational position is also remotely controlled in order to reverse the polarity of the laser (right or left-handed) according to the spin state of ^{3}He. Next, the laser profile is reshaped using lenses to match that of the ^{3}He cell. The laser is reflected using two mirrors to illuminate the ^{3}He cell parallel to the neutron beam. The transmission spectrum of the ^{3}He cell is measured by an optical spectrum analyzer (OSA) using a partial reflection mirror and an optical fiber. A schematic of the laser optical system is shown in figure 2. Si wafers are used as substrates in the full and partial reflection mirrors in the neutron beamline in order to minimize unwanted neutron scattering.

The POLANO ^{3}He NSF including the optical and magnetic systems is shown in figure 9. All the optical components as well as the coils, oven, and ^{3}He cell are enclosed in a laser shield box for safety.

IOP Conf. Series: Journal of Physics: Conf. Series **862** (2017) 012011 doi:10.1088/1742-6596/862/1/012011

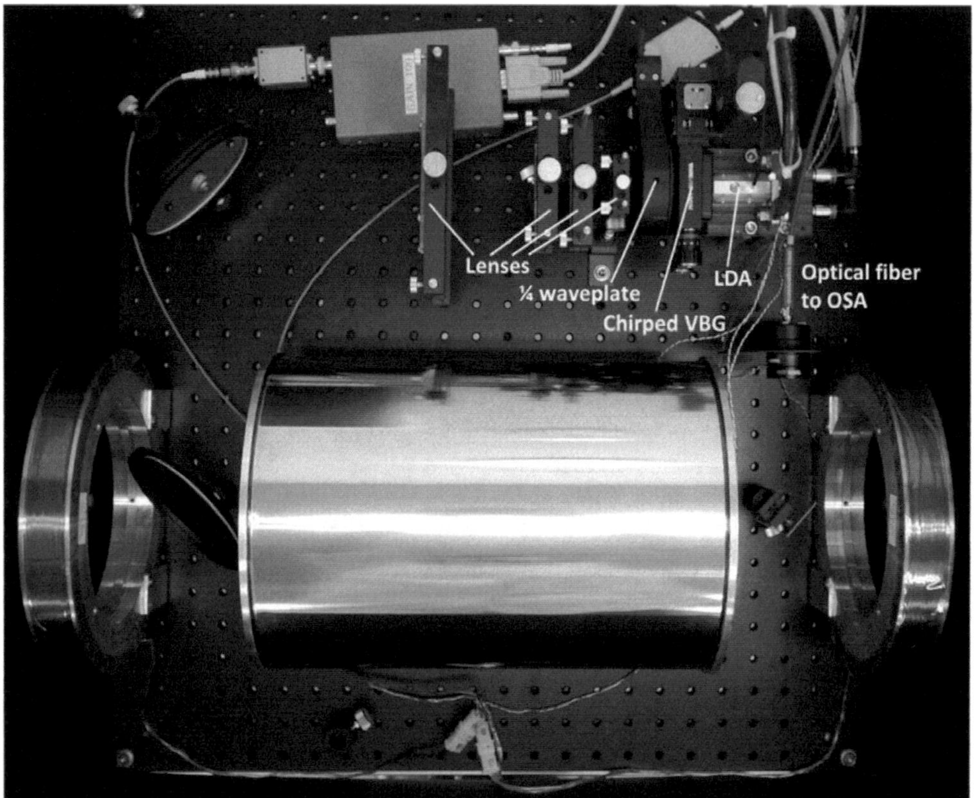

Figure 9 POLANO ^{3}He NSF viewed from the top. All components are fixed on a 60 cm × 60 cm optical table and enclosed in a laser shield box for safety.

2.5. Safety measures

The POLANO ^{3}He NSF system is equipped with interlocks for safety. The interlock system monitors the temperatures of the LDA, ^{3}He cell, polyimide heater, and solenoid as well as the flow rate and hydraulic pressure of the coolant water for the LDA. The heater voltage and current are continuously monitored to detect heater burnout. The interlock system automatically shuts down the laser and heater current when the parameters exceed acceptable levels. Physical guards and interlock switches are also included in the system to prevent injuries due to the laser and high operating temperatures.

3. Optical pumping test

An optical pumping test was performed for the POLANO ^{3}He NSF in an off-site laboratory. A ^{3}He cell inside the oven was kept at 240 °C for continuous optical pumping. The polarity of the laser was reversed every 15 minutes along with the ^{3}He spins using AFP NMR. Figure 10 shows the variation of the peak height of the measured AFP NMR signals that are proportional to the ^{3}He polarization over four days. Positive and negative signals were alternately observed corresponding to the change of the spin states from down to up and vice versa. The ^{3}He polarization increased for approximately 30 hours from the start of the optical pumping, after which it reached a maximum. The time constant of the polarization build-up was 5.5 hours. The optical pumping was continued for an additional 60 hours, during which the ^{3}He polarization stayed at a constant value until the end of the experiment.

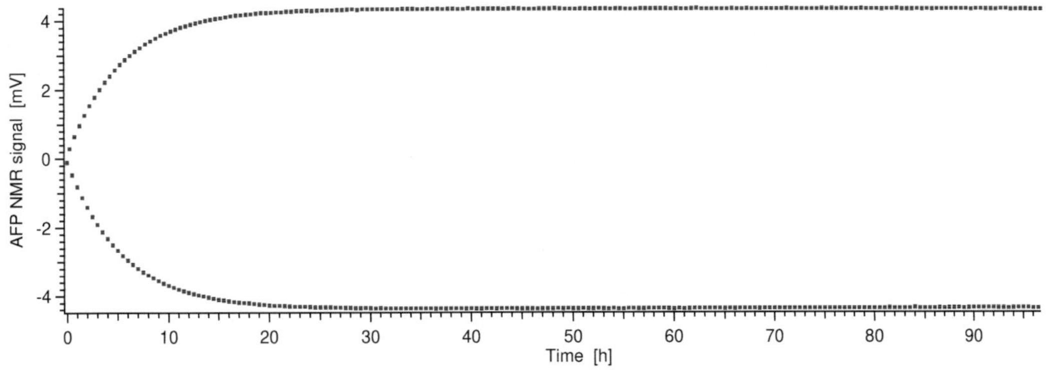

Figure 10 AFP NMR signals measured every 15 minutes during an optical pumping test of the POLANO ^{3}He NSF. Positive and negative signals correspond to the changes of the ^{3}He spin states from down to up and vice versa. A ^{3}He cell with a diameter of 60 mm and a length of 60 mm filled with ^{3}He at 2.3 atm was used.

4. Summary and plans for future upgrade

We have developed an *in-situ* polarized ^{3}He NSF that will be installed in POLANO for incident neutron beam polarization. Polarized neutron beams with diameters of 50 mm and energies up to 100 meV will be provided by this ^{3}He NSF at the first stage of the POLANO operation. A continuous optical pumping test performed off-site demonstrated a successful operation of the POLANO ^{3}He NSF. We have performed similar tests with different ^{3}He cells and have not encountered any problems so far. The ^{3}He polarizations measured using Rb electron paramagnetic resonance shifts for several ^{3}He cells were found to be 70–75 % [10]. We will perform further optical pumping tests for longer periods to investigate the durability of the ^{3}He NSF before final installation in POLANO.

The usable neutron beam energy and diameter will be increased with larger ^{3}He cells in the near future. According to figure 3, an effective ^{3}He thickness of approximately 50 atm·cm is required to polarize neutrons with an energy of 200 meV. Taking into consideration the safety issues pertaining to the use of the ^{3}He cells at high pressure and large cell volumes, the ^{3}He gas pressure at room temperature and thicknesses should lie between 2.5–5 atm and 10–20 cm, respectively. The cell diameter will also have to be increased to 80 mm in order to accommodate the complete neutron beam profile. Further studies are necessary to fabricate such large cells that can withstand higher gas pressures. In the current design, only one end of the ^{3}He cell is illuminated by the laser. The ^{3}He polarization can be increased by illuminating both ends of the ^{3}He cell. Increasing the laser power will also be beneficial for ^{3}He cells with larger volumes. Designing of the double laser system is currently under progress and will constitute the next upgrade of the POLANO ^{3}He NSF.

References

[1] Y. Nambu *et al.*, "Magnetism of the triangular antiferromagnet $NiGa_2S_4$ and introduction of POLANO," in these proceedings, T. Yokoo *et al.*, EPJ Web of Conf. 83, 03018 (2015).
[2] T. Ino *et al.*, J. Phys.: Conf. Ser. 711, 012011 (2016).
[3] A. Abragam, The Principles of nuclear Magnetism, Oxford University Press (1961).
[4] E. Babcock *et al.*, Phys. Rev. Lett. 91, 123003 (2003).
[5] G. D. Cates, S. R. Schaefer, and W. Happer, Phys Rev. A 37, 2877 (1988); L. D. Schearer and G. K. Walters, Phys. Rev. 139, A1398 (1965).
[6] K. Ohoyama *et al.*, J. Phys.: Conf. Ser. 711, 012010 (2016).
[7] T. Ino *et al.*, submitted to Rev. Sci. Instrum.
[8] T. Ino *et al.*, J. Phys.: Conf. Ser. 340, 012006 (2012).
[9] W. C. Chen *et al.*, J. Appl. Phys. 116, 014903 (2014); B. Chann *et al.*, Opt. Lett. 31, 1253

IOP Conf. Series: Journal of Physics: Conf. Series **862** (2017) 012011 doi:10.1088/1742-6596/862/1/012011

(2006); C. Moser and G. Steckman, Photon. Spectra 36 (6), 82 (2005); B. L. Volodin *et al.*, Opt. Lett. 29, 1891 (2004).
[10] M. V. Romalis and G. D. Cates, Phys. Rev. A 58, 3004 (1998).

Thermal stabilization of neutron Larmor diffractometers

T Keller[1,2], **F Tralmer**[1,2]

[1]Max Planck Institute For Solid State Research, D 70569 Stuttgart
[2]Max Planck Society Outstation at the FRM II, D 85748 Garching

E-mail: t.keller@fkf.mpg.de

Abstract. We report on the design of a support unit for the radio frequency (RF) coils of a Larmor diffractometer (LD) eliminating fluctuations of the Larmor phase resulting from thermal expansion of the support structures. The key component defining the spacing between the RF coils is a Zerodur bar with a very low thermal expansion coefficient ($\alpha = 7 \times 10^{-8}\,\mathrm{K}^{-1}$). This support unit will allow for LD measurements on the 10^{-6} accuracy level even if the ambient temperature is fluctuating.

1. Introduction

Neutron Larmor diffraction (LD) is a high resolution diffraction technique based on Larmor precession of neutron spins. One key property of LD is that the resolution is decoupled from the monochromatization and divergence of the neutron beam. At present, LD is available at TRISP at the FRM II and at FLEXX at HZB. At both instruments a resolution $\Delta d/d \sim 10^{-6}$ is achieved, further improvement by one order of magnitude seems feasible with an optimized design. Material properties measured by LD include the thermal expansion under pressure and at low temperature [1, 2], the spread of lattice spacing resulting from micro strains or structural phase transitions [3, 4], accurate absolute lattice parameters [5], and structural and magnetic domain sizes [6].

Rekveldt first pointed out the appropriate field geometry shown in Fig. 1 [7, 8]: A uniform DC magnetic field is placed up- and downstream the sample with boundaries oriented parallel to the diffracting lattice planes. The Larmor phase $\phi = \omega_L \times t$ accumulated by a neutron spin after passing both fields is proportional to the Larmor frequency $\omega_L = \gamma B_0$, with $\gamma = 2.916\,\mathrm{kHz/Gauss}$, and the time $t = 2L/v_\perp$ the neutron spends in the field, where the latter is given by the velocity component $v_\perp = \pi\hbar/(m d_{\mathrm{hkl}})$ perpendicular to the field boundary. Thus the Larmor phase $\phi = 2m/(\pi\hbar)\omega_L L d_{\mathrm{hkl}}$ only depends on the field integral $J_0 = 2\omega_L L$ and the lattice spacing d_{hkl}. Thermal expansion is measured by tracking the phase shift $\Delta\phi(T)/\phi = \Delta d_{\mathrm{hkl}}/d_{\mathrm{hkl}}$.

The Larmor diffractometers at FLEXX and TRISP use the resonance spin echo technique (NRSE). The PDs are formed by pairs of RF spin flippers $C_1 - C_2$ and $C_3 - C_4$ (Fig. 1), the surfaces of these flippers define the boundaries of the precession regions [9, 10]. Phase shifts are measured by scanning the RF flipper C_4 along the beam axis. The resolution is both limited by counting statistics and by instrumental phase instabilities $\Delta\phi_{\mathrm{instr}}$, resulting from temporal fluctuations of the field integral ΔJ_0. ΔJ_0 arises from fluctuations of the frequency ω_{RF} applied to the spin flippers, or the spacing L between the coils. The design target value is $\Delta J_0/J_0 < 10^{-6}$

IOP Conf. Series: Journal of Physics: Conf. Series **862** (2017) 012012 doi:10.1088/1742-6596/862/1/012012

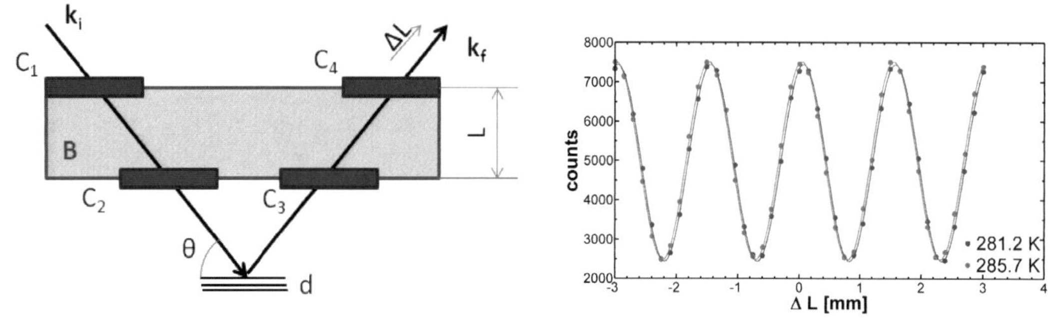

Figure 1. (left) Sketch of a Larmor diffractometer. An initially polarized neutron beam crosses the uniform field $\boldsymbol{B}$ twice (polarizer and analyzer are not shown). The boundaries of $\boldsymbol{B}$ are oriented parallel to the lattice planes d_{hkl}. In the resonance spin echo configuration (NRSE), instead of a uniform field $\boldsymbol{B}$ four RF spin flippers $C_1 - C_4$ are used, and there is no field along the flight path between the flippers and in the sample region. (right) LD raw data measured at TRISP on germanium (220), $k = 1.92\,\text{Å}^{-1}$, by scanning the position the RF flipper C_4 at two temperatures. The data points were taken at identical ΔL values and are vertically shifted by the phase shift. $\Delta\phi/\phi = \Delta d_{220}/d_{220} = 3.05(14) \times 10^{-5}$ induced by thermal expansion is consistent with $\alpha_{\mathrm{Ge}} = 6 \times 10^{-6}\,\mathrm{K}^{-1}$.

for the existing LDs and 10^{-7} for next generation instruments. ω_{RF} can easily be stabilized to the level 10^{-8} by using stable reference oscillators, such as oven controlled quartz or rubidium clocks.

The main source of instability is then the linear thermal expansion of the structures carrying the coils, an aluminum profile in the case of TRISP and FLEXX. The thermal expansion of the Al with a coefficient $\alpha_{\mathrm{Al}} = 2.31 \times 10^{-5}\,\mathrm{K}^{-1}$ directly affects the field integral and thus the phase, such that $\Delta\phi/\phi = \alpha_{\mathrm{Al}} \times \Delta T$. At TRISP in the experimental hall of the FRM II, the ambient temperature is very stable and drifts typically less that 0.1 K within 12 h, corresponding to $\Delta\phi/\phi < 2.3 \times 10^{-6}$. Only if the temperature drift is slow (< 0.2 K/h), the whole support structure is thermalized and the phase drift can be corrected by measuring the ambient temperature. Faster temperature drifts make measurements beyond the 10^{-5} level impossible. Besides this large linear thermal expansion of the support structures, there are certainly additional thermal drift mechanisms acting on the effective coil spacing L and thus on the Larmor phase, including bending of the support structure and bending of the DC coil wires, which leads to a curved coil surfaces and eventually to an effective shift of the coil center. These mechanisms are probably smaller than the linear expansion effect. Fluctuations of the DC current I_0 enter only in second order as $\delta\phi \sim (\delta I_0/I_0)^2$. With stabilized power supplies $\delta I_0/I_0 \sim 10^{-4}$, and $\delta\phi \sim 10^{-8}$ can be neglected. However, if there is as asymmetry in the RF flippers, resulting, for example, from an offset between the centers of the RF and DC coils along the beam path, current fluctuations might also enter the phase in first order and could become significant.

There are several strategies to minimize these drifts: (i) Exact measurement of the spacings $C_1 - C_2$ and $C_3 - C_4$, for example by a laser interferometer, and active compensation of the drift. As the coils are sitting on rotation units, it is difficult to find adequate points for attaching reference mirrors. (ii) Constructing the support with low thermal expansion materials. As the support must be non-magnetic to avoid stray fields disturbing the Larmor phase, Invar ($\alpha = 1.5 \times 10^{-6}\,\mathrm{K}^{-1}$) is not usable. Granite can be machined and is used for constructing precise and stable structures and tables, but the thermal expansion is too large for our application

IOP Conf. Series: Journal of Physics: Conf. Series **862** (2017) 012012 doi:10.1088/1742-6596/862/1/012012

($\alpha = 5 \times 10^{-6}\,\mathrm{K}^{-1}$). The Zerodur ceramic (Schott) shows excellent stability, the type 0 material has a very low $\alpha = 0 \pm 2 \times 10^{-8}\,\mathrm{K}^{-1}$. The construction of large structures from Zerodur is possible, but in the present case precluded by the high prize. We thus decided to keep the supporting aluminum structure and to compensate the linear expansion of the support by defining the spacing between the coils by a thin Zerodur bar.

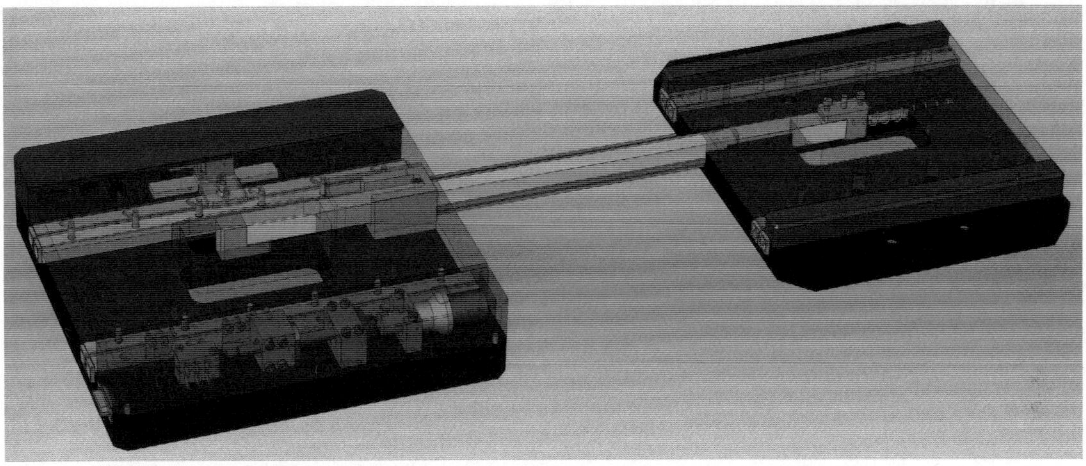

Figure 2. Support unit for RF coils $C_1 - C_2$ or $C_4 - C_3$.

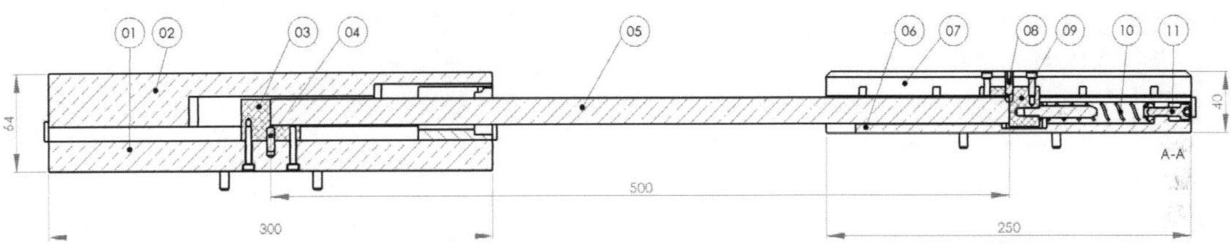

Figure 3. Longitudinal cut through the unit (side view). The center of the base plate (1) defines the reference point. The carriage (2) with motor drive and optical linear encoder allows for translational scanning of one RF coil. The post (3) is attached to the base plate with the pivot pin (4) to provide a suspension point (reference point) for the Zerodur bar (5). The right unit includes a base plate (6), and a carriage (7) suspended with linear bearings on (6). The post (9) is connected to (7) via the pivot pin (6). The spring (10) with setscrew (11) presses (7) against the Zerodur bar (5).

The design is shown in Figs. 2 and 3. The key component is the Zerodur bar ($20 \times 20 \times 500\,\mathrm{mm}^3$) keeping the distance between the centers of the base plate of the left unit and the upper part of the right unit constant. The end faces of the bar rest on flat posts being attached to pivot pins in the left bottom and right upper plates. The latter is suspended on a precise linear bearing (Schneeberger R6-250) and is pressed against the Zerodur bar by a spring (60 N). Thus the right carriage can translate with respect to the base to compensate thermal expansion of the aluminum support structure (not shown). An additional motor driven translation stage incorporated in the left unit is used to scan the spacing between the coils to measure the spin echo phase as shown in Fig. 1 (right). The pitch and yaw angles over the translation range of ± 25 mm are $< 70\,\mu$rad and $< 20\,\mu$rad, respectively. Over short translation distances, both angles

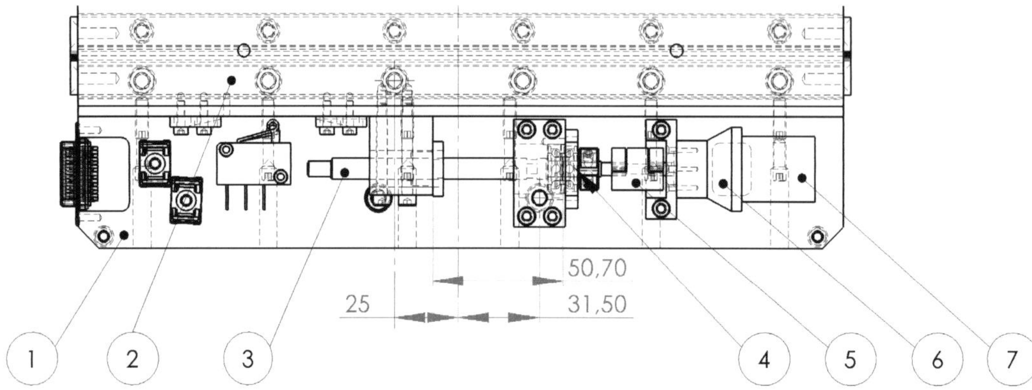

Figure 4. Left translation unit: (1) Aluminum base plate. (2) Linear ball bearing (Schneeberger R6-250) (3) Ball screw (Hipp KGT-F-08-01-132-74-O-IT1, slope 1 mm, dynamic load 850 N, material 1.1213 Cf53, $\alpha = 11.5 \times 10^{-6}$ /K. (4) Fixed bearing. (6,7) Stepping motor Phytron ZSS 25.200.0,6 with HarmonicDrive gear HD 05/80. Linear encoder Heidenhain LIP 481, precision $< 0.15\,\mu$m (not shown).

are $< 10\,\mu$rad. Special care was taken to avoid drifts in this translation by thermal expansion of the drive screw (Fig. 4). First, the position is measured by a precise linear encoder (Heidenhain LIP 481, error $< 0.15\,\mu$m) directly attached to the translation unit, thus that thermal expansion of the ball screw (Fig. 4, (3)) is directly monitored. Second, the expansion of the screw is compensated by a asymmetric mount to the carriage, thus that expansion of the screw and of the carriage compensate.

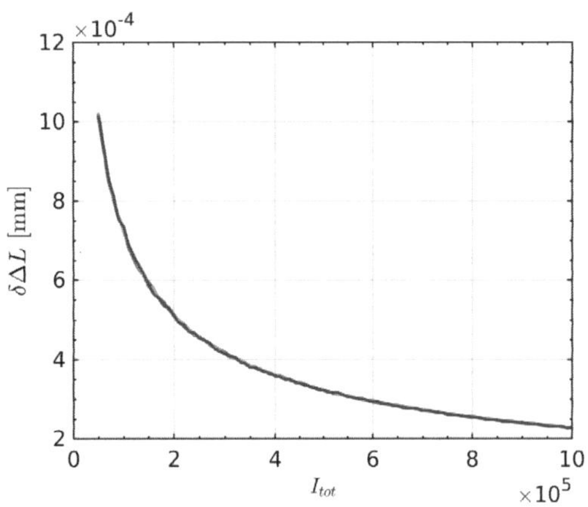

Figure 5. Phase shift *vs.* total counts (eq. 1) for $\Delta L = 0.5$ mm, ($\lambda = 4$ Å, $\omega_{\rm RF} = 2\pi \times 500$ kHz), $P = 0.5$.

In summary, we designed a support for the RF flippers of a LD with passive compensation of the linear thermal expansion by using a Zerodur standard. The unit is easy to manufacture in institute workshops and eliminates the dominating phase instability resulting from thermal expansion of support structures. We tested the stability by repeated measurements of the Larmor phase on the aforementioned Ge (220) Bragg reflection. On a time interval of 12 h, the phase shows random fluctuations in the order of $\Delta\phi/\phi \sim 1 \times 10^{-6}$, corresponding to an effective variation $\delta L \sim 1\,\mu$m of the total length $L = 1$ m, where the accuracy was not limited by counting statistics. Operation on a 10^{-6} accuracy level will thus be possible in experimental facilities with fluctuating ambient temperature.

For the design of future Larmor diffractometers with relative accuracies around 10^{-7}, first the aforementioned additional drift mechanisms resulting from bending of the support and heating of the coils have to be quantified and reduced. If we assume, that these drifts are independent of the spacing of the coils, the resolution can be increased by a factor 4 using the present coil and support design by increasing the coil spacing from presently 0.5 m to 2 m. Further the frequency applied to the coils can be increased. Presently we operate the coils at the relatively low frequency of 200 kHz to avoid heating. The temperature in the center of the beam window, where cooling of the coil is worst, rises only by 3 K after switching the coil on. Increasing the frequency by a factor 2.5 to 500 kHz will induce additional thermal drifts, and probably will require a modified coil design with reduced bending of the wires.

Finally we briefly discuss counting statistics. The statistical accuracy is estimated via the derivative of the cosine-shaped signal in eq. 1. The phase error $\delta\Delta L$ is

$$\delta\Delta L = \Delta L_{\text{period}}/(\pi P\sqrt{2I_{\text{total}}}) \qquad (1)$$

where ΔL_{period} is the period of the cosine, P is the polarization and I_{total} the total number of counts (the sum of all data points). Eq. 1 is shown in Fig. 5 for typical parameters. An error $\delta\Delta L < 0.4\,\mu$m is required for the resolution 10^{-7}, which is obtained for $I_{\text{total}} > 3\times10^5$. There is the question on how many signal periods should the total counts be distributed? The number of signal periods has an influence if the period ΔL_{period} is no a fixed parameter in the fit and if the phase shift is far from $\Delta L = 0$, which happens usually when the phase travels due to thermal expansion of the lattice d_{hkl}. In this case the phase error $\delta\Delta L$ is also affected by the error in ΔL_{period}. The latter is minimized by increasing the number of signal periods. At TRISP, it turned out as a good compromise to use 4 signal periods, which fit in the limited translation range of ΔL even for the shortest wavelengths.

Acknowledgments
We thank the workshop of the Max Planck Institute for manufacturing the devices with excellent precision, and Manfred Ohl for coordinating and supervising the work.

References

[1] Pfleiderer C, Boni P, Keller T, Rossler U K and Rosch A 2007 *Science* **316** 1871–1874
[2] Niklowitz P G, Pfleiderer C, Keller T, Vojta M, Huang Y K and Mydosh J A 2010 *Phys. Rev. Lett.* **104**(10) 106406
[3] Inosov D S, Friemel G, Park J T, Walters A C, Texier Y, Laplace Y, Bobroff J, Hinkov V, Sun D L, Liu Y, Khasanov R, Sedlak K, Bourges P, Sidis Y, Ivanov A, Lin C T, Keller T and Keimer B 2013 *Phys. Rev. B* **87**(22) 224425
[4] Lu X, Tseng K F, Keller T, Zhang W, Hu D, Song Y, Man H, Park J T, Luo H, Li S, Nevidomskyy A H and Dai P 2016 *Phys. Rev. B* **93**(13) 134519
[5] Repper J, Keller T, Hofmann M, Krempaszky C, Petry W and Werner E 2010 *Acta Materialia* **58** 3459–3467
[6] Náfrádi B, Keller T, Hardy F, Meingast C, Erb A and Keimer B 2016 *Phys. Rev. Lett.* **116**(4) 047001
[7] Rekveldt M T 1998 *Journal of Applied Physics* **84** 31–37
[8] Rekveldt M T, Keller T and Golub R 2001 *Europhysics Letters* **54** 342–346
[9] Golub R and Gähler R 1987 *Phys. Lett. A* **123** 43–48
[10] Gähler R and Golub R 1988 *Journal de Physique* **49** 1195–1202

IOP Conf. Series: Journal of Physics: Conf. Series **862** (2017) 012013 doi:10.1088/1742-6596/862/1/012013

Magnetic state of Nb(1-7nm)/$Cu_{30}Ni_{70}$ (6nm) superlattices revealed by Polarized Neutron Reflectometry and SQUID magnetometry

Yu. Khaydukov[1,2,7], R. Morari[3,4], D. Lenk[3], V. Zdravkov[3,4], D.G. Merkel[5], B.-K. Seidlhofer[6], C. Müller[3], H.-A. Krug von Nidda[3], T. Keller[1,2], R. Steitz[6], A. Sidorenko[4], S. Horn[3], R. Tidecks[3], and B. Keimer[1]
[1]Max-Planck-Institut für Festkörperforschung, Stuttgart, Germany
[2] Max Planck Society Outstation at the FRM-II, Garching, Germany
[3]Institut für Physik, Universität Augsburg, Augsburg, Germany
[4]D. Ghitsu Institute of Electronic Engineering and Nanotechnologies ASM, Kishinev, Moldova
[5]Wigner Research Centre for Physics, Budapest, Hungary
[6]Helmholtz-Zentrum für Materialien und Energie, Berlin, Germany
[7]Skobeltsyn Institute of Nuclear Physics, Moscow State University, Moscow, Russia

E-mail: y.khaydukov@fkf.mpg.de

Abstract. We report results of a magnetic characterization of $[Cu_{30}Ni_{70}(6nm)/Nb(x)]_{20}$ (x=1÷7nm) superlattices using Polarized Neutron Reflectometry (PNR) and SQUID magnetometry. The study has shown that the magnetic moment of the structures growths almost linearly from H = 0 to H_{sat} = 1.3kOe which can be interpreted as evidence of antiferromagnetic (AF) coupling of the magnetic moments in neighbouring layers. PNR, however, did not detect any in-plane AF coupling. Taking into account the out-of-plane easy axis of the $Cu_{30}Ni_{70}$ layers, this may mean that only the out-of-plane components of the magnetic moments are AF coupled.

1. Introduction

Hybrid superconducting/ferromagnet (S/F) heterostructures are intensively studied objects due to their interesting and promising properties [1]. At the moment the main research is concentrated on the study of simple S/F bilayers and S/F/S, F/S/F, and S/F/F trilayers (see [2] and references therein). However, we may expect that both superconducting ([3]-[11]) and magnetic ([12]-[14]) properties of a more complex S/F systems, such as $[S/F]_n$ superlattices, will differ from those of their constituent elements (S/F bilayers or S/F/S-, F/S/F -trilayers). A difference between the behaviour of the constituent elements and the superlattice is especially expected when the thicknesses of the layers become comparable with the correlation length of superconductivity, ξ_S, and magnetism, ξ_F, in the respective layers [12]-[15]. In a sense such superlattices can be considered as metamaterials assembled from "atoms" of S/F bilayers.

Our main goal is the design and fabrication of SF metamaterials with unique superconducting and magnetic properties arising from the competition of magnetic and superconducting ordering. As the building blocks of this metamaterial we propose to use Nb/CuNi bilayers investigated in detail before [16]-[19].

IOP Conf. Series: Journal of Physics: Conf. Series **862** (2017) 012013 doi:10.1088/1742-6596/862/1/012013

There are several reasons for this choice of materials, like very small but non-vanishing solubility of CuNi and Nb (yielding smooth interfaces), good electrical contact and relatively high superconducting correlation lengths $\xi_{S,F} \sim$ 10nm [17]. The aim of this investigation is to search for a possible antiferromagnetic (AF) coupling of $Cu_{30}Ni_{70}$ layers through the Nb spacer. Such an AF coupling can give rise to unique superconducting and ferromagnetic properties of SF metamaterials [13]-[15]. A similar AF coupling has been previously observed by Polarized Neutron Reflectometry (PNR) in Fe/Nb systems [20] for thicknesses of the Nb spacers of d_{Nb} = 1.3nm, 1.7nm, 2.4nm, 2.7nm, 3nm.

2. Sample preparation

A series of periodic structures (PS) with nominal composition $Si/[Nb(d_{Nb})/Cu_{30}Ni_{70}(6nm)]_{20}/Si$ ($d_{Nb} = 1 \div 7$nm) were prepared at room temperature using the automatic magnetron sputtering device UNIVEX at Augsburg University. This machine allows an automatic fabrication of layered structure with specified parameters. All targets are mounted in independent chambers (*target-cells*) and the sample, fixed on the sample holder, is transferred between the target-cells by a robotic manipulator (see Fig. 1).

The typical base pressure in the chambers is 3×10^{-7}, 6×10^{-7} and 5×10^{-6} mbar for target-cells, transfer and load-lock chambers, respectively. Pure argon (99.999%, "Linde AG") at a pressure of 8×10^{-3} mbar was used as sputter gas. Three targets, Si, Nb and $Cu_{30}Ni_{70}$ (100 mm in diameter), were pre-sputtered for 10 – 15 minutes to remove contaminations before deposition of the samples and 1 minute before each layer deposition. The {111} silicon substrate was etched in pure argon plasma for 5 minutes. After the sample holder was transferred into the cleaned Si-target chamber, the Si-cell was hermetically closed and a 6 nm buffer silicon layer was deposited. The growth rate of the Si film was about 0.4 nm/sec. After deposition the waste gas was rapidly pumped out and the pressure in the target-cell and transfer chamber equalized (6×10^{-7} mbar). Next the load lock was opened and the manipulator picked up the sample holder. Then the sample holder was alternatingly transferred to the Nb target-cell and $Cu_{30}Ni_{70}$ target-cell for deposition of the superlattice structure. To avoid oxidation of the structure, the Si-cap film was grown on the top of the layered structure. The processes of pressure leveling between target-cell and transfer chamber and also sealing/opening of the load lock were repeated for every layer, thus ensuring a clean atmosphere in the transfer chamber during fabrication of the structures. The time spent for gas pumping, pressure leveling process and sample transferring between two target-cells was about 25-30 seconds.

The deposition rates for Nb and $Cu_{30}Ni_{70}$ are 0.4 and 0.5 nm/sec, respectively. The thicknesses of the layers were controlled by the time of the ignited target exposition and checked later using neutron reflectometry. For this purpose, the chambers of the UNIVEX are equipped with a movable shutter between target and sample holder. While pure Nb was sputtered using a pulsed power module, the $Cu_{30}Ni_{70}$ and Si – layers deposition and substrate etching were performed using an AC power module.

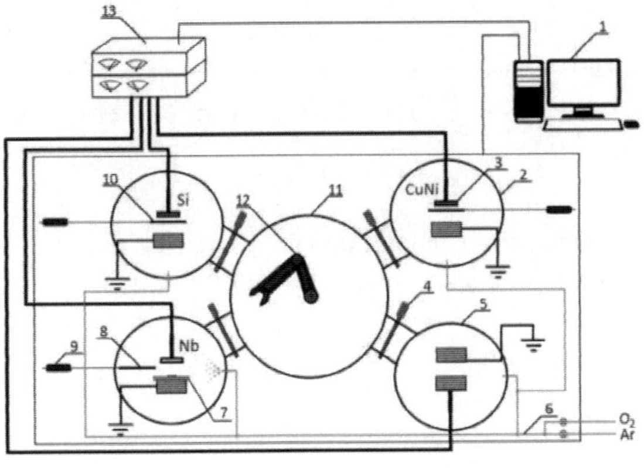

Figure. 1. Magnetron sputtering device UNIVEX. 1 - Computer (PC) for controlling the UNIVEX; 2-Target-cell or target chamber; 3 - Target; 4 - load lock controlled by PC;5 - Load-lock chamber; 6- Gas pipe; 7 -Sample holder and silicon substrate; 8 - Movable shutter (opened position); 9 - Electric motor for shutter moving; 10 - Movable shutter (closed position); 11 - Transfer chamber; 12 - Robotic manipulator; 13 - AC and Pulsed power modules for magnetron

IOP Conf. Series: Journal of Physics: Conf. Series **862** (2017) 012013 doi:10.1088/1742-6596/862/1/012013

The structural properties of the superlattices were characterized using neutron reflectometry. The necessity of using neutron reflectometry instead of the more widely available X-ray reflectometry is due to the extremely low optical X-ray contrast for layers of CuNi and Nb [18],[19]. Figure 2 shows the non-polarized reflectivity curve of the sample with nominal d_{Nb} = 2nm measured at room temperature at the GINA reflectometer [21]. Two peaks at positions $Q_1 = 0.54$ nm^{-1} and $Q_2 = 1.04$ nm^{-1} can be identified and assigned to Bragg reflection from the periodic structure with period D. The positions of the peaks are given by $Q_n \approx 2\pi n/D$ (n = 1,2...) and their presence gives evidence for a high repeatability of the bilayer thickness within the superlattice. We were able to reproduce the experimental curve with the scattering length density (SLD) depth profile depicted in the inset to the Fig.1. According to the fit, the nuclear SLD of $Cu_{30}Ni_{70}$ and Nb layers are $\rho_{CuNi} = 7.8\text{x}10^{-4}$ nm^{-2} and $\rho_{Nb} = 4.3\text{x}10^{-4}$ nm^{-2}. The rms roughness of the $Cu_{30}Ni_{70}$/Nb and Nb/$Cu_{30}Ni_{70}$ interfaces is obtained as 0.6 and 1.2 nm, respectively. The obtained thickness of $Cu_{30}Ni_{70}$ layer d_{CuNi} = 6.3nm is close to the nominal value. However thickness of Nb film d_{Nb} = 6.0nm turned out to be much higher than nominal one. Similar disagreement was also observed for the sample with nominal d_{Nb} = 3nm. For the rest of the samples nominal thickness was within 10% accuracy close to nominal. The reason of such disagreement at $d_{Nb} < 4$nm will be discussed elsewhere.

In order to demonstrate sensitivity of the model to repeatability of the bilayer thickness we calculated model reflectivity for the same SLDs but with 10% variation of the thickness of each layer around mean d_{CuNi} = 6.3nm and d_{Nb} = 6.0nm (see red dashed curve in Fig. 2). One can see that such deviation of real thickness of each layer will lead to the smearing of the 1st Bragg peak which was not observed on the experiment.

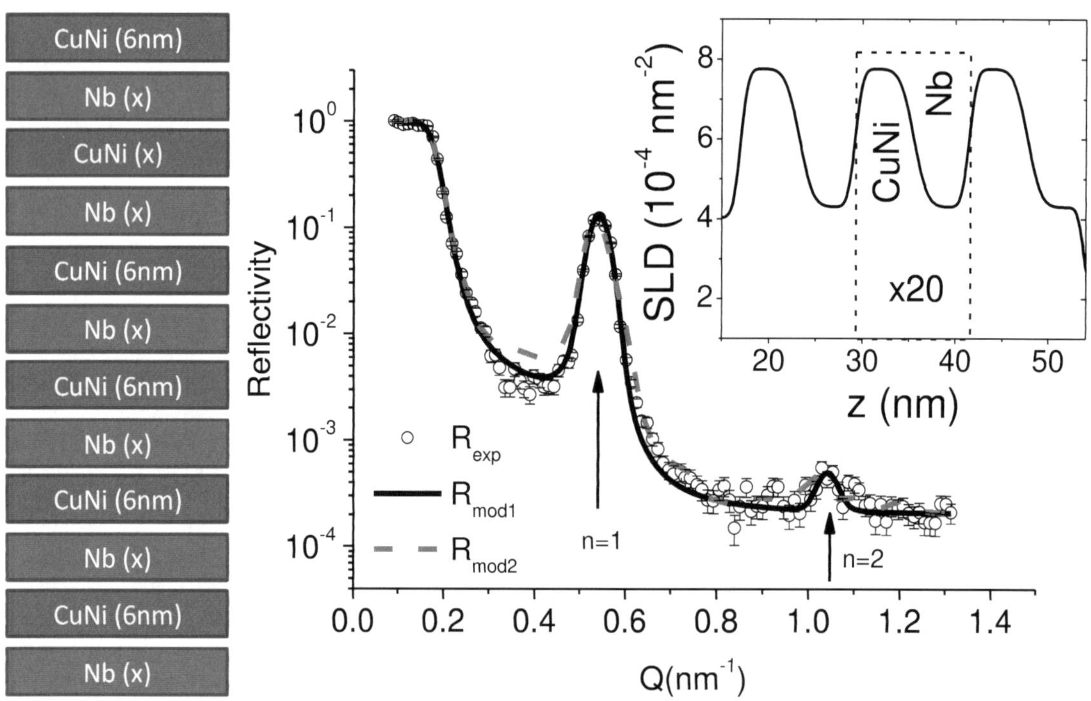

Figure 2. Sketch of the structure (left) and room– temperature data (dots) measured at the GINA reflectometer. Lines show model curve for 20 bilayers with the same thickness (solid black) and with 10% deviation around mean values (dashed red). The inset shows part of the SLD depth profile. One "unit cell" of the periodic CuNi/Nb structure is shown by the dashed rectangular.

3. Magnetic properties

3.1. Single copper-nickel layer

In order to better understand magnetic behaviour of complex $Cu_{30}Ni_{70}$/Nb superlattices we first have characterized magnetic properties of single $Cu_{30}Ni_{70}$ layers. Figure 3a shows the hysteresis loop of the $Cu_{30}Ni_{70}$(23.5nm) film measured by Superconducting Quantum Interference Device (SQUID) in the Max-Planck Institute for Solid State Research (Stuttgart). The curve is characterized by the coercive field $H_c \approx 150$ Oe and a remanent magnetization which is only 25% of the saturation magnetization. Such a shape of the hysteresis loop is typical for copper-nickel films with the easy axis turned out of plane of the sample [22]. In order to calculate the direction of the easy axis relative to the sample plane, β, we used the following expression for the magnetic energy of the film

$$E(\alpha) = K_{anys} \sin^2(\beta - \alpha) - M_S H \cos(\alpha), \quad (1)$$

where K_{anys} - anisotropy energy, α, β - the angles between the magnetic moment and the easy axis and direction of the external field H (or sample surface, see sketch in Fig. 3a), respectively, M_S is the saturation magnetization. The value of M_S can be found from the saturation magnetic moment m_{sat} as $M_S = m_{sat}/(d_{CuNi} \times A)$, where d_{CuNi} and A are the thickness of the $Cu_{30}Ni_{70}$ layer and the sample area, respectively. However, taking into account possible mis-calibration of the film thickness during deposition we decided to use Polarized Neutron Reflectometry (see below) which measures directly the magnetic contrast. Minimizing this energy by varying α for every magnetic field allows to draw a model dependence $m_{sat} \cos(\alpha)$ vs. H (see red curve) and estimate the direction of the easy axis as $\beta \approx 80°$.

The saturation magnetization was measured in a PNR experiment on the same sample. The reflectivity curves R^+ and R^- were measured at NREX reflectometer [23] at $T = 150$K in a magnetic field $H = 4.5$kOe (Fig. 3b). Since both temperatures $T = 150$K and $T = 10$K are far below the Curie temperature $T_m = 295$K (Fig. 3d) we believe that comparison of PNR and SQUID data are possible. Oscillations of the reflectivity curves are caused by the interference over the thickness of the $Cu_{30}Ni_{70}$ film and allowed us also to calibrate the thickness of the $Cu_{30}Ni_{70}$ layer for further preparation of periodic structures. Presence of spin asymmetry $(R^+ - R^-)/(R^+ + R^-)$ (Fig. 3c) is related to the magnetic moment in the film. Fitting the experiment to a model allows us to extract thickness of the layer $d_{CuNi} = 23.4 \pm 0.2$nm, nuclear SLD of the single film $\rho_{CuNi} = (8.3 \pm 0.1) \times 10^{-4}$ nm^{-2} and saturation magnetization $4\pi M_{sat} = 1.7 \pm 0.3$kG.

Using the values obtained from PNR for ρ_{CuNi} and M_{sat} we may estimate the magnetic moment per one unit cell atom as

$$m[\mu_B] = M_{sat}[G]/(N\,[cm^{-3}]\,\mu_B), \quad (2)$$

where N - is the packing density of $Cu_{30}Ni_{70}$ layer, $\mu_B = 9.27 \times 10^{-21}$ erg/G - is the Bohr magneton. The density can be derived from the nuclear SLD, which can be written as

$$\rho_{CuNi} = N \times [C_{Ni}\, b_{Ni} + (1 - C_{Ni})\, b_{Cu}], \quad (3)$$

where $C_{Ni} = 0.7$ -concentration of nickel atoms, $b_{Ni} = 10.3$ fm, $b_{Cu} = 7.7$ fm are neutron scattering lengths for nickel and copper atoms, respectively. Using Eqs. (2) and (3) a magnetic moment of 0.17 μ_B/u.c. or 0.24 μ_B/Ni atom can be estimated from experimentally obtained values of M_{sat} and ρ_{CuNi}. This value is in agreement with the previously reported 0.27 μ_B/atom for $C_{Ni} = 0.67$ [24].

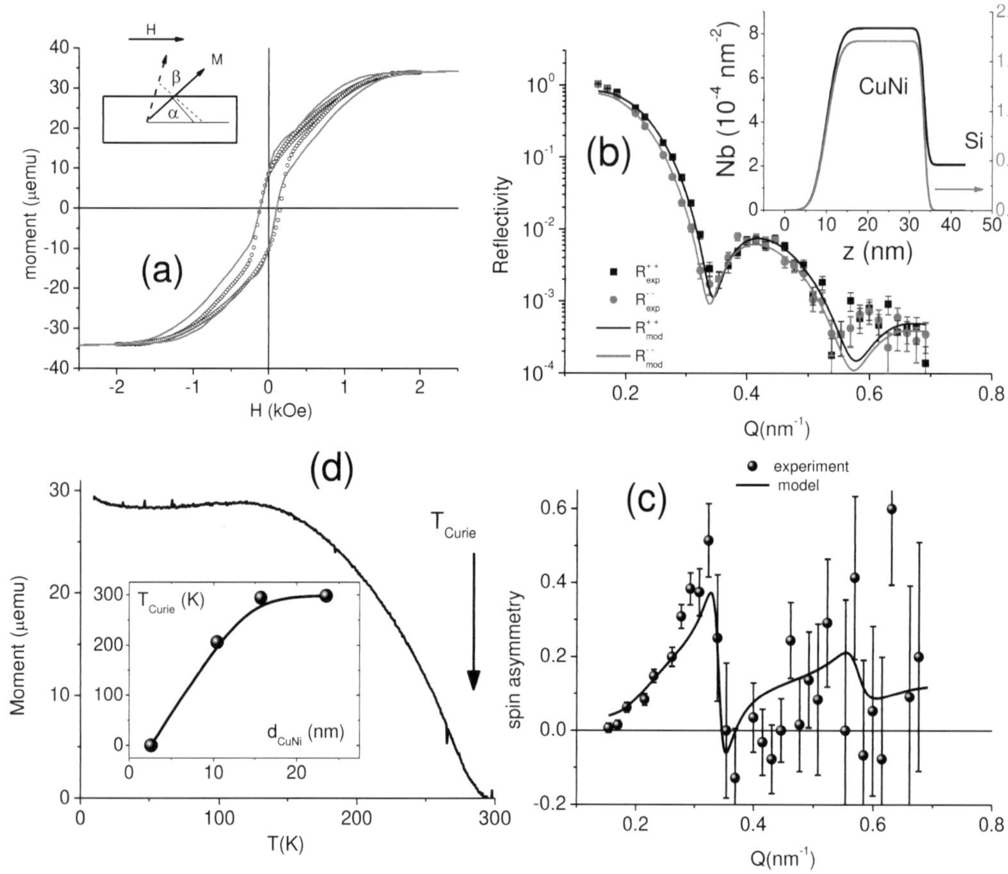

Figure 3. (a) Hysteresis loop for the $Cu_{30}Ni_{70}$(23.5nm) layer measured at T = 10K. (b) Low temperature PNR curves measured at T = 150K and H = 4.5kOe. The error bars if not seen are below the dot size. (c) Spin asymmetry $(R^+ - R^-)/(R^+ + R^-)$ correspondent to data depicted in (b). (d) Temperature dependence of the magnetic moment measured at H = 1kOe. The vertical arrow shows the position of the Curie temperature. Inset shows the dependence of the Curie temperature of films on their thickness.

3.2. Periodic structure

The next step is the characterization of the magnetic properties of the periodic structures. Fig. 4a shows the field dependence of the magnetic moment of the sample with d_{Nb} = 5nm measured by SQUID magnetometry at a temperature of 13K. The magnetic moment growths almost linearly from zero at H = 0 till its saturation value at $H > H_{sat}$ = 1kOe. One of the explanations of the linear growth of the magnetic moment is an antiferromagnetic (AF) coupling of the magnetic moments in the neighbouring layers [24]. We have observed that *all* measured samples with different thickness of Nb spacer exhibit similar curves. In order to prove the presence of in-plane AF coupling we have performed low-temperature PNR experiments. The measurements were conducted at the angle dispersive (λ = 0.466 nm) reflectometer V6 [26]. The protocol of the measurements was as follows. The sample was cooled down to T = 15K in a magnetic field H = 5kOe. At this temperature the field was released and several reflectivity curves were measured at fields $H < H_{sat}$. Fig. 4b shows PNR curves of the sample with d_{Nb} = 5nm measured in a magnetic field H = 400 Oe at T = 15K. Similar to the curve depicted in Fig. 2, the curve is characterized by the presence of the n = 1 Bragg peak from the superstructure. Similar to Fig. 3b, Bragg peaks of different polarization are split due to the presence of the magnetic moment in $Cu_{30}Ni_{70}$ layers. Inset to Fig. 4b shows the field dependence of the spin asymmetry of the first Bragg peak $S_1 = (R^+ - R^-)/(R^+ + R^-)$. One can see that this dependence correlates with the field dependence of the magnetic moment depicted in Fig. 4a. A similar behaviour was observed for the sample with nominal d_{Nb} = 3nm.

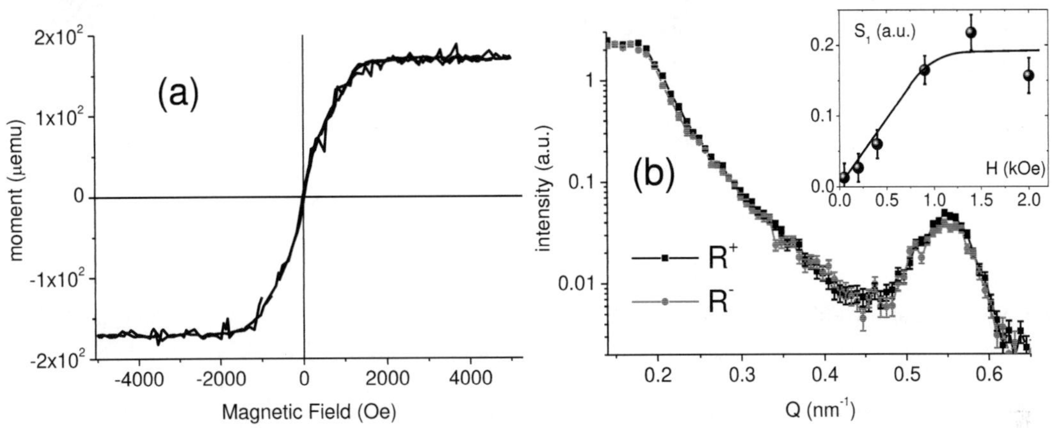

Figure 4. (a) Hysteresis loop measured by SQUID magnetometry at T = 13K on the sample with d_{Nb} = 5nm. (b) PNR curves measured at T = 15K in a magnetic field H = 400 Oe on the same sample. Dots in the inset show the field dependence of the spin asymmetry of the first Bragg peak. The solid line is shown to guide an eye.

4. Discussion and conclusion

Our measurements on single $Cu_{30}Ni_{70}$ films have shown that the magnetic easy axis of the films is out of plane (OOP). This means that the magnetic moments of the $Cu_{30}Ni_{70}$ layers in small magnetic fields will tend to align out of plane. If an AF coupling exists in the system it would try to align the magnetic moments in the neighbouring layers antiparallel to each other. Such a magnetic configuration and its reflectivity curves are shown in Fig. 5a. One can see that in this case there will be the $n = 1/2$ peak due to the doubling of magnetic period in comparison to the structural one. Such peak was indeed seen in PNR experiments with Fe/Nb systems [20]. Another feature is the absence of a splitting at the $n = 1$ peak. We may conclude that this model does not describe our data. Moreover this magnetic configuration leads to a strong increase of the OOP stray field which, in turn, increases the magnetostatic energy. Hence we consider another model, where only out of plane components of the magnetization of neighbouring copper-nickel layers are AF coupled (Fig. 5b). Such a configuration allows decreasing the magnetostatic energy. Since PNR is only sensitive to the in-plane magnetic moment, therefore this configuration will not produce an $n = 1/2$ peak, but will lead to the splitting of curves at the $n = 1$ peak. This model agrees with PNR data depicted in Fig. 4b.

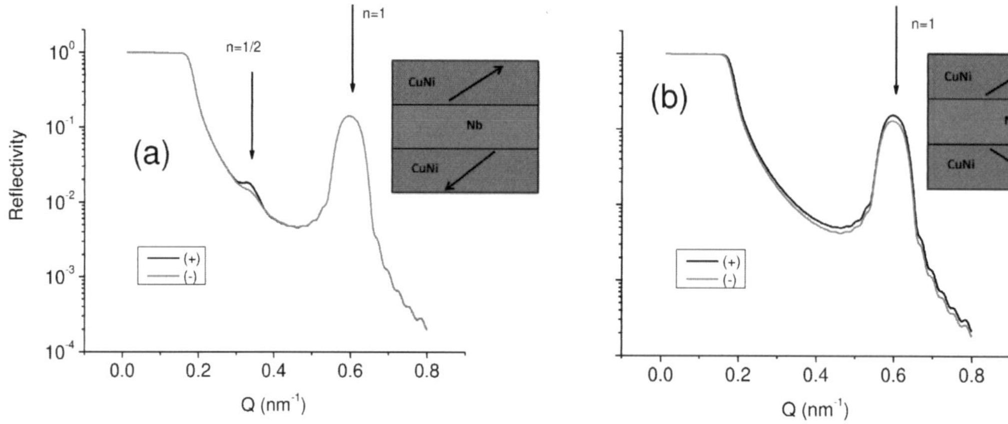

Figure 5. Model reflectivity curves for different magnetic configuration depicted in the corresponding insets

In conclusion we report results of the magnetic characterization of $[Nb(d_{Nb})]/Cu_{30}Ni_{70}(6nm)]_{20}$ (d_{Nb} =1÷7nm) superlattices using SQUID magnetometry and polarized neutron reflectometry. The study has shown that in all structures the magnetic moment growths almost linearly from $H = 0$ to H_{sat} = 1.3kOe which may indicate presence of antiferromagnetic coupling of the moments in neighbouring layers. PNR however did not detect any AF coupling in-plane. Taking into account an out-of-plane easy axis of the $Cu_{30}Ni_{70}$ layers, this probably indicates that only the out of plane components of magnetic moments are AF coupled. Further depth and out-of-plane sensitive techniques like X-ray magnetic scattering are required to answer this question explicitly.

The authors would like to thank A. Petrzhik for assistance in measurement of magnetic moment of single $Cu_{30}Ni_{70}$ layers and T. Kraus and W. Reiber for their help in preparation of the samples. This work is based upon experiments performed at the NREX instrument operated by Max-Planck Society at the Heinz Maier-Leibnitz Zentrum (MLZ), Garching, Germany and partially supported by the DFG collaborative research center TRR80 and DFG grant No HO 955/9-1.

References

[1] Buzdin A 2005 *Rev. Mod. Phys.* **77** 935
[2] Lenk D *et al.* 2016 *Beilstein J. Nanotechnol.* **7** 957
[3] Chien C L and Reich D H 1999 *J. Magn. Magn. Mater.* **200** 83
[4] Verbanck G, Potter C, Schad R, Beilen P, Moshchalkov V and Bruynseraede Y 1994 *Physica C* **235–240** 3295
[5] Koorevaar P, Coehoorn R and Aarts J 1995 *Physica C* 248 61
[6] Mattson J, Potter C, Conover M, Sowers C and Bader S 1997 *Phys. Rev.* B **55** 70
[7] Verbanck G, Potter C, Metlushko V, Schad R, Moshchalkov V and Bruynseraede Y 1998 *Phys. Rev.* B **57** 6029
[8] Prischepa S, Cirillo C, Bell C, Kushnir V, Aarts J, Attanasio C and Kupriyanov M 2008 *JETP Lett.* **88** 431
[9] Armenio A A, Cirillo C, Iannone G, Prischepa S and Attanasio C 2007 *Phys. Rev.* B **76** 024515
[10] Huang S, Liang J-J, Tsai T, Lin L, Lin M, Hsu S and Lee S 2008 *J. Appl. Phys.* **103** 07C704
[11] Cirillo C, Bell L C, Iannone G, Prischepa S, Aarts J and Attanasio C 2009 *Phys. Rev.* B **80** 094510
[12] Sá de Melo C A R. 2000 *Phys. Rev.* B **62** 12303
[13] Proshin Y, Izyumov Y and Khusainov M 2001 *Phys. Rev.* B **64** 064522
[14] Halterman K and Valls O 2004 *Phys. Rev.* B **69** 014517

IOP Conf. Series: Journal of Physics: Conf. Series **862** (2017) 012013 doi:10.1088/1742-6596/862/1/012013

[15] Bakurskiy S, Kupriyanov M, Baranov A, Golubov A, Klenov N,. Soloviev I 2015 *JETP Lett.* **102** 586-593

[16] Zdravkov V, Sidorenko A, Obermeier G, Gsell S, Schreck M, Müller C, Horn S, Tidecks R and Tagirov L 2006 *Phys Rev. Lett.* **97** 057004

[17] Zdravkov V *et al.* 2010 *Phys. Rev.* B **82** 054517

[18] Khaydukov Y *et al.* 2015 *J Supercond Nov Magn* **28** 1143

[19] Khaydukov Y, Morari R, Soltwedel O, Keller T, Christiani G, Logvenov G, Kupriyanov M, Sidorenko A and Keimer B 2015 *J. of Appl. Phys.* **118** 213905

[20] Rehm C, Nagengast D, Klose F, Maletta H and Weidinger A 1997 *Europhys. Lett.* **38** 61-66

[21] Bottyán L, Merkel D, Nagy B, Füzi J, Sajti S, Deák L, Endrőczi G, Petrenko A and Major J 2013 *Rev. Sci. Instrum.* **84** 015112

[22] Ruotolo A, Bell C, Leung C and Blamire M 2004 *J. Appl. Phys.* **96** 512

[23] http://www.mlz-garching.de/nrex

[24] Rusanov A, Boogaard R, Hesselberth M, Sellier H and Aarts J 2002 *Physica* C **369** 330

[25] Granberg P, Isberg P, Svedberg E, Hjörvarsson B, Nordblad P and Wäppling R 1998 *J. Magn. Magn. Mater.* **186** 154

[26] Paul A, Krist T, Teichert A and Steitz R 2011 *Physica* B **406** 1598–1606

IOP Conf. Series: Journal of Physics: Conf. Series **862** (2017) 012014 doi:10.1088/1742-6596/862/1/012014

The effect of amphiphilic polymers with a continuous amphiphilicity profile on the membrane properties in a bicontinuous microemulsions studied by neutron scattering

Helge F.M. Klemmer[1], Henrich Frielinghaus[2], Jürgen Allgaier[3], Michael Ohl[4], Olaf Holderer[2]

[1] Department of Chemistry, University of Cologne, Cologne, Germany

[2] Jülich Centre for Neutron Science (JCNS) at Heinz Maier-Leibnitz Zentrum (MLZ), Forschungszentrum Jülich GmbH, Garching, Germany

[3] Jülich Centre for Neutron Science JCNS and Institute for Complex Systems ICS, Forschungszentrum Jülich GmbH, Jülich, Germany

[4] Jülich Centre for Neutron Science (JCNS) at SNS, Forschungszentrum Jülich GmbH, Garching, Germany

o.holderer@fz-juelich.de

Abstract. Microemulsion systems consisting of oil, water and surfactant have been studied with neutron scattering techniques. The amount of surfactant needed to form a microemulsion can be dramatically reduced by the addition of small amounts of amphiphilic block copolymers (boosting effect). Here, we studied the influence of block copolymers with gradually changing amphiphilicity from hydrophilic to hydrophobic. Small angle neutron scattering (SANS), neutron spin echo spectroscopy (NSE) and phase diagram measurements in combination give access to the elastic properties of the membrane. The underlying NSE experiments for this interpretation rely on smallest changes of the relaxation curves (of ca. 1% steps) for still small changes of the bending rigidity (of ca. 10% steps). This high reliability of the experiments conducted at the SNS-NSE displays the accuracy of the instrument itself and the latest developments of the evaluation software, which were necessary to interpret such tiny changes of the bending rigidity reliably.

1. Introduction

The combination of neutron spin echo spectroscopy (NSE) with phase diagram measurements and small angle neutron scattering (SANS) is a very powerful combination for the determination of the elastic properties of microemulsions and the determination of the bending rigidity, κ, and the saddle splay modulus, $\bar{\kappa}$ [1]. The efficiency of microemulsions (i.e. thermodynamically stable water-oil-surfactant mixtures) can be strongly increased (the so called "boosting effect") by the addition of amphiphilic diblock copolymers with a hydrophobic and a hydrophilic block. The diblock copolymers are incorporated into the membrane and make it stiffer by increasing the bending rigidity [2,3]. This "boosting effect" of the microemulsion efficiency could be used to reduce the amount of surfactant needed in applications. At high surfactant concentrations, lamellar phases are formed instead of isotropic bicontinuous phases. Recently, a new type of copolymers has been developed with a gradual

IOP Conf. Series: Journal of Physics: Conf. Series **862** (2017) 012014 doi:10.1088/1742-6596/862/1/012014

change of polarity, i.e. instead of one hydrophilic and one hydrophobic part (similar to the surfactant structure but with a much higher molecular weight), a smooth transition from the hydrophobic to the hydrophilic side has been realized [4]. This new class of "tapered" polymers also behaves in a similar way as the diblock copolymers in the sense that a significantly smaller amount of surfactant can be used to emulsify a given amount of water and oil. But contrary to the diblock copolymers, the tapered polymers suppress at high surfactant concentrations the formation of a lamellar phase instead of favoring it. Also at high surfactant concentrations, an isotropic phase is therefore preserved, which might be of interest for some applications. The elastic properties have been analyzed by a combination of SANS, NSE and phase diagram measurements [5]. Here we show how the strong requirements in measurement precision for the NSE experiments have been satisfied at the SNS-NSE spectrometer at the SNS (Oak Ridge) [6] and how this helps to deduce the elastic properties of the microemulsion together with SANS and phase diagram measurements.

2. Experimental results

The synthesis of the tapered polymers has been described in detail elsewhere [4]. The tapered polymers were made from polybutylene (BO) and polyethylene (EO) blocks, where the copolymer P(BO/EO)22 had a molecular weight of 22 kg/mol. Microemulsions were mixed from water and decane (bulk contrast for SANS: D_2O/h-decane, film contrast for NSE: D_2O/d-decane) with a water/decane ratio of 1, and with the nonionic surfactant tetraethyleneglycol-mono-*n*-decylether ($C_{10}E_4$).

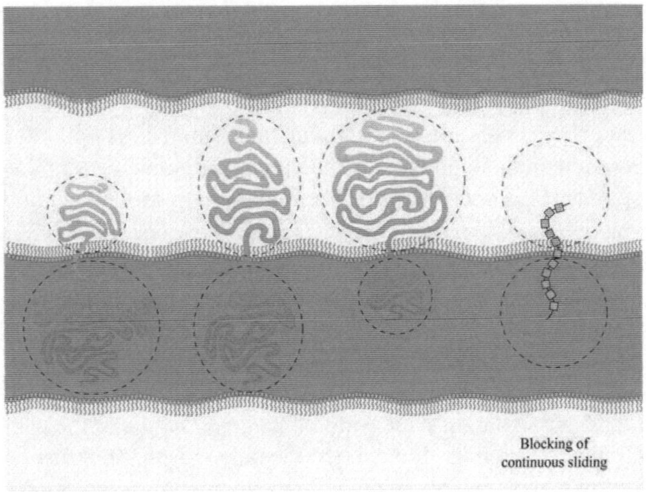

Fig. 1: Sketch of a surfactant membrane layer with tapered diblock copolymer decoration. The gradual change of amphiphilicity allows to slide the anchoring point along the contour of the polymer chain.

The structure of microemulsions has been investigated with SANS experiments and phase diagram measurements (Fig. 2). The typical correlation peak at ~0.02 Å^{-1} shows the repeating distance of the structure. The width of the peak corresponds inversely to the correlation length ξ according to the well known Teubner-Strey formula $I(q) \sim 1/(q^4 - 2(q_0^2 - \xi^{-2})\, q^2 + (q_0^2+\xi^{-2})^2)$ [7]. The asymptotic behavior shows the typical Porod dependence (although the Teubner-Strey formula would give the same behavior, it is only valid around the peak position). A series of SANS experiments allowed to determine the variation of ξ for varying d from ξ=132 A (δ=0) to 314 A (δ=0.3) [5]. Together with the structural length scale, the renormalized bending rigidity κ_R can be determined [3] and the sensitivity parameter Ξ (the change of bending rigidity with surface coverage of the membrane by diblock polymers) can be determined [5]. The saddle splay modulus comes from the position of the so called fish tail point in phase diagrams (minimum amount of surfactant needed to emulsify a given amount of oil and water, see [5]). The intermediate scattering function measured by NSE spectroscopy (Fig. 3) allows a direct determination of κ, and also the determination of the sensitivity parameter Ξ.

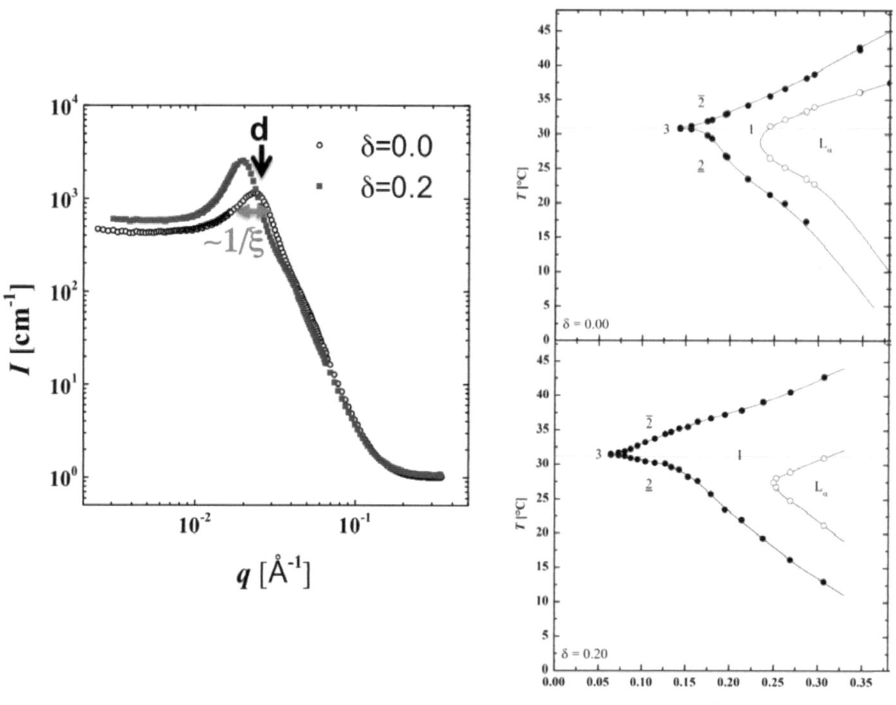

Fig. 2: Left: SANS scattering curves for bicontinuous microemulsions with no additives and with 20% tapered polymer contents in the membrane [5]. Right: Phase diagrams of the same samples: the pure microemulsion (top) and the one with added polymers (bottom), showing the different phase regions (1: one phase bicontinuous, L_α: lamellar phase, 2: oil- or water excess phase). It shows the boosting effect by the shift of the one phase region to lower surfactant concentrations γ (i.e. to the left) and the beginning of lamellar phase suppression.

3. Discussion

Combining SANS, NSE and phase diagram measurements gives access to the elastic properties of the membrane. Phase diagram measurements showed that the lamellar phase is suppressed to a large extend in favor of a bicontinuous phase [5]. Here, we focus on the high precision of the NSE experiments needed to detect tiny differences in the resulting relaxation rate. Varying the fraction of tapered copolymers in the membrane, δ, leads to a small but systematic variation of the relaxation rate, as depicted in Fig. 3. The intermediate scattering function has been fitted with a stretched exponential function according to the Zilman-Granek theory [8], $S(q,\tau) = \exp(-(\Gamma\tau)^{\beta})$. This theory applies only for local membrane patch fluctuations, i.e. it is valid at high q ($q>4\, q_0$ with q_0 from the Teubner-Strey fits, roughly the correlation peak position). There, the stretching exponent takes the value β=2/3 (Top right of Fig. 3). At lower q closer to the correlation peak (e.g. the top left plot of Fig. 3) the structural influences change the stretching exponent (here to 0.84). The corresponding relaxation rates (for large q) as a function of δ are shown in Fig. 4. The bending rigidity κ can be extracted from the relaxation rate either by numerically evaluating the height correlation function of Zilman and Granek instead of the simplified version of the stretched exponential function [9], or by converting the reduced relaxation rate (Γ corrected for the average solvent viscosity η and the third power of the inverse length scale, q^3) taking into account the correlation length ξ from SANS. The dependence of κ on both, ξ and Γ, is the reason that rather small changes of Γ can result in significant changes of κ. A graphical procedure for this evaluation procedure is shown in Ref. [10]. The bending rigidity κ increases in the present case from ~0.8 k_BT for δ=0 in agreement to the results reported in Refs [9,10] to 1.8 k_BT for the highest δ.

The determination of the sensitivity parameter Ξ lead to the interesting result, that NSE showed that the influence of the tapered copolymers on the bending rigidity is very similar to the case of the simple diblock copolymers (i.e. with a very well defined anchoring point) studied in Ref. [9]. In both cases, the bending rigidity increased by ~50%. On the other hand, SANS showed a significantly reduced value of Ξ i.e. a much smaller influence on the bending rigidity in the case of tapered polymers. A detailed evaluation can be found in Ref. [5]. In both cases, the added polymers are very similar in structure and size. The typical end-to-end distance of a block of PEO or PEP with a molecular weight of 10 kg/mol is around 10 nm. The domain size d (~2π/q) is around 20 nm in the studied microemulsions. The difference between NSE and SANS results can be rationalized by the different time scales of the measurements. In the case of NSE, the dynamics is probed on the nanosecond time scale. There, the anchoring points of the tapered copolymers stay at the same position and the diblock copolymers act in a very similar way as the non-tapered equivalents as a "booster" for the microemulsion. SANS on the other hand probes the static structure, i.e. an average over very long time scales. There the tapered copolymers can slide through the membrane, the anchoring point moves to one side of the copolymer chain or to the other. On the different time scales, the copolymers have therefore different signatures on the membrane conformation with an apparent stronger membrane fluctuation due to the possibly different block sizes of the copolymer on both sides. This is also the reason for the suppression of the lamellar phase of the microemulsion.

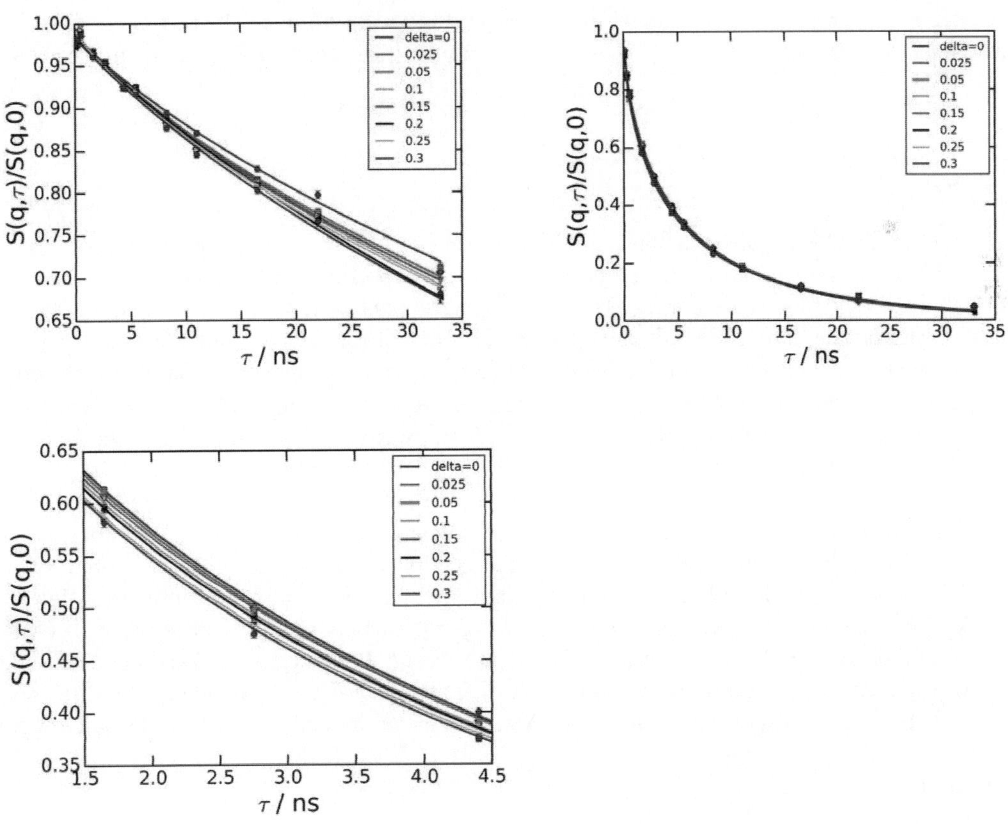

Fig. 3: S(q,t) from NSE experiments for a) q=0.06 (top left) and b) q=0.17 (top right) as well as a zoom of b) (bottom left, delta increases from top to bottom of the curves) at intermediate Fourier times. Tiny concentration variations of the tapered polymer lead to small but nonetheless systematic variations in the relaxation rate.

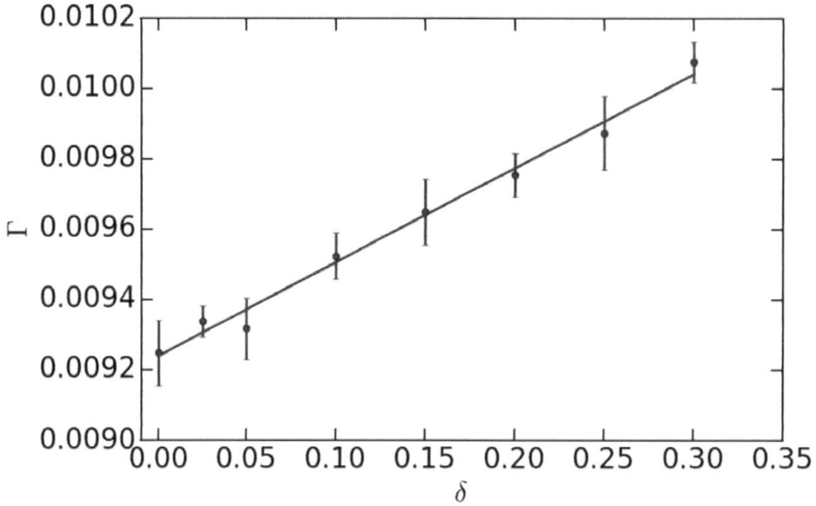

Fig. 4: Relaxation rate Γ as a function of tapered polymer concentration δ. The rate is given as the dimensionless quantity $\Gamma = (\Gamma_0 / q^3)\ (\eta/k_BT)$ corrected for the q-dependence and viscosity (see [10]).

4. Conclusion

Small changes in the elastic constants have significant influence on macroscopic properties of complex fluids such as microemulsions. High precision NSE experiments are needed to provide the relaxation rate with the required accuracy. The stable operation and magnetic shielding of the SNS-NSE instrument provides a magnetically very stable environment for such measurements. As an example, the reason of the suppression for lamellar phases by tapered copolymers has been studied. Probing the bending rigidity κ on different time scales, i.e. the nanosecond time scale with NSE and the seconds time scale with the static SANS experiments, allowed to observe the effect of a slowly sliding anchoring point of the tapered copolymer, where NSE probes the situation where the copolymer is at the same place in the relaxation time scale of S(q,τ), while the static structure of SANS probes different local chain configurations where the anchoring point slides across the surfactant membrane. This sliding behavior of the tapered polymers is the reason for the suppression of the lamellar phase up to high surfactant concentrations. A microscopically small effect results therefore in a macroscopically useful change of phase behavior. Many scientific questions concerning thermal fluctuations deal with small effects as in our case the small change in surfactant concentration, or as an other example domain motions in proteins, which are observed as a change in the diffusion constant [11]. The present experiments show that NSE is capable of dealing whith such small effects.

5. References

[1] Holderer O. et al, Soft Matter **9**, 2308-2313 (2013)
[2] Endo H., et al., Phys. Rev. Lett. **85**, 102-105 (2000)
[3] Gompper G., et al, Europhys. Lett. **56**, 683-689 (2001)
[4] Zhang W., et al., Macromolecules **46**, 3931 – 3938 (2013)
[5] Klemmer H.F.M., et al., in preparation

[6] Ohl M., Nucl. Instr. Meth. A **696**, 85 – 99 (2012)
[7] Teubner, M, Strey, R., J. Chem. Phys. **87**, 3195 – 3200 (1987)
[8] Zilman A.G., Granek R., Phys. Rev. Lett **77**, 4788 – 4791 (1996)
[9] Mihailescu M., et al., J. Chem. Phys. **115**, 9563 – 9577 (2001)
[10] Monkenbusch M., et al., J.Phys.: Condens. Matter, **17**, S2903 – S2909 (2005)
[11] Stadler A., et al., J. Am. Chem. Soc., **136**, 6987 – 6994 (2014)

IOP Conf. Series: Journal of Physics: Conf. Series **862** (2017) 012015 doi:10.1088/1742-6596/862/1/012015

A framework for modeling polarized neutron scattering from NMR spin-modulated systems

M Kotlarchyk, G M Thurston

School of Physics and Astronomy, Rochester Institute of Technology, Rochester, New York, 14623, USA

E-mail: mnksps@rit.edu,gmtsps@rit.edu

Abstract. In this work we study the potential for utilizing the scattering of polarized neutrons from nuclei whose spin has been modulated using nuclear magnetic resonance (NMR). From first principles, we present an in-depth development of the differential scattering cross-sections that would arise in such measurements from a hypothetical target system containing nuclei with non-zero spins. In particular, we investigate the modulation of the polarized scattering cross-sections following the application of radio frequency pulses that impart initial transverse rotations to selected sets of spin-1/2 nuclei. The long-term aim is to provide a foundational treatment of the scattering cross section associated with enhancing scattering signals from selected nuclei using NMR techniques, thus employing minimal chemical or isotopic alterations, so as to advance knowledge of macromolecular or liquid structure.

1. Introduction

In this work we continue the investigation of the potential for manipulating nuclear spins by nuclear magnetic resonance (NMR) techniques in order to enhance scattering of polarized neutrons from selected nuclei, a strategy that was proposed and described previously by Buckingham [1]. That work broadly considered the overall feasibility and utility of such experiments. If scattering from NMR-selected nuclei can be robustly accomplished, it could serve as an extremely useful probe of soft-matter and other liquid systems. Our goal here is to develop a sound, foundational framework for calculating the neutron scattering cross-sections associated with such experiments. Establishing these cross sections will be a basis for designing and evaluating experimental strategies to measure the desired signals, which are likely to be very small. The needed framework should be rigorous, yet flexible enough to address a variety of experimental strategies.

The present project was motivated by two pressing needs of very long standing. One is the need to create methods of more directly *measuring* the probabilistic, molecular orientation-dependent structure of liquids and liquid mixtures. While beautiful theoretical techniques exist to model liquid structure and other properties in terms of intermolecular potentials [2, 3, 4], the uncertainties presently associated with accurately knowing the intermolecular potentials in the first place [5, 6] call for the development of ways to actually *measure* the intermolecular potentials in undisturbed liquid samples. Another motivating need is that of supplementing current methods of measuring the structure of biological macromolecules.

The basic physical idea behind the present work is to use NMR to take advantage of the nuclear spin-dependent cross sections for neutron scattering. Can one "light up" the nuclei of

IOP Conf. Series: Journal of Physics: Conf. Series **862** (2017) 012015 doi:10.1088/1742-6596/862/1/012015

one's choice using NMR to prepare the spin states, then use neutron scattering to see how far apart the selected nuclei are? A key challenge is that very small differences in neutron scattering signals would be associated with the common, very small degrees of nuclear spin polarization. As pointed out in Reference [1], a typical polarization of the sample at room temperature is only on the order of 10^{-5}. Can small-angle neutron scattering be made sensitive enough to detect these signals?

The present work is intended to lay the groundwork for making *quantitative* evaluations of the sensitivity of strategies proposed to achieve useful NMR-modulated, spin-polarized neutron scattering. By having suitable expressions, built from first principles, for the fundamental scattering cross sections, one can eventually be in a position to quantify the signal-to-noise ratio that would be expected from proposed experimental designs. To start assembling the needed elements we have made a relatively simple choice of NMR modulation, as we describe below.

2. Static scattering cross-section and spin-space decoupling approximation

We begin with a useful form for the ensemble-average elastic cross section for the scattering of polarized neutrons from a spin-polarized target system [7, 8],

$$\frac{d\sigma^{ss'}}{d\Omega}\left(\vec{Q}\right) = \left\langle\left\langle \sum_{i,j} \left(\hat{b}_i^{ss'}\right)^{\dagger}\left(\hat{b}_j^{ss'}\right) e^{-i\vec{Q}\cdot\vec{r}_i} e^{+i\vec{Q}\cdot\vec{r}_j} \right\rangle\right\rangle \tag{1}$$

where $\vec{Q} = \vec{k} - \vec{k}'$ is the wavevector transfer between the initial neutron wavevector $\vec{k}$ and the final wavevector $\vec{k}'$, $\vec{r}_j$ denotes the position of target nucleus j, and $\hat{b}_j^{ss'}$ is the scattering-length operator for target nucleus j that connects the incident neutron spin state s with the final neutron spin state s'. In Eq. 1, we have denoted the ensemble average by double brackets (<< ... >>) to emphasize the fact that it is an average over *both* the spins and the positions of the nuclei.

Denoting the polarization-state of the neutron $|s\rangle$ by either $|+\rangle$, for spin up, or by $|-\rangle$, for spin down, the scattering length operators for the jth nucleus can be written in terms of its z-component, raising, and lowering angular momentum operators $\hat{I}_z$, $\hat{I}_+$, and $\hat{I}_-$, respectively, as follows [7]:

$$\hat{b}_j^{++} = \left(A + B\hat{I}_z\right)_j; \quad \hat{b}_j^{--} = \left(A - B\hat{I}_z\right)_j; \quad \hat{b}_j^{+-} = \left(B\hat{I}_+\right)_j; \quad \hat{b}_j^{-+} = \left(B\hat{I}_-\right)_j. \tag{2}$$

in which, for any one particular target nucleus, the constants A and B can be expressed in terms of the two scalar scattering lengths b^+ and b^-, which correspond to the cases where the total spin quantum numbers, t, of the neutron-nucleus system are $t_+ = I + \frac{1}{2}$ and $t_- = I - \frac{1}{2}$, respectively. Specifically,

$$A_j = \frac{(I_j + 1)\, b_j^+ + I_j b_j^-}{2I_j + 1} \quad \text{and} \quad B_j = \frac{b_j^+ - b_j^-}{2I_j + 1}, \tag{3}$$

in which I_j is the total spin quantum number for nucleus j. The utility of the expressions in Eqs. 2 and 3 stems from the fact that b^+ and b^- values have been determined experimentally and tabulated for various nuclei [7, 8].

In the following we develop the cross section given by Eq. 1 when the spin states of the nuclei respond to an applied magnetic field, to various NMR pulses, and to other factors that affect their degree of polarization. For simplicity in the present treatment we assume that the responses of different nuclei to the imposed external fields are independent of one another.

IOP Conf. Series: Journal of Physics: Conf. Series **862** (2017) 012015 doi:10.1088/1742-6596/862/1/012015

Technically, as will be developed below, this is equivalent to assuming that the density operators for the nuclear spin states factor into products of density operators for each nucleus, and that the nuclear spin states are uncorrelated with their *relative* positions. However, it is important to note that spin states of different nuclei can nevertheless be closely related to one another, owing to their commonality of response to applied fields and other sample conditions.

By making the decoupling approximations just described, for the time being we postpone the more complicated description of polarized neutron scattering that will be needed when nuclear spins are linked by one or a few covalent bonds by scalar, or J-coupling, and when nuclear spins are directly coupled by through-space spin-spin interactions. Also, for the present purpose, we neglect the coupling with the surroundings ("lattice") that will lead to longitudinal and transverse relaxation of the nuclear spin polarizations. Due in part to the spatial proximity needed for the coupling effects, the associated spin and position variables have correlations that the present analysis does not consider.

Consistent with the present decoupling approximation, we rewrite Eq. 1 to distinguish between the spin average, denoted by an overline $\overline{(...)}$ and the spatial average, denoted by <>.

$$\frac{d\sigma^{ss'}}{d\Omega}\left(\vec{Q}\right) = \left\langle \sum_{i,j} \overline{\left(\hat{b}_i^{ss'}\right)^{\dagger}\left(\hat{b}_j^{ss'}\right)} e^{+i\vec{Q}\cdot(\vec{r}_j-\vec{r}_i)} \right\rangle \tag{4}$$

At this point, it is interesting to note that each of the indices i and j enters into *both* the spin *and* the spatial averages that appear in Eq. 4. As a consequence, despite the assumption that the spin states of different nuclei are independent of their relative positions, the results of evaluating the spin averages of the cross sections, for various types of polarized neutron scattering from an NMR-modulated sample, can in principle lead to methods of gaining information about the relative positions $\vec{r}_{ji} = \vec{r}_j - \vec{r}_i$ of NMR-selected nuclei [1].

We briefly remark on a possibly puzzling fact, namely, that we anticipate that *a posteriori* knowledge of *actual* spatial correlations between nuclear positions can in principle be gained from analysis of the dependence of polarized neutron scattering on their joint spin states. It is important to note that these *deduced* spatial correlations, which result from the interactions of the nuclei with incident neutrons, do not contradict the *physical* assumption of decoupling of target nuclear spin states from their relative positions.

We now consider further the spin-averaged scattering-length product that appears in Eq. 4. This can be expressed in terms of the joint spin-state density-operator for spins i and j, $\hat{\rho}_{\mathrm{ij,spin}}$:

$$\overline{\left(\hat{b}_i^{ss'}\right)^{\dagger}\left(\hat{b}_j^{ss'}\right)} = Tr\left[\hat{\rho}_{\mathrm{ij,spin}}\left(\hat{b}_i^{ss'}\right)^{\dagger}\left(\hat{b}_j^{ss'}\right)\right] \tag{5}$$

Again consistent with our present neglect of J and direct coupling, we assume, for different nuclei, that the joint spin-state density-operator can be written as a product of spin-density operators for spins i and j: $\hat{\rho}_{\mathrm{ij,spin}} = \hat{\rho}_{\mathrm{i,spin}}\hat{\rho}_{\mathrm{j,spin}}$, in which $\hat{\rho}_{\mathrm{i,spin}}$ is the single-nucleus density matrix for nucleus i, and has the property $Tr[\hat{\rho}_{\mathrm{i,spin}}] = 1$. On the other hand, if $i = j$, we have $\hat{\rho}_{\mathrm{ij,spin}} = \hat{\rho}_{\mathrm{i,spin}}$. Equivalently, $\hat{\rho}_{\mathrm{ij,spin}} = (\hat{\rho}_{\mathrm{i,spin}} - \hat{\rho}_{\mathrm{i,spin}}\hat{\rho}_{\mathrm{j,spin}})\delta_{\mathrm{ij}} + \hat{\rho}_{\mathrm{i,spin}}\hat{\rho}_{\mathrm{j,spin}}$. Consequently,

$$\begin{aligned}\overline{\left(\hat{b}_i^{ss'}\right)^{\dagger}\left(\hat{b}_j^{ss'}\right)} &= Tr\left[\hat{\rho}_i\hat{\rho}_j\left(\hat{b}_i^{ss'}\right)^{\dagger}\left(\hat{b}_j^{ss'}\right)\right] = Tr_{(i)}\left[\hat{\rho}_i\left(\hat{b}_i^{ss'}\right)^{\dagger}\right]Tr_{(j)}\left[\hat{\rho}_j\left(\hat{b}_j^{ss'}\right)\right]\ (i \neq j)\\ &= Tr_{(i)}\left[\hat{\rho}_i\left(\hat{b}_i^{ss'}\right)^{\dagger}\left(\hat{b}_i^{ss'}\right)\right]\ (i = j),\end{aligned} \tag{6}$$

in which we have omitted the "spin" designation from the density operators for clarity, and in which $Tr_{(i)}$ represents a "reduced" trace over the spin states for nucleus i alone. The two cases can be written together as

$$\overline{(\hat{b}_i^{ss'})^\dagger(\hat{b}_j^{ss'})} = \Big\{Tr_{(i)}\Big[\hat{\rho}_i(\hat{b}_i^{ss'})^\dagger(\hat{b}_i^{ss'})\Big] - Tr_{(i)}\Big[\hat{\rho}_i(\hat{b}_i^{ss'})^\dagger\Big]Tr_{(j)}\Big[\hat{\rho}_j(\hat{b}_j^{ss'})\Big]\Big\}\delta_{ij} + Tr_{(i)}\Big[\hat{\rho}_i(\hat{b}_i^{ss'})^\dagger\Big]Tr_{(j)}\Big[\hat{\rho}_j(\hat{b}_j^{ss'})\Big]. \tag{7}$$

However, because terms for which $i = j$ are the only ones that contribute to the portion in curly brackets, we can replace j by i in that term to obtain

$$\overline{(\hat{b}_i^{ss'})^\dagger(\hat{b}_j^{ss'})} = \Big\{Tr_{(i)}\Big[\hat{\rho}_i(\hat{b}_i^{ss'})^\dagger(\hat{b}_i^{ss'})\Big] - Tr_{(i)}\Big[\hat{\rho}_i(\hat{b}_i^{ss'})^\dagger\Big]Tr_{(i)}\Big[\hat{\rho}_i(\hat{b}_i^{ss'})\Big]\Big\}\delta_{ij} + Tr_{(i)}\Big[\hat{\rho}_i(\hat{b}_i^{ss'})^\dagger\Big]Tr_{(j)}\Big[\hat{\rho}_j(\hat{b}_j^{ss'})\Big]. \tag{8}$$

We now define coherent and incoherent cross sections as follows:

$$(b_{ij}^{ss'})^2_{\text{coh}} = Tr_{(i)}\Big[\hat{\rho}_i(\hat{b}_i^{ss'})^\dagger\Big]Tr_{(j)}\Big[\hat{\rho}_j(\hat{b}_j^{ss'})\Big] \tag{9}$$

$$(b_i^{ss'})^2_{\text{inc}} = Tr_{(i)}\Big[\hat{\rho}_i(\hat{b}_i^{ss'})^\dagger(\hat{b}_i^{ss'})\Big] - Tr_{(i)}\Big[\hat{\rho}_i(\hat{b}_i^{ss'})^\dagger\Big]Tr_{(i)}\Big[\hat{\rho}_i(\hat{b}_i^{ss'})\Big], \tag{10}$$

in terms of which

$$\overline{(\hat{b}_i^{ss'})^\dagger(\hat{b}_j^{ss'})} = (b_i^{ss'})^2_{\text{inc}}\delta_{ij} + (b_{ij}^{ss'})^2_{\text{coh}} \tag{11}$$

Inserting Eq. 11 into Eq. 4 puts the differential scattering cross-section in the form

$$\frac{d\sigma^{ss'}}{d\Omega}\left(\vec{Q}\right) = \left[(b_{\text{inc}}^{ss'})^2 + \left\langle \sum_{ij}(b_{ij}^{ss'})^2_{\text{coh}}e^{i\vec{Q}\cdot\vec{r}_{ji}}\right\rangle\right] \tag{12}$$

in which $\vec{r}_{ji} = \vec{r}_j - \vec{r}_i$ and $(b_{\text{inc}}^{ss'})^2 = \sum_i (b_i^{ss'})^2_{\text{inc}}$. Eq. 12, in combination with the expressions in Eqs. 2, 3, 9, and 10, allows for calculation of both the $\vec{Q}$-independent incoherent cross section and the coherent polarized neutron scattering cross-section from a spin-polarized target, all within the present spin-space decoupling approximation.

3. Spin-density operator for a prototypical NMR-SANS experiment

We now use Eq. 12 to calculate the cross-sections resulting from a prototypical, candidate NMR spin-modulation scenario. This requires a model of the spin-density operators of the system needed to evaluate Eqs. 9 and 10. In the present work we restrict our analysis of the static cross section to density operators that are independent of time, which, nevertheless, we will need to have evolved forward in time starting from an equilibrium or other steady-state situation. The density operator at time $t = 0$ represents the probabilities p_+ and p_- that a spin-1/2 nucleus will initially be spin-up and spin-down, respectively, along along the positive z-axis ($p_+ + p_- = 1$). An NMR pulse will subsequently be applied to the sample, and we need to calculate how the density operator changes as a result.

$$\hat{\rho}(0) = \begin{pmatrix} p_+ & 0 \\ 0 & p_- \end{pmatrix} = \begin{pmatrix} p_+ & 0 \\ 0 & (1-p_+) \end{pmatrix} = \frac{1}{2}\hat{\mathbb{I}} + P\left(\frac{1}{2}\right)\begin{pmatrix} 1 & 0 \\ 0 & -1 \end{pmatrix} = \frac{1}{2}\hat{\mathbb{I}} + P\hat{I}_z \tag{13}$$

where $\hat{\mathbb{I}}$ is the identity operator and $P = (+1)p_+ + (-1)p_- = 2p_+ - 1$ is the polarization state of the target nucleus ($-1 \leq P \leq +1$). This form of $\hat{\rho}$, where the off-diagonal elements are zero, is applicable if the phases of the quantum-mechanical spin states of the target nuclei are randomly distributed, and would hold, for example, for a statistical mixture of up and down states in thermal equilibrium.

We now find the single-nucleus density operator $\hat{\rho}_i$ immediately after a radio-frequency (RF) pulse that is applied to a nucleus that is initially in polarization state P_i, as described by the density operator in Eq. 13. The derivation follows a pattern whose physical basis is given in a number of books, such as [8, 9, 10, 11, 12].

We now briefly describe the nature of the derivation, which will be detailed more completely elsewhere. For nucleus j, one first transforms the density operator $\hat{\rho}$ in the laboratory frame to the corresponding density operator $\hat{\rho}_R$ in a frame that is rotating at the RF frequency ω. We assume that the RF frequency is tuned near the relevant Larmor resonant frequency of the nucleus, $\omega_{0,i} = \gamma_i H_0(1-\sigma_i)$, in which γ_i is the gyromagnetic ratio of the ith nucleus, H_0 is the strength of the static magnetic field applied to the sample, taken to be in the z-direction, and σ_i is the chemical shift of nucleus i. This transformation to the rotating frame takes the form $\hat{\rho}_R = \hat{R}^{-1}\hat{\rho}\hat{R}$, in which $\hat{R} = \exp(i\omega t \hat{I}_z)$. In the rotating frame, the effective Hamiltonian is time-independent, allowing for the treatment below.

At time $t = 0$, clearly $\hat{\rho}_R(0) = \hat{\rho}(0)$. At the end of the pulse, which we take to have duration t_p, $\hat{\rho}_R(t_p) = \hat{T}^{-1}\hat{\rho}_R(0)\hat{T}$, in which $\hat{T} = \exp(i\hat{H}_{\text{eff}}t_p/\hbar)$ and in turn, $\hat{H}_{\text{eff}} \approx -\gamma_j \hbar H_{\text{RF}}(1-\sigma_j)\hat{I}_x$, in which H_{RF} is the strength of the radio-frequency magnetic field and $\hat{I}_x$ is the x-component angular momentum operator for nucleus j; this form of $\hat{H}_{\text{eff}}$ assumes that the pulse is in the negative x-direction in the rotating frame. The effect of the pulse can be summarized by the so-called "spin-flip" frequency, $(\omega_{\text{flip}})_i = \gamma_i H_{\text{RF}}(1-\sigma_i)$; in particular, as a result of the pulse, the direction of the expected nuclear spin polarization rotates by an angle $\theta_i = (\omega_{\text{flip}})_i t_p$ from the z-axis. Finally, one transforms the density operator back to the laboratory frame, that is, the frame in which we are calculating neutron scattering cross sections, with use of $\hat{\rho}(t_p) = \hat{R}\hat{\rho}_R(t_p)\hat{R}^{-1}$.

Immediately following the pulse, the density operator in the laboratory frame can be shown by the full derivation to become

$$\hat{\rho}_i = \frac{1}{2}\hat{\mathbb{I}}_i + P_i \sin\theta_i\Big[(\hat{I}_x)_i \sin(\omega t_p) + (\hat{I}_y)_i \cos(\omega t_p)\Big] + P_i \cos\theta_i (\hat{I}_z)_i \tag{14}$$

or, in terms of the raising and lowering operators for nucleus i, and using $\omega t_p = (\omega/(\omega_{\text{flip}})_i)\theta_i$,

$$\hat{\rho}_i = \frac{1}{2}\hat{\mathbb{I}}_i + \frac{i}{2}P_i \sin\theta_i\Big[e^{-i(\omega/(\omega_{\text{flip}})_i)\theta_i}(\hat{I}_-)_i - e^{+i(\omega/(\omega_{\text{flip}})_i)\theta_i}(\hat{I}_+)_i\Big] + P_i \cos\theta_i (\hat{I}_z)_i. \tag{15}$$

Use of the raising and lowering operators, as in Eq. 15, greatly simplifies the evaluation of the traces over spin states needed to evaluate the scattering cross-sections in Eqs. 9 and 10.

4. Polarized scattering cross-sections for a prototypical NMR-SANS experiment

We now envision a hypothetical scenario, depicted schematically in Fig. 1, in which a collection of nuclei has been prepared with NMR so that it is well-modeled by the spin density operator given in Eq. 15. The sample is then quickly removed from the magnetic field H_0, and polarized neutron scattering is performed during the time interval within which the selected nuclei in the sample remain sufficiently polarized. The sample is then returned to the NMR apparatus, and the cycle is repeated. As mentioned above, we postpone quantitative consideration of the relaxation effects that will affect the evolution of the density operator subsequent to times when Eq. 15 is a good model.

We now use Eq. 15 to calculate the neutron scattering cross sections for the various cases of incident neutron spin ($s = \pm$) and detected neutron-spin ($s' = \pm$). By keeping the subscript i on the density operator for each of the distinct target nuclei, including the initial rotation angle θ_i,

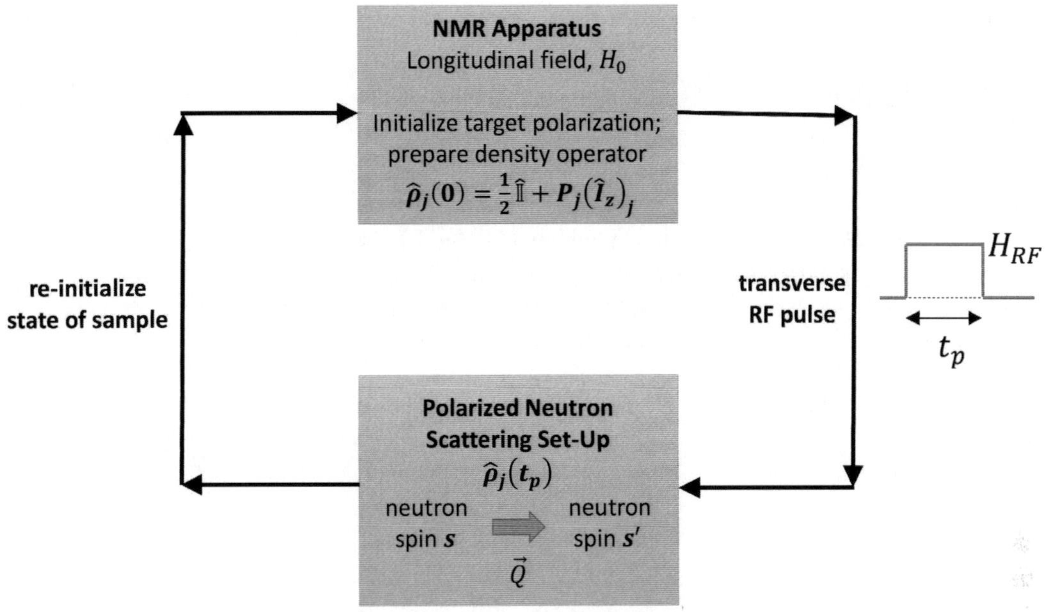

Figure 1. A possible measurement scenario; see text.

we retain the ability of the formalism to model distinct scattering cross sections that correspond to nuclei of different chemical elements, or to nuclei that have differing chemical shifts, or that have undergone distinct NMR preparations.

For $s = +$ and $s' = +$, or the $(++)$-case, from Eqs. 2, 9, and 10, we need to evaluate

$$Tr_{(j)}[\hat{\rho}_j(\hat{b}_j^{++})] = Tr_{(j)}[\hat{\rho}_j(A_j + B_j(\hat{I}_z)_j)] \tag{16}$$

and

$$Tr_{(i)}[\hat{\rho}_i(\hat{b}_i^{++})^\dagger(\hat{b}_i^{++})] = Tr_{(i)}[\hat{\rho}_i(A_i + B_i(\hat{I}_z)_i)(A_i + B_i(\hat{I}_z)_i)] \tag{17}$$

Upon evaluating the result of substituting the expression for $\hat{\rho}_j$ from Eq. 15 into Eq. 16, evaluating the trace, then using Eq. 9, one obtains

$$(b_{ij}^{++})^2_{\rm coh} = (A_i + \frac{1}{2}B_iP_i\cos\theta_i)(A_j + \frac{1}{2}B_jP_j\cos\theta_j). \tag{18}$$

By similarly evaluating the trace in Eq. 17 and using that result together with the one from Eq. 16 in Eq. 9, one obtains

$$(b_i^{++})^2_{\rm inc} = \frac{1}{4}B_i^2(1 - P_i^2\cos^2\theta_i) \tag{19}$$

The details of this derivation will be published elsewhere. Upon inserting Eqs. 18 and 19 into Eq. 12, the $(++)$ cross section takes the form

$$\frac{d\sigma^{++}}{d\Omega}\left(\vec{Q}\right) = \frac{1}{4}\sum_i B_i^2(1 - P_i^2\cos^2\theta_i) + \left\langle \sum_{ij}(A_i + \frac{1}{2}B_iP_i\cos\theta_i)(A_j + \frac{1}{2}B_jP_j\cos\theta_j)e^{i\vec{Q}\cdot\vec{r}_{ji}} \right\rangle \tag{20}$$

Because of the similar forms of the respective scattering length operators (Eq. 2), the $(--)$ cross section can be obtained by replacing B_i in the $(++)$ cross section of Eq. 20 with $-B_i$, giving

$$\frac{d\sigma^{--}}{d\Omega}\left(\vec{Q}\right) = \frac{1}{4}\sum_i B_i^2(1-P_i^2\cos^2\theta_i) + \left\langle \sum_{ij}(A_i - \frac{1}{2}B_iP_i\cos\theta_i)(A_j - \frac{1}{2}B_jP_j\cos\theta_j)e^{i\vec{Q}\cdot\vec{r}_{ji}} \right\rangle \tag{21}$$

We find that the second terms on the right-hand sides of Eqs. 20 and 21, when specialized to θ_i and $\theta_j = 0$ or π, are consistent with the result obtained by Buckingham, as specified in the second term on the right-hand side of Eq. 4 of reference [1], with use of Eq. 2 of that paper; we note that reference 1 considered only the effect of π-pulses. However, we also find that the first, or incoherent terms in Eqs. 20 and 21 differ from the corresponding term in Eq. 4 of [1]. Details will be given in another publication.

Eqs. 20 and 21 reduce correctly to the cross sections for neutron scattering from an unpolarized target by setting the polarizations P_i, which are normally very small, to 0. In that case, one has

$$(b_{ij}^{++})^2_{\text{coh}} = (b_{ij}^{--})^2_{\text{coh}} = A_iA_j \quad \text{and} \quad (b_i^{++})^2_{\text{inc}} = (b_i^{--})^2_{\text{inc}} = \frac{1}{4}B_i^2 \quad (\text{for } P_i = P_j = 0) \tag{22}$$

giving

$$\frac{d\sigma^{++}}{d\Omega}\left(\vec{Q}\right) = \frac{d\sigma^{--}}{d\Omega}\left(\vec{Q}\right) = \frac{1}{4}\sum_i B_i^2 + \left\langle \sum_{ij} A_iA_je^{i\vec{Q}\cdot\vec{r}_{ji}} \right\rangle \quad (\text{for } P_i = P_j = 0). \tag{23}$$

For N unpolarized target nuclei of a single spin $\frac{1}{2}$ species, Eq. 23 reduces further to the expected result for the no spin-flip case for scattering from an unpolarized target [7, 8, 2]

$$\begin{aligned}\frac{d\sigma^{++}}{d\Omega}\left(\vec{Q}\right) = \frac{d\sigma^{--}}{d\Omega}\left(\vec{Q}\right) &= \frac{1}{3}Nb_{\text{inc}}^2 + \left\langle \sum_{ij} A^2e^{i\vec{Q}\cdot\vec{r}_{ji}} \right\rangle \\ &= \frac{1}{3}Nb_{\text{inc}}^2 + NA^2\left\langle \frac{1}{N}\sum_{ij} e^{i\vec{Q}\cdot\vec{r}_{ji}} \right\rangle \\ &= \frac{1}{3}Nb_{\text{inc}}^2 + NA^2S(\vec{Q})\end{aligned} \tag{24}$$

in which $S(\vec{Q})$ is the static structure factor, and in which we have used the definition [8]

$$b_{\text{inc}}^2 = I(I+1)B^2 = \frac{3}{4}B^2. \tag{25}$$

For the (+-) case, i.e. neutron spin-flip scattering, evaluation of the needed traces gives

$$\left(b_{ij}^{+-}\right)^2_{coh} = \frac{1}{4}B_iB_jP_iP_j\sin\theta_i\sin\theta_j \tag{26}$$

and

$$\left(b_i^{+-}\right)^2_{inc} = \frac{1}{2}B_i^2\left[1 - P_i\cos\theta_i - \frac{1}{2}P_i^2\sin^2\theta_i\right]. \tag{27}$$

Again, details of this derivation will be published elsewhere. Using Eqs. 26 and 27 in Eq. 12 gives the scattering cross section

$$\frac{d\sigma^{+-}}{d\Omega} = \frac{1}{2}\sum_i B_i^2\left(1 - P_i\cos\theta_i - \frac{1}{2}P_i^2\sin^2\theta_i\right) + \frac{1}{4}\left\langle \sum_{i,j} B_iB_jP_iP_j\sin\theta_i\sin\theta_je^{i\vec{Q}\cdot\vec{r}_{ji}} \right\rangle \tag{28}$$

For (-+) neutron spin-flip scattering, one has

$$\left(b_{i,j}^{-+}\right)_{coh}^{2} = \frac{1}{4}B_iB_jP_iP_j\sin\theta_i\sin\theta_j \tag{29}$$

and

$$\left(b_{i}^{-+}\right)_{inc}^{2} = \frac{1}{2}B_i^2\left[1 + P_i\cos\theta_i - \frac{1}{2}P_i^2\sin^2\theta_i\right]. \tag{30}$$

giving

$$\frac{d\sigma^{-+}}{d\Omega} = \frac{1}{2}\sum_i B_i^2(1 + P_i\cos\theta_i - \frac{1}{2}P_i^2\sin^2\theta_i) + \frac{1}{4}\left\langle \sum_{i,j} B_iB_jP_iP_j\sin\theta_i\sin\theta_j e^{i\vec{Q}\cdot\vec{r}_{ji}}\right\rangle \tag{31}$$

For the zero-polarization case, Eqs. 28 and 31 both give the expected result for spin-flip scattering, which is solely incoherent:

$$\frac{d\sigma^{+-}}{d\Omega} = \frac{d\sigma^{-+}}{d\Omega} = \frac{1}{2}\sum_i B_i^2; \tag{32}$$

for a single species the cross-sections in Eq. 32 become $(2/3)Nb_{\text{inc}}^2$, where we have used Eq. 25.

5. Obtaining structure factors of NMR-selected nuclei

With use of the cross sections detailed in Eqs. 20, 21, 28, and 31, we now consider possible experiments designed so that sums and differences of their results may give useful information about NMR-selected partial structure factors.

We note first that an experiment that measures the difference between the two no-spin-flip cross sections gives structural information for which the polarization of only one of the nuclei in each pair of nuclei enters:

$$\left[\left(\frac{d\sigma^{++}}{d\Omega}\right) - \left(\frac{d\sigma^{--}}{d\Omega}\right)\right] = 2\langle \sum_{i\neq j} A_jB_iP_i\cos\theta_i\cos(Q\cdot\vec{r}_{ji})\rangle \tag{33}$$

A difference experiment that uses the two spin-flip cross sections yields a constant background:

$$\begin{aligned}\left[\left(\frac{d\sigma^{+-}}{d\Omega}\right) - \left(\frac{d\sigma^{-+}}{d\Omega}\right)\right] = & \frac{1}{2}\langle \sum_i B_i^2\big[(1 + P_i\cos\theta_i - \frac{1}{2}P_i^2\sin^2\theta_i) \\ & -(1 - P_i\cos\theta_i - \frac{1}{2}P_i^2\sin^2\theta_i)\big]\rangle \\ = & \langle \sum_i B_i^2P_i\cos\theta_i\rangle \end{aligned} \tag{34}$$

For the no-spin-flip cases, if one applies a $\pi/2$-pulse to all types of nuclei, the scattering cross-sections are not sensitive to the nuclear polarization states, being given by an expression that matches that in Eq. 23:

$$\left[\left(\frac{d\sigma^{++}}{d\Omega}\right)\right]_{\theta_i=\theta_j=\pi/2} = \left[\left(\frac{d\sigma^{--}}{d\Omega}\right)\right]_{\theta_i=\theta_j=\pi/2} = \frac{1}{4}\sum_i B_i^2 + \langle \sum_{i,j} A_iA_je^{i\vec{Q}\cdot\vec{R}_{ij}}\rangle \tag{35}$$

We now consider the *sum* experiment for the no spin-flip scattering cases:

$$\begin{aligned}\Big[\Big(\frac{d\sigma^{++}}{d\Omega}\Big)+\Big(\frac{d\sigma^{--}}{d\Omega}\Big)\Big] &= \frac{1}{2}\sum_i B_i^2\big[(1-P_i^2\cos^2\theta_i) \\ &+\Big\langle\sum_{ij}\big[(A_i+\frac{1}{2}B_iP_i\cos\theta_i)(A_j+\frac{1}{2}B_jP_j\cos\theta_j)e^{i\vec{Q}\cdot\vec{r}_{ji}} \\ &+(A_i-\frac{1}{2}B_iP_i\cos\theta_i)(A_j-\frac{1}{2}B_jP_j\cos\theta_j)e^{i\vec{Q}\cdot\vec{r}_{ji}}\big]\Big\rangle \\ &= \frac{1}{2}\sum_i B_i^2(1-P_i^2\cos^2\theta_i) \\ &+\Big\langle\sum_{ij}(2A_iA_j+\frac{1}{2}B_iB_jP_iP_j\cos\theta_i\cos\theta_j)e^{i\vec{Q}\cdot\vec{r}_{ji}}\Big\rangle \qquad (36)\end{aligned}$$

Comparison of Eqs. 36 and 35 suggests that the following combination of the *sum* experiment with the no-spin-flip $\pi/2$-pulse experiment yields

$$\begin{aligned}\Big\{\frac{1}{2}\Big[\Big(\frac{d\sigma^{++}}{d\Omega}\Big)+\Big(\frac{d\sigma^{--}}{d\Omega}\Big)\Big]-\Big(\frac{d\sigma^{aa}}{d\Omega}\Big)_{\theta_i=\theta_j=\pi/2}\Big\} &= \frac{1}{4}\Big(\Big\langle\sum_{ij}B_iB_jP_iP_j\cos\theta_i\cos\theta_j e^{i\vec{Q}\cdot\vec{r}_{ji}}\Big\rangle \\ &-\sum_i B_i^2P_i^2\cos^2\theta_i\Big); \qquad (37)\end{aligned}$$

here the superscript 'aa' indicates the fact that the $\theta = \pi/2$ experiment is a no spin-flip experiment that could be either ++ or −−, or indeed the average of the two. Provided adequate signal strength, this or closely-related experiments could prove very useful indeed, because the structure factor incorporates the polarization of *each* nucleus of a pair, and thus points to the possibility of NMR control of which *pairs* of nuclei contribute to the scattering.

Finally, we consider another possible scenario: suppose one wants to "light up" scattering from a particular spin-$\frac{1}{2}$ nuclear species of just *one* type, call it α, e.g. ^{1}H, ^{15}N, or ^{13}C. To do so one could apply a sequence of shaped pulses that cause all the *other* types, β, to have $\theta_\beta = \pi/2$, so that $\cos\theta_\beta = 0$ (these "lie down"), while having little effect on the α-type nuclei. When the results of such experiments are combined as in Eq. 37, only terms having $B_i = B_\alpha$ will contribute. For simplicity we also assume that $P_i = P_\alpha$ and $\theta_i = \theta_j = \theta_\alpha$ for all remaining i and j of type α. Under these assumptions, Eq. 37 becomes

$$\begin{aligned}\Big\{\frac{1}{2}\Big[\Big(\frac{d\sigma^{++}}{d\Omega}\Big)+\Big(\frac{d\sigma^{--}}{d\Omega}\Big)\Big]-\Big(\frac{d\sigma^{aa}}{d\Omega}\Big)_{\theta_i=\theta_j=\pi/2}\Big\} &= \frac{1}{4}\Big(\Big\langle\sum_{ij}B_\alpha^2P_\alpha^2\cos^2\theta_\alpha e^{i\vec{Q}\cdot\vec{r}_{ji}}\Big\rangle \\ &-\sum_i B_\alpha^2P_\alpha^2\cos^2\theta_\alpha\Big) \\ &= \frac{1}{4}N_\alpha B_\alpha^2P_\alpha^2\cos^2\theta_\alpha\Big[S_{\alpha\alpha}(\vec{Q})-1\Big] \qquad (38)\end{aligned}$$

in which

$$S_{\alpha\alpha}(\vec{Q}) = \frac{1}{N_\alpha}\Big\langle\sum_{ij}e^{i\vec{Q}\cdot(\vec{r}_\alpha)_{ji}}\Big\rangle \qquad (39)$$

is the structure factor of the α-type nuclei, and N_α is the total number of such nuclei. We anticipate that the pulses needed to prepare experiments to which Eq. 38 is applicable can readily be produced because of the widely differing Larmor frequencies of the different nuclei. In principle, one could also imagine such a scheme applied to nuclei that have different chemical shifts, provided that sufficiently selective preparation pulses are practical, given other constraints on the measurement process.

6. Summary and Conclusions

In summary, we have developed first-principles expressions for the four types of polarized differential cross-sections that would arise in the scattering of polarized neutrons from nuclei whose spins have been modulated using NMR. In particular, we have considered an NMR pulse that rotates the expectation value of the nuclear spins through a chosen angle, starting from the direction along which both the nuclei and the incident neutrons are initially polarized. For each type of polarized scattering, we have derived generalized structure factors that explicitly show the linear and quadratic dependences of the scattering cross sections on the degree of polarization of each of the sample nuclei. In doing so we have also identified the incoherent and coherent contributions to the cross sections, which agree with the standard ones in the case of an unpolarized sample. Finally, we have considered potentially useful combinations of the four polarized scattering-experiment types that, in principle, could be used to measure NMR-selected partial structure factors.

Further work is needed in order to investigate the feasibility of NMR-modulated neutron scattering. First, it will be useful to calculate the polarized scattering cross sections that would result from use of selective NMR pulse shaping and sequencing. Second, it is important to find effective ways of optimizing methods of weak-signal detection that take full advantage of (Q, ω)-space, as well as temporal considerations relative to timing of pulses. Third, the formalism needs to be generalized and extended in order to robustly handle target T1/T2 relaxation considerations, as well as alternate experimental scenarios, such as performing neutron scattering while the target remains in the NMR fields. Fourth, it would be interesting to develop scattering cross-sections when the spin-decoupling approximations used here are removed.

References

[1] Buckingham A 2003 *Chemical Physics Letters* **371** 517–521
[2] Hansen J P and McDonald I 2006 *Theory of Simple Liquids* 3rd ed (Amsterdam: Elsevier Academic Press)
[3] Gray C G and Gubbins K E 1984 *Theory of Molecular Fluids, Volume 1: Fundamentals* (*International Series of Monographs on Chemistry* no 9) (Oxford: Oxford University Press)
[4] Gray C G, Gubbins K E and Joslin C G 2011 *Theory of Molecular Fluids, Volume 2: Applications* (*International Series of Monographs on Chemistry* no 10) (Oxford: Oxford University Press)
[5] Ninham B W and Lo Nostro P 2010 *Molecular Forces and Self Assembly: In Colloid, Nano Sciences and Biology* (Cambridge: Cambridge University Press)
[6] Israelachvili J N 2011 *Intermolecular and Surface Forces* 3rd ed (Amsterdam: Elsevier Academic Press)
[7] Squires G L 1996 *Introduction to the Theory of Thermal Neutron Scattering* (Mineola: Dover (Cambridge University Press, 1978))
[8] Chen S H and Kotlarchyk M 2007 *Interactions of Photons and Neutrons with Matter* 2nd ed (New Jersey: World Scientific)
[9] Slichter C P 1996 *Principles of Magnetic Resonance* 3rd ed vol 1 (Berlin: Springer Verlag)
[10] Goldman M 1988 *Quantum Description of High-Resolution NMR in Liquids* (Oxford: Oxford University Press)
[11] Cavanagh J, Fairbrother W J, Palmer III A G, Rance M and Skelton N J 2007 *Protein NMR Spectroscopy: Principles and Practice* 2nd ed (Amsterdam: Elsevier Academic Press)
[12] Levitt M H 2001 *Spin Dynamics: Basics of Nuclear Magnetic Resonance* (Chichester: John Wiley & Sons)

IOP Conf. Series: Journal of Physics: Conf. Series **862** (2017) 012016 doi:10.1088/1742-6596/862/1/012016

Neutron multimonochromator-bipolarizer based on magnetic multilayer Fe/Co and new scheme for the total neutron polarization analysis

V.G. Syromyatnikov[1,2], **Kyaw Zaw Lin**[1]

[1] Department of Physics, St. Petersburg State University, Ulyanovskaya, 1, Petrodvorets, St. Petersburg 198504, Russia

[2] Neutron Research Department, Petersburg Nuclear Physics Institute, National Research Center "Kurchatov Institute", Orlova Roscha, Gatchina, St. Petersburg 188300, Russia

syromyatnikov_vg@pnpi.nrcki.ru

Abstract. In this paper, we present a new neutron-optical element, Neutron Multimonochromator-Bipolarizer (NMB). It consists of a multimultilayer structure made of 12 periodic multilayer Fe/Co magnetic nanostructures whose period increases with distance from the substrate. Results are presented of calculations of the reflection coefficients from the NMB. We propose a new scheme of the total neutron polarization analysis for the time-of-flight method in the reflectometry. In this scheme, double NMB is used as a polarizer and there is no spin-flipper before the sample. NMB can be used in polarized neutron reflectometry, in SESANS, and for research of low-angle and inelastic scattering of polarized neutrons.

1. Introduction

In recent years the number of experiments on the study of surfaces and multilayer nanostructures by neutron reflectometry has significantly increased. In recent years, the number has increased significantly of neutron-reflectometry experiments on surfaces and multilayer nanostructure. Polarized neutron reflectometry [1] plays a unique role in the study of interphase boundaries and multilayer nanostructures. The latter are used increasingly in the electronics industry and in creating unique scientific equipment such as neutron multilayer monochromators, supermirror polarizers, and analyzers.

There is a number of works devoted to multilayer periodic nanostructures consisting of alternating layers of two different materials [2-9]. There is a Bragg peak of the 1st order on a curve of a reflectivity of a neutron beam from such a nanostructure. The magnitude of this peak is proportional to the square of the difference between the neutron-optical potentials of materials of the structure. If one of these materials is magnetic and another one is nonmagnetic, such a structure with suitable characteristics of materials can serve as a neutron multilayer monochromator-polarizer. The difference between neutron-optical potentials of the layers in the saturating magnetic field can be made sufficiently great for one spin component of the beam and negligibly small for another one. As a result, the reflectivity of the Bragg peak is close to 1 and to 0, in the former and in the latter cases,

IOP Conf. Series: Journal of Physics: Conf. Series **862** (2017) 012016 doi:10.1088/1742-6596/862/1/012016

respectively (see, e.g., [8, 9]). Therefore, the beam reflected from such a structure is monochromatic and polarized.

In [10], we studied a periodic magnetic multilayer Fe / Co nanostructure. In this structure, both materials are magnetic. Parameters of materials of layers for the saturating magnetic field are presented in Table 1.

Table 1. Parameters of materials of iron and cobalt layers. Here SLD (+) and SLD (-) are Scattering Length Density for spin-up and spin-down neutrons, correspondingly; σ_{abs} is the cross-section of absorption of neutrons by nuclei of material.

Parameters	**Fe**	**Co**
d (Å)	100	110
$SLD\ (+)\ \left(10^{10}\ cm^{-2}\right)$	13.13	6.45
$SLD\ (-)\ \left(10^{10}\ cm^{-2}\right)$	2.93	-1.98
$\sigma_{abs}\left(10^{-24}\ cm^{2}\right)$	*2.56*	*37.18*

A structure consisting of 20 pairs of alternating layers was created on a polished glass substrate. Thicknesses of the iron and cobalt layers were 100 Å and 110 Å, respectively. Reflection coefficients for (+) and (-) spin components of the neutron beam as a function of the momentum transfer were measured in such nanostructure for several values of the in-plane external magnetic field. Measurements were carried out using the time-of-flight method on neutron reflectometer NR-4M [11] (the reactor WWR-M PNPI NRC "KI"). The corresponding experimental data and theoretically obtained curves (solid lines) are shown in Fig. 1 for the saturating magnetic field H = 472 Oe [10]. The value of the reflectivity smaller than one below the critical edge and low values of the Bragg reflections of the 1st order are due to the absorption in cobalt layers and the roughness of the layers. The roughness of layers was 35 Å.

It follows from Fig. 1 that intensities of Bragg peaks of the 1st order are close for (+) and (-) spin component but their positions are shifted relative to one another. Furthermore, it was also found that intensities of these Bragg peaks are weakly dependent on the magnetic field. These unusual properties of Fe / Co nanostructures can be explained by the fact that differences are significant between the neutron optical potentials of iron and cobalt layers for (+) and (-) spin components of the beam [10]. Let us consider this point in more detail. As it is well known, in the kinematic approximation, the reflection coefficient of the Bragg peak of the 1st order of the multilayer monochromator-polarizer consisting of alternating layers of two kinds of materials depends, in particular, on the square of the difference between scattering length densities of adjacent layers [4]. It appears that this difference is slightly dependent on the magnetic field applied to the Fe / Co structure [10]. Therefore, Bragg peaks of the 1st order are virtually independent of the applied magnetic field in the (ideal) Fe / Co structure.

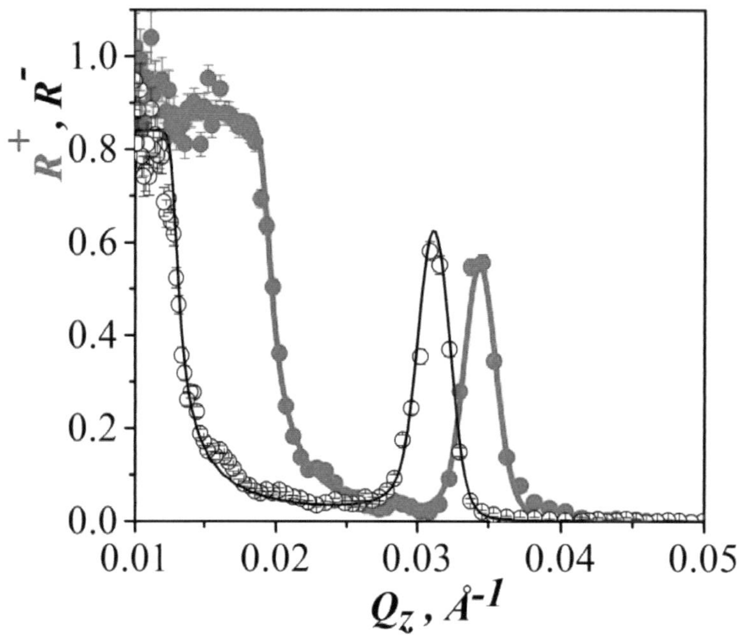

Figure 1. Reflection coefficients of both spin components of the neutron beam as a function of the normal component of the momentum transfer Q_z for the scattering from the multilayer periodic magnetic nanostructure Fe / Co. Experimental data and results of theoretical calculations (solid lines) are shown for the saturating in-plane magnetic field $H = 472$ Oe.

2. Calculations of the neutron multimonochromator-bipolarizer

After analysis of experimental results for the multilayer periodic magnetic Fe / Co nanostructures [10], one of the authors of the present work (V.S.) suggested to use these unique properties of Fe / Co nanostructures to polarize and monochromize a neutron beam with opposite polarizations for two close monochromatic wavelengths. In this case, measurements of intensities I^+ and I^- reflected from the sample can be obtained without the spin flipper. If one uses a set of similar structures with properly chosen different periods, one measurement would give dependences of both I^+ and I^- on the momentum transfer that would reduce the time of the experiment. To realize this idea, one of the authors of the present paper (V.S.) proposed a new neutron-optical element, Neutron Multimonochromator-Bipolarizer (NMB). NMB has a multimultilayer structure that consists of twelve periodic multilayer magnetic Fe/Co nanostructures. Each periodic multilayer magnetic nanostructure has a considerable number of identical bilayers and its own thickness of the bilayer (or the period value). The structure is schematically shown in Fig. 2 of a NMB with twelve periodic magnetic Fe/Co nanostructures in which period values increase with distance from the substrate. Iron and cobalt layers in each periodic structure have the same thickness. This eliminates Bragg peaks of the 2nd order for each periodic structure. Each structure has two Bragg peaks of the 1st order on the reflectivity curve for (+) and (-) spin components with the reflection coefficients close to 1. Values of structures periods

Figure 2. Scheme of multimultilayer structure of NMB made of twelve periodic multilayer magnetic Fe/Co nanostructures whose periods increase with distance from the substrate.

have been selected so that the Bragg peak of the first order for (+) spin component of each structure and the Bragg peak for (-) spin component of the neighboring structure (with larger period) to be as close to each other as possible. Thus, the reflection coefficient for this NMB should show a set of densely standing alternating Bragg peaks of the 1st order.

The scheme of such NMB is presented in Fig. 3a. Schemes of distribution of neutron-optical potentials for periodic Fe/Co nanostructures in the NMB for (+) and (-) neutron spin components are presented in Figs. 3b and 3c, correspondingly. As it is seen from Figs. 3b and 3c, there is a large difference of neutron-optical potentials for both spin components of the beam for all structures in NMB. Therefore, on curves of reflectivities of both spin components from each periodic structure depending on the momentum transfer, at sufficient number of pairs of layers in each structure, almost identical Bragg peaks of first order should be observed which are slightly displaced from each other and which intensities are close to 1.

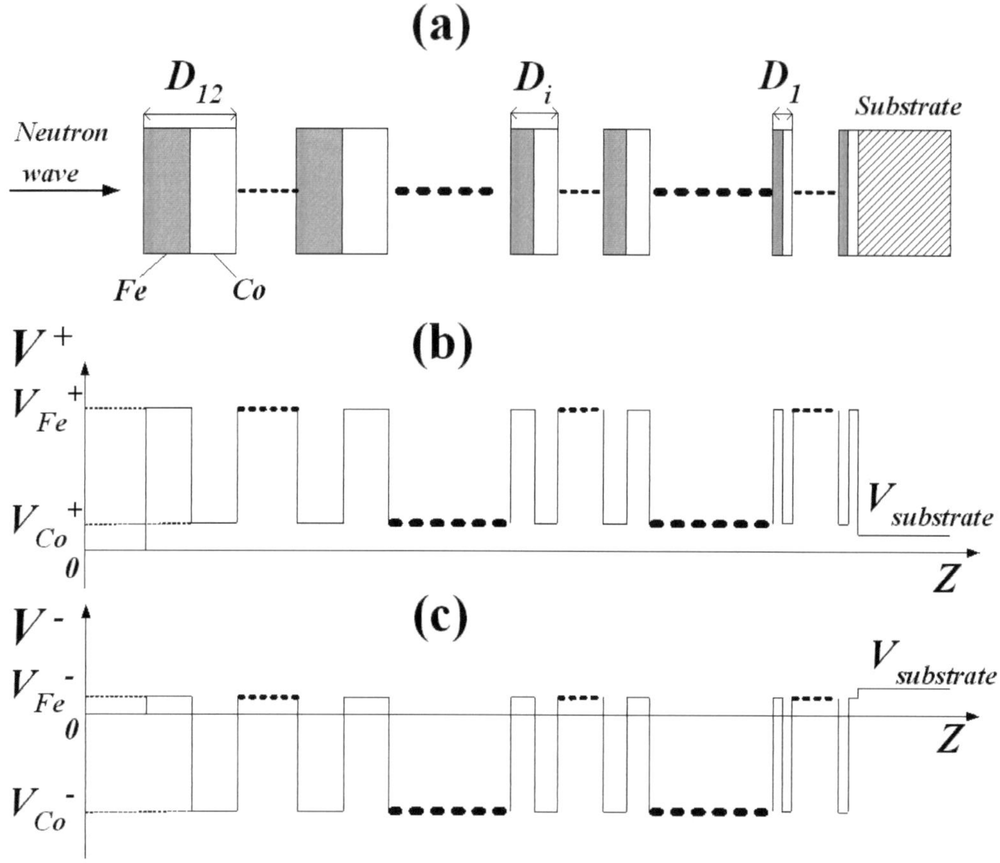

Figure 3. Scheme of periodic Fe/Co structures in NMB (a); scheme of distribution of neutron-optical potentials for periodic Fe/Co structures in NMB for (+) (b) and for (-) (c) neutron spin components.

Standard Parratt32 program was used for calculations, which is one of the most effective tools for studying of multilayered nanostructures. Coefficients of reflection from each of twelve multilayer periodic magnetic nanostructures are presented in Fig. 4, which were obtained as results of calculations. The absorption of neutrons in the materials of the layers of structures was not taken into account. Periods of nanostructures in this NMB have the following values: D_1=*50 Å* (a), D_2=*53 Å* (b), D_3=*56 Å* (c), D_4=*59 Å* (d), D_5=*62 Å* (e), D_6=*66 Å* (f), D_7=*72 Å* (g), D_8=*78 Å* (h), D_9=*88 Å* (i), D_{10}=*98 Å* (j), D_{11}=*118 Å* (k), D_{12}=*146 Å* (l). Such periods provide a set of densely standing alternating Bragg peaks of the 1st order with opposite spins (see above). Each multilayered nanostructure of this NMB contains 245 bilayers. Layer thicknesses are the same in each structure. The structure with period D_1 is evaporated on the glass substrate.

As is seen from Fig. 4, only Bragg peaks of odd orders are visible in each nanostructure for both spin components of the beam. For Bragg peaks of the 1st order, coefficients of reflection are close to 1 for both spin components. The absence of peaks of even orders is explained by the fact that thicknesses of layers in every period are identical [9].

In Fig. 5, we present calculated coefficients of reflection from the NMB consisting of twelve periodic magnetic Fe/Co nanostructures with periods D_1=*50 Å*, D_2=*53 Å*, D_3=*56 Å*, D_4=*59 Å*, D_5=*62 Å*, D_6=*66 Å*, D_7=*72 Å*, D_8=*78 Å*, D_9=*88 Å*, D_{10}=*98 Å*, D_{11}=*118 Å*, D_{12}=*146 Å*. The

IOP Conf. Series: Journal of Physics: Conf. Series **862** (2017) 012016 doi:10.1088/1742-6596/862/1/012016

number of bilayers in all periodic structures remains the same and equals to 245. Calculations were carried out without taking into account a structure imperfection and the absorption of neutrons in materials of layers. These dependences show 13 pairs of peaks located close to each other, which reflection coefficients are close to 1. Red and black curves correspond to R^+ and R^-, respectively. The number of peak pairs corresponds to the number of periods of NMB. Peaks in a pair are Bragg peaks of the 1st order of the corresponding period for (+) and (-) spin components. Peaks in the thirteenth pair are Bragg peaks of the 3rd order for (+) and (-) spin components produced by the structure with period D_{12}.

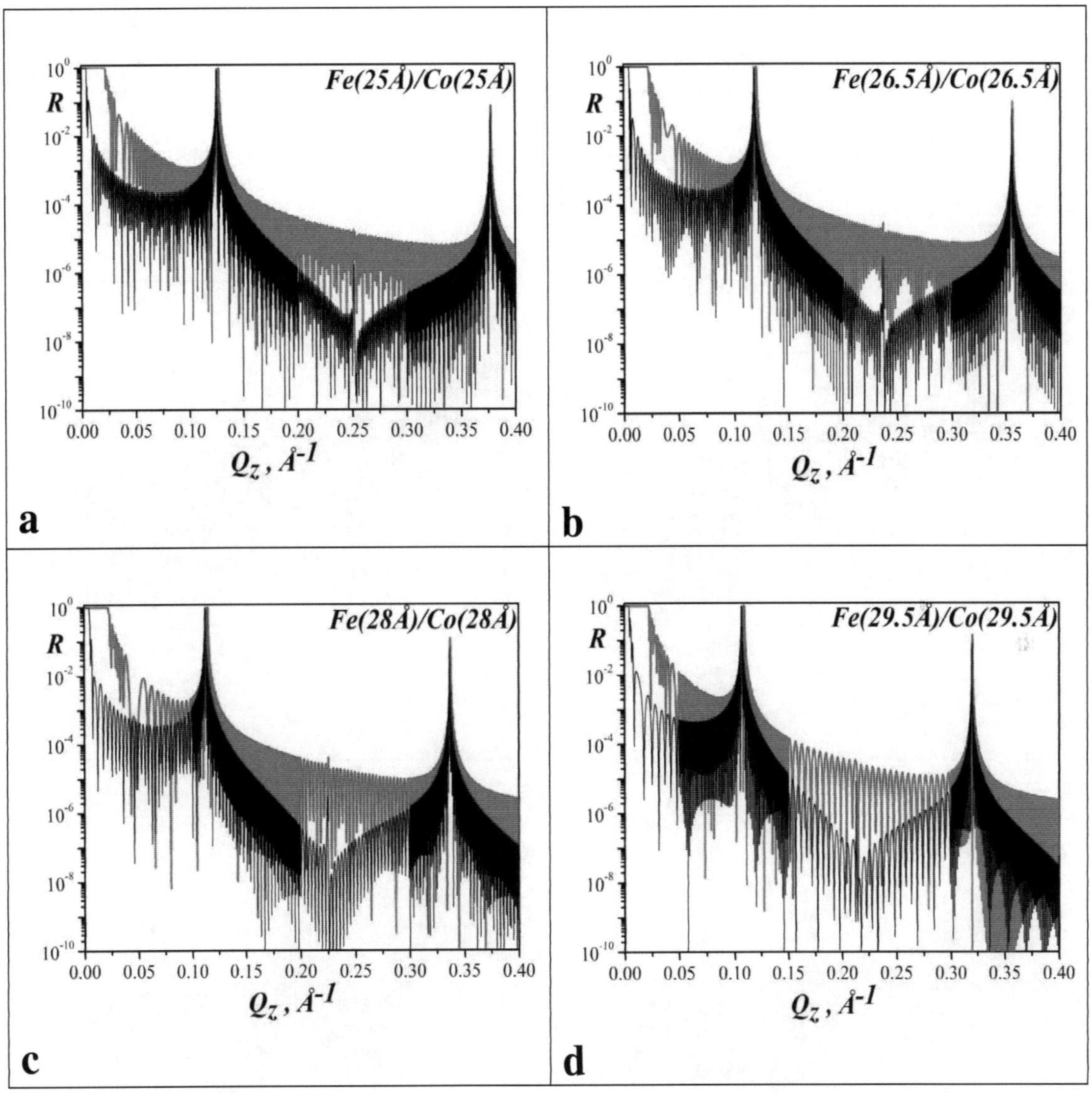

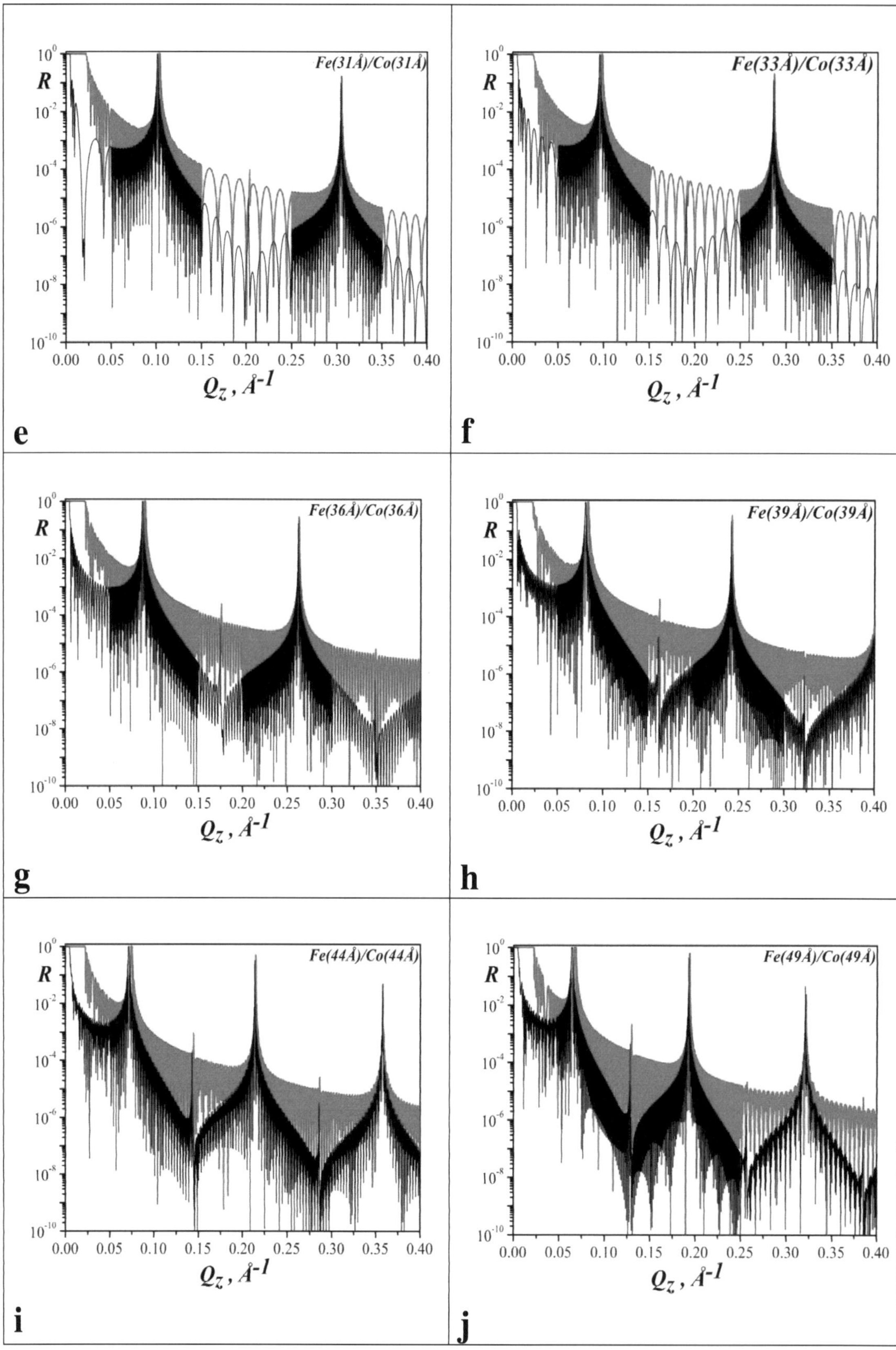
Fe(31Å)/Co(31Å)
Fe(33Å)/Co(33Å)
Fe(36Å)/Co(36Å)
Fe(39Å)/Co(39Å)
Fe(44Å)/Co(44Å)
Fe(49Å)/Co(49Å)
R
Qz, Å-1
e
f
g
h
i
j

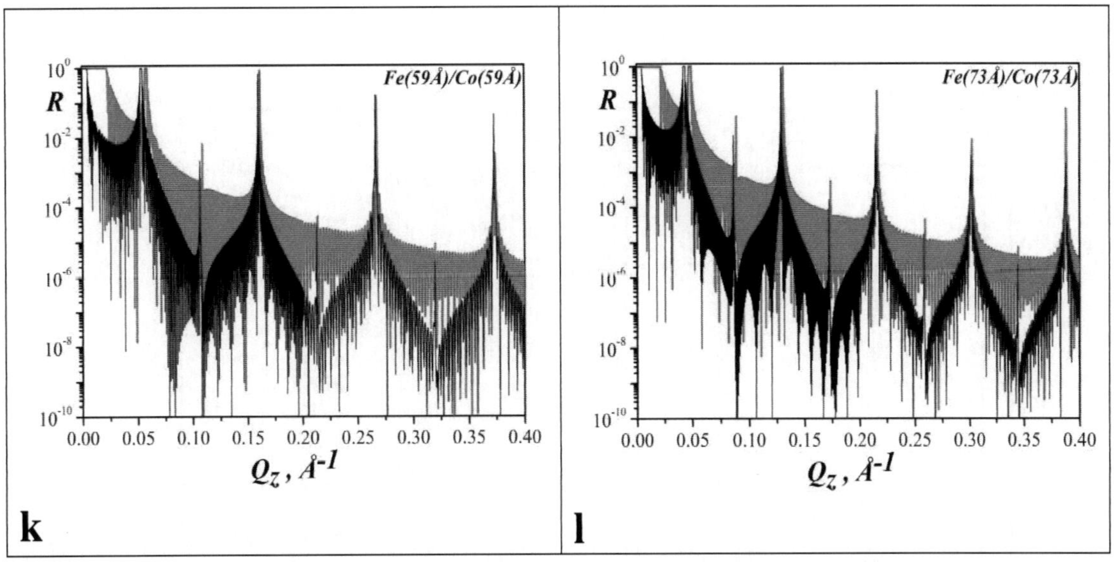

Figure 4. Calculated reflection coefficients of both spin components of the neutron beam from each of twelve multilayered periodic magnetic nanostructures as functions of the normal component of the momentum transfer. Periods of nanostructures in this NMB are D_1=*50* Å (a), D_2=*53* Å (b), D_3=*56* Å (c), D_4=*59* Å (d), D_5=*62* Å (e), D_6=*66* Å (f), D_7=*72* Å (g), D_8=*78* Å (h), D_9=*88* Å (i), D_{10}=*98* Å (j), D_{11}=*118* Å (k), D_{12}=*146* Å (l).

As appears from Fig. 5, a small overlapping occurs of peak wings of neighboring peak pairs. Let us consider this point in more detail. In Figs. 6a and 6b, we show calculated reflectivities from the nanostructure of NMB with period D_7=72 Å near Bragg peaks of the 1st order for single (a) and double (b) reflections. Calculations were carried out without taking into account the absorption of neutrons in materials of layers. Notice that at single reflection (Fig. 6a) a small overlapping of peaks is observed, there is practically no overlapping at double reflection (Fig. 6b), whereas peaks heights are the same in both cases. Therefore, the degree of polarization of Bragg peaks for this period will be significantly higher after double reflection than after single reflection. This statement is valid also for all other structures of this NMB.

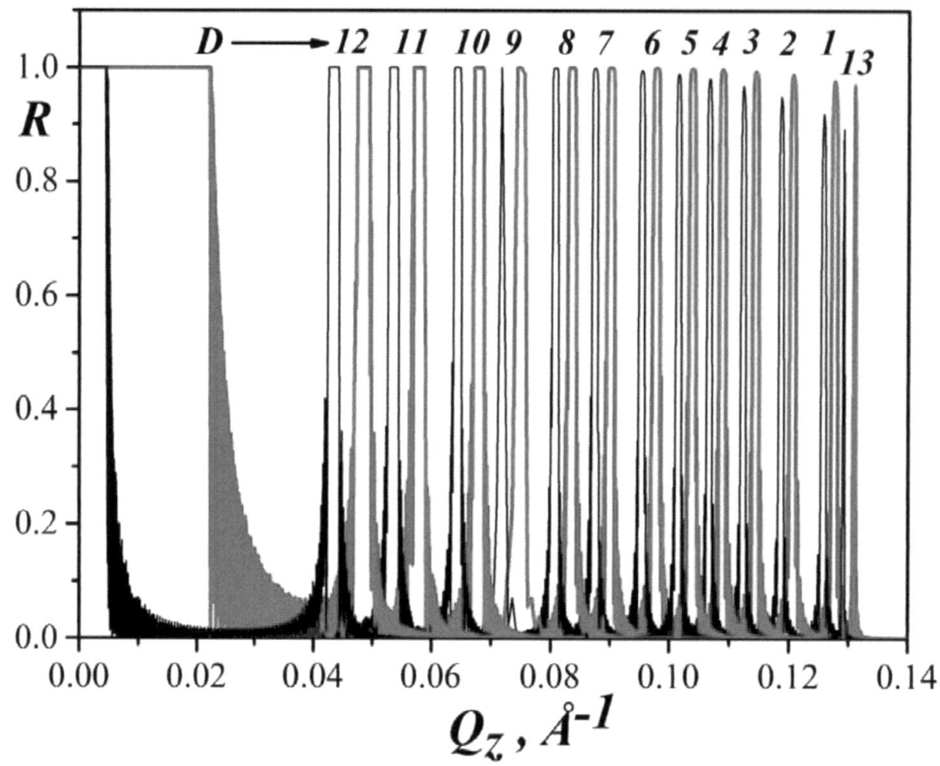

Figure 5. Calculated reflection coefficients of both spin components of the beam as functions of the normal component of the momentum transfer of the NMB consisting of twelve periodic magnetic Fe/Co nanostructures with periods D_1=50 Å, D_2=53 Å, D_3=56 Å, D_4=59 Å, D_5=62 Å, D_6=66 Å, D_7=72 Å, D_8=78 Å, D_9=88 Å, D_{10}=98 Å, D_{11}=118 Å, D_{12}=146 Å. R^+ and R^- are shown in red and black, respectively. Each of the twelve periodic Fe / Co structures contains 245 bilayers.

Parameters are presented in Table 2 of Bragg peaks of the 1st and the 3rd orders which are shown in Fig. 5.

Let us introduce coefficient of attenuation A_i of a monochromatic neutron beam when reflected from the *i*-th periodic structure of NMB. We define this coefficient as

$$A_i = \exp(-\gamma_i),$$

$$\text{where } \gamma_i = \frac{8\cdot\pi\cdot\overline{N}\cdot\overline{\sigma}}{\lambda_0\cdot Q_{zB1i}}\cdot\sum_{k=i}^{12} n_k\cdot d_k \quad, \quad 1\le i\le 12,$$

where Q_{zB1i} is the value of the normal component of the momentum transfer corresponding to the Bragg reflection of the 1st order from the *i*-th periodic structure of NMB (*i* is counted from the substrate), n_k is the number of bilayers in the *k*-th periodic structure ($k \ge i$), d_k is the thickness of bilayer in the *k*-th periodic structure, $\overline{N}\cdot\overline{\sigma} = (N_{Fe}\cdot\sigma_{abs\,Fe} + N_{Co}\cdot\sigma_{abs\,Co})/2 = 1.78\ cm^{-1}$, N_{Fe} and

IOP Conf. Series: Journal of Physics: Conf. Series **862** (2017) 012016 doi:10.1088/1742-6596/862/1/012016

N_{Co} are numbers in 1 cm^3 of iron and cobalt nuclei, respectively, $\sigma_{abs\,Fe}$ and $\sigma_{abs\,Co}$ are cross-sections of absorption of neutrons by iron and cobalt nuclei, correspondingly, and $\lambda_0 = 1.8$ Å.

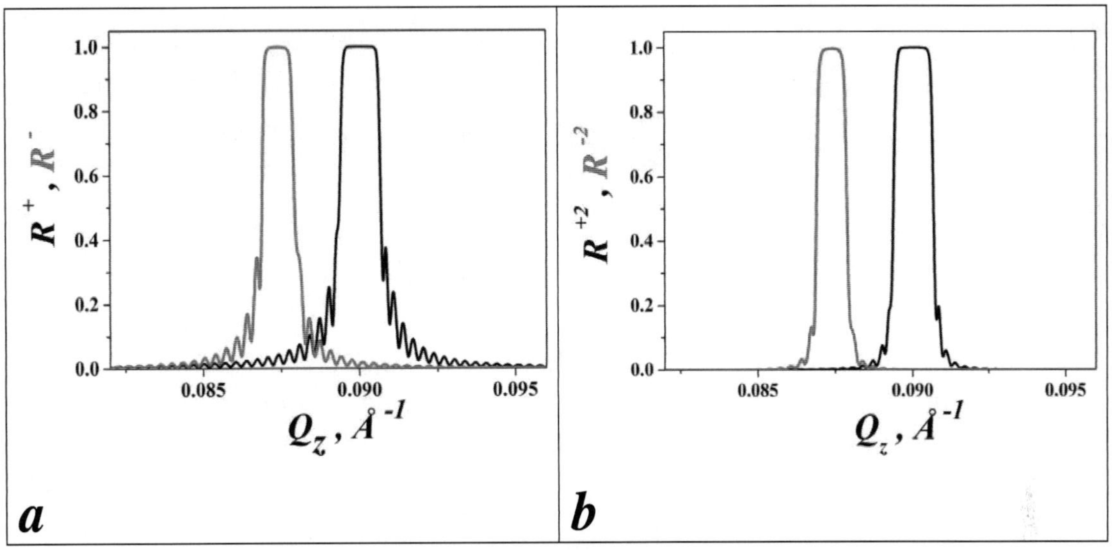

Figure 6a, b. Calculated reflectivities from the nanostructure of NMB with period D_7 = 72 Å near Bragg peaks of the 1st order for single (a) and double (b) reflections.

Table 2. Parameters of Bragg peaks of the 1st and the 3rd orders shown in Fig. 5. Here *i* is the number of the periodic structure in NMB. For *i* = *1, 2, ... 12*, parameters of Bragg peaks of the 1st order are specified for structures with periods D_1 - D_{12}. For *i* = 13, parameters of Bragg peaks of the 3rd order are specified for the structure with period D_{12}. *d* is the value of period, $R^{\pm}_{B1i,B313}$ are reflectivities of Bragg peaks for (+) and (-) spin components of the neutron beam, $Q^{\pm}_{zB1i,B313}$ are positions of Bragg peaks produced by *i*-th structure, $\left(\frac{\Delta Q_{zB1i,B313}}{Q_{zB1i,B313}}\right)^{\pm}$ are relative widths of Bragg peaks of the first and the third orders, $P_{Q^+_{zB1i,B313}}$ are polarizing efficiencies of Bragg peaks of the first and the third orders given by $P_{Q^+_{zB1i,B313}} = [R^+_{B1i,B313}(Q^+_{B1i,B313}) - R^-_{B1i,B313}(Q^+_{B1i,B313})]/[R^+_{B1i,B313}(Q^+_{B1i,B313}) + R^-_{B1i,B313}(Q^+_{B1i,B313})]$; A_{1i} and A_{2i} are coefficients of attenuation of monochromatic neutron beams, when the number of bilayers in all 12 periodic structures of NMB are the same and equal to 245 and 100, respectively.

i	*d* (Å)	$R^+_{B1i,B3}$	$Q^+_{zB1i,B3}$ (Å^{-1})	$R^-_{B1i,B3}$	$Q^-_{zB1i,B313}$ (Å^{-1})	$\left(\frac{\Delta Q_{zB1i,B313}}{Q_{zB1i,B313}}\right)^+$	$\left(\frac{\Delta Q_{zB1i,B313}}{Q_{zB1i,B313}}\right)^-$	$P_{Q^+_{zB1i,B313}}$	A_{1i}	A_{2i}
1	*50*	*0.98*	*0.128*	*0.92*	*0.126*	*0.0085*	*0.0068*	*0.952*	*0.64*	*0.83*
2	*53*	*0.98*	*0.121*	*0.95*	*0.119*	*0.0099*	*0.0076*	*0.999*		
3	*56*	*0.98*	*0.114*	*0.96*	*0.112*	*0.0099*	*0.0081*	*0.951*		

4	*59*	*0.99*	*0.109*	*0.98*	*0.107*	*0.0107*	*0.0090*	*0.949*		
5	*62*	*0.99*	*0.104*	*0.99*	*0.102*	*0.0115*	*0.0094*	*0.949*		
6	*66*	*1*	*0.098*	*0.99*	*0.095*	*0.0127*	*0.0105*	*0.987*		
7	*72*	*1*	*0.090*	*1*	*0.087*	*0.0133*	*0.0106*	*0.958*	*0.67*	*0.85*
8	*78*	*1*	*0.084*	*1*	*0.081*	*0.0149*	*0.0114*	*0.987*		
9	*88*	*1*	*0.075*	*1*	*0.072*	*0.0187*	*0.0182*	*0.969*		
10	*98*	*1*	*0.068*	*1*	*0.064*	*0.0280*	*0.0218*	*0.960*		
11	*118*	*1*	*0.058*	*1*	*0.053*	*0.0308*	*0.0346*	*0.959*		
12	*146*	*1*	*0.048*	*1*	*0.043*	*0.0577*	*0.0462*	*0.959*	*0.83*	*0.93*
13	*146*	*0.97*	*0.131*	*0.89*	*0.129*	*0.0027*	*0.0019*	*0.99*	*0.93*	*0.97*

As follows from Table 2, Bragg peaks of the first order for all periodic structures and Bragg peaks of the 3rd order for the structure with N = 12 have high reflectivities R^{+}_{B1} > 0.97 and R^{-}_{B1} > 0.89. The relative widths of Bragg peaks are in range $(\Delta Q_{B1}/Q_{B1})^{\pm}$ = 0.002 – 0.058. The polarizing efficiency for all Bragg peaks is high $P_{Q^{+}_{zB1,B3}} \geq 0.95$. Coefficient of attenuation A_i depends significantly on the number of bilayers in periodic structures of NMB. Then, A_{2i} is greater than A_{1i} in all considered structures. This value can be increased further by reflecting the neutron beam twice from a given NMB, as it is discussed above by the example of the structure with period D_7 = 72 Å. Thirteen pairs of Bragg peaks presented in Fig. 5 including peaks of the 3rd order cover the considerable range of the momentum transfer Q_Z = (0.043 - 0.131) Å^{-1}, where $Q_{Z\max}/Q_{Z\min} \cong 3.05$.

3. The new scheme of the total neutron polarization analysis

Four-modes neutron reflectometer NR-4M [11] at the WWR-M reactor of Petersburg Nuclear Physics Institute (PNPI NRC KI) was constructed using the traditional scheme of the total neutron polarization analysis, where the main components in the monochromatic polarization mode are monochromator-polarizer, two spin-flippers (one before and one after the sample), sample, analyzer, and detector. The scheme of neutron reflectometer NR-4M in polarization modes is shown in Fig. 7.

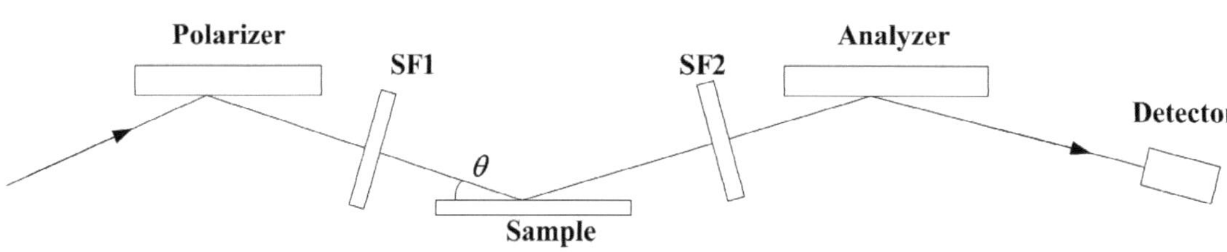

Figure 7. Scheme of neutron reflectometer NR-4M in polarization modes [11].

In the monochromatic polarization mode of neutron reflectometer NR-4M, the mirror multilayer monochromator-polarizer is used which is based on the pair Co/Ti [7]. Scheme of the double mirror monochromator-polarizer based on the Co / Ti pair is shown in Fig. 8a.

IOP Conf. Series: Journal of Physics: Conf. Series **862** (2017) 012016 doi:10.1088/1742-6596/862/1/012016

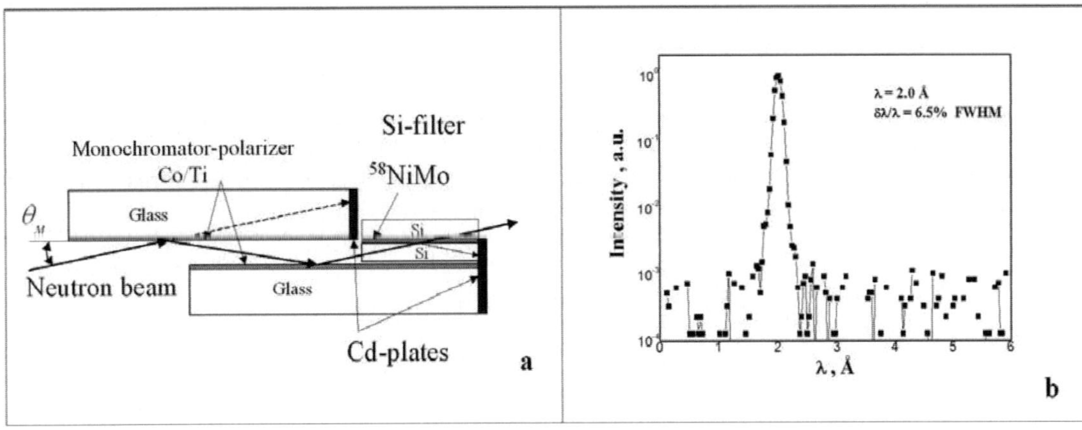

Figure 8. (a) Scheme of mirror neutron monochromator-polarizer based on Co / Ti in the unit of the beamformer of neutron reflectometer NR-4M, (b) Intensity of (+) spin component of the beam as a function of wavelength at the exit of the monochromator-polarizer after double reflection from the monochromator-polarizer and after passing of the beam through the filter.

The neutron beam with a broad spectral distribution falls on the mirror monochromator-polarizer Co / Ti at the glancing angle θ_m. Due to the Bragg diffraction on periodic Co / Ti structure [7], monochromatic neutrons are reflected from it with wavelength λ and with spins parallel to the magnetic field. Neutrons with this wavelength and with spins opposite to the magnetic field are not practically reflected. They pass through the whole structure, through the glass substrate, and they are absorbed in a cadmium plate which is glued to the substrate edge. By setting the appropriate angle θ_m, one can obtain the desired wavelength λ of the monochromatic beam passing through the unit. As is seen from Fig. 8a, the double reflection of neutrons and the silicon filter are used in the monochromator-polarizer Co / Ti. This is done in order to improve the polarization of the reflected neutron beam and to increase the degree of its monochromatization. In Fig. 8b, the dependence on the wavelength is shown of the intensity of (+) spin component of the beam at the exit of the monochromator-polarizer that is tuned to wavelength of $\lambda = 2\,\text{Å}$. Parasitic contributions to the reflected intensity amount to only 0.2% from non-monochromatic neutrons after double reflection from the monochromator-polarizer Co / Ti and after passing through the silicon filter. The polarization of the produced monochromatic beam is very high: the flipping ratio at the peak is 250. The flipping ratio is defined as $I^+(\lambda)/I^-(\lambda)$. In Fig. 8b, only $I^+(\lambda)$ is shown. The highly polarized monochromatic beam with a wavelength λ formed thereby passes through the first collimation system, the first spin-flipper, and falls on the sample at the glancing angle θ. The reflected beam from the sample with the intensity $\boldsymbol{I}$ falls on the supermirror analyzer after passing through the second spin-flipper. The reflected beam from the analyzer is recorded by the detector. In this scheme of the total polarization analysis, four intensities ($\boldsymbol{I^{++}}$, $\boldsymbol{I^{+-}}$, $\boldsymbol{I^{-+}}$, $\boldsymbol{I^{--}}$) are measured in four measurements. Here, signs "+" and "-" correspond to situations when the spin-flipper is switch on and switch off, respectively. When the spin-flipper is switched on, the neutron spin flips. When it switched off, the beam passes through the spin-flipper without a spin flip. The first and the second signs "+" or "-" indicate the state of the first and of the second spin-flipper, correspondingly. Four values $I^{++}(Q_z)$, $I^{+-}(Q_z)$, $I^{-+}(Q_z)$, $I^{--}(Q_z)$ (where $Q_z = \dfrac{4\pi \sin\theta}{\lambda}$) can be measured by carrying out such measurements (the total polarization analysis) in the wide range of glancing angles θ of the beam on the sample. It is possible to obtain a reflectivity matrix for each value of Q_z from these four intensities.

In the supermirror polarization mode of neutron reflectometer NR-4M [11], the conventional total neutron polarization analysis scheme is also used. The main set-up components are supermirror polarizer, two spin-flippers (one before and one after the sample), the sample, supermirror analyzer and detector. In Fig. 9a, we show the supermirror polarization mode of NR-4M reflectometer, in which a double polarizing CoFeV / TiZr supermirror is used as a polarizer.

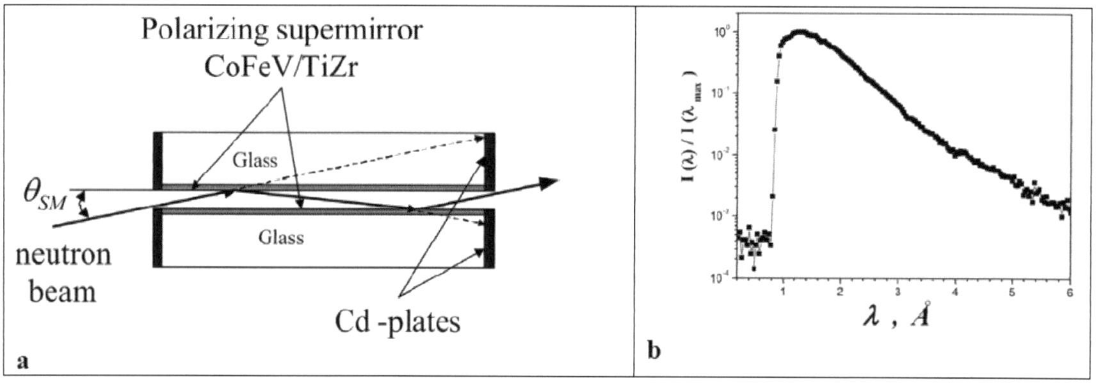

Figure 9. (a) Scheme of CoFeV / TiZr supermirror neutron polarizer based in the unit of the beamformer of neutron reflectometer NR-4M. (b) Relative intensity (as a function of the wavelength) of (+) spin component of the beam at the exit of the supermirror polarizer after double reflection from the polarizer.

The neutron beam with a broad spectral distribution falls on the *CoFeV/TiZr* suppermirror [12] at the glancing angle θ_{SM}. In Fig. 9b, the relative intensity is presented as a function of the neutron wavelength. In the spectral range λ = 0.88 - 4.7 Å used in this reflectometer, the polarization is very high. Its average value over the spectrum is 0.99. Formed thereby highly polarizing neutron beam with a wide range of wavelength λ passes through the first collimation system, through the first spin-flipper and falls on the sample at the glancing angle θ. The reflected beam from the sample with intensity $I(\lambda)$, after passing through the second spin-flipper, falls on the supermirror analyzer. The reflected beam from the analyzer is recorded by the detector. In this scheme of the total polarization analysis, four spectral intensities ($I^{++}(\lambda)$, $I^{+-}(\lambda)$, $I^{-+}(\lambda)$, $I^{--}(\lambda)$) are measured during four measurements. Carrying out such measurements for a glancing angle θ of the beam containing neutrons with a wide range of wavelength λ, it is possible to obtain four dependences $I^{++}(Q_z)$, $I^{+-}(Q_z)$, $I^{-+}(Q_z)$, $I^{--}(Q_z)$ (where $Q_z = \dfrac{4\pi \sin\theta}{\lambda}$) within the total polarization analysis.

We propose the new scheme of the total neutron polarization analysis for the reflectometer, where the above NMB based on multilayer periodic Fe / Co nanostructures is used as a monochromator-polarizer. The new scheme of the total neutron polarization analysis is shown in Fig. 10 for the reflectometer without the spin-flipper before the sample.

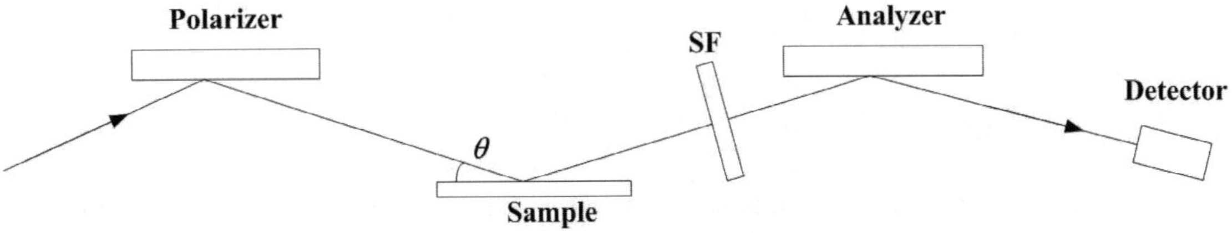

Figure 10. The new scheme of the total neutron polarization analysis for the reflectometer without the spin-flipper before the sample.

There is no spin-flipper before the sample because NMB works as a polarizer in this scheme, which reflects neutrons of both (+) and (-) spin components of the beam. The polarizer is shown in Fig. 11 for the new scheme that is created in the form of the double NMB.

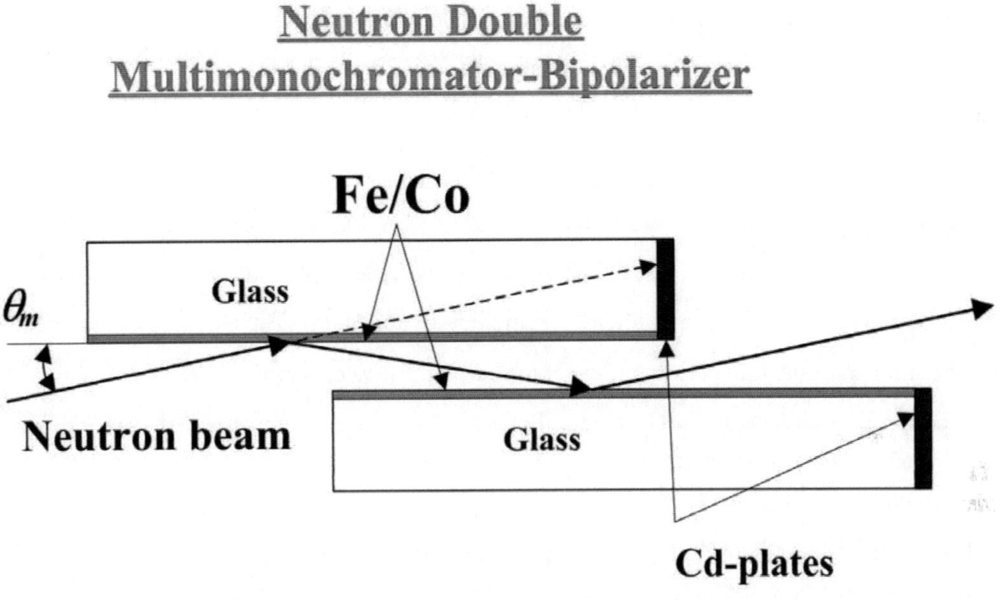

Figure 11. The polarizer for the new scheme of the total neutron polarization analysis for a reflectometer created in the form of the double NMB.

The unpolarized neutron beam with a broad spectral distribution falls on the first mirror of the NMB at angle θ_m. The intensity of the reflected beam shows a set of polarized monochromatic peaks with monotonically growing wavelength. Peaks wavelengths are very close to each other. The neutron polarization is opposite in adjacent peaks. Then, the resulting beam is reflected from the second mirror under the same angle θ_m. The output beam has the same direction, as that falling on the NMB, just as in the cases of Co/Ti monochromator-polarizer and CoFeV/TiZr supermirror considered above. The second reflection allows to make wings of all peaks more sharp. This reduces partial overlapping of wings of the neighboring peaks that allows to increase the polarization of all peaks. At the same time, peaks intensities decrease slightly since reflectivities of peaks are close to 1 (see Table 2). Thus, in this reflectometer, the beam which is twice reflected from NMB and which passes through the collimation system falls on the sample at the glancing angle θ. As it is noted above, it is possible to represent the beam intensity $I(\lambda)$ in the form of the set of monochromatic peaks with monotonically growing wavelengths. In this set, neighboring peaks have opposite polarizations. As noted above, the use of thirteen pairs of Bragg peaks (Fig. 5) leads to the covering of a large interval of the momentum transfer $Q_z = 0.043 - 0.131\ \text{Å}^{-1}$, where $\frac{Q_{z\max}}{Q_{z\min}} \cong 3.05$. Therefore, after reflection of the neutron beam at angle θ_m from NMB, one obtains the set of 13 pairs of monochromatic peaks. For this set, the ratio of the maximal wavelength to the minimal one is $\frac{\lambda_{\max}}{\lambda_{\min}} \cong 3.05$, as for the momentum transfer because $Q_z = \frac{4\pi\sin\theta}{\lambda}$. Bragg peaks of the 3rd order were not considered at the time-of-flight measurements for all structures except for the structure with period D_{12}. Besides, Bragg peaks of higher orders are not considered as well. Thus, the set of highly polarized monochromatic peaks with opposite polarizations allows to omit the spin-flipper before the sample. Indeed, the intensity of the beam reflected from the sample can be represented in the form of two functions: $I_R^+(\lambda)$ and $I_R^-(\lambda)$ that can be obtained in one measurement. Whereas the conventional scheme requires two measurements at two states of the spin flipper to obtain these functions. The beam reflected from the sample passes through the spin-flipper after the sample, it is then reflected from the analyzer and is registered in the detector. At the same time, we obtain $I_R^{++}(\lambda)$, $I_R^{-+}(\lambda)$ using $I_R^+(\lambda)$ (when the spin-flipper after the sample is switched off) and we find $I_R^{+-}(\lambda)$, $I_R^{--}(\lambda)$ using $I_R^-(\lambda)$ $I_R^+(\lambda)$ (when the spin-flipper after the sample is switched on). Thus, in the new scheme of the total polarization analysis, it is possible to receive four dependences $I_R^{++}(\lambda)$, $I_R^{-+}(\lambda)$, $I_R^{+-}(\lambda)$ and $I_R^{--}(\lambda)$ performing only two measurements. Let us consider the beam reflected from the standard supermirror analyzer for two states of the spin-flipper after the sample (it is switched off (+) and switched on (-)). One obtains $I_R^{++}(\lambda)$ and $I_R^{+-}(\lambda)$ from $I_R^+(\lambda)$ and one obtains $I_R^{-+}(\lambda)$ and $I_R^{--}(\lambda)$ from $I_R^-(\lambda)$. On the other hand, four measurements are required corresponding to four states of two spin-flippers in the case of the traditional total polarization analysis for obtaining four intensities $I_R^{++}(\lambda)$, $I_R^{+-}(\lambda)$, $I_R^{-+}(\lambda)$ and $I_R^{--}(\lambda)$.

4. Discussion and Conclusions

To conclude, we propose the new neutron-optical element - Neutron Multimonochromator-Bipolarizer (NMB). It is based on a multimultilayer structure consisting of 12 periodic Fe/Co multilayers which are evaporated on the substrate. The period of these nanostructures increases with distance from the substrate. We present results of calculations of reflectivities of both spin components of a beam from NMB as functions of the momentum transfer.

It is shown that the absorption is quite large of the neutron beam in layers of cobalt during the reflection from the Co/Fe NMB. However, it follows from calculations that the beam attenuation decreases upon decreasing of the number of bilayers. Therefore, it is required to minimize the number of bilayers for each periodic structure of NMB. At the same time, the reflectivity at the Bragg peak of the first order should be close to one. For example, for the structure with the period of 146 Å, the number of bilayers of 245 is apparently almost 8-10 times greater than it is required, because 30-40 bilayers are sufficient for the reflectivity to be close to unity at the Bragg peak of the first order.

We propose the new scheme of the total neutron polarization analysis in reflectometry with application of double NMB as a polarizer. In this scheme, there is no spin-flipper before the sample. The beam consisting of 13 pairs of monochromatic peaks falls on the sample. Peaks in each pair have opposite polarization. Wavelengths at which peaks arise are close to each other and fill the range $(\lambda_{\min}, \lambda_{\max})$, where $\frac{\lambda_{\max}}{\lambda_{\min}} \cong 3.05$. Only two measurements are required to find four intensities $I_R^{++}(\lambda)$, $I_R^{+-}(\lambda)$, $I_R^{-+}(\lambda)$ and $I_R^{--}(\lambda)$.

It is possible to make a compact polarizer based on NMB and to place it at the sample unit in magnetic field. Then, it is possible to remove also the second spin-flipper when using another NMB as the analyzer (instead of the standard supermirror) which is similar to that in the beamformer before the sample. In this case, one can obtain four intensities during two measurements by performing them for two angles of this analyzer which differ from each other by a small angle $\Delta\theta$. Using such a polarizer and such an analyzer, it is possible to create a very compact cell at the sample unit to perform the total polarization analysis without using of spin flippers.

Use of the Neutron Multimonochromator-Bipolarizer is possible in such time-of-flight neutron techniques as polarized neutron reflectometry, SESANS, for research of low-angle and inelastic scattering of polarized neutrons.

Acknowledgements

The work was supported by the Federal target program of Ministry of Education and Science of Russian Federation (project No. RFMEFI61614X0004). Authors are grateful to M. V. Avdeev and Yu. V. Nikitenko for interest in work.

References

[1] Nikitenko Yu V, Syromyatnikov V G 2013 *Reflectometry of polarized neutrons (M.: Fizmatlit)* 224 p (in russian)
[2] Schoenborn B P, Caspar D L D, Kammerer O F 1974 *J. Appl. Cryst* **7** 508
[3] Lynn J W, Kjems J K, Passell L, Saxena A M, Schoenborn B P 1976 *J Appl Cryst* **9** 454
[4] Saxena A M, Schoenborn B P 1977 *Acta Cryst* **A33** (5) 805
[5] Gukasov A G, Deriglazov V V, Kezerashvili V Ya, Kudriashov V A, Krutov G A, Peskov B G, Syromyatnikov V G, Trunov V A, Kharchenkov V P, Shchebetov A F 1979 *ZhETF* **77** (5) 1720 (in russian)
[6] Kezerashvili V Ya, Schebetov A F, Peskov B G, Pleshanov N K, Soroko Z N, Syromyatnikov V G 1987 *Journal of Technical Physics* **57** (7) 1372 (in russian)
[7] Syromyatnikov V G, Menelle A, Soroko Z N, Schebetov A F 1998 *Physica B* **248** (1–4) 355
[8] Kyaw Zaw Lin, Syromyatnikov V G 2016 *Vestnik of Saint Petersburg State University* series 4 **3** (61) N1 4 (in russian)
[9] Kyaw Zaw Lin, Syromyatnikov V G 2016 *Journal of Surface Investigation: X-ray, Synchrotron and Neutron Techniques* **10** (4) 687
[10] Kyaw Zaw Lin, Syromyatnikov V G 2017 *Journal of Surface Investigation: X-ray, Synchrotron and Neutron Techniques* **11** (1) 20
[11] Syromyatnikov V G et al 2005 *Preprint of Petersburg Nuclear Physics Institute (NRC "Kurchatov Institute")* N 2619 Gatchina 47 p
[12] Schebetov A F, Pleshanov N K, Syromyatnikov V G et al 1996 *Journal of Physical Society of Japan* **65** supplement A 195

IOP Conf. Series: Journal of Physics: Conf. Series **862** (2017) 012017 doi:10.1088/1742-6596/862/1/012017

Accessible length scale of the in-plane structure in polarized neutron off-specular and grazing-incidence small-angle scattering measurements

R Maruyama[1], T Bigault[2], A R Wildes[2], C D Dewhurst[2], T Saerbeck[2], D Honecker[2], D Yamazaki[1], K Soyama[1] and P Courtois[2]

[1] J-PARC Center, Japan Atomic Energy Agency, 2-4 Shirakata, Tokai, Ibaraki 319-1195, Japan
[2] Institut Laue Langevin, 71 avenue des Martyrs, 38042 Grenoble, France

E-mail: ryuji.maruyama@j-parc.jp

Abstract. Polarized neutron off-specular and grazing-incidence small-angle scattering measurements are useful methods to investigate the in-plane structure and its correlation of layered systems. Although these measurements give information on complementary and overlapping length scale, the different characteristics between them need to be taken into account when performed. In this study, the difference in the accessible length scale of the in-plane structure, which is one of the most important characteristics, was discussed using an Fe/Si multilayer together with simulations based on the distorted wave Born approximation.

1. Introduction

Layered magnetic structures exhibit interesting and important magnetic properties which are not present in the bulk, such as exchange coupling between layers, giant magnetoresistance, and tunnel magnetoresistance [1, 2, 3]. These anomalous magnetic properties arise from the structure in the out-of-plane and/or in-plane directions reduced to the nm range. To elucidate those mechanisms and to open the possibility of new applications, it is important to obtain information on the in-plane magnetic structure.

Off-specular scattering (OSS) and grazing-incidence small-angle scattering (GISAS) using polarized neutrons are unique and powerful techniques to observe correlations of small magnetic objects in layered systems. Since there is a difference between the OSS and GISAS in the accessible range and resolution of the lateral component of the momentum transfer $\boldsymbol{q}$, one has to verify if the length scale of the in-plane structure of the sample matches the in-plane $\boldsymbol{q}$-range and resolution [4]. The in-plane component of the coherence volume can be estimated by the in-plane $\boldsymbol{q}$-resolution. The interpretation of the data can be different if the lateral dimension of the in-plane structure of the sample is sufficiently small or large compared with the in-plane component of the coherence volume in that dimension. OSS only measures a limited range of the in-plane component of $\boldsymbol{q}$ due to the restriction in the scattering geometry, making difficult the access to the lateral correlation length smaller than one μm. This can be overcome by GISAS, which allows measurements to a larger in-plane component of $\boldsymbol{q}$ and hence to a smaller lateral correlation length [5]. This study has mainly discussed the difference in the accessible in-plane

IOP Conf. Series: Journal of Physics: Conf. Series **862** (2017) 012017 doi:10.1088/1742-6596/862/1/012017

$\boldsymbol{q}$-range between the OSS and GISAS using polarized neutrons to illustrate the limitation of the accessible lateral correlation length for both measurements.

2. Results

Let us assume a scattering geometry where the sample surface is parallel to the (x, y) plane and neutrons incident on the sample at incident angle α_i are scattered at exit angles of α_f and $2\theta_f$ in the planes within and perpendicular to the (x, z) plane [6]. The scattering intensity dependence on q_x is measured in OSS, whereas that on q_y observed in the GISAS. In this study, the OSS and GISAS data were analyzed in the framework of the distorted wave Born approximation (DWBA) given by Toperverg *et al.* [4, 7, 8].

Figure 1(a) shows the measured polarized neutron OSS image in the I_{-+} spin-flip channel for an Fe/Si multilayer of 30 bilayers with a thickness of 10 nm, which was fabricated using the DC-magnetron sputtering [9] [1]. The measurement was performed on the D17 polarized neutron reflectometer installed at the high-flux reactor at the Institut Laue Langevin (ILL) [10, 11]. The monochromatic mode of operation with an average wavelength of 0.555 nm and wavelength spread of 4.0% in the full width at half maximum (FWHM) was used. The incident beam was collimated to keep the angular resolution less than 3.6%. Further detail of the measurement can be found in Ref. [6]. The spins in the sample are assumed to be aligned within the sample plane and to be perpendicular to $\boldsymbol{q}$ because of $q_x \ll q_z$ in the OSS. Since the sample was magnetized to 70% of saturation under an external field of 4.6×10^3 A/m, the magnetic scattering is seen in both of the spin-flip and non-spin-flip channels. The lateral correlation length for the magnetic scattering, $\xi_{\parallel}$, corresponding to the length of an area in which the spins are aligned to the same direction, can be determined by the scattering intensity dependence on q_x. The intensity profile under the condition of $\alpha_i + \alpha_f =$ const., which is approximately considered as $q_z =$ const., is useful because it does not depend on the z-correlation length [12]. The measured and simulated OSS profiles along the broken line on Fig. 1(a) are shown in Fig. 1(b), where $\alpha_i + \alpha_f = 2.5^\circ$ ($q_z \approx 0.49$ nm^{-1}) is satisfied. The simulated profile does not vary for small values of $\xi_{\parallel}$ less than 250 nm, which makes it difficult to determine $\xi_{\parallel}$ and the error of it. Since $\boldsymbol{q}$ is oriented close to parallel to z-axis in the OSS, q_x is smaller than a few percent of $|\boldsymbol{q}|$ for most cases. Although the q_x-range can be increased by increasing q_z, this causes another problem of the reduction in the scattering intensity. In general, it is difficult to obtain sufficient scattering intensity over full q_x-range for $q_z > 0.5$ nm^{-1}. This limits the accessible q_x-range to $|q_x| < 0.01$ nm^{-1} in the OSS. Let us consider the q_x-dependence of the structure factor in the simulation to understand this. Since the coupled areas with uniform orientation of the spins are modeled as rectangular boxes with lengths of $2\xi_{\parallel}$ within the plane and a height corresponding to the layer thickness, the structure factor is approximately given by the Lorentzian function, $1/\{1 + (q_x \cdot \xi_{\parallel})^2\}$ [13, 8, 6]. Figure 2 shows the Lorentzian profiles for different values of $\xi_{\parallel}$. The FWHM of the Lorentzian function becomes comparable to the accessible q_x-range with decreasing $\xi_{\parallel}$. This means that only a small part of the full peak profile is measured, making the determination of $\xi_{\parallel}$ more ambiguous. The condition of $q_{x,\max} \cdot \xi_{\parallel} \gg 1$ needs to be taken into account when performing the OSS measurement, where $q_{x,\max}$ is the maximum value of $|q_x|$ in the measurement. For this case with $q_{x,\max} = 0.01$ nm^{-1}, the GISAS measurement is needed to obtain quantitative information on the in-plane structure with a length scale smaller than 500 nm. The in-plane component of the coherence volume, $l_x \approx 2\pi/\delta q_x$, was estimated as $l_x \gtrsim 10$ μm because δq_x was roughly given by $\delta q_x \lesssim 5 \times 10^{-4}$ nm^{-1} (5% of q_x) for this OSS measurement. Since l_x is much larger than the length of the coupled area, the magnetic scattering due to the fluctuation of the spins can be expected.

Figure 3(a) shows the measured polarized neutron GISAS image in the I_{-} channel for the

[1] Figure 1(a) is reprinted from [6], with permission from Elsevier.

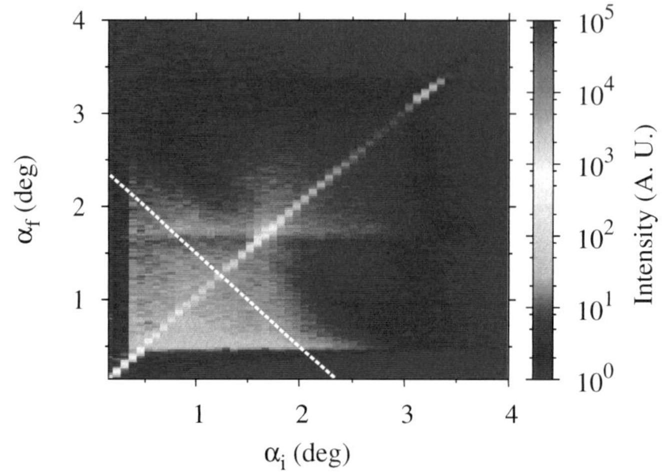

(a)

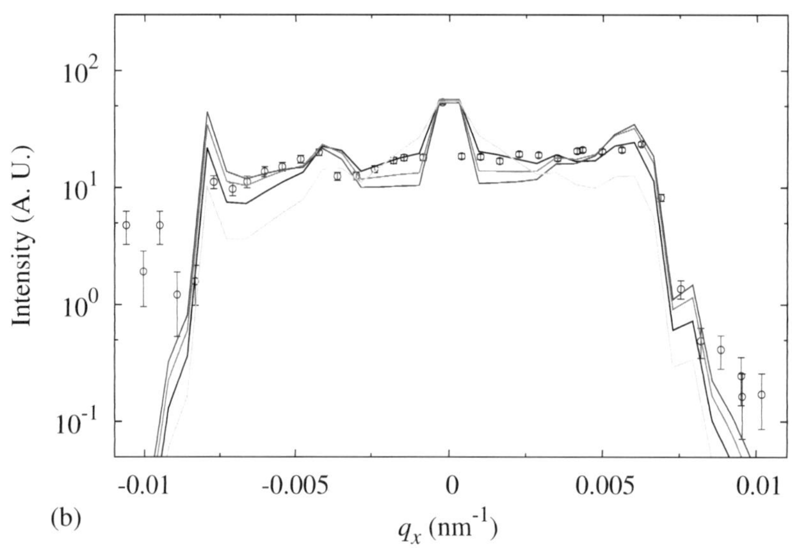

(b)

Figure 1. (a) Measured polarized neutron OSS contour plot of an Fe/Si multilayer. (b) Measured (symbols) and simulated (lines) OSS profiles along the broken line in (a), where $\alpha_i + \alpha_f = 2.5°$ ($q_z \approx 0.49$ nm^{-1}) is satisfied. Green, black, red, and blue line represents the simulated profile with $\xi_{\parallel}$ =500, 250, 125, and 50 nm, respectively.

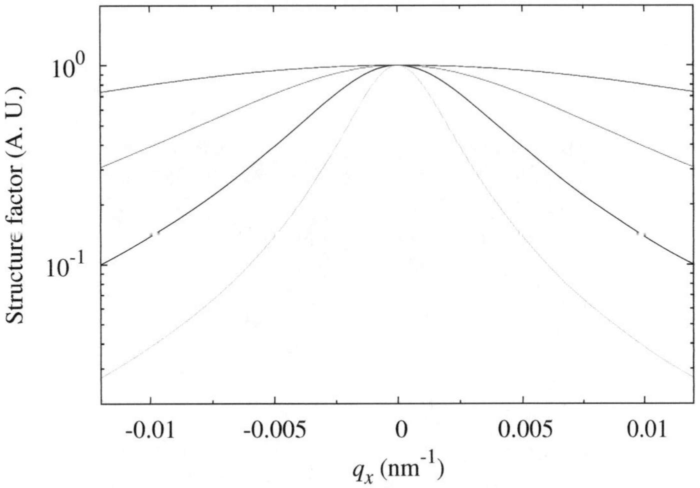

Figure 2. Structure factors proportional to the Lorentzian function, $1/\{1 + (q_x \cdot \xi_{||})^2\}$. Green, black, red, and blue line indicates the profile with $\xi_{||} = 500$, 250, 125, and 50 nm, respectively.

same sample [2]. The measurement without polarization analysis was performed on the D33 small angle neutron scattering (SANS) instrument at the ILL [14, 15]. The monochromatic mode of operation with an average wavelength of 0.80 nm and wavelength spread of 10% in FWHM was used. The beam divergence was less than 1.95 mrad. Further detail of the measurement can be found in Ref. [6]. In the GISAS, α_i was chosen to 3.8° to include the intense magnetic scattering observed in the OSS image Fig. 1(a) at $\alpha_i = 2.6°$ and $\alpha_f = 1.9°$, taking into account the difference in wavelength between the two techniques. The values of the angle given in the parentheses are the corrected angles for a wavelength of 0.80 nm. The spin-flip scattering is considered as dominant at $\alpha_f = 2.7°$ in Fig. 3(a). The measured and simulated profiles at $\alpha_f = 2.7°$ in Fig. 3(a) are shown in Fig. 3(b). The determination of $\xi_{||}$ in the GISAS is simpler and more reliable than that in the OSS since the intensity profile along $q_y(2\theta_f)$ does not depend on the Fresnel coefficients in the layers. The simulated peak profile becomes sharp with increasing $\xi_{||}$ and the peak broadening due to the scattering is very small for $\xi_{||} = 1000$ nm, which would be close to the upper limit of the accessible range of $\xi_{||}$ for the GISAS measurement using a pin-hole-type SANS instrument because a well-collimated beam with a relatively long wavelength was used in this measurement. In addition, the in-plane component of the coherence volume, $l_y \approx 2\pi/\delta q_y$, is estimated to be $l_y \lessapprox 1000$ nm because δq_y is in the order of 10% of q_y. It is no longer possible to discuss the length scale of the in-plane structure with $\xi_{||} \geqq 1000$ nm using this GISAS setup. For an in-plane structure with larger length scale, more reliable parameter would be obtained by the OSS.

3. Conclusions

The accessible range of the in-plane component of $\boldsymbol{q}$ and the relation between the in-plane component of the coherence volume and length scale of the in-plane structure of the sample for the OSS and GISAS measurements with polarized neutrons were discussed using the measured data for the Fe/Si multilayer with the simulation based on the DWBA. $\xi_{||}$ for the magnetic scattering of this sample was determined to be 325 ± 80 nm, where the result of the GISAS was more reliable [6]. To precisely determine $\xi_{||}$, the condition of $q_{x,\max} \cdot \xi_{||} \gg 1$ has to be satisfied

[2] Figure 3(a) is reprinted from [6], with permission from Elsevier.

IOP Conf. Series: Journal of Physics: Conf. Series **862** (2017) 012017 doi:10.1088/1742-6596/862/1/012017

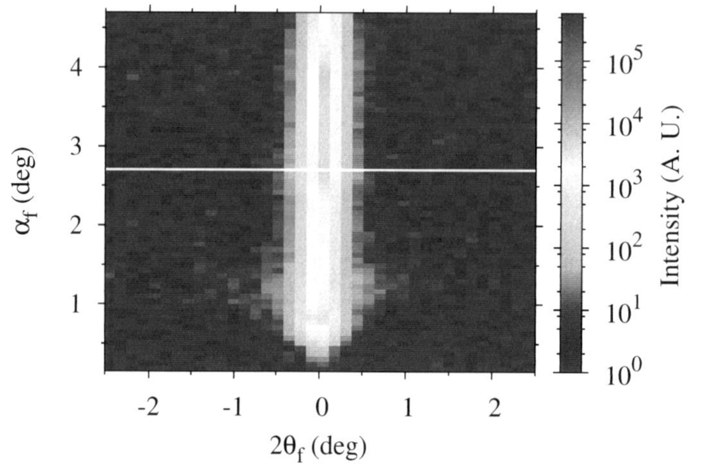

(a)

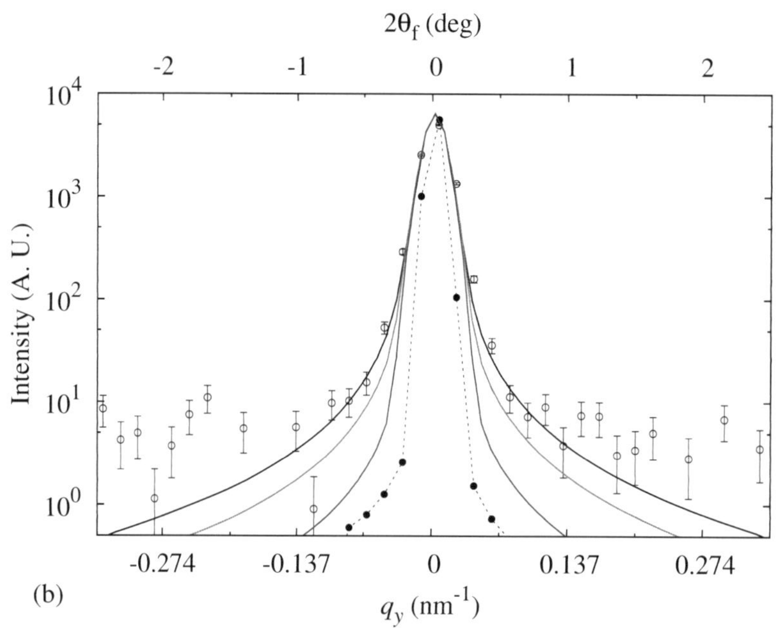

(b)

Figure 3. (a) Measured polarized neutron GISAS image of an Fe/Si multilayer. (b) Measured (○) and simulated (lines) GISAS profiles along the solid line in (a). Black, red, and blue line represents the simulated profile with $\xi_{\parallel}$ =325, 500, and 1000 nm, respectively. Direct beam profile (●) is plotted with scaling. Statistical errors for the direct beam profile are smaller than the size of the symbols.

for the OSS (usually $\xi_{\parallel} \gtrapprox 100$ nm), whereas $\xi_{\parallel} \lessapprox 1000$ nm is needed for the GISAS. For the overlapping length scale of $100 \lessapprox \xi_{\parallel} \lessapprox 1000$ nm, the accuracy of the obtained parameter can be improved by performing both OSS and GISAS measurements.

Acknowledgments

The authors thank the ILL for providing the beam time and support. This work was supported in part by JSPS KAKENHI Grant number 24561049 and Sasakawa Grants for Scientific Fellows from the Japan Science Society.

References

[1] Grünberg P, Schreiber R, Pang Y, Brodsky M B and Sowers H 1986 *Phys. Rev. Lett.* **57** 2442
[2] Baibich M N, Broto J M, Fert A, Nguyen Van Dau F, Pertroff F, Etienne P, Creuzet G, Friederich A and Chazelas J 1988 *Phys. Rev. Lett.* **61** 2472
[3] Binasch G, Grünberg, Saurenbach F and Zinn W 1989 *Phys. Rev.* B **39** 4828
[4] Zabel H, Theris-Bröhl K and Toperverg B P 2007 *Polarized neutron reflectivity and scattering from magnetic nanostructures and spintronic materials Handbook of Magnetism and advanced magnetic materials* ed Kronmüller H and Parkin S (Chichester: John Wiley & Sons Ltd.) pp 1237
[5] Salditt T, Metzger T H and Peisl J 1994 *Phys. Rev. Lett.* **73** 2228
[6] Maruyama R, Bigault T, Wildes A R, Dewhurst C D, Soyama K and Courtois P 2016 *Nucl. Instrum. Methods Phys. Res.* A **819** 37
[7] Toperverg B P 2002 *Appl. Phys.* A **74 (Suppl.)** S1560
[8] Kentzinger E, Rücker U, Toperverg B, Ott F and Brückel T 2008 *Phys. Rev.* B **77** 104435
[9] Høghøj P, Anderson I, Siebrecht R, Graff W and Ben-Saidane K 1999 *Physica* B **267-268** 355
[10] Cubitt R and Fragneto G 2002 *Appl. Phys.* A **74 (Suppl.)** S329
[11] Andersen A H, Cubitt R, Humblot H, Jullien D, Petoukhov A, Tasset F, Schanzer C, Shah V R and Wildes A R 2006 *Physica* B **385-386** 1134
[12] Maruyama R, Yamazaki D, Ebisawa T and Soyama K 2009 *J. Appl. Phys.* **105** 083527
[13] Kentzinger E, Rücker U, Toperverg B, Brücker T 2003 *Physica* B **335** 89.
[14] Dewhurst C D 2008 *Meas. Sci. Technol.* **19** 034007
[15] Dewhurst C D 2012 *Nucl. Instrum. Methods Phys. Res.* A **683** 16

IOP Conf. Series: Journal of Physics: Conf. Series **862** (2017) 012018 doi:10.1088/1742-6596/862/1/012018

Simulation and optimization of a new focusing polarizing bender for the diffuse neutrons scattering spectrometer DNS at MLZ

K Nemkovski[1], A Ioffe[1], Y Su[1], E Babcock[1], W Schweika[2] and Th Brückel[2]

[1]Jülich Centre for Neutron Science (JCNS) at Heinz Maier-Leibnitz-Zentrum (MLZ), Forschungszentrum Jülich GmbH, Lichtenbergstr. 1, 85748 Garching

[2]Jülich Centre for Neutron Science (JCNS) und Peter Grünberg Institut (PGI), JARA-FIT, Forschungszentrum Jülich GmbH, Lichtenbergstr. 1, 524258 Jülich, Germany

k.nemkovskiy@fz-juelich.de

Abstract. We present the concept and the results of the simulations of a new polarizer for the diffuse neutron scattering spectrometer DNS at MLZ. The concept of the polarizer is based on the idea of a bender made from the stack of the silicon wafers with a double-side supermirror polarizing coating and absorbing spacers in between. Owing to its compact design, such a system provides more free space for the arrangement of other instrument components. To reduce activation of the polarizer in the high intensity neutron beam of the DNS spectrometer we plan to use the Fe/Si supermirrors instead of currently used FeCoV/Ti:N ones. Using the VITESS simulation package we have performed simulations for horizontally focusing polarizing benders with different geometries in the combination with the double-focusing crystal monochromator of DNS. Neutron transmission and polarization efficiency as well as the effects of the focusing for convergent conventional C-benders and S-benders have been analyzed both for wedge-like and plane-parallel convergent geometries of the channels. The results of these simulations and the advantages/disadvantages of the various configurations are discussed.

1. Introduction

DNS is a versatile diffuse neutron scattering instrument with polarization analysis operated by JCNS at the research reactor FRM II (Heinz Maier-Leibnitz Zentrum) [1, 2]. It is used for studies of highly frustrated spin systems, strongly correlated electrons, emergent functional materials and soft condensed matter.

The general layout of the DNS is shown in figure 1. During the last years the instrument has been considerably upgraded. In particular, supermirror coating of the neutron guide (1) has been changed from m=1.2 to m=2. Respectively, the flux for the short-wavelength range has been considerably enhanced. The available short wavelength range has been extended from 2.4 Å to 1.5 Å. In order to deal with increased neutron flux, the monochromator (2) shielding has been replaced and improved. In addition, a neutron velocity selector (3) has been installed and successfully commissioned. The velocity selector allows suppressing the high-order contamination of the monochromatized beam coming from double-focusing PG monochromator (2). Alternatively, the selector can be used for selecting a shorter wavelength by the PG004 reflection with better resolution and without moving the secondary spectrometer. For the purpose of inelastic scattering experiments recently we have successfully installed a new high-frequency disc chopper (5) enabling the time-of-flight operation mode (in commissioning).

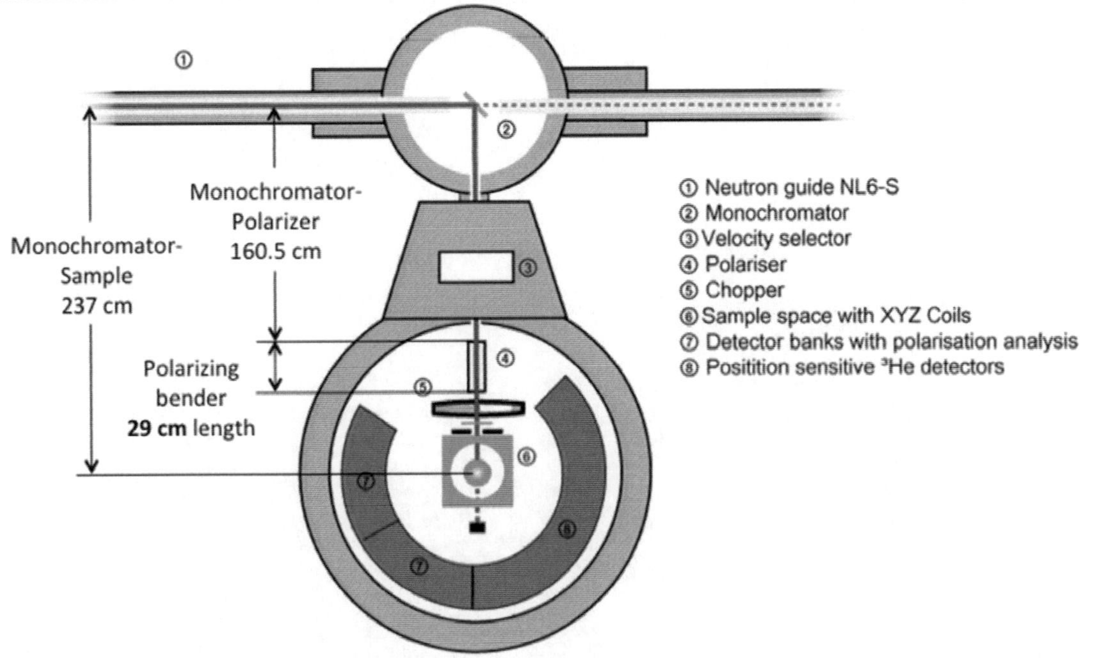

Figure 1. General layout of the diffuse neutron scattering spectrometer DNS at MLZ.

The polarization analysis at DNS is realized by a polarizer (4), XYZ-coils (6) with a Mezei-flipper and a polarization analyzer installed in the front of the detector bank (7). The currently used polarizer is a focusing Schärpf-type bender [3,4] of 29 cm length. It is composed of a stack of boron-containing glass sheets separated by wedge-type Al spacers. The spacers are essentially hollow sheets for optimal transmission, with only 6 mm Al material in trajectory of the neutrons. The glass sheets are coated on both sides by polarizing FeCoV/Ti:N supermirrors with a Gd:Ti anti-reflection layer underneath [5, 6]. To match the extended wavelength range, in the next step of the instrument upgrade we plan to change the existing polarizer.

The concept of the new polarizer is based on the idea of a solid-state bender composed of a number of silicon wafers with a double-side supermirror polarizing coating separated by absorbing spacers [7]. With this approach one can design very compact devices [7-10], which allows us to free space necessary for other instrument elements in the sample area. Additionally, to reduce the activation of the polarizer we plan to use the Fe/Si supermirrors [11] instead of currently used FeCoV/Ti:N ones. The optimization of the parameters of such a polarizing bender is the subject of this paper.

2. Analytical relations between bender parameters

Different geometries of benders impose certain constraints and relations between bender parameters, which can be easily obtained analytically and are presented below:

2.1. C-bender

In the case of a simple C-bender with plane-parallel channels (see figure 2a), the minimum length allowing avoiding the direct sight of view is given by

$$L_1^C = \sqrt{8Rd} \quad (1),$$

where R is the radius of curvature and d is the thickness of the channel ($R \gg d$) [12]. If the bender length L exceeds L_1^C, each neutron is reflected inside the channel at least once.

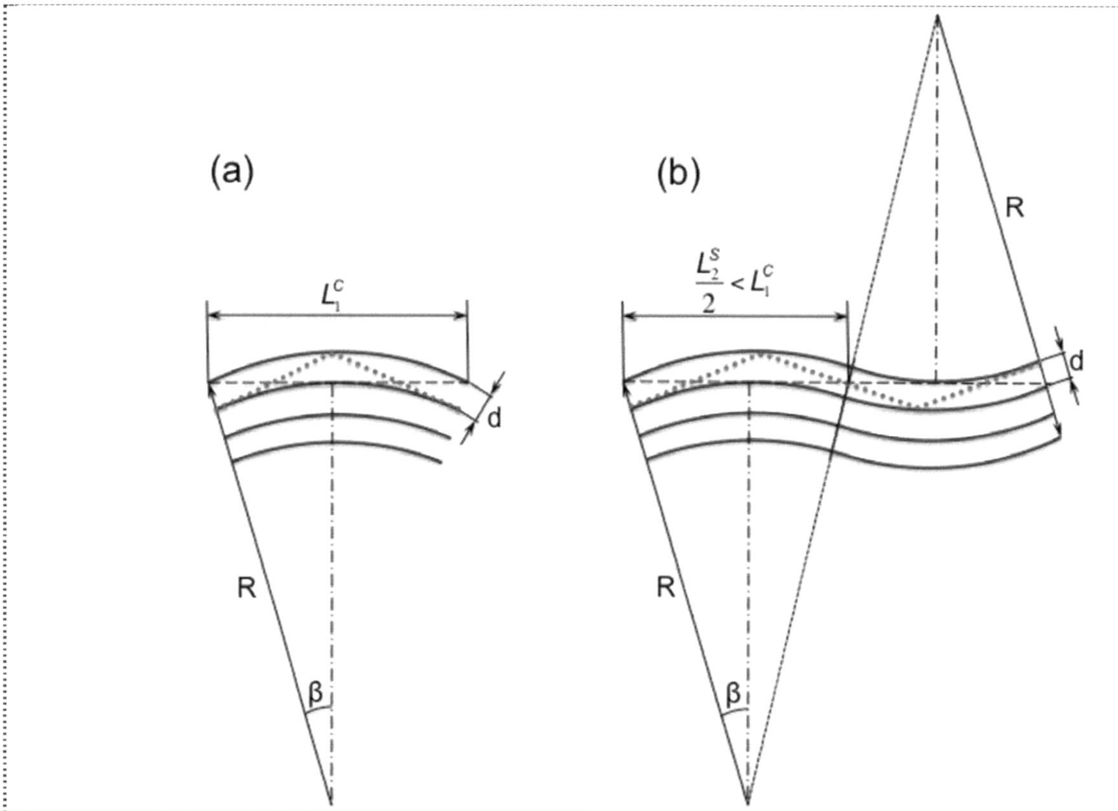

Figure 2. Geometry of neutron transport for a C-bender (a) and S-bender (b).

Another important parameter is the characteristic angle:

$$\beta = \sqrt{\frac{2d}{R}} \quad (2).$$

The value of 2β is sometimes referred as the bending angle of C-bender. It is also closely related to the characteristic shortest wavelength transmitted by the bender:

$$\lambda^* = \frac{\beta}{\gamma_c m} = \frac{1}{\gamma_c m}\sqrt{\frac{2d}{R}} \quad (3),$$

where $\gamma_c = 0.0017$ rad/Å is the critical angle of Ni coating for λ=1 Å, m is the index of the supermirror coating. Here one has to note, that for the C-bender there is no real cut-off value. Neutrons with shorter wavelength $\lambda<\lambda^*$ still can be transmitted by the C-bender (the so-called garland reflections), though their amount diminishes fast.

The equations (1-3) can be generalized to the case of convergent or divergent C-bender:

$$L_1^C = \sqrt{2Rd_1} + \sqrt{2Rd_2} \quad (4),$$

$$\beta = \sqrt{\frac{d_1}{2R}} + \sqrt{\frac{d_2}{2R}} \quad (5),$$

IOP Conf. Series: Journal of Physics: Conf. Series **862** (2017) 012018 doi:10.1088/1742-6596/862/1/012018

where d_1 and d_2 is width of the channel at the entrance and the exit, respectively ($|d_1 - d_2| \ll d$). By combining equations (3) and (5) one can see that for the convergent bender λ^* shifts towards a higher value, whereas for divergent bender it is just opposite.

2.2. *S-bender*

In the S-bender (see figure 2b) each neutron reflects inside the channel at least twice. The corresponding minimum length is given by:

$$L_2^S = 2(1+\sqrt{2})\sqrt{Rd} \qquad (6),$$

The characteristic angle β coincides with the one for the C-bender (see Eq. 2), however, here it is not anymore related to the deflection of the polarized beam with respect to the incoming beam: for an S-bender incoming and outgoing beams are parallel. However, β is still an important parameter that defines the transmitted wavelength range. Though in the case of an identical curvature and channel width the value of λ^* is the same both for C- and S-benders, the S-bender provides a sharp cut-off that strictly limits the minimum wavelength value:

$$\lambda_{cut-off} = \frac{2}{3}\lambda^* \qquad (7).$$

The effect of the wavelength cut-off is illustrated in figure 3.

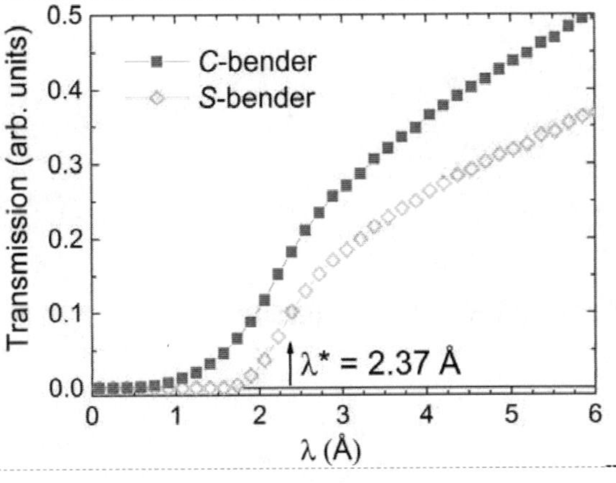

Figure 3. Transmission of the polarizing *not-convergent C-* and *S*-bender as a function of neutron wavelength as obtained from VITESS simulation. The parameters of both benders are based on the present-day geometry: d = 0.8 mm, R = 13500 mm. L = 300 and 510 mm for C- and S-bender, respectively. Transmission is normalized to the value for C-bender at λ = 6 Å.

The generalization for the case of convergent or divergent bender can be carried out in the same way as it is done in the section 2.1 (equations 4-5).

3. Monte-Carlo simulations

To optimize the parameters of the future polarizing bender for DNS, we have performed a series of Monte-Carlo simulations using the VITESS package [13]. Our model includes the cold source, the neutron guide, the double-focusing crystal monochromator (7x5 pieces of PG, 2.5 cm × 2.5 cm each, 0.55° mosaicity) and different kinds of polarizing benders. The main parameters for the options discussed in this paper are listed in Table 1.

The goal of simulations is to optimize the neutron beam parameters, flux and polarization, at the sample position. The neutron source has been simulated using the standard module "source_FRM2" from the VITESS library. In this module a simplified analytical description of the moderator characteristics is used, so that the resulted neutron spectrum slightly differs from the real one. Therefore, all intensity-related parameters are shown below in arbitrary units. In turn, the neutron wavelength dependences those are not normalized to the incoming flux, though being rather close to the expected in reality, are not completely accurate.

The reflectivity of the supermirror coating for the desired *m*-values is obtained by extrapolating the available data from Swiss Neutronics [14]. For solid state benders with Fe/Si coating, the position of the threshold at small Q in the spin-down reflectivity has been additionally corrected for the shift in the critical reflection angle caused by the difference in the refraction indices of the supermirror and the material of the bender channel (supermirror substrate), because the reflectivity curves in [14] have been obtained for neutrons incident from vacuum whereas in the case of the solid-state bender it is silicon). More details on the influence of the neutron-transmitting material on the reflectivity can be found in [11].

The neutron transmission of single-crystalline silicon has been taken into account according to data from [15]. The effect of the Laue extinction (the loss of the transmitted beam intensity by Bragg scattering in Si wafers) in the wavelength range of interest (i.e. 2-6 Å) is limited by the few narrow dips around 2.42, 2.79, and 5.06 Å, respectively [8]. During the measurements these particular wavelength values can be easily avoided. Therefore, for simplicity of the simulation we neglected the effect of Laue extinction.

Table 1. Main parameters of the simulated benders.

Style	Present-day air-gap C-bender	Solid-state C-bender, with wedge-type channels	Solid-state C-bender, with plane-parallel channels	Solid-state S-bender, with wedge-type channels	Solid-state C-bender, with plane-parallel channels
Length (mm)	290	55	55	95	95
Radius of curvature (mm)	13500	236	236	236	236
Channel width at the entrance (mm)	0.80	0.157	0.150	0.162	0.150
Channel width at the exit (mm)	0.57	0.150	0.150	0.150	0.150
Thickness of spacer between channels at the entrance (mm)	—	—	0.007	—	0.012
Thickness of absorbing layer (mm)	0.21	0.002	0.002	0.002	0.002
Supermirror coating	FeCoV/Ti:N m=2.7	Fe/Si m=3.5	Fe/Si m=3.5	Fe/Si m=3.5	Fe/Si m=3.5
Thickness of supermirror coating (mm)	0.003	0.003	0.003	0.003	0.003
Number of channels	44	271	271	265	265

Here we do not consider the air-gap C-bender with the Fe/Si coating instead of the FeCoV/Ti:N, because, first, it would not allow us to achieve a compact design – one of main objectives for a new polarizer and, second, having the same transmission for the desired spin state, it does not provide sufficiently high polarization for the whole required wavelength range.

The reason for the latter is the following. For a given wavelength, the maximal reflection angle for neutrons transmitted through the channel of bender is limited by the critical angle of the supermirror coating. The minimal angle is determined by the half of the bending angle β defined in equation 5 (some neutrons reflected by the angle smaller than $\beta/2$ still can pass through due to the garland reflections). On the other side, below critical reflection angle of Fe(↓)/Si the reflectivity for the Fe/Si coating is equal for both spin states that results in a low polarization at small incident angles (see figure 4). For neutrons incident to the supermirror from Si, the region of low polarization extends up to $\Theta \approx 0.8\gamma_C\lambda$. However, for neutrons incident from vacuum (or from air) it is increased to $\Theta \approx 0.8\gamma_C\lambda$. Obviously, the consequences of such shift are most severe for long wavelength neutrons. For instance, neutrons with $\Theta/(\gamma_C\lambda) = 0.5$ will be still well polarized (about 90%) when being incident from Si, and will not be polarized at all when being incident from vacuum. Indeed, in the latter case neutrons with wrong spin component will admix to the transmitted beam reducing its overall polarization. Simulations show that the beam polarization drops below 80% for $\lambda = 6$ Å - the upper limit of our wavelength range of interest.

To fix this problem one can decrease the radius of curvature by 2.5-3 times (to shift up the value of $\beta/2$, see equation 5) with the simultaneous increase of the critical angle of the supermirror coating to approximately 4.5 (to keep the value of λ^*, see equation 3). However, the length of such a bender will be about 170 mm, so that it will not be as compact as the solid state benders listed in table 1.

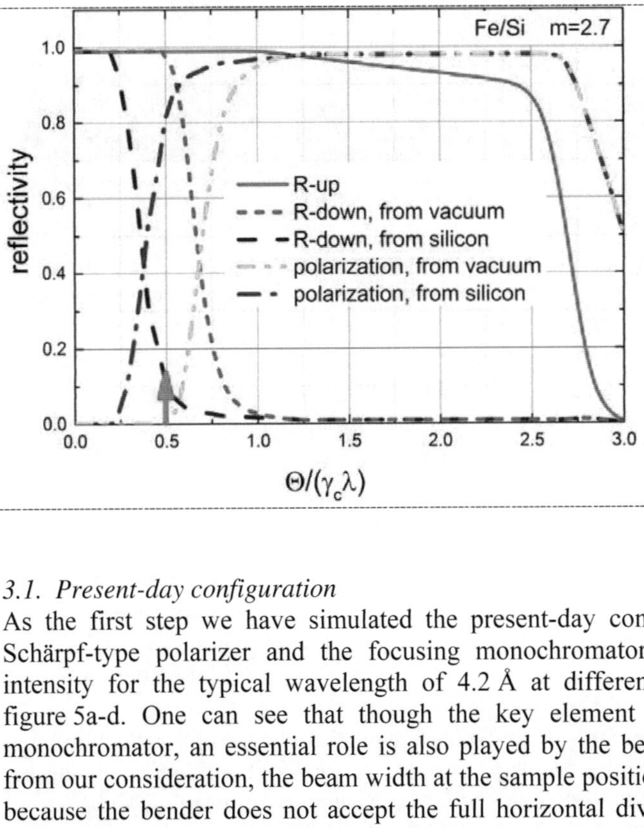

Figure 4. Model reflectivity for spin-up $|\uparrow\rangle$ and spin-down $|\downarrow\rangle$ states and polarization of m=2.7 Fe/Si supermirror vs. the normalized reflection angle. Reflectivity for the $|\downarrow\rangle$ component and polarization are shown for the neutron beam incidence from vacuum and silicon. The arrow indicates the position of the characteristic angle $\beta/2$ for $\lambda = 6$ Å.

3.1. Present-day configuration

As the first step we have simulated the present-day configuration of DNS, a combination of an Schärpf-type polarizer and the focusing monochromator. The distribution of the neutron beam intensity for the typical wavelength of 4.2 Å at different positions along the beam is shown in figure 5a-d. One can see that though the key element of the focusing system is the focusing monochromator, an essential role is also played by the bender. If we exclude the polarizing bender from our consideration, the beam width at the sample position (frame (d)) appears several times larger, because the bender does not accept the full horizontal divergence provided by the monochromator. Interestingly, at the same time the bender serves as the additional wavelength filter, because the most horizontally diverging neutrons also have the largest deviation in the wavelength. As a result (compare figures 5(e) and (f)), the bender reduces the wavelength spread and the beam divergence nearly by factor of 2 (at $\lambda = 4.2$ Å). This effect is most pronounced for long wavelengths and less important for the short ones.

The important consequence of this analysis is that for the proper optimization of the polarizing bender it is crucial to consider it in the combination with the monochromator.

Further results for the present-day configuration (transmission, polarization and quality factor as a function of wavelength) are discussed in the next section in the comparison with other bender options.

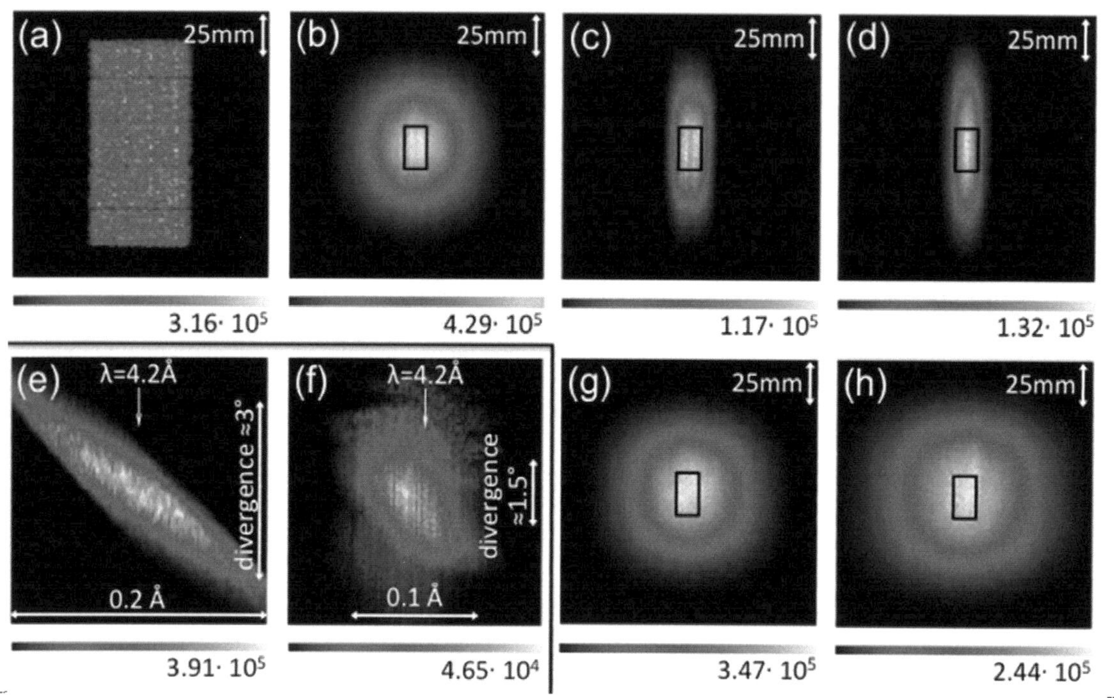

Figure 5. VITESS simulation of the present-day configuration of DNS (see text) at $\lambda = 4.2$ Å.
(a-d): Distributions of the neutron intensity across the beam at the (a) positions after monochromator, (b) entrance to the polarizing bender, (c) exit of the polarizing bender, (d) sample position.
(e,f): Neutron beam divergence vs. wavelength at the entrance (e) and the exit (f) of polarizing bender.
(g,h): Neutron intensity at the same positions as (c) and (d), respectively, but without bender.
The black rectangular in frames (a-d) and (g,h) corresponds to the image of one PG crystal of the monochromator.

3.2. Optimization of parameters of a new bender

To find the optimal supermirror coating index m for a new bender we have used the analytical relations (3) and (5). Using the parameters given in Table 1, one can obtain the characteristic wavelength for the present-day bender: $\lambda^* = 2.38$ Å. With a new polarizing bender we plan to expand the available wavelength range down to ~2 Å, which requires $m \geq 3.2$ for the same geometry. To be on the safe side, we have chosen slightly higher value of m=3.5.

The next important aspect is the arrangement of the bender channels. The current air-gap bender uses convergent wedge-type channels (see figure 6a). However, for the compact solid-state bender the production of silicon wedges of the necessary size is technologically very complicated. Therefore, as the main option we consider a bender with plane-parallel Si channels and spacers in between (figure 6b). Obviously, the part of the neutron beam entering the bender between Si wafers will be lost according to the ratio of the channel width to the spacer thickness. However, for a given inclination angle between wafers the spacer thickness is in turn proportional to the bender length, so that in the case of the compact bender the losses should not be really significant.

For the purpose of quantitative comparison we have carried out the simulations for the currently installed air gap bender, and C- and S-benders of different geometries shown in figure 6: using wedge-type Si wafers (a), plane-parallel Si wafers separated by spacers at the entrance (b).

The distance from focusing monochromator to bender, as well as the total entrance width and the angle of total bender convergence are fixed according to the given DNS layout: 1605 mm, 45 mm and 2°, respectively. The width of the bender channel, i.e. the silicon wafers' thickness, is 150μm according to the present-day technological possibilities. The radius of curvature and the length of benders are scaled from the present-day geometry using equations (4, 5). Parameters used for simulations are listed in Table 1.

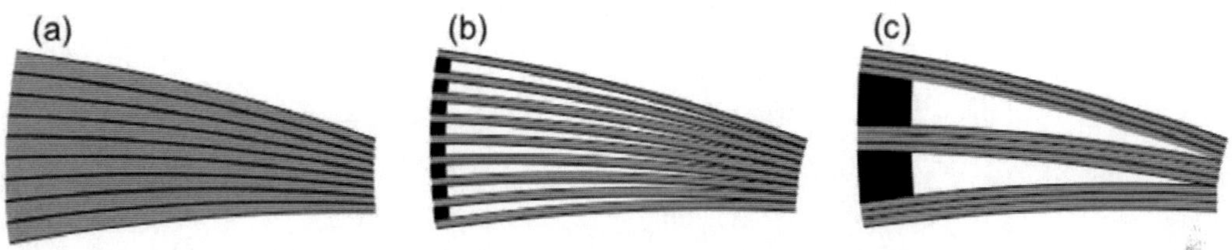

Figure 6. Possible channel arrangements for the convergent C-bender: wedge-type channels (a), plane-parallel channels with spacers at the entrance between each channel (b), channels grouped in blocks with spacers between blocks (c). (The gaps are largely exaggerated in this schematic picture.)

The results of the simulations are shown in figure 7. Here $P = \frac{N_\uparrow - N_\downarrow}{N_\uparrow + N_\downarrow}$ is the polarization of the outgoing neutron beam, $T = \frac{N_{total}^{out}}{N_{total}^{in}}$ - total transmission of initially unpolarized beam through the bender, $Q=P^2T$ - the figure of merit (also called the quality factor) commonly used for the characterization of polarizing devices [16-18]. However, for weak signals $Q=P^2T$ is not a good criteria anymore and one should better use the neutron beam with a high polarization.

The first important conclusion is that for the selected bender parameters the difference between benders with wedge-type and plane-parallel channels is negligible. From pure geometrical considerations one could expects the difference in transmission of 4.5% for C-benders, and 7.4% for S-benders. However, real numbers are even smaller (less than 3.5% and 5%, respectively), because the wedge-type channels accept smaller beam divergence (the same effect manifests itself as the reduction of the characteristic wavelength discussed in Section 2).

Among all analyzed options, the best polarization of the neutron beam is provided by the solid-state S-bender, owing to the highest number of reflections in the channels. The polarization of the solid-state C-bender is about 1% (2% for long wavelengths) worse than for the present-day bender, but still at a high level above 97%. At the same time, the transmission of the present-day is considerably lower than the one for C- and S-benders. Moreover, the transmission of the C-bender is 30% higher than for the S-bender. Therefore, the figure of merit $Q=P^2T$ is dominated by the transmission and demonstrates the clear superiority of the solid-state C-bender.

Beside the polarization and transmission of the benders, it is also important to watch the neutron intensity at the sample position. Figure 8 shows the distribution of the neutron flux across the beam for the above-discussed benders at three characteristic wavelength values of 6, 4.2 and 2 Å. The wavelength dependencies of the peak width, amplitude and integrated intensity are presented in the figure 9.

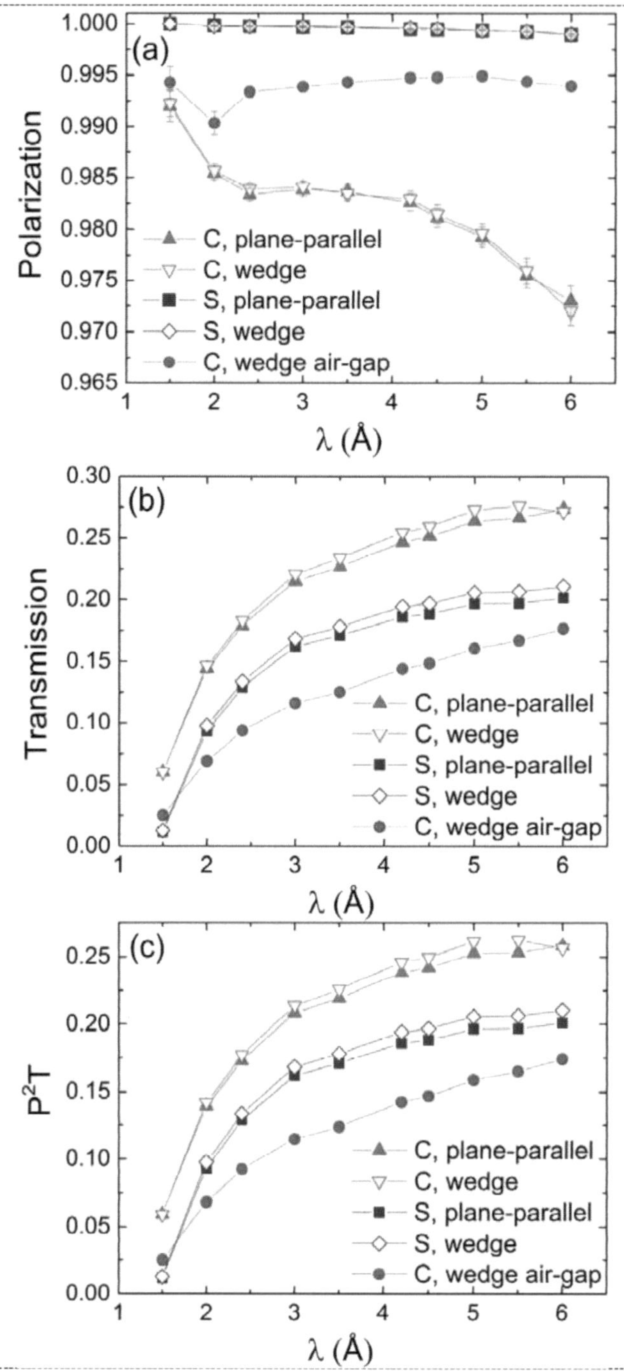

Figure 7. Main results of the simulation for different solid-state benders (see details in the text) in comparison with present-day air-gap bender: (a) transmission T, (b) polarization P, (c) quality factor P^2T. Where not shown, the error bars are within the size of the symbol. The transmission has been calculated as the ratio of the number of neutrons at the exit to the *total* number of neutrons (i.e. the both spin states) at the entrance of the polarizing bender.

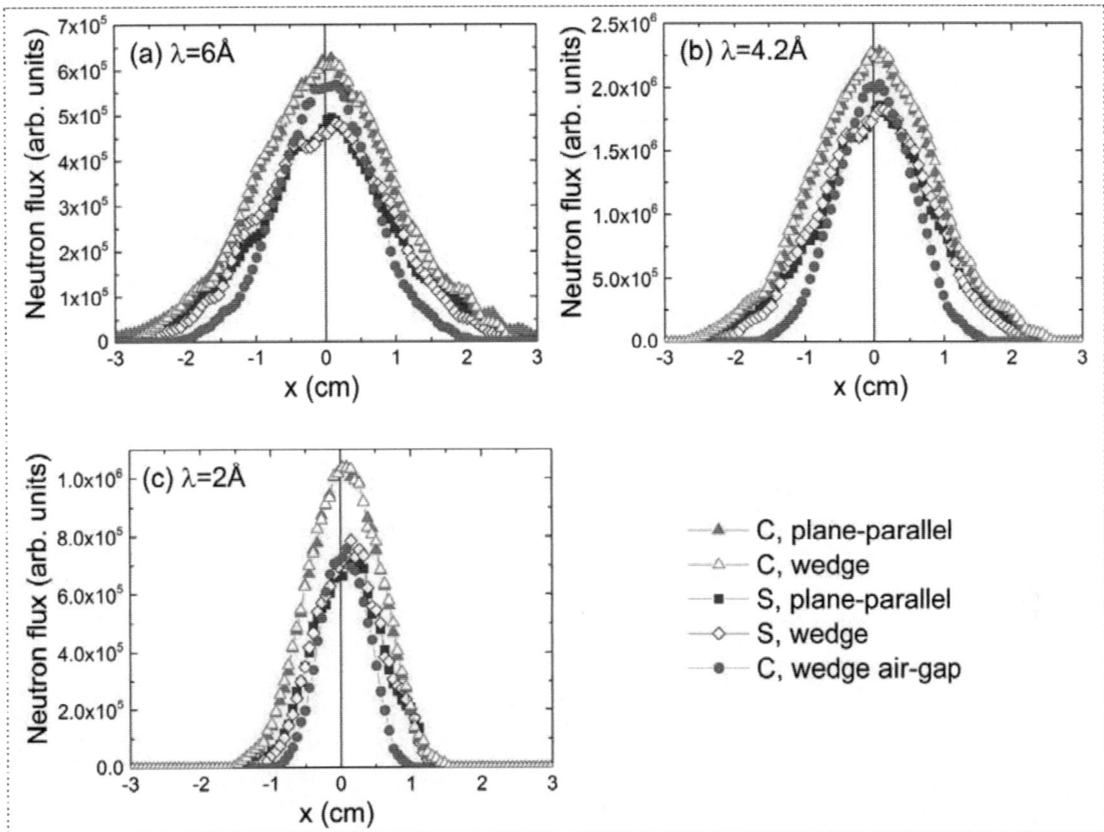

Figure 8. Neutron flux at the sample position for different benders at $\lambda = 6$ Å (a), $\lambda = 4.2$ Å (b) and $\lambda = 2$ Å (c). Neutron flux is integrated over the full beam height and shown as a distribution across the beam along horizontal x-axis perpendicular to the beam direction.

One can see that considered benders provide different beam size at the sample position. The smallest one can be achieved with present-day bender because it accepts the least angle divergence due to the lower value of $m = 2.7$, and has the largest length (in certain sense the polarizing bender serves also as a neutron guide). For the shortest solid-state C-benders with $m = 3.5$ the beam at the sample position is ~30% wider, whereas the S-benders with intermediate length demonstrate the values of the beam width between the present-day air-gap and solid-state C-bender. The effect of accepted divergence is clearly seen in the difference between solid-state benders of the same type with wedge-type and plane-parallel channels (for the bender with wedge-type channels the accepted divergence is higher). On the other side, for all bender we observe the increase of the beam size at the sample governed by the growth of the incoming beam divergence along with increase of the wavelength.

Due to the large beam width, the numbers of the intensity gain obtained above for the solid-state benders might be overestimated. To get more accurate numbers one has to analyze carefully beam distribution at the sample position. In agreement with transmission results, the highest peak amplitude is realized for the solid-state C-benders (figure 9b). Notably, this value is strongly dependent on the neutron wavelength. It has a clear maximum around 3-3.5 Å, corresponding to the spectral maximum in the flux coming to the monochromator.

In order to exclude the uncertainty related to the spectrum of the neutron source, in figure 9c we present the amplitude normalized to the incoming flux. Here, after the sharp increase at the short wavelengths, the curve is nearly constant, with some decay at long wavelengths because of the beam spreading over a larger area due to higher divergence. The gain in respect to the air-gap bender is almost 40% at λ = 2 Å but only 10% at 6 Å.

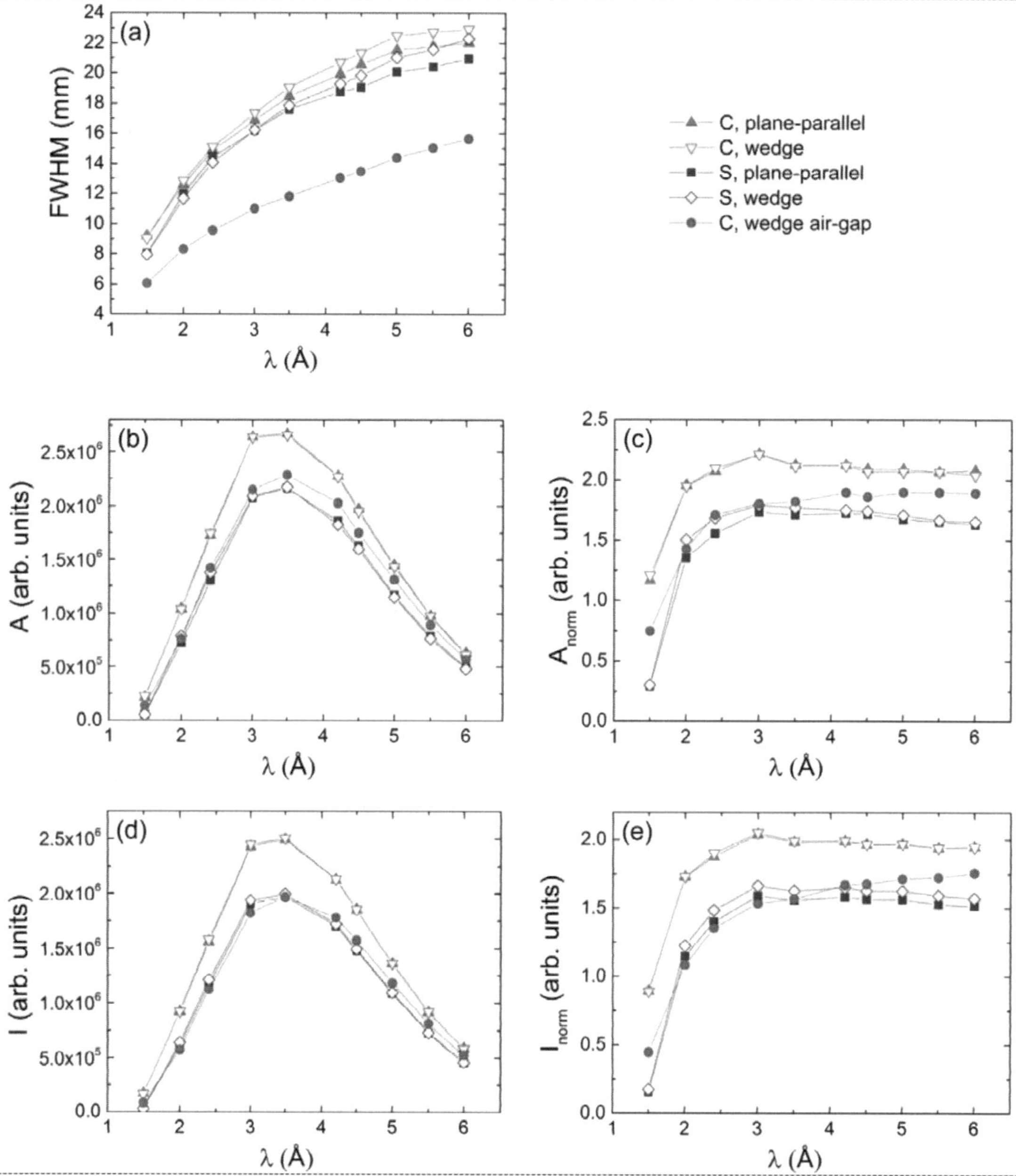

Figure 9. Parameters of the neutron flux distribution at the sample position as a function of neutron wavelength (neutron flux is integrated over the full beam height) for different benders: (a) the width (FWHM) of the distribution across the beam along horizontal x-axis perpendicular to the beam direction, (b) peak amplitude, (c) peak amplitude normalized to the wavelength dependence of the incoming flux, (d) intensity integrated over the range of 10mm in horizontal direction, (e) intensity integrated over the range of 10mm in horizontal direction and normalized to the wavelength dependence of the incoming flux.

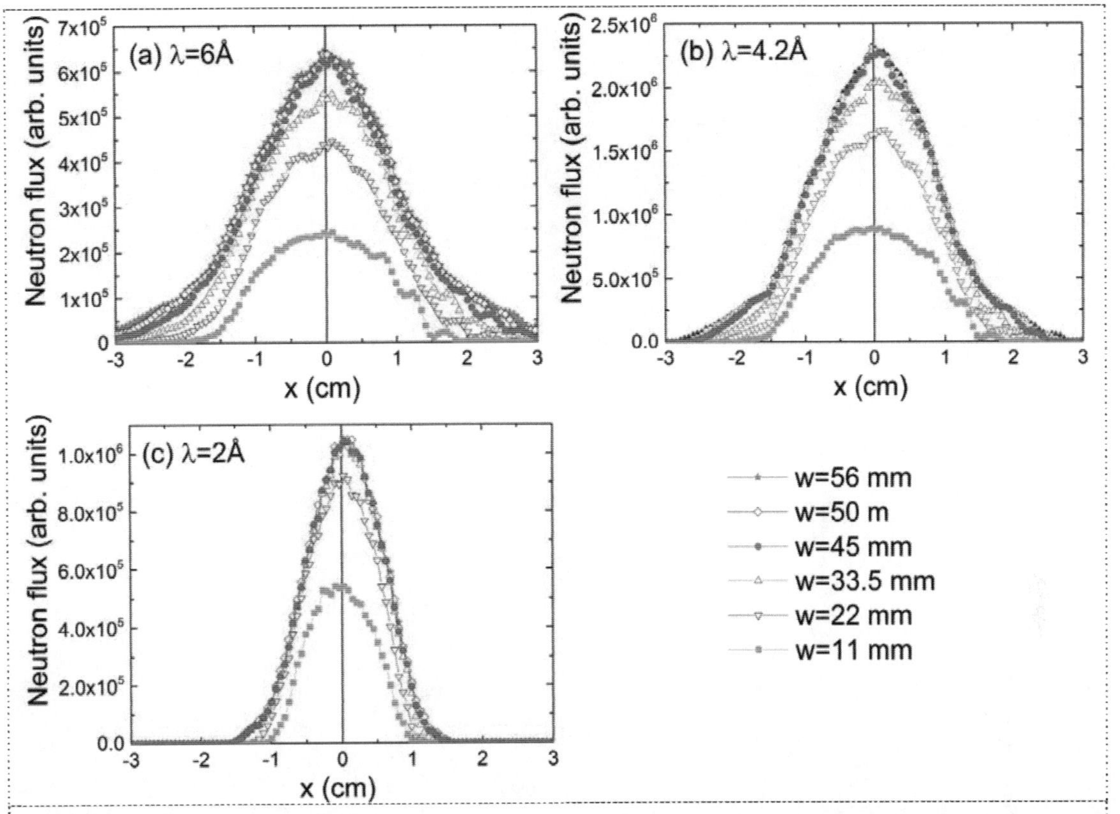

Figure 10. Influence of the width of C-bender with plane-parallel channels on the neutron flux at the sample position for $\lambda = 6$ Å (a), $\lambda = 4.2$ Å (b) and $\lambda = 2$ Å (c). Neutron flux is integrated over the full beam height and shown as a distribution across the beam along horizontal x-axis perpendicular to the beam direction.

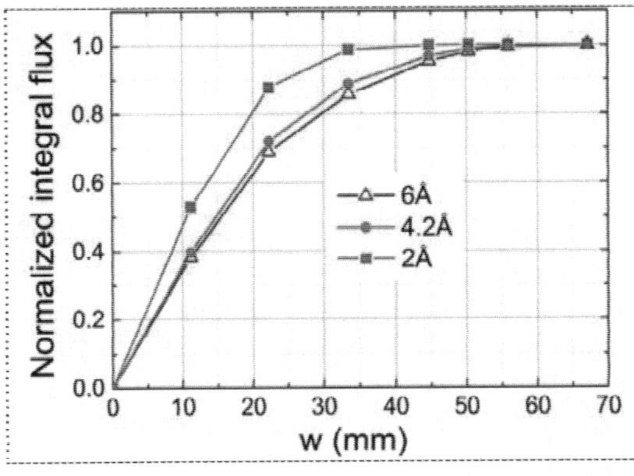

Figure 11. Dependence of the integral neutron flux at the sample position on the width of the C-bender with plane-parallel channels for several typical wavelength values. Neutron flux is integrated over the full beam height in vertical direction and over the range of 10 mm in horizontal direction, and then normalized for each wavelength to the unity at the maximum considered bender width.

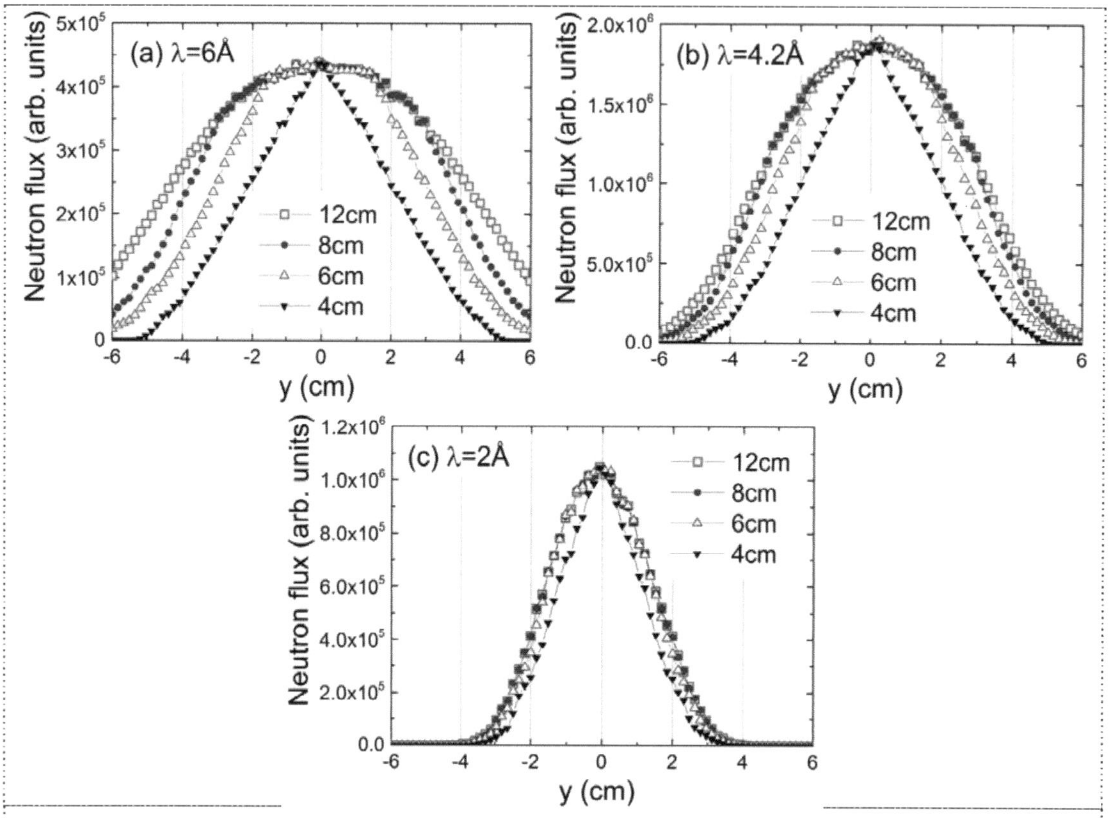

Figure 12. Influence of the height of C-bender with plane-parallel channels on the neutron flux at the sample position for λ = 6 Å (a), λ = 4.2 Å (b) and λ = 2 Å (c). Neutron flux is integrated over the full beam width and shown as a distribution across the beam along vertical y-axis perpendicular to the beam direction.

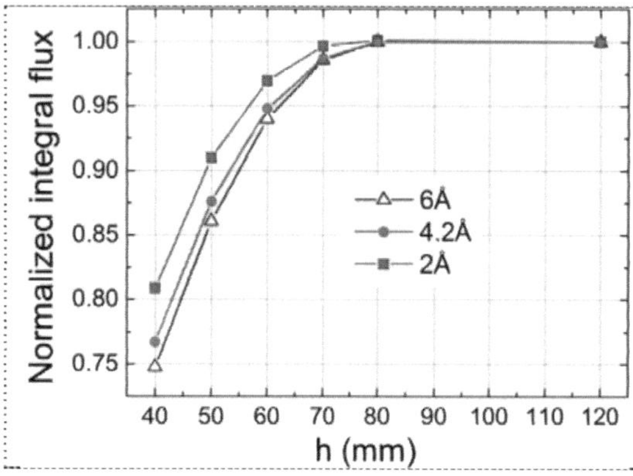

Figure 13. Dependence of the integral neutron flux at the sample position on the height of the C-bender with plane-parallel channels for several typical wavelength values. Neutron flux is integrated over the full beam width in horizontal direction and over the range of 60 mm in vertical direction, and then normalized for each wavelength to the unity at the maximum considered bender height.

IOP Conf. Series: Journal of Physics: Conf. Series **862** (2017) 012018 doi:10.1088/1742-6596/862/1/012018

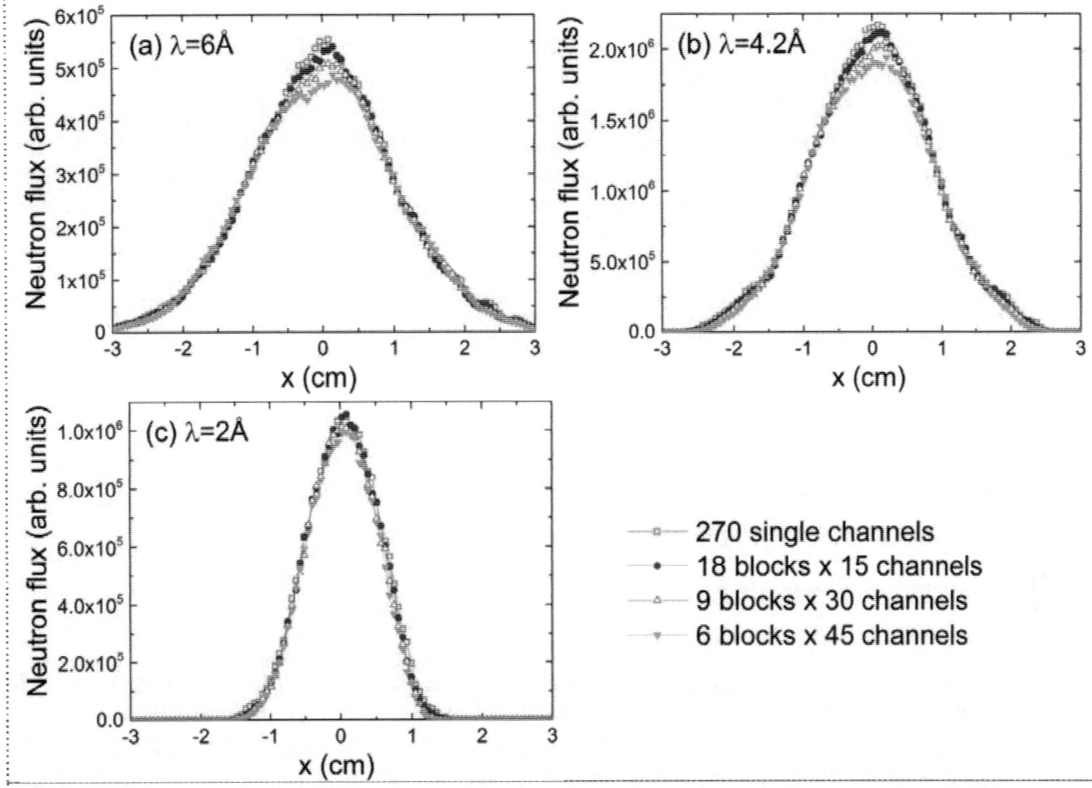

Figure 14. Neutron flux distributions at the sample position for the C-bender with the channel grouping. Neutron flux is integrated over the full beam height and shown as a distribution across the beam along horizontal x-axis perpendicular to the beam direction.

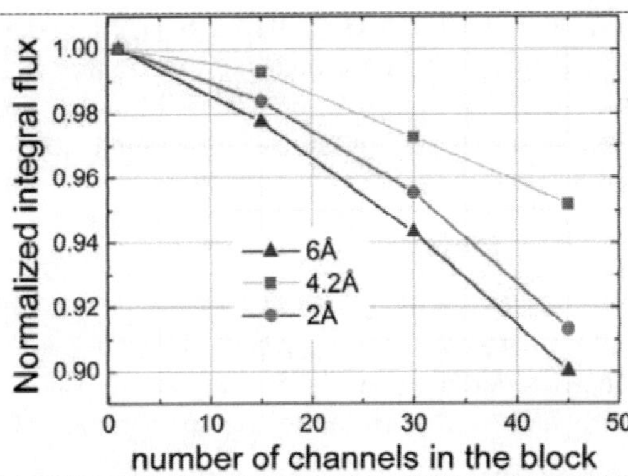

Figure 15. Dependence of the integral neutron flux at the sample position on the number of channels in the block (the number of channels is fixed to be 270). The neutron flux is integrated over the full beam height and 10 mm width, and normalized to unity for the case of the bender with single channels (figure 6b)

To get more accurate estimations, one has to integrate the intensity around the center of the beam over the typical sample width (10 mm at DNS). It is especially important for the short wavelength limit, where the present-day DNS bender cannot provide large enough beam size at the sample position. The larger beam width allows for the increase of the integral flux at the sample. Finally, for a typical experiment with the solid-state C-bender the gain in the integral intensity varies from 60% at $\lambda = 2$ Å to 10% at $\lambda = 6$ Å. In the long wavelength limit integration does not change the numbers because of the already sufficiently large beam size resulted from a large beam divergence at the exit from the neutron guide.

For the solid-state S-benders, despite of some gain in total transmission in respect to the present-day bender, the neutron flux at the sample is quite close to the present-day configuration, with some small gain or loss, depending on the neutron wavelength (see figure 9e). Thus, we can undoubtedly confirm the superiority of the solid-state C-bender.

For the practical realization of such bender, it is necessary to find the optimal width and height as the larger sizes leads to higher costs as well as extra complications in manufacturing. The influence of the bender width w on the neutron flux distribution at the sample position is presented in the figure 10. The dependence of the intensity at the 10 mm wide sample is shown in figure 11. One can see that for the bender with w=45mm intensity losses of about 5% are observed for longer wavelengths 4.2 Å and 6Å, whereas for short wavelength of 2 Å the losses are only about 1%. Indeed, the bender with a width of 45 mm is chosen for DNS as a reasonable compromise between the performance and costs.

A similar analysis has been carried out for the bender height h (figure 12). The reduction of the bender height gives rise to the cropping of the wings of the intensity distribution and subsequently to the intensity losses (see figure 13). However, since the maximum sample height at DNS is around 60mm (though usually does not exceed 40 mm), the height of the bender can be reduced down to 80 mm without any losses of the flux. Note, that the bender with h=40 mm limits the divergence of the beam delivered by the focusing monochromator and therefore the corresponding transmission profiles have a triangular shape that is typical for a slit collimation.

On the same reasons as for the width and the height, costs and complexity of manufacturing, it seems reasonable to reduce the number of spacers by combining several silicon wafers in a single block without using spacers, and place spacers between such blocks (see figure 6c). Below we present the optimization of these parameters for the C-bender with plane-parallel channels.

The results of simulations for the C-bender consisting of 270 channels (wafers) grouped in 18 blocks by 15 channels, 9 blocks by 30 channels, 6 blocks by 45 channels are presented in figures 14, 15. In general, the influence of grouped wafers is similar to the reduction of the bender width, with the most pronounced effect at the long wavelength. With 18 blocks by 15 wafers the integrated intensity is reduced by 2% in the worst case of $\lambda = 6$ Å, which is still acceptable.

4. Conclusion

The performance of the focusing polarizing benders illuminated by a convergent neutron beam reflected from focusing crystal monochromators has been considered. In sense of total transmission, the solid state S-bender is found to be inferior to the solid state C-bender due to the shorter neutron path in silicon, but still better than the Schärpf-type air-gap bender. However, taking into account the real sample size, our simulations unambiguously demonstrate the superiority of the focusing solid-state C-bender with respect to both Schärpf-type and solid-state S-benders.

We have analyzed the performance of a C-bender with a reduced number of spacers, when several silicon wafers are combined in the blocks, which may significantly simplify the manufacturing process. Though it will generally result in extra losses in the transmission of the bender, the losses may be tolerated if the number of wafers in the blocks is not too large; e.g. grouping together 15 wafers reduces the bender transmission only by 2%.

Finally, we consider the solid-state C-bender with Fe/Si coating m=3.5 as the optimal solution for a new polarizer at DNS. Such bender is much more compact than the present-day air-gap Schärpf-type bender (55 mm against 290 mm) and provides the intensity gain from 10% to 60% at the 10mm sample (depending on the wavelength) by the cost of slightly lower polarization (97-98.5% against 99.5%).

Acknowledgments
We would like to thank S. Manoshin (JINR, Russia), M. Khaneft, S. Mattauch, N. Violini, Z. Salhi (all JCNS) and Th. Krist (HZB) for useful discussions.

References
[1] Schweika W, Böni P 2001 Physica B **297** 155-159
[2] Su Y, Nemkovskiy K and Demirdiş S 2015 JLSRF **1** A27 1-4
[3] Schaerpf O 1989 Physica B **156&157** 639-646
[4] Hahn W, Schweika W 1997 Physica B **234** 1165-1167
[5] Böni P, 1996 J. Neutron Res. **5** 63-70
[6] Böni P, Clemens D, Senthil Kumar M, Pappas C 1999 Physica B **267-268** 320-327
[7] Krist Th, Kennedy S J, Hicks T J, Mezej F 1998 Physica B **241-243** 82-85
[8] Stunault A, Andersen K H, Roux S, Bigault T, Ben-Saidane K, Rønnow H M 2006 Physica B **385–386** 1152–1154
[9] Krist Th, Rucker F, Brandl G, Georgii R 2013 Nucl. Instr. Meth. Phys. Res. A **698** 94-97
[10] Shah V R, Washington A L, Stonaha P, Ashkar R, Kaiser H, Krist T, Pynn R 2014 Nucl. Instr. Meth. Phys. Res. A **768** 157-163
[11] Majkrzak C F, Nunez V, Copley J R D, Ankner J F, and Greene G C 1992 Proc. SPIE **1738** 90
[12] Hayter J B, Penfold J, Williams W G 1978 J. phys. E: Sci. Instrum. **11** 454-458
[13] http://www.helmholtz-berlin.de/forschung/oe/em/transport-phenomena/projekte/vitess/index_en.html
[14] http://www.swissneutronics.ch/index.php?id=24
[15] Freund A K 1983 Nucl. Instr. Meth. **213** 495-501
[16] William W G 1973 Rutherford Appleton Laboratory Report **RL-73-034**
[17] Gussen L D 1998 J. Neutr. Res. **7** 15-37
[18] Goossens D J, Cussen L D 2002 Nucl. Instr. Meth. A **481** 475-492

IOP Conf. Series: Journal of Physics: Conf. Series **862** (2017) 012019 doi:10.1088/1742-6596/862/1/012019

Polarisation analysis on the LET time-of-flight spectrometer

G J Nilsen, J Košata, M Devonport, P Galsworthy, R I Bewley, D J Voneshen, R Dalgliesh and J R Stewart

ISIS Neutron and Muon Facility, Rutherford Appleton Laboratory, Science and Technology Facilities Council, Didcot OX11 0QX, UK

E-mail: goran.nilsen@stfc.ac.uk

Abstract. We present a design for implementing uniaxial polarisation analysis on the LET cold neutron time-of-flight spectrometer, installed on the second target station at ISIS. The polarised neutron beam is to be produced by a transmission-based supermirror polariser with the polarising mirrors arranged in a "double-V" formation. This will be followed by a Mezei-type precession coil spin flipper, selected for its small spatial requirements, as well as a permanent magnet guide field to transport the beam polarisation to the sample position. The sample area will contain a set of holding field coils, whose purpose is to produce a highly homogenous magnetic field for the wide-angle ^{3}He analyser cell. To facilitate fast cell changes and reduce the risk of cell failure, we intend to separate the cell and cryostat from the vacuum of the sample tank by installing both in a vessel at atmospheric pressure. When the instrument upgrade is complete, the performance of LET is expected to be commensurate with existing and planned polarised cold neutron spectrometers at other sources. Finally, we discuss the implications of performing uniaxial polarisation analysis only, and identify quasi-elastic neutron scattering (QENS) on ionic conducting materials as an interesting area to apply the technique.

1. Introduction

Longitudinal neutron polarisation analysis grants the ability to separate the components of the neutron scattering cross section, thereby permitting the isolation of weak magnetic or structural signals, as well as giving direct access to the directional components of the sample magnetisation [1]. While the technique has been applied successfully on a wide range of instruments at both reactor and spallation sources, its extension to wide-angle time-of-flight spectrometers has proved challenging. Indeed, despite 10 years passing since the first proof of principle [2], the technique is still not in routine use. The reasons for this are manifold. Firstly, time-of-flight spectroscopy remains a flux limited technique, requiring large samples and long counting times, and it is only recently that polarisers and analysers with a sufficient transmission to permit experiments on reasonable timescales have become available. Second is the issue of analysing the scattered neutron polarisation over the large solid angles covered by the detectors on direct geometry time-of-flight spectrometers. Two options for achieving this exist: channeled supermirror-based devices and polarised ^{3}He filters. For the former, acceptance of beam divergence from large samples, the outgoing divergence from the device, and shadowing of the detectors by the absorbing blades all complicate data collection and analysis. The cost of a large supermirror array can also be prohibitive. For the latter, maintaining the ^{3}He polarisation over a large

IOP Conf. Series: Journal of Physics: Conf. Series **862** (2017) 012019 doi:10.1088/1742-6596/862/1/012019

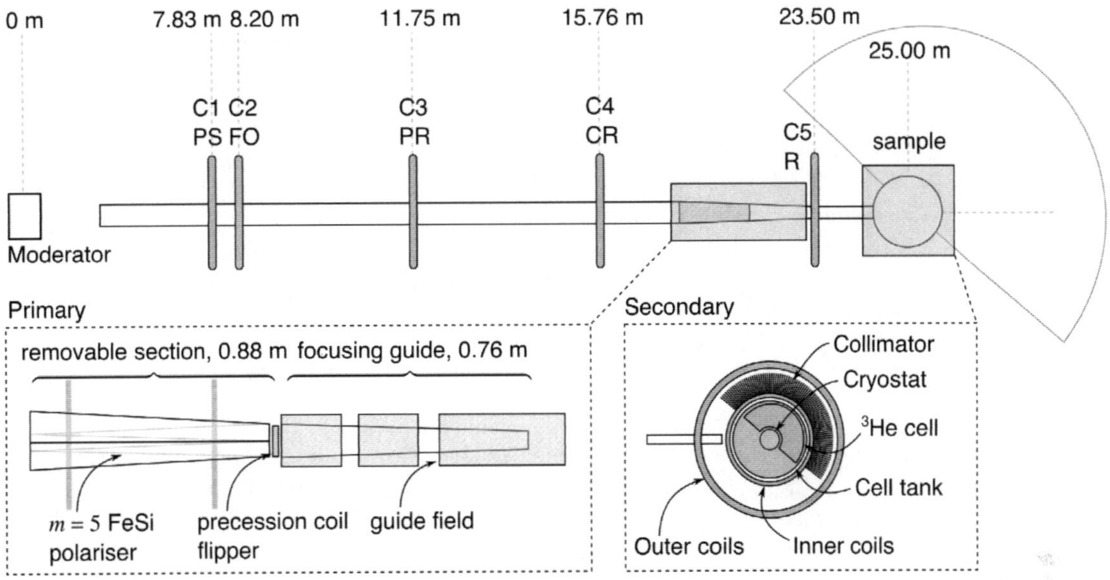

Figure 1. Schematic of the LET instrument viewed vertically. Five sets of disc choppers (C1-C5) select one or more neutron wavelength (Pulse Removal – PR, Contaminant Removal – CR), shape the pulse (Pulse Shaping – PS, Resolution – R), and eliminate frame overlap (Frame Overlap – FO). The magnified regions show the positions of the polariser and flipper and the analyser, respectively. Once polarised (see Section 3.1 and Figure 2), the neutron beam can be flipped by a precession coil flipper (see Section 3.2). It is then transported to the sample by guide fields and a set of coils (see Section 3.4), before being analysed by a wide-angle ^{3}He spin filter analyser (see Section 3.3). Scattering from the analyser cell and the sample environment is eliminated by a radial oscillating collimator.

filter volume presents a significant challenge. There is currently a large impetus to overcome these issues, with several polarised time-of-flight instruments under construction (PANTHER at ILL, TOPAS at FRMII [3]), under commissioning (MACS at NIST [4], HYSPEC at SNS [5], POLANO at JPARC [6, 7, 8]), or already in operation (D7 at ILL [11] and DNS at FRMII [10]]).

The LET spectrometer is a direct geometry time-of-flight spectrometer located on the second target station at the ISIS pulsed neutron and muon source in the United Kingdom. It is installed on a coupled hydrogen moderator, which provides a neutron spectrum in the wavelength range $\lambda = 2 - 10$ Å[12]. A set of five disc choppers allow for so-called repetition rate multiplication, whereby the time-frame between neutron pulses is efficiently filled by measuring spectra at several different incident energies E_i [12, 13]. The neutron flux at the sample position at a resolution of $\Delta E/E_i = 2\%$ and $\lambda = 5$ Å, corresponding to the maximum in the wavelength spectrum, is $\Phi = 4.4 \times 10^4$ n cm^{-2} s^{-1}, close to those of CNCS at the Spallation Neutron Source [14] and IN5 at the Institut Laue-Langevin [15] for the same conditions. A unique feature of LET among cold time-of-flight spectrometers is its large detector coverage, approaching π steradians, making it ideally suited for collecting four-dimensional $S(\mathbf{Q}, \omega)$ maps of excitations in single crystals. Its excellent energy resolution, as low as 1% at at $E_i = 1$ meV, also makes it an excellent choice for quasi-elastic neutron scattering (QENS) on materials ranging from biological systems to polymers to inorganic ion conductors. The key parameters of the instrument are summarised in Table 1.

IOP Conf. Series: Journal of Physics: Conf. Series **862** (2017) 012019 doi:10.1088/1742-6596/862/1/012019

Table 1. Key parameters for the LET spectrometer.

Parameter	Value
E_i	$0.8 - 20$ meV ($\lambda = 2 - 10$ Å)
$\Delta E/E_i$	$0.5 - 4\%$
Φ (5 Å, 2%)	$\Phi = 4.4 \times 10^4$ n cm^{-2} s^{-1} (unpolarised)
Beam size	20×40 mm^2
Detectors	^{3}He PSD, $-40° \leq \theta \leq 140°$, $-30° \leq \gamma \leq 30°$
Polarizer	$m = 5$ FeSi channeled "double-V" cavity
Flipper	ramped precession coil (Mezei)
Analyser	wide-angle ^{3}He spin filter

We present here the design for a longitudinal polarisation analysis option on LET, the construction of which is currently in progress. We plan to overcome the difficulties highlighted above by choosing the simplest possible implementation of polarisation analysis: the beam will be polarised by a solid state super-mirror device, and flipping will be performed using a Mezei-type precession coil π-flipper [16]. Uniaxial polarisation analysis will be achieved using a wide-angle Si-windowed ^{3}He spin filter cell, to be contained in a vessel also holding the sample environment. The paper is structured as follows: in section 2, we will give an overview of the history and the aims of the project. Then, we will provide a more detailed description of the simulation and design of the polarised neutron components in section 3. The expected performance will be given in section 4, along with a discussion of the possible scientific applications and the relative merits of three-directional (xyz) versus uniaxial polarisation analysis. We will conclude in section 5.

2. Project overview

Longitudinal polarisation analysis, in which only the longitudinal components of the beam polarisation with respect to the external field are measured [1], requires means to polarise, analyse, and manipulate the spin state of the neutron beam; the classic implementation on a triple-axis spectrometer comprises a polariser and a flipper before the sample, followed by a set of coils to rotate the beam polarisation into three orthogonal directions, x, y, and z. Following scattering from the sample, the beam then passes through another flipper and an analyser before being detected. This setup permits measurement of the 12 (longitudinal) cross sections of the type $\sigma_{\pm\alpha\pm\alpha}$ and $\sigma_{\pm\alpha\mp\alpha}$, where $\alpha = \{x, y, z\}$. These have the general form:

$$\begin{aligned} \sigma_{\pm\alpha\pm\alpha} &= \sigma^{nuc} + \sigma^{mag}_{||\alpha\perp\mathbf{Q}} + \frac{1}{3}\sigma^{si} \pm \sigma^{nm} \\ \sigma_{\pm\alpha\mp\alpha} &= \sigma^{mag}_{\perp\alpha\perp\mathbf{Q}} + \frac{2}{3}\sigma^{si} \pm \sigma^{chi} \end{aligned} \tag{1}$$

where σ^{nuc} is the nuclear (and isotope incoherent) cross section, σ^{si} the spin incoherent cross section, and σ^{mag} the magnetic cross section, which is sensitive to both the direction of the incident polarisation and the scattering vector $\mathbf{Q}$: non spin-flip processes $\sigma_{\pm\alpha\pm\alpha}$ measure the component of the Fourier-transformed magnetisation parallel to the polarisation and perpendicular to $\mathbf{Q}$, while spin-flip processes probe the component perpendicular to both the polarisation and $\mathbf{Q}$. Finally, σ^{chi} and σ^{nm} are terms which arise from non-collinear structures and nuclear-magnetic interference, respectively. Separation of the cross section components from these cross sections is generally possible.

IOP Conf. Series: Journal of Physics: Conf. Series **862** (2017) 012019 doi:10.1088/1742-6596/862/1/012019

On a multi-detector spectrometer, the second flipper is usually dispensed with as a result of the difficulty of flipping over a wide scattering angle, meaning that the final polarisation analysis system consists of a polariser, a flipper, a guide field, and an analyser. As such, only the six of the twelve cross sections *i.e.* $\sigma_{+\alpha+\alpha}$, $\sigma_{+\alpha-\alpha}$, etc. can typically be measured. Nonetheless, if certain assumptions are made on the form of the magnetic scattering, separation of the cross section components is still possible using either so-called xyz (three-directional) polarisation analysis [9, 11], or the more recent 10−point technique, devised for instruments which integrate over the detector height [17].

LET was designed with a polarisation analysis option from the outset; the beam was originally polarised by a ^{3}He spin filter, with the ^{3}He gas polarised *in situ* via the spin exchange optical pumping (SEOP) method [18]. The spin filter and associated optics were mounted on a trolley on rails to facilitate easy changes between the polarised and unpolarised configurations. The ^{3}He gas in the polariser, and hence also the neutron beam, could furthermore be flipped by applying an adiabatic fast passage NMR pulse to the spin filter cell. Analysis of the scattered beam was to be achieved using wide-angle ^{3}He analyser cells filled remotely by a metastable exchange optical pumping (MEOP) filling station. However, test experiments with the SEOP polariser showed that its maximum ^{3}He polarisation $P_{He} = 65\%$, combined with the removal of approximately 1 m of guide, resulted in large losses in flux, in excess of 90% across the wavelength band of LET. As a result of the limitations of the SEOP polariser, and because the moderator spectrum on LET is colder than originally envisaged, it was decided to replace the original polariser by a supermirror-based device, which exhibits superior performance at all but the shortest wavelengths. This device consists of a two-channel tapering guide with each channel containing a pair of polarising FeSi supermirrors arranged in a "V" formation, and its design is detailed in Section 3.1.

The replacement of the ^{3}He polariser also means that flipping has to be performed directly on the neutron beam. The length of the new polariser, as well as the large stray field produced by the permanent magnets used to saturate the supermirrors, severely restricts the alternatives in this respect. Indeed, the default option for a white beam with a large cross-section, the adiabatic radio-frequency (RF) flipper, consumes too much space and is sensitive to field inhomogeneities. As such, we choose a much simpler precession coil (or Mezei) flipper, which can be made less than 10 mm thick and is only sensitive to field gradients of $\mathcal{O}(10^{-2}|B|)$. Despite the robustness of the device, the large stray field produced by the polariser still requires careful shielding, and since a precession coil only flips a single wavelength for a given current, the current needs to be ramped in time. Solutions to both of these issues will be detailed in Section 3.2

The polarised beam is to be transported to the sample by a set of permanent magnet guide fields followed by a set of coils oriented perpendicular to the beam direction; in other words, the magnetic field throughout the instrument will point in the z (up) direction, meaning that non-adiabatic rotation of the neutron polarisation is less of a concern for the present implementation, as compared to cases where full three-direction polarisation analysis is attempted. The coils will also generate a homogenous holding field across the volume of the ^{3}He spin filter analyser cell, as demonstrated in Section 3.4.

The ^{3}He spin filter analyser presents a significant challenge; previous tests of wide-angle composite Si/pyrex cells for use at 3 atm resulted in a high failure rate. To mitigate the risk of cell failure and to simplify their manufacture, we will reduce the operating pressure to ~ 1 atm, and maintain the cells (as well as the sample environment) in 1 atm of inert gas by placing them in a vacuum-tight vessel, to be installed inside the LET sample tank. This leads to cell dimensions which are compatible with both the wavelength band of LET and the homogenous region of the holding field, as shown in Section 3.3. We will aim to reduce the background generated by the cell by using polished Si windows for all parts of the cell which transmit the neutron beam. Any remaining scattering from the cell and the sample environment will be removed by a radial oscillating collimator.

The layout of the new polarisation analysis option for LET is summarised in Figure 1. The supermirror polariser and precession coil flipper are located ahead of the final set of disc choppers, approximately 3 m before the sample position. In the sample area, the coils will be suspended on a rack which also supports the inner vessel, which is to hold the ^{3}He analyser cell and the sample environment. The collimator will be installed on existing supports in the sample tank.

3. Design studies

In this section, we will provide a more detailed description of the components either under design or construction. Some components, like the radial oscillating collimator, have not yet been designed, and we therefore defer their description to a future paper.

3.1. Polariser

With recent advances in polarising supermirror technology, it has become possible to polarise a broad neutron wavelength band in a relatively short distance. This is crucial for LET, where only 882 mm is available to accommodate both the polariser and the flipper. To determine whether this space is sufficient to provide a reasonable neutron polarisation and transmission, we consider a transmission-based device consisting of a tapered supermirror guide containing two pairs of polarising supermirrors arranged in a "double-V" formation (Figures 1,2). The walls of the device are linearly tapered by 0.29° horizontally and 0.22° vertically, like the focusing guide section the polariser replaces, with the cross section dimensions at the entrance an exit being 63.95(H)×40.00(W) mm^2 and 57.96(H)×32.15(W) mm^2. Furthermore, it is considered as coated with $m = 2$ supermirror on the sides and $m = 4$ on the top and bottom walls, as compared to $m = 2/3$ (top/sides) and $m = 4/4$ in the preceding and following guide sections. The lower m-number on the side walls of the polariser ensures the absorption of reflected neutrons of the undesired spin state. An m-number of 5 was assumed for the polarising supermirrors; this both permits a short device, and is readily attainable using FeSi deposited on an Si substrate. Three device lengths – 732, 782 and 832 mm – were simulated to find a compromise between leaving sufficient space for the flipper and polariser performance. Finally, the necessity of dividing the device into two channels was considered by simulating it with and without a vertical $m = 2$ double-sided centre wall. The simulations were carried out for the entire LET instrument from the moderator to the sample position using the McStas package.

The criteria for acceptable performance were a transmission T of at least 35% (70% of neutrons with the correct polarisation state, here $|\downarrow\rangle$ with respect to the guide field) and a neutron polarisation P_n of $> 95\%$ across the wavelength band $3 < \lambda < 9$ Å^{-1}. This could be achieved for a device of length 782 mm, leaving approximately 70 mm for the flipper when vacuum flanges and windows are included. While a longer device yields slightly better polarisation and transmission, we consider the stray field in excess of what can be easily shielded at the flipper position (see Section 3.2). The neutron polarisation P_n and transmission T for a polariser length of 782 mm are plotted versus λ in Figure 3, along with the conventional quality factor P_n^2T. Results are shown for both the design with (two-channel) and without (single-channel) the $m = 2$ dividing wall. Starting with the polarisation, the two-channel device clearly performs better than the single-channel version, with the latter showing dips in $P_n(\lambda)$ at approximately 5 Å and 7 Å. The origin of these features are reflections of neutrons with angles of incidence on the back face of the polarising supermirror $n\phi_v$ where n is an even integer and ϕ_v is the angle of the polarising supermirrors [19, 20]. These processes are largely eliminated by the dividing wall.

Despite the significantly improved $P_n(\lambda)$ of the two-channel configuration, P_n^2T for the two devices is quite similar, owing to the higher transmission of the single-channel case. Nonetheless, as the differences in transmission are small, we prefer the former due to its more homogenous

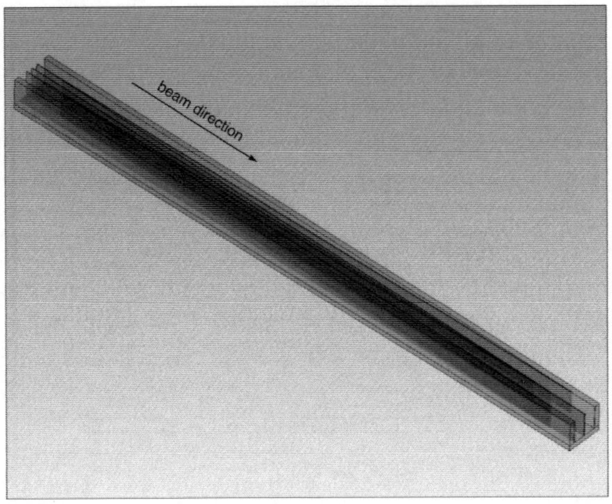

Figure 2. Horizontal cross section of the supermirror polariser, with the holding field assembly and housing omitted. The device consists of a horizontally and vertically focusing guide with $m = 2$ and $m = 4$ coatings on the sides and top/bottom, respectively, linearly tapered walls, $m = 5$ FeSi supermirrors arranged in a "double-V" formation, and a dividing wall to create two channels. The device is 782 mm long.

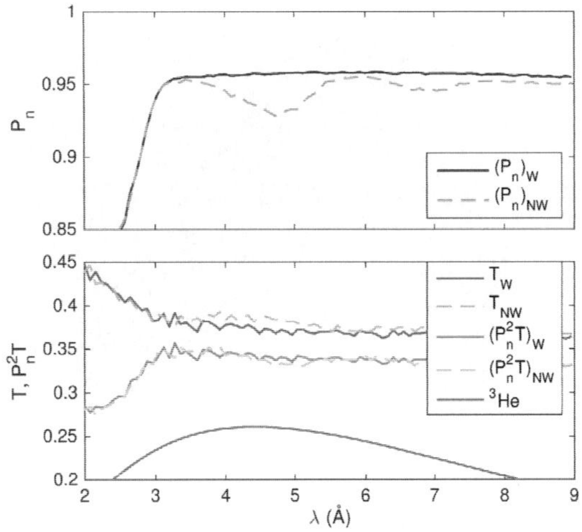

Figure 3. The neutron polarisation P_n (upper panel), transmission T, and quality factor P_n^2T (lower panel) versus wavelength λ for designs of the polariser with (W, solid lines) and without (NW, dashed lines) a dividing wall coated in $m = 2$ supermirror. The design with the wall in place gives a smooth P_n for all wavelengths. The performance of a ^{3}He polariser with gas polarisation $P_{He} = 75\%$ is shown for comparison.

performance in λ. To compare the performances of the present design with the previous ^{3}He-based polariser, we calculate P_n^2T for the latter assuming a very high ^{3}He polarisation of 75%. Even at the maximum P_n^2T, ^{3}He falls well short of the solid state device. Furthermore, this simple comparison disregards the losses due to removing 880 mm of guide, compared to just 100 mm.

Finally, the FeSi mirrors require a relatively large and homogenous magnetic field to bring them close to saturation. This will be produced by columns of $Nd_2Fe_{14}B$ permanent magnets sandwiched between soft Fe yokes on the top and bottom of the device. These supply a homogenous field of $\sim$ 400 G across the mirrors. However, they also lead to a significant stray field, estimated to be $\sim$ 100 G at the flipper position. The implications of this will be discussed in the following section.

3.2. Flipper and guide field

As mentioned previously, the length of the polariser precludes flipper designs which occupy more than 80 mm along the beam direction. When the space constraint is combined with the necessity to operate in the large stray field produced by the permanent magnets around the polariser, even fewer possibilities remain. One of the simplest flipper designs, which is additionally both compact and relatively robust towards stray fields, is the precession coil, or Mezei, flipper [16], which uses a non-adiabatic change of field direction and Larmor precession to flip the beam. Despite its advantages, this type of flipper only flips a single neutron wavelength, and would therefore have to be ramped in synchronisation with the ISIS neutron pulse should the option of repetition rate multiplication be maintained. A precession coil flipper also places material in

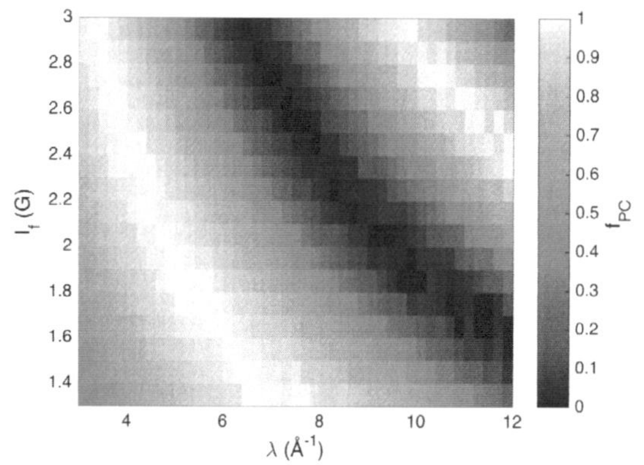

Figure 4. Flipping efficiency f_{PC} versus wavelength λ and flipping current I_f for a flipper-analyser distance of $L_{af} = 20$ cm. The expected $f_{PC}(\lambda) \propto \cos^2(\lambda)$ is obeyed for all B_f. The first ridge of high f_{PC} corresponds to a π flip, and the following ridge 3π.

the beam, which generates both attenuation and small angle scattering. The latter is, however, mitigated by the 2 m of guide between the flipper and the sample position.

In order to test the suitability of a precession coil flipper for flipping a white beam in the vicinity of a strong magnetic field, we installed a prototype device on the LARMOR instrument, also at TS2, ISIS. The beam was polarised by a "V"-cavity polariser, similar to the one intended for LET, then flipped by the first flipper, an adiabatic RF device. The beam dimensions were defined to be 30×30 mm^2 at the flipper position by two sets of slits. The precession coil flipper, consisting of two coils wound, respectively, along the guide field (the compensation coil) and perpendicular to both this direction and the beam axis (the flipping coil), was installed in a soft iron housing adjacent to the analyser. The latter was a supermirror device consisting of a single FeSi mirror magnetised by $Nd_2Fe_{14}B$ magnets, which result in a magnetic field $|B| = 300$ G at the exit of the polariser. The transmitted beam was measured for all configurations of both RF and precession coil flippers on and off, with the flipper at three positions along the beam axis; $L_{af} = 45$ mm, 100 mm, and 200 mm from the analyser entrance. The flipping efficiency of the precession coil and RF flippers for each configuration was then extracted from the formula [21, 22]:

$$f_{PC} = \frac{I^{00} - I^{01} - I^{10} + I^{11}}{2(I^{00} - I^{10})} \tag{2}$$

$$f_{RF} = \frac{I^{00} - I^{01} - I^{10} + I^{11}}{2(I^{00} - I^{01})} \tag{3}$$

where I^{ab} are intensities integrated over the entire detector measured for all configurations of RF/precession coil (a/b) flipper on/off $(0/1)$. The RF flipper was found to be close to 100% efficient across the entire wavelength band (not shown). At $L_{af} = 45$ mm, where the field at the downstream face of the flipper was 50 G, the beam was found to be nearly depolarised at the detector (not shown). This was ascribed to steep field gradients in the region of the flipper. Moving the flipper away to 100 mm or 200 mm, however, performance was found to be acceptable, with peak f_{PC} values ranging between $94 - 99\%$ for the former and $98 - 100\%$ for the latter.

In order to quantify this more precisely, we turn to the wavelength-dependence of f_0. To fit the curves, we recall some simple results, which, although relatively well known, are not often shown in the literature. Firstly, the time-evolution of the neutron wave-function in the direction perpendicular to the beam and the initial neutron polarisation (which is $|\downarrow\rangle$ for a transmission-based polariser) is

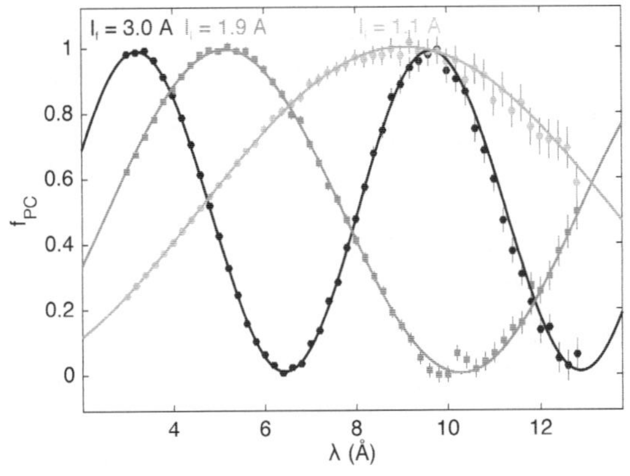

Figure 5. Wavelength-dependence of the precession coil flipper efficiency f_{PC}. Curves are shown for three values of I_f, with fits to equation 6 indicated by solid lines.

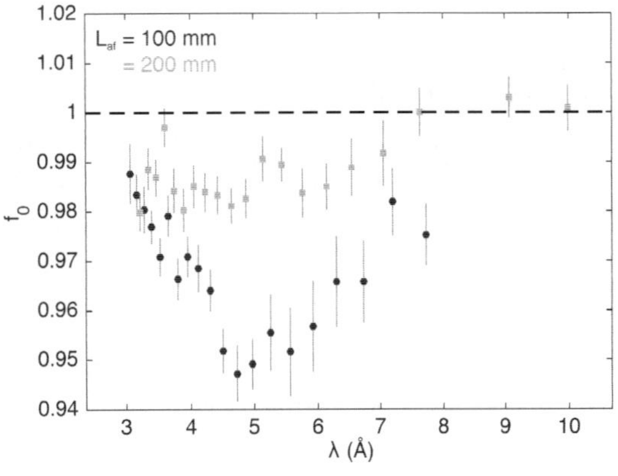

Figure 6. Wavelength-dependence of the absolute flipper efficiency f_0 extracted from fits to equation 6. Stray fields cause the distribution of points to shift to the left for $L_{af} = 100$ mm.

$$\begin{pmatrix} \psi_1(t) \\ \psi_2(t) \end{pmatrix} = \exp\left(\frac{-i\omega_L t}{2}\sigma_x\right)\begin{pmatrix} 0 \\ 1 \end{pmatrix} \tag{4}$$

$$= \begin{pmatrix} i\sin\left(\frac{\omega_L t}{2}\right) \\ \cos\left(\frac{\omega_L t}{2}\right) \end{pmatrix}, \tag{5}$$

where σ_x is the x component of the Pauli matrices and $\omega_L = -\gamma_N B_f$ is the Larmor precession frequency of the neutrons in the field B_f generated by the flipper coil. The probability of finding a spin up after passing through the flipper is thus $P(\psi_1) = |i\sin\left(\frac{\omega_L t}{2}\right)|^2 = \sin^2\left(\frac{\omega_L t}{2}\right)$. Substituting $t_f = \lambda m_N L_f/h$, where t_f is the time spent in the flipper and L_f is its thickness, and including the absolute flipper efficiency f_0 (with $(1 - f_0)/2$ defined as the depolarisation due to the flipper), the wavelength-dependence of the efficiency at a particular B_f is expressed as

$$f_{PC}(\lambda) = \frac{1}{2} - f_0\left[\frac{1}{2} - \cos^2\left(\frac{\gamma_N m_N B_f L_f \lambda}{2h}\right)\right] \tag{6}$$

$$= \frac{1}{2} - f_0\left[\frac{1}{2} - \cos^2(C B_f \lambda)\right], \tag{7}$$

where the constants have their usual meanings, and, assuming the flipping coil approximates a solenoid, $B_f = 4\pi n[\text{cm}^{-1}]I_f[\text{A}]/10$, where n is the winding density of the coil and I_f is the current. As can be seen from the fits to this expression for $L_{af} = 200$ mm, shown in Figure 5, agreement is excellent at all values of B_f. The effect of moving the flipper from $L_{af} = 200$ mm to $L_{af} = 100$ mm is twofold: the efficiencies f_0 (Figure 6) show a broad dip down to 94% around $\lambda = 5$ Å, and there is an offset between the expected peak and actual peaks in λ. This is probably due to components of the analyser stray field along the flipping field adding to that generated by the coils.

The above measurements and analysis show that a precession coil is a viable option as a broad-band flipper, but also that it has to be shielded carefully from stray fields. In addition, it is clear that to flip a white beam (or discrete wavelengths in the case of repetition rate multiplication mode), the current through the flipper needs to be both synchronised with the ISIS pulse and ramped in time. Taking the derivative of Equation 6 with respect to λ, the maximum flipping efficiency is given by the condition $\lambda_{\max} = \pi/2CB_f$, and the form of the ramp is found to be

$$I_f(t) = \frac{5}{2}\frac{L_{mf}}{\gamma_N n L_f t}, \tag{8}$$

where L_{mf} is the distance from the moderator to the flipper. Some preliminary tests using a signal generator amplified by a power supply and triggered by the ISIS pulse have indicated that excellent performance ($f_0 > 97\%$) at all wavelengths is possible (not shown).

Returning to the issue of stray fields, the field at the exit of the LET polariser will exceed that of the analyser in our test experiment, and L_{af} will only be $\sim$ 7 cm. To solve this problem, a model of the flipper and its environs was built in Radia [23], which calculates the magnetic field generated by an assembly of magnetic (both current-carrying and permanent) components using analytical approximations. The model comprised the holding field environment for the polariser, a μ-metal screen (relative permeability $\mu = 5 \times 10^4$) with an aperture to reduce the stray field at the flipper position, the flipper itself, several movable permanent magnets to fine tune the magnetic field, and the three existing soft iron guide field housings following the flipper (Figure 7).

The criteria used to optimise the model were threefold: ensuring adiabatic passage of the neutrons by avoiding sharp gradients in the angle θ (corresponding to the polar angle between the z and x, y components of the magnetic field), minimising θ in the region immediately before flipping (denoted θ_{pf}), and maximising the field homogeneity in the flipping region to allow for complete guide field B_g cancellation and accurate rotation of the neutron spin. An additional limitation was imposed by the ohmic heating of the compensation coil combined with its proximity to the $Nd_2Fe_{14}B$ guide magnets. Increasing the winding density of the copper wire to avoid overheating results in a higher neutron flux loss. On the other hand, aluminium wire can be used for its transparency to neutrons, but in turn suffers from lower electrical conductivity and deterioration under high temperatures, limiting the current. Designs where the guide field exceeded a set threshold ($\sim$ 60 G) were therefore disregarded.

The best solution was found by a combination of algorithmic optimisation for a simplified model containing only the flipper and polariser, as well as manual movement of the flipper and magnets for the full model. This solution includes four cube-shaped $Nd_2Fe_{14}B$ magnets (24x24x24 mm^3), magnetised along the guide field and placed near the corners of the soft iron housing of the guide field section immediately following the flipper. Combined with position and strength adjustments of the guide magnets further down the beam, a near-complete cancellation of the stray fields was achieved in the simulation. The remaining variation in the z-component of the field B_z (around 3 G, as to compared to the total field $|B| = 58$ G) along the z-axis in the flipper will be remedied by slight adjustments of the winding density of the compensation coil. To test the validity of our simulations, a partial model of the flipper was constructed (Figure 7, bottom left), and its magnetic field mapped using a Hall probe. The results of these measurements, which are shown in Figure 7, agree almost quantitatively with the corresponding simulation, giving confidence that the final device will perform as expected.

To estimate the setup's performance, the standard deviation of B_f across the beam cross-section δB_f was calculated. As the Larmor frequency scales linearly with B_f, the ratio $\delta B_f / B_f$ corresponds to the fractional error in the rotation angle. As shown above, the expectation value of the rotated polarisation state (here $|\uparrow\rangle$) then depends primarily on the cosines squared of the total rotation and the field direction ahead of the flipper:

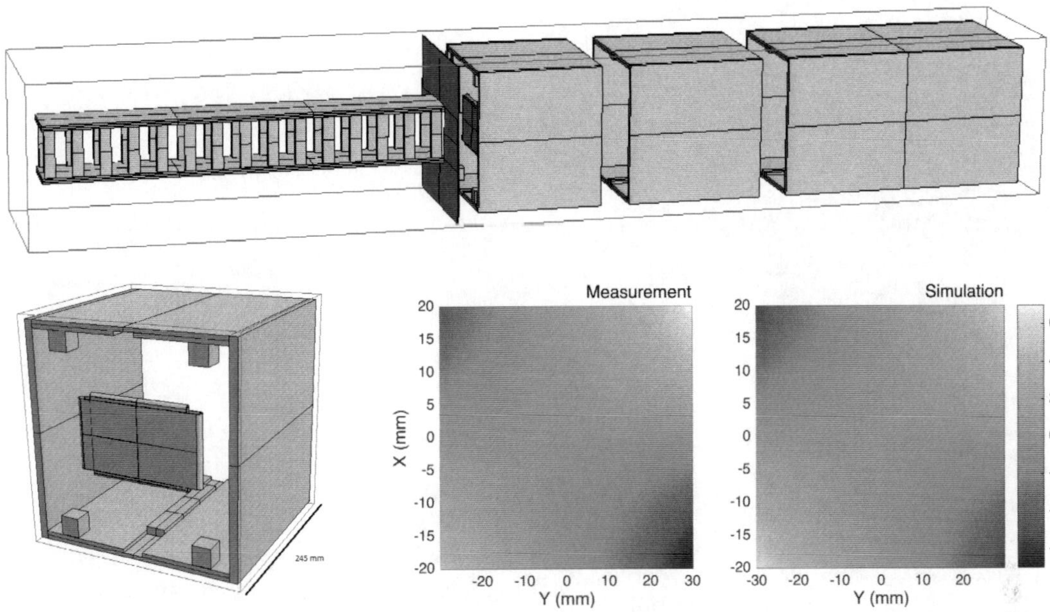

Figure 7. (top) A side view of the model used to optimise the magnetic field environment around the flipper. The polariser is on the left, separated from the flipper by a μ-metal screen. The guide field section immediately following the flipper contains several permanent magnets to compensate the stray fields at the flipper position (bottom left). The best solution yields a flipping efficiency > 99%. (bottom right) Comparison of measured and simulated component of the magnetic field parallel to the flipping field around the beam centre for a model of the flipper containing only the surrounding guide field section and magnets. The slight discrepancy is due to a small spatial offset of the Hall probe.

$$P(\psi_2) \simeq \cos^2(\theta_{pf}) \cos^2\left[\pi \frac{\delta B_f}{B_f}\right]. \tag{9}$$

The values obtained from the simulations are $\theta_{pf} = 0.035$ and $\delta B_f = 0.03$ G, which gives expectation values $P(\psi_2) > 0.998$ across the desired wavelength range. The losses due to non-adiabatic passage are negligible and are therefore disregarded.

3.3. ^{3}He spin filter cells

To analyse the scattered beam on LET, we require a device capable of covering the large solid angle subtended by the detector banks. As mentioned in the introduction, wide-angle solid state analysers are associated with considerable complications in both data collection and analysis, and ^{3}He spin filters therefore remain the preferred option, despite the fact that these only perform optimally for a single wavelength. The dimensions of the ^{3}He spin filter cells are a compromise between the dimensions of the sample environment, the minimum wavelength λ_{min} admitted by the polariser (see Section 3.1), and the maximum admissible pressure in the cell. For $\lambda_{min} = 3$ Å, the pressure length pl required to maximise the quality factor $P_n^2 T$, where

$$P_n = \tanh(0.0733 pl \lambda P_{He}) = \tanh(OP_{He}) \tag{10}$$

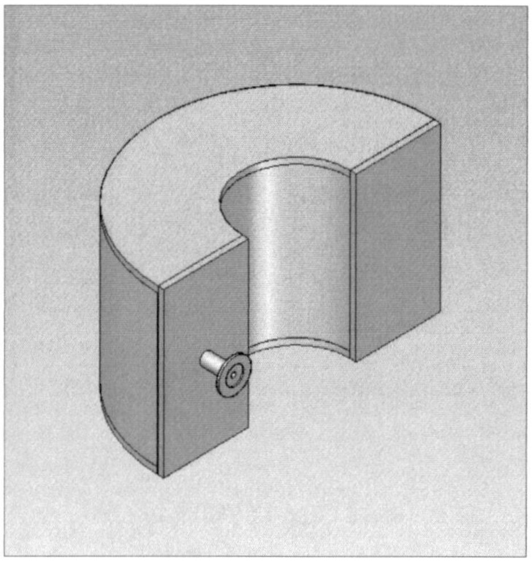

Figure 8. Three dimensional model of the envisaged ^{3}He spin filter cell. The cell covers 180° in the horizontal plane and $\gamma = \pm 30.2°$ in the vertical direction.

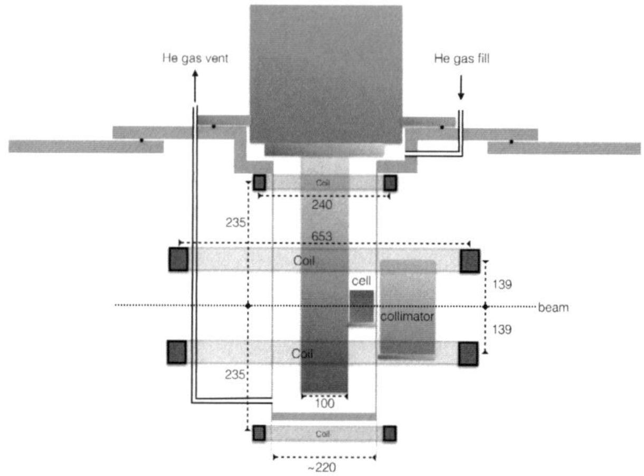

Figure 9. A side view of the sample area, including the coils, cell, collimator, and inner tank. The latter will maintain the ^{3}He spin filter and the sample environment under 1 atm of inert gas, such as ^{4}He or Ar.

and

$$T = \exp(-O)\cosh(OP_{He}) \tag{11}$$

is ~ 9 bar cm. The inner radius is determined by the maximum radius of the sample environment, 40 mm for the cryostat we are currently adapting for exclusive use with polarised ^{3}He. For a cell with 5 mm walls operating at atmospheric pressure, the outer radius would thus be at least 140 mm (allowing 5 mm clearance between the cell and cryostat tail), and the height at least 200 mm. As will be shown in the following section, it is difficult ensure a homogenous magnetic field (and hence long ^{3}He field gradient lifetime $T_{1,grad}$) across the volume filled by a cell of these dimensions, especially given the spatial constraints the LET sample space imposes on the holding field coils. On the other hand, previous experience with glued cells suggests that these may be operated at up to 1 bar overpressure. Provided the cell can be contained in a vessel at atmospheric pressure, this allows us to reduce the gas length to 6 cm, much more amenable to long cell lifetimes. The latter may be achieved by the use of an inner vacuum vessel (Figure 9), also containing the sample environment, which fits inside the LET sample tank. The vessel will be filled with a gas with small neutron scattering and absorption cross sections (*e.g* ^{4}He or Ar) to reduce attenuation of the beam. An additional advantage of this arrangement is that the sample tank vacuum does not need to be broken, which should facilitate rapid cell changes.

Regarding the spin filter cell material, single crystal silicon is both compatible with both long lifetimes and a low background at small scattering vector Q, provided care is taken with respect to the orientation of the panels. This is a large advantage over cells constructed of quartz or GE180 glass, where the background is broadly distributed in Q [24], and centred around the elastic line, hence potentially impeding studies at low neutron energy transfers. Given that the cell will intercept the direct beam, an absorbing cadmium strip will be fitted on the inner radius of the cell to cover the beam cross section. This will, however, restrict the scattering angle θ in the horizontal plane to $|\theta| \geq 11°$. A three dimensional model of the proposed ^{3}He spin filter analyser cell is shown in Figure 8.

The ^{3}He gas is to be polarised in the FLYNN filling station, installed at ISIS in 2011. FLYNN operates using the metastable exchange optical pumping (MEOP) method, whereby

IOP Conf. Series: Journal of Physics: Conf. Series **862** (2017) 012019 doi:10.1088/1742-6596/862/1/012019

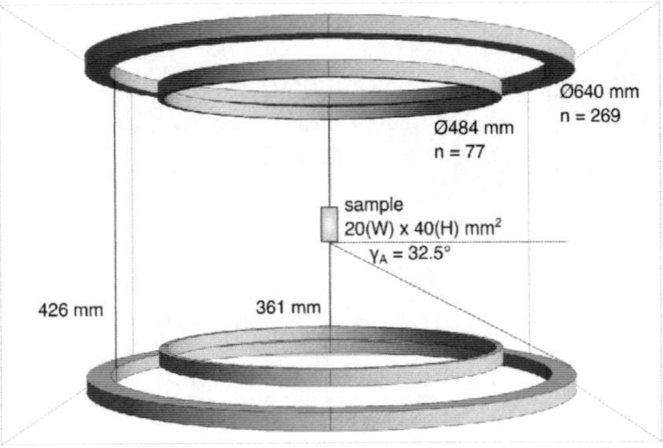

Figure 10. Optimised coil geometry including numbers of windings for the four coil configuration. The vertical opening of the coils is ±32.5°. n is the number of windings of copper wire on each set of coils.

metastable ^{3}He atoms generated by a high-voltage discharge are directly polarised by circularly polarised laser light, before transferring their polarisation to ground state atoms via collisions [25]. Following recent improvements, FLYNN is now producing polarised ^{3}He gas with an excellent $P_{He} \sim 75\%$.

3.4. Holding field

As discussed in the previous section, polarisation analysis on LET will be achieved using wide-angle ^{3}He spin filters. Due to the large size of the spin filter cells (60 mm path length, at least 170 mm tall), the magnetic field environment around the cells must be carefully designed to minimise ^{3}He polarisation losses from magnetic field gradients. These losses may be quantified by the inverse relaxation time (or lifetime) $1/T_{1,grad}$

$$\frac{1}{T_{1,grad}} = \frac{D}{p}\left[\left(\frac{\nabla B_x}{|B|}\right)^2 + \left(\frac{\nabla B_x}{|B|}\right)^2\right] \tag{12}$$

$$= \frac{D}{p}\left(\frac{\nabla B_t}{|B|}\right)^2 \tag{13}$$

where D/p is the ^{3}He gas diffusion constant, which is $6700/p$ cm^{-2} hr^{-1} at a pressure of p atm [26]. The terms in the brackets represent the square of the transverse field gradient $\nabla B_t = \nabla\sqrt{B_x^2 + B_y^2}$ normalised to the norm of the field $|B|$. Thus, to achieve a 150 hour lifetime for a cell at $p = 1$ atm, the maximum allowable average transverse field gradient over the cell is approximately 1×10^{-3} cm^{-1}.

The exceptionally large detector coverage on LET presents another issue; designing a coil-set with a horizontal opening of 180° is trivial, but covering $\gamma_{max} = \pm 30°$ in the vertical direction without shadowing the detector is more difficult. For example, the well-known Barker and Braunbek coil geometries [27] have fixed openings close to ±10°.

In order to optimise the coil geometry, we again employed the Radia software for Mathematica [23]. Three configurations were simulated; a simple pair of Helmholtz coils, four coils arranged in two symmetric pairs, and six coils arranged in three symmetric pairs. All coil configurations but the Helmholtz configuration were optimised using the internal Mathematica simulated annealing algorithm. The figure of merit to minimise was

$$F = w_B|B_c - B_i| + w_G\left(\frac{\nabla B_t}{|B|}\right)_{max} + w_A\left(\frac{\gamma_A}{\gamma_{max}} - 1\right), \tag{14}$$

IOP Conf. Series: Journal of Physics: Conf. Series **862** (2017) 012019 doi:10.1088/1742-6596/862/1/012019

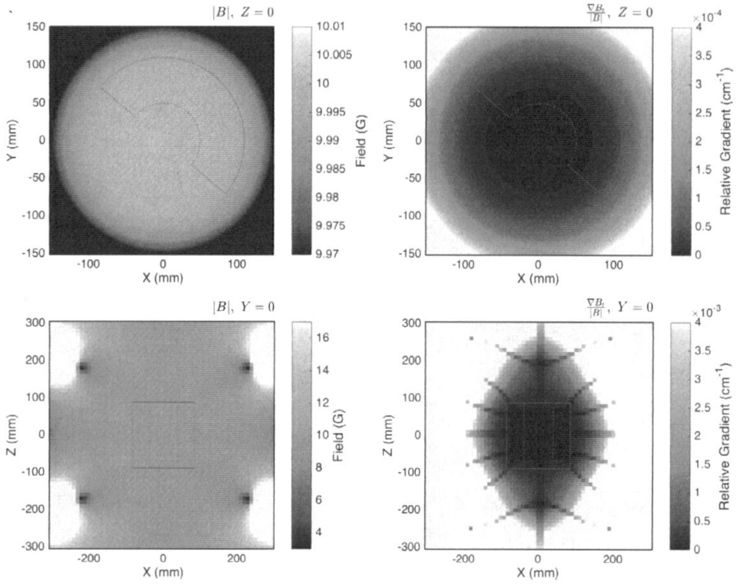

Figure 11. Field and field gradient maps for the optimised four coil configuration. The left panels show the absolute field in the XY (top) and XZ (bottom) planes, while the right panels show the relative gradient of the transverse component of the field. The red lines indicate the position of the cell, with the inner and outer walls denoted by dashed and solid lines, respectively.

where B_c is the field at the centre of the cell, B_i is the desired field, here 10 G, $(\nabla B_t/|B|)_{max}$ is the maximum relative transverse gradient along either the inner or outer wall of the cell, and γ_A is the opening angle of the coils, as defined in Figure 10. The weights w_B, w_G, and w_A were adjusted for each model to yield an acceptable compromise between the three terms.

The best solution was found for four coils arranged in two symmetric pairs; the main pair with 269 windings at a separation of 426 mm, and the compensation pair with 77 windings and the current reversed at 361 mm (Figure 10). The figure of merit was $F = 2.0$ as compared to 16.4 for the second best model, the six coil configuration, and 27.4 for the third best model, the Helmholtz configuration. This is largely a consequence of the fact that the four-coil configuration allows a full vertical view of the LET detector, *i.e.* $\gamma > \gamma_{max}$. In addition, the magnetic field profiles (Figure 11) confirm that the field gradients for this model are excellent, with the entirety of the cell held in a gradient of $< 5 \times 10^{-4}$ cm^{-1}. Indeed, the maximum transverse gradient observed is 4.3×10^{-4} cm^{-1}, corresponding to a $T_{1,grad}$ in the region of 800 hr. However, adiabatic transport of neutron polarisation from the polariser to the sample position remains to be simulated.

4. Expected performance and applications

The guide system on LET has recently been upgraded with a focusing snout to provide a smaller, more divergent beam at the sample position. The resulting gain in flux is a factor 3 versus the previous version of the instrument; the unpolarised flux at $\lambda = 5$ Å and 2% resolution is now 4.4×10^4 n cm^{-2} s^{-1}, while for a relaxed resolution of 4%, this increases to 8×10^4 n cm^{-2} s^{-1}. The high flux, coupled with the exceptionally low instrumental background, make LET ideally suited for polarisation analysis. Indeed, the high transmission of the polariser will deliver a polarised flux at the sample position of 1.5×10^4 n cm^{-2} s^{-1} at 5 Åand 2%, similar to the unpolarised pre-upgrade instrument. Additional losses will of course be incurred in the ^{3}He analyser. For example, if the gas pressure is matched to an incident wavelength of $\lambda = 7$ Å$(E_i = 1.67$ meV), the peak transmission will be 33% at the elastic line (assuming an unpolarised scattered beam), but only 24% at $0.6E_i$. As such, in cases where inelastic features are of interest, it may be desirable to adjust the optimum wavelength to that of the scattered rather than the incident beam.

IOP Conf. Series: Journal of Physics: Conf. Series **862** (2017) 012019 doi:10.1088/1742-6596/862/1/012019

Other cold time-of-flight instruments with current or planned polarisation analysis capabilities include MACS, NEAT, and HYSPEC. Initial experiments on HYSPEC and MACS, both of which currently operate in uniaxial mode, have shown that polarised experiments are viable on a reasonable timescale, typically one week. On these instruments, polarisation and analysis are achieved using two ^{3}He spin filter cells and a Heusler reflector and supermirror array, respectively. Contrary to ^{3}He, the transmission of a supermirror-based device is typically flat for long neutron wavelengths, perhaps making supermirror-based instruments more suitable for inelastic studies over a broad energy range. Nevertheless, the high combined transmission of the polariser and analyser will make LET perform at a commensurate level with the above instruments once complete.

We finish by debating the relative merits of performing uniaxial and three-directional polarisation analysis on a wide-angle time-of-flight spectrometer. At the outset, we again stress that polarised inelastic neutron scattering is a flux limited technique and that three directional polarisation analysis measurements typically take at least three times longer than uniaxial ones. It is thus hard to envisage completing a three-directional measurement in a reasonable time-frame given current instrument performance. Furthermore, inelastic neutron scattering with three-directional polarisation analysis has typically been applied to single crystals of superconductors (*e.g.* to separate anisotropic spin fluctuations), conventional and quantum magnets (*e.g.* to distinguish longitudinal and transverse magnons), and materials with hybrid excitations (*e.g.* electromagnons). A triple-axis spectrometer is superior to a time-of-flight spectrometer for addressing most these problems, as the excitations probed are typically localised in $(\mathbf{Q}, \omega)$ space (quantum magnets with continuum excitations are excepted). On the other hand, time-of-flight spectroscopy with uniaxial polarisation analysis can complement triple axis spectroscopy by surveying large regions of $(\mathbf{Q}, \omega)$ space for interesting points to probe in more detail with three-directional polarisation analysis. In addition, for certain systems, uniaxial polarisation provides sufficient information to separate the spatial components of the magnetisation. For example, uniaxial polarisation analysis has been successfully applied to uncover pinch points and other diffuse features in rare earth pyrochlores like $Ho_2Ti_2O_7$ and $Tb_2Ti_2O_7$ [28]. We expect future polarised single crystal measurements on LET to be divided between these applications.

For powder samples, it is also debatable how much extra information can be gained by performing full three-directional polarisation. If little or no spin incoherent scattering is present, as in many metal oxide systems, a full separation of the scattering cross section is straightforward from a uniaxial measurement. In some other instances, energy resolution can help separate the cross section components. For example, in a metal-organic magnet, the low-energy dynamics is dominated by quasi-elastic scattering from protons, and there is very little coherent nuclear contribution. The magnetic scattering can then be isolated from the spin incoherent by a linear combination of the spin-flip and non-spin-flip cross sections. Where three-directional polarisation analysis is desirable, as in, for example molecular nano-magnets, the large amount of incoherent scattering from the protons in the organic ligands often makes polarisation analysis considerably less time-efficient than sample deuteration.

Non-magnetic samples, of course, also only require uniaxial polarisation analysis. Recent work on D7 at the Institut Laue-Langevin has identified the promise of polarised quasi-elastic neutron scattering (PQENS), a rarely used technique thus far [29]. Here, polarised neutrons can be used both to separate weak signals from dilute diffusing species in ionic conductors as well as isotopic labelling of functional groups in biological or polymer systems. LET has excellent resolution in energy transfer (down to 8 μeV at E_i = 1 meV) and can cover both several decades in time (5×10^{-13} s – 3×10^{-9} s) as well as a broad range in Q using repetition rate multiplication, making it well suited for PQENS. A particularly promising avenue to explore is diffusion of protons, Li^+ and Na^+ in inorganic battery and fuel cell materials.

5. Conclusion

We have presented design studies for the components comprising the future polarised option on the LET spectrometer, located at the ISIS Neutron and Muon Facility. The polarised neutron beam will be produced by a relatively short "V"-cavity polariser containing $m = 5$ FeSi supermirrors. Flipping will be achieved using a simple precession coil device, selected due to its small size and its robustness to stray field. The wide-angle ^{3}He analyser cell, to be constructed of Si, will be maintained in a homogenous magnetic field defined by two pairs of coils wound along the z direction. Fabrication of several of the components has already begun, and completion of the hardware is anticipated in 2017, and the final performance should be similar to instruments at other sources. We have also discussed the relative merits of uniaxial and three directional polarisation analysis, and come to the conclusion that the former will be sufficient for most applications. A major challenge will be the development of software to correct and analyse the large datasets generated by time-of-flight spectrometers.

Acknowledgements

We thank M. Schneider (Swiss Neutronics) for suggesting the polariser dividing wall, for the original simulations of the magnetic field environment around the polariser, and for helpful discussions.

References

[1] Moon R M, Riste T, Koehler W C 1969 *Phys. Rev.* **181** 920
[2] Stewart J R, Andersen K H, Babcock E, Frost C D, Hiess A, Jullien D, Stride J A, Barthelemy J-F, Marchal F, Murani A P, Mutka H, Schober H 2006 *Physica B* **385-386** 1142
[3] Voigt J, Soltner H, Babcock E, Aldus R J, Salhi Z, Gainov R R, and Bruckel T 2015 *EPJ Web of Conferences* **83** 03016
[4] Rodriguez J A, Adler D M, Brand P C, Broholm C, Cook J C, Brocker C, Hammond R, Huang Z, Hundertmark P, Lynn J W, Maliszewskyj N C, Moyer J, Orndorff J, Pierce D, Pike T D, Scharfstein G, Smee S A, and Vilaseca R 2008 *Meas. Sci. Technol.* **19** 034023
[5] Winn B, Filges U, Garlea V O, Graves-Brook M, Hagen M, Jiang C, Kenzelmann M, Passell L, Shapiro S M, Tong X, and Zaliznyak I 2015*EPJ Web of Conferences* **83** 03017
[6] Yokoo T, Ohoyama K, Itoh S, Suzuki J, Nanbu M, Kaneko N, Iwasa K, Sato T J, Kimura H, and Ohkawara M 2014 *J. Phys.: Conf. Ser.* **502** 012046
[7] Ino T, Ohoyama K, Yokoo T, Itoh S, Ohkawara M, Kira H, Hayashida H, Sakai K, Hiroi K, Oku T 2016 *J. Phys.: Conf. Ser.* **711** 012011
[8] Ohoyama K, Yokoo T, Itoh S, Nanbu M, Iwasa K, Ohkawara M, Kaneko N, Ino T, Hayashida H, Oku T, Kira H, Tasaki S, Takeda M, Kimura H, and Sato T J 2016 *J. Phys.: Conf. Ser.* **711** 012010
[9] Scharpf O and Capellmann H 1993 *phys. stat. sol. (a)* **135** 359
[10] Su Y, Nemkovski K, and Demirdis S 2015 *Journal of large-scale facilities* **1** A27
[11] Stewart J R, Deen P P, Andersen K H, Schober H, Barthelemy J-F, Hillier J M, Murani A P, Hayes T, and Lindenau B *J. Appl. Cryst.* **42** 69
[12] Bewley R I, Taylor J W, and Bennington S M 2011 *Nucl. Instr. and Meth. A* 637 128
[13] Russina M and Mezei F, 2009 *Nucl. Instr. and Meth. A* **604** 624
[14] Ehlers, G, Podlesnyak A A, and Kolesnikov A I, 2016 *Rev. Sci. Instrum.* **87** 093902
[15] Ollivier J and Mutka, H, 2011 *J. Phys. Soc. Jpn.* SB003
[16] Hayter J B 1978 *Z. Physik B* **31** 117
[17] Ehlers G, Stewart J R, Wildes A R, Deen P P, and Andersen K H, 2013 *Rev. Sci. Instrum.* **84** 093901
[18] Beecham C J, Boag S, Frost C D, McKetterick T J, Stewart J R, Andersen K H, Bentley P M, Jullien D 2011 *Physica B* **406** 2429
[19] Gilles R, Ostermann A, and Petry W 2007 *J. Appl. Cryst.* **40** S428
[20] Boni P, Munzer W, and Ostermann A 2009 *Physica B* **404** 2620
[21] Wildes A R 1999 *Rev. Sci. Instrum* **70** 4241
[22] Wildes A R 2006 *Neutron News* **17** 17
[23] Chubar O, Elleaume P, and Chavanne J 1998 *J. Synchrotron Rad.* **5** 481
[24] Babcock E, Salhi Z, Pistel P, Simeoni G, and Ioffe A 2014 *J. Phys.: Conf. Series* **528** 012018
[25] Colegrove F D, Schearer L D, and Walters G K 1961 *Phys. Rev.* **13** 2561

[26] Cates G D, Schaefer S R, and Happer W 1988 *Phys. Rev. A* **37** 2877
[27] Robinson P R 1983 *J. Phys. E: Sci. Instrum.* **16** 39
[28] Fennell T, Deen P P, Wildes A R, Schmalzl K, Prabhakaran D, Boothroyd A T, Aldus R J, McMorrow D F, and Bramwell S T 2009 *Science* **326** 415
[29] Burankova T, Hempelmann R, Wildes A R, and Embs J P 2014 *J. Phys. Chem. B* **118** 14452

IOP Conf. Series: Journal of Physics: Conf. Series **862** (2017) 012020 doi:10.1088/1742-6596/862/1/012020

Optimization of a polarizer device for SANS-2 instrument at the PIK reactor

K A Pavlov[1,2], P I Konik[1,2], V G Syromyatnikov[1,2], S V Grigoriev[1,2] and E V Moskvin[1,2]

[1] Petersburg Nuclear Physics Institute NRC KI, Gatchina, 188300 Russia
[2] Saint Petersburg State University, St. Petersburg, 199034 Russia

E-mail: orbita.pk@lns.pnpi.spb.ru

Abstract. A polarizer of V-cavity type for small angle diffractometer was considered. The device performance was simulated using McStas ray tracing package with the input parameters for supermirrors of PNPI home production (CoFe/TiZr, $m = 2.13$). Two replaceable devices with lengths 0.75 m and 1.8 m are proposed to cover the wavelength range 4–25 Å with beam polarization not less than 95%.

Introduction

A small-angle neutron scattering diffractometer SANS-2 was transferred to Petersburg Nuclear Physics Institute (PNPI) from Helmholtz Zentrum Geesthacht (HZG), Germany, after the FRG reactor was shutted down. At the moment the instrument is being reconstructed for installation at the PIK reactor, Gatchina, Russia. Among other parts, a new polarizer for SANS-2 has to be developed. When operated at FGR reactor SANS-2 used the reflection-type bender as a polarizer. That device provided inconvenience in use since the switching to polarized neutrons mode caused movement and alignment of the whole 40 m long instrument. Furthermore, the transmission rate was also an issue: the bender provided transmission not more than 70% due to losses in glass walls. Thus the construction of a new transmission polarizer is required. The polarizer of V-cavity type [1, 2, 3] was chosen for simulations since a single mirror appeared to be too long for assembly and adjusting considering a spectral range specified, and bender type devices couldn't increase the transmission compare to the old device.

The reason why we consider the device based on polarizing supermirrors is the following. Compare to the ^{3}He polarizers [4] the supermirrir polarizer is relatively simple in construction and needs no service when being used. The technology of polarizing supermirrors production is available at the moment at PNPI. The m-values available at the moment are not extremely high, but these coatings were never applied for the wide spectral range optics before, and the aim of this work is to set whether is it possible to implement.

1. Input parameters

The neutron ray-tracing package McStas [5, 6] was used to calculate the flux and polarization after the polarizer. A sketch of a model used is shown in the figure 1. As the input parameters we used a real spectrum of HEC-3 reactor beam tube equipped with cold neutron source (as a source see Fig. 1) and a model of neutron guide being now developed for SANS-2 ($m = 1$, cross-section 30×30 mm^2). To simulate the instrument optical scheme we used a velocity selector component and a collimation line consisted of 8 exchangeable sections of 2 m length corresponding to SANS-2 layout.

Figure 1. A sketch of model used in simulations.

Corresponding to McStas reflectivity description [6], a number of parameters is to be set: critical Q-vector Q_c, supermirror index m, supermirror regime reflectivity slope α and mQ_c drop width W. These values were obtained from the best fit of the experimental curves for CoFe/TiZr multilayer structures, produced by PNPI [7, 8]. These parameters are given in the table 1. In this table $R^{\pm}$ stands for two components of polarization, and R^{+}_{dbl} is the reflectivity of the double coated (on both sides) supermirror we used in simulations. To take into account the reflections from both sides, we added the contribution to the intensity the fraction which was once transmitted and then reflected. Mathematically this simple assumption holds

$$R^{+}_{dbl} = R^{+} + (1 - R^{+})R^{+}, \tag{1}$$

and as a result it sufficiently changes the slope α (see third raw in table 1). Different slope gives new improved magnitude of the cut-off reflectivity $R_m = 0.90$ instead of $R_m = 0.75$ for single coated supermirror. For the negative component R^{-} there are no changes since a reflectivity curve has no sloping region corresponding to multilayer structures: the reflectivity is 0.99 up to Q_c and then drops dramatically.

Table 1. McStas parameters for CoFe/TiZr supermirror

Pol. comp.	Q_c[Å^{-1}]	m	R_0	α [Å]	W [Å^{-1}]
R^{+}	0.0219	2.13	0.995	13.87	0.00187
R^{-}	0.0102	1	0.99	0	0.001
R^{+}_{dbl}	0.0219	2.13	0.995	3.43	0.00187

The polarizer should provide polarization no less than 95% with the maximal transmission in the whole spectral range which is used by the instrument. SANS-2 is planned to operate in the wavelength range from 4.5 to about 20 – 25 Å. Covering such a wide range with one optical polarizer with in-house coating $m = 2.13$ was found impossible as it is shown below, so we came to two exchangeable polarizers for short and long wavelength ranges. For optimization of the monochromatic beam we used a velocity selector model adjusted to 4.5 Å and to 10 Å with 10% resolution.

IOP Conf. Series: Journal of Physics: Conf. Series **862** (2017) 012020 doi:10.1088/1742-6596/862/1/012020

2. Results and discussions

First we optimized the short wavelength range polarizer. Figure 2 shows the dependence of polarization versus the polarizer length for five different beam collimations. The minimum

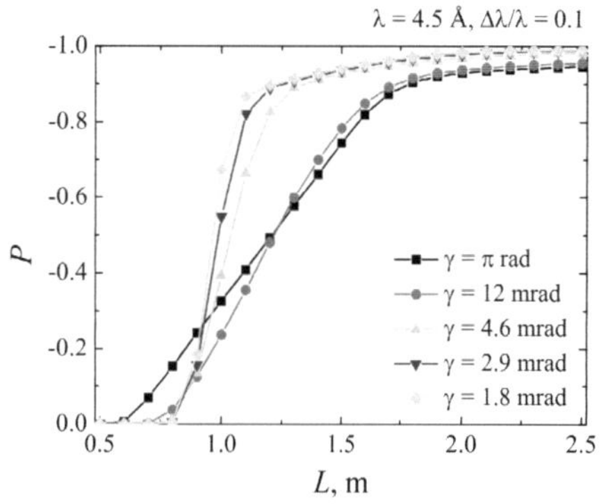

Figure 2. Polarizer length dependence of polarization for different beam collimations for $\lambda = 4.5$ Å.

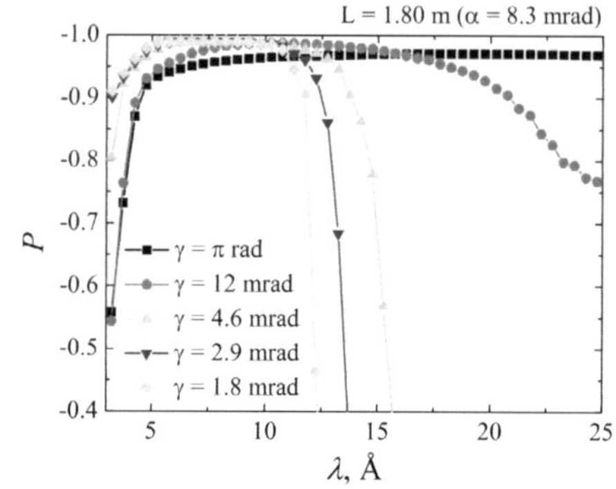

Figure 3. The polarization spectral dependence of 1.8 m long polarizer.

polarizer length corresponds to the value of $L = 1.8$ m where all the curves saturate. This length is directly coupled to the inclination angle of supermirror plate. The corresponding angle is 8.3 mrad. Figure 3 shows the spectral dependence of polarization. It is well seen that such polarizer provides good polarization in the wavelength range from about 4.5 Å to about 11 Å for all the values of the beam divergence.

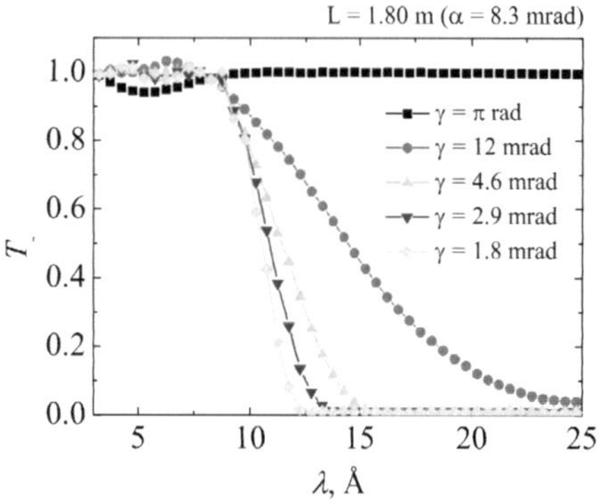

Figure 4. Spectral dependence of the transmission of R^- component of polarization.

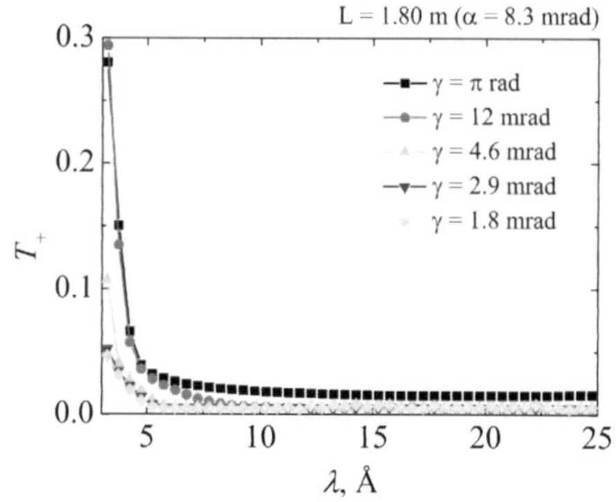

Figure 5. Spectral dependence of the transmission of R^+ component of polarization.

Figures 4 and 5 show transmission spectral dependencies for R^- and R^+ components respectively. As it is clearly seen from these figures, this polarizer performs well in the range from about 4.5 to about 10 Å. A limit at shorter wavelengths is induced by the R^+ reflectivity drop-off at high Q-values: the short wavelength neutrons are not reflected at all and the polarization degree in the transmitted beam is low. In the opposite way, the long-wavelength bound is caused by the full-reflectivity area in silicon: both spin components are reflected and the transmission of the device drops to zero. It is not the case for the non-collimated beam ($\gamma = \pi$ rad), since in divergent beam there still are trajectories with falling angle exceeding critical angle. Therefore

IOP Conf. Series: Journal of Physics: Conf. Series **862** (2017) 012020 doi:10.1088/1742-6596/862/1/012020

it is very important to optimize optics considering individual instrument beam requirements and setting divergence bounds is essential.

Exactly the same procedure was done for the second polarizer optimized for the long wavelength range (from about 10 to about 20 Å). Corresponding pictures are figures 6-9.

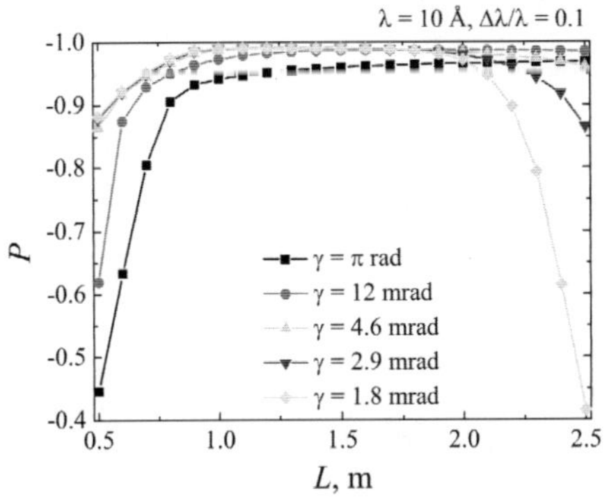

Figure 6. Polarizer length dependence of polarization for different beam collimations for $\lambda = 10$ Å.

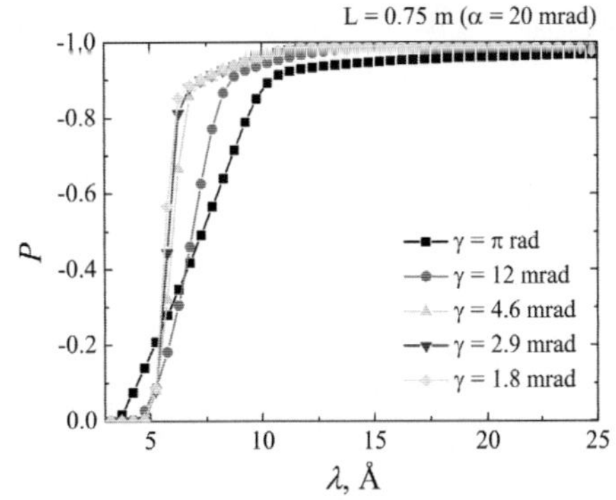

Figure 7. The polarization spectral dependence of 0.75 m long polarizer.

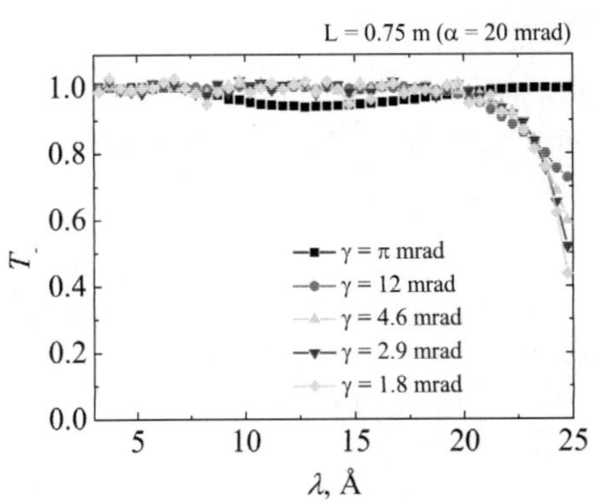

Figure 8. Spectral dependence of the transmission of R^- component of polarization.

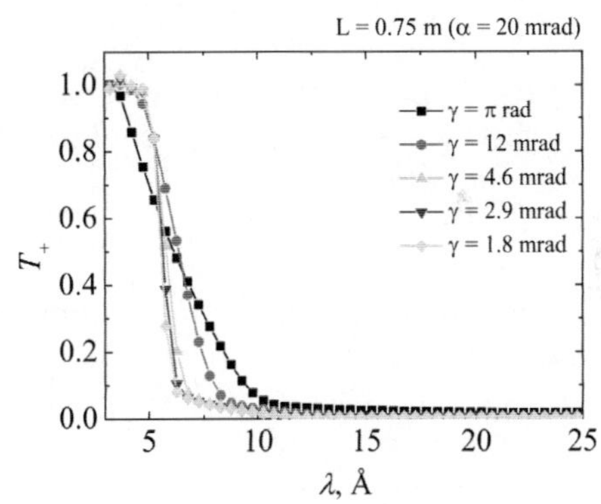

Figure 9. Spectral dependence of the transmission of R^+ component of polarization.

The optimal length of the second polarizer found to be 0.75 m. This corresponds to the supermirror inclination angle of 20 mrad.

3. Conclusions

Using the coatings considered (production available at PNPI at the moment) doesn't allow to make a single polarizer for a wide spectral range. Our calculations show that to cover the wavelength range from about 4.5 to about 25 Å two exchangeable devices are needed. The first one (figures 3, 4, 5) operates in a range from about 4.5 to about 10 Å, the second one (figures 7, 8, 9) fits the range from about 10 up to about 25 Å. Both polarizers provide good beam polarization — not less than 0.95 for any neutron wavelength in the range $4.5 \div 25$ Å. The transmission of negative spin component is also satisfactorily high: not less than 0.95 for the major part of the operating range and not less than 0.7 at the range edges. It is worth to note that the operation of PNPI in-house supermirror coatings was first simulated and shown to be applicable in optics for wide spectral range.

In principle it is possible to cover entire range of wavelengths with the only one polarizer, but for this purpose one need the supermirror with higher m coefficient. We chose as example a supermirror with $m = 5.5$ commercially available at the moment and we performed simulations using parameters provided by SwissNeutronics [9].

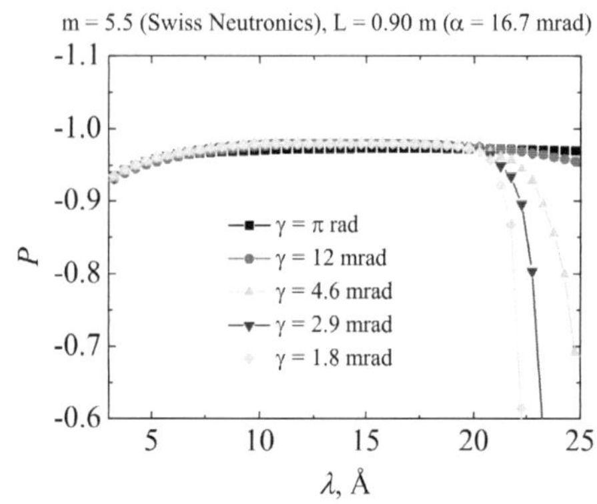

Figure 10. Polarization spectral dependence for V-cavity type polarizer with $m = 5.5$.

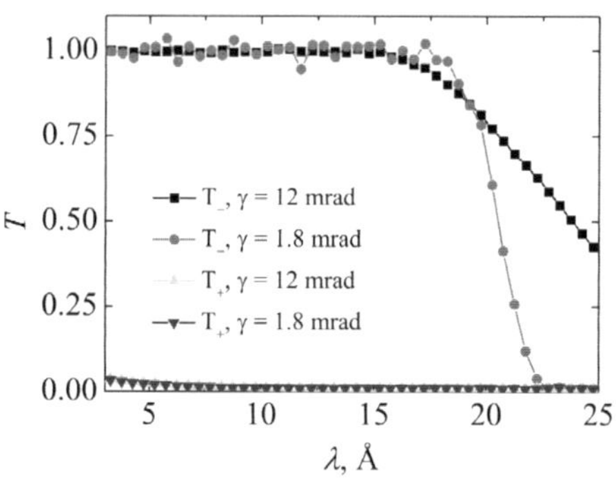

Figure 11. Transmission spectral dependence for V-cavity type polarizer with $m = 5.5$.

Figures 10-11 show the performance of V-shaped polarizer of 0.9 m length. As one can see from these pictures, such a device could perform well in the entire desired wavelength range (from 4.5 to about 20 Å).

One thing should be pointed out. The polarization degree for polarizer with high-m supermirror is few percent less compare to the one based on the lower m coating. This is caused by the lower reflectivity in supermirror region (from Q_c to mQ_c) for R^+ component. Higher m obviously makes this region wider, but the reflectivity here is lower. The transmission for negative spin component (T_-) supposed to be around 1, and it remains the same for both types of supermirrors (see fig. 4 for $m = 2.13$ and fig. 11 black squares and red circles for $m = 5.5$). For positive spin component (T_+) transmission should be as low as possible, but for supermirror with $m = 5.5$ it increases compare to the one with $m = 2.13$ (see fig. 5, and fig. 11 blue and green triangles). The higher T_+ component becomes, the lower polarization it gives. In our case it is not the most important factor since we initially established the restriction for polarization at 0.95, but in some applications of supermirror polarizers it could be significant.

Acknowlegement

The authors are grateful to PNPI researchers N.K. Pleshanov and V.A. Ulyanov for fruitful discussions.

The work was supported by Russian Federation Ministry of Education and Science (Federal target program, project RFMEFI61614X0004).

References

[1] Krist T, Lartigue C and Mezei F 1992 *Physica B* **180–181** 1005–1006
[2] Keller T, Krist T, Danzig A, Keiderling U, Mezei F and Wiedenmann A 2000 *Nuclear Instruments and Methods in Physics Research A* **451** 474–479
[3] Dewhurst C 2008 *Measurement Science and Technology* **19**

[4] Babcock E, Boag S, Becker M, Chen W C, Chupp T E, Gentile T R, Jones G L, Petukhov A K, Soldner T and Walker T G 2009 *Phys. Rev. A* **80**(3) 033414
[5] Lefmann K and Nielsen K 1999 *Neutron News* **10** 20–23
[6] Willendrup P, Farhi E, Knudsen E, Filges U, Lefmann K and Stein J 2016 *User and Programmers Guide to the Neutron Ray-Tracing Package McStas, Version 2.3*
[7] Chen B, Huang C, Li X, Pleshanov N, Syromyatnikov V and Schebetov A 2006 *Physica B* **385–386, part 1** 663–666
[8] Pleshanov N, Peskov B, Schebetov A, Syromyatnikov V, Chen B, Huang C and Li X 2007 *Physica B: Condensed Matter* **397** 62–64
[9] http://www.swissneutronics.ch

Neutron multilayer-backed flipper: experiment, prospects

N K Pleshanov
Petersburg Nuclear Physics Institute, NRC "Kurchatov Institute",Gatchina, St. Petersburg, 188300, Russia.

E-mail: pnk@pnpi.spb.ru

Abstract. The efficiency of a multilayer-backed neutron mirror flipper for monochromatic beams was found to be 97.5±0.5%. Such mirror flippers can combine monochromatization of a polarized beam with flipping neutron spins. Supermirror-backed flippers are shown to be more advantageous with initially monochromatic beams, with a gain in the outcome intensity up to 4 times. Prospects in improving mirror spin turners, incl. flippers, are discussed. Performance of spin turners is analyzed. Scheme of a hyperpolarizer for monochromatic beam mode reflectometers is considered, its working parameters are estimated.

1. Introduction

Neutron spin manipulation optics recently proposed [1,2] may essentially widen the functionality of neutron optics and contribute to the development of tools for obtaining information about the objects under study. Here the term *neutron spin optics* (NSO) is used to designate the new possibilities. Among the advantages of NSO over the existing spin manipulation devices [3] are

- compactness (miniaturization is practically unlimited),
- zero-field option (no external fields are required, guide fields are optional),
- multi-functionality (beam spectrum, beam divergence and spin manipulations can be handled at the same time).

NSO devices may play an important role in developing alternative schemes of measurements, esp. with small samples, which are often of special interest for neutron studies. Therefore, the probing of the basic principles and elaboration of the basic elements for NSO are quite essential.

2. Design

The necessary condition of building mirror spin turners is the equality of the reflectivities of neutrons with the opposite spins:

$$R_+ = R_-. \qquad (1)$$

Then the change of the spin orientation of neutrons under reflection can be described as precession about the fields in magnetic layers. A multilayer-backed spin turner [1] for monochromatic beams is the simplest for realization. Observation of the spin flipping for neutrons reflected from such a mirror flipper was recently reported [4]. It is designed (Fig. 1) as a magnetic layer weakly reflecting neutrons on top of a non-magnetic multilayer (ML). At large q (at a Bragg peak) the layers of Co, Fe and their alloys weakly reflect neutrons and may be used to build mirror spin turners. Spins inclined to the magnetic induction vector $\mathbf{B}_{\text{in}}$ in the magnetic layer undergo Larmor precession twice, before and after reflection of neutrons from ML. The precession angle depends on q, so the spread of precession angles is defined by the Bragg peak width, and the efficiency of multilayer-backed spin turners is increased for multilayers with smaller d-spacings.

According to the Larmor precession law, the spin rotates about $\mathbf{B}_{in}$ by an angle

$$\varphi = CB_{in}\lambda \frac{2d_m}{\sin\theta} \quad (2)$$

($C = 4\pi m_n|\mu_n|/h^2 \cong 4.63 \cdot 10^{-4}$ T^{-1}nm^{-2}) proportional to B_{in}, λ and the total (before and after reflection from ML) path length in the magnetic layer with a thickness d_m. The Bragg condition taking account of the refraction in ML can be written as

$$4\pi \sin\theta = \lambda q_B, \quad (3)$$

and one obtains the precession angle at the Bragg peak maximum ($q = q_B$):

$$\varphi = 8\pi C B_{in} d_m / q_B. \quad (4)$$

To turn spins of the Bragg-reflected neutrons by an angle $\varphi = \pi$, the required thickness of the magnetic layer can be found from (4) as

$$d_m(\pi) = q_B/(8CB_{in}). \quad (5)$$

When $\mathbf{B}_{in}$ is perpendicular to the spins of the incident neutrons, the layer will work as a flipper for neutrons reflected with q near q_B.

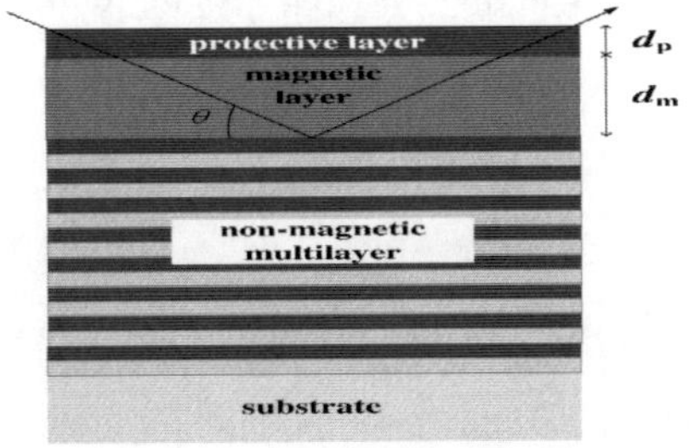

Figure 1. The design of a mirror flipper: non-magnetic multilayer coated with a magnetically anisotropic and protective layers.

The magnetic layer can be made from a CoFe alloy used in production of polarizing CoFe/TiZr supermirrors. From the nuclear and magnetic potentials found for CoFe layers [5], one can determine the saturation magnetization $B_s = 2.0$ T. Owing to pronounced easy-axis anisotropy, the residual magnetization of the layers is quite high. Assuming that it is equal to $0.95B_s$, we find from (5) for q_B= 0.5 nm^{-1} that $d_m(\pi) = 71.1$ nm.

To produce a mirror flipper, a periodic structure of 20 pairs of quarter-wavelength non-magnetic NiMo(6.96 nm) and Ti(6.17 nm) layers was sputtered onto two glass substrates at the sputtering machine LUNA (PNPI, Gatchina, Russia). Then the NiMo target was replaced with a CoFe target, and two layers, $Co_{70}Fe_{30}$(71.1 nm) and Ti(30 nm), were sputtered onto one of the samples. The additional Ti layer protects [6] the CoFe layer from oxidation and prevents deterioration of its magnetic properties.

3. Experiments and analysis

Experiments with the mirror flipper were carried out at the neutron reflectometer NR-4M (beam 13, reactor WWR-M, Gatchina) with the maximum spectral density for the polarized beam at a wavelength 0.130 nm. The spin-up (+) and spin-down (-) neutron reflectivities $R_\pm(q)$ for the mirror flipper in a field sufficiently strong to magnetically saturate the CoFe layer are represented in Fig. 2. One can see that the precession condition $R_+ = R_-$ is well fulfilled for the Bragg peak with $\Delta q/q$=7.5% (FWHM) in the q-scale. However, the peak maximum is not at the designed value 0.5 nm^{-1}, but at a value 0.48 nm^{-1} corresponding to thicker layers in ML.

In the low q region the spin-up and spin-down reflectivities are quite different (see Fig. 2). It can be used to study the reversal of the CoFe layer magnetization by measuring integral reflectivities at a small glancing angle. Then the integral polarizing efficiency

$$P_{int}(B) = [R_{off}^{int}(B) - R_{on}^{int}(B)]/[R_{off}^{int}(B) + R_{on}^{int}(B)] \quad (6)$$

obtained from the reflectivities, measured with the flipper f_1 switched off and on as a function of the applied field B, represents a hysteresis loop. The hysteresis loop obtained with the field **B** along the e.a. is represented in Fig. 3-a; the field was reduced from 60 mT to 0, then the layer was oppositely magnetized in the field -60 mT and then measurements were resumed in positive fields from 1 to 60 mT. Note that the magnetization reversal starts about 12 mT.

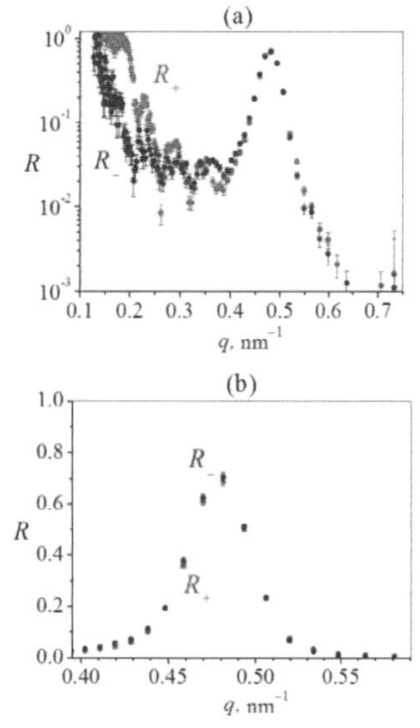

Figure 2. (a) The spin-up (+) and spin-down (-) neutron reflectivities $R_\pm$ for the mirror flipper (Fig. 1) are represented as functions of q. The TOF data are obtained with the sample at θ = 4.95 mrad in a field 47 mT.
(b) The same $R_\pm$ in the linear scale for the q-range with the Bragg peaks.

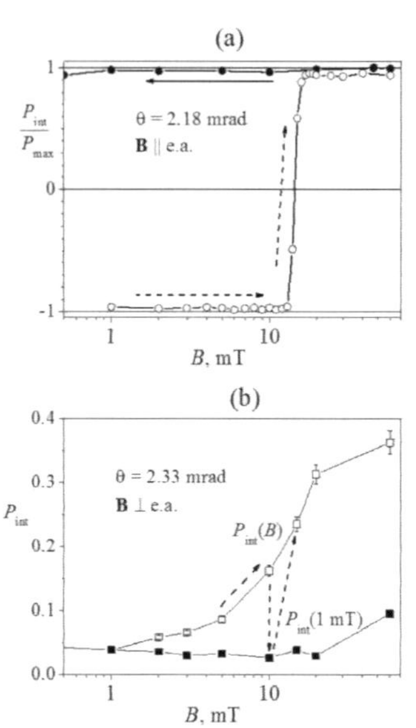

Figure 3. (a) The dependence $P_{int}(B)$, normalized to P_{max}= max(P_{int}), obtained with a white beam of neutrons reflected at θ = 2.18 mrad from the mirror flipper in the field **B** || easy axis (in the CoFe layer).
(b) The dependence $P_{int}(B)$ obtained with a white beam of neutrons reflected at θ=2.33 mrad from the mirror flipper in the field **B** ⊥ easy axis (in the CoFe layer). Previously the sample was magnetized in B=60 mT along the easy axis and turned in B=0 about its surface normal by 90°. The field B changed as shown with the arrows: each rise from 1 mT was followed by a drop to 1 mT.

The sample magnetized in B=60 mT was turned in B=0 about its surface normal by 90°. The measurements obtained with the field **B** ⊥ e.a. are represented in Fig. 3-b. The field B changed as follows: 0 → 1 mT → 2 mT→ 1 mT → 3 mT→ 1 mT → ... As a consequence, the lower points (filled squares) in Fig. 3-b correspond to measurements in a field 1 mT after applying the field given by the field axis. One can infer that no oppositely magnetized domains form in the fields up to 20 mT. The layer behaves as a single-domain magnetic spring. It means that the mirror spin flipper, in principle, can work in guide fields up to 20 mT.

In the standard method of measuring the flipper efficiencies, two flippers are switched off and on, and the 4 intensities measured with an analyzer give the efficiencies of both. A mirror spin flipper would change both intensity and direction of the beam, violating the requirement that the source and the analyzer in the measurements be the same. So the scheme is modified. Since two multilayer samples were prepared in the same sputtering process and only one was later coated with the magnetic film, both can be used, one imitating the "flipper-off" state, and the other flipping spins. The samples were adjusted on the sample unit and could be moved into and out of the beam. The same glancing angles mean the same neutron paths, therefore the efficiencies of flipper 2 and the analyzer will be identical in each set of measurements. The designed-in flipper after the sample unit was found [4] to be more efficient with small fields at the sample, than that before the sample unit. Therefore, it was used for testing the mirror flipper in measurements with the analyzer, and the mirror flipper was the first flipper in the modified scheme of measuring the flipper efficiencies.

Normalized TOF intensities (Fig. 4-a) measured for the non-magnetic sample at $\theta = 7.27$ mrad in a field 3 mT with the analyzer and flipper 2 switched off and on show that the Bragg-reflected neutrons are sufficiently polarized, even though the polarization guiding is not optimal with low fields at the sample unit of the reflectometer. Normalized TOF intensities (Fig. 4-b) measured for the mirror flipper at the same glancing angle show that the polarization of the Bragg-reflected neutrons is reversed. The mirror works as a flipper, indeed.

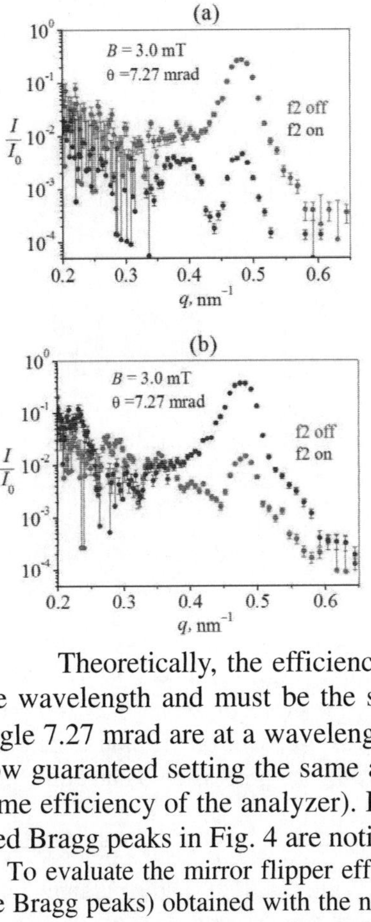

Figure 4. The TOF neutron intensities I measured with (a) the non-magnetic sample and (b) the mirror flipper at $\theta = 7.27$ mrad in a field 3 mT with the analyzer and flipper 2 switched off and on are represented as functions of q. The intensity in each TOF channel is normalized with the direct beam intensity I_0 in the respective channel.

Theoretically, the efficiency of the mirror flipper in a given field does not depend on the wavelength and must be the same at the Bragg peak. The Bragg peaks for the glancing angle 7.27 mrad are at a wavelength $\lambda_B = 0.189$ nm. A narrow slit on the single detector window guaranteed setting the same angle for the reflection from the analyzing supermirror (the same efficiency of the analyzer). Because of the additional reflection and the slit, the normalized Bragg peaks in Fig. 4 are noticeably lower than the Bragg reflectivities in Fig. 2.

To evaluate the mirror flipper efficiency, the integral Bragg peak intensities (integrated only over the Bragg peaks) obtained with the non-magnetic ML ($I_{\mathrm{off,off}}$, $I_{\mathrm{off,on}}$) and the mirror flipper ($I_{\mathrm{on,off}}$, $I_{\mathrm{on,on}}$) are further used. The standard equations lead to efficiencies exceeding 1 and other contradictions, indicating that the Bragg peak reflectivities R_0 (for the non-magnetic ML) and R_F (for the mirror flipper) are not equal. So the standard equations have been generalized [7] for $\rho = R_0/R_F \neq 1$:

$$f_1 = \frac{1}{2}\left(1 + \rho\frac{I_{\mathrm{on,on}} - I_{\mathrm{on,off}}}{I_{\mathrm{off,off}} - I_{\mathrm{off,on}}}\right), \quad f_2 = \frac{1}{2}\left(1 + \frac{\rho I_{\mathrm{on,on}} - I_{\mathrm{off,on}}}{I_{\mathrm{off,off}} - \rho I_{\mathrm{on,off}}}\right). \qquad (7)$$

Regretfully, the reflectivities R_0 and R_F were not measured during the experiments. So a new approach was developed. The following equations have been shown [7] to hold:

$$f_1 = \frac{1}{2}\left[1 + \frac{|\sigma_F|/\sigma_0}{1+(P_0P_A/\sigma_0 - 1)(|\sigma_F| + \sigma_0)/(1+\sigma_0)}\right], \quad (8)$$

$$f_2 = \frac{1+(P_0P_A)^{-1}}{1+\sigma_0^{-1}}, \quad (9)$$

where P_0P_A is the product of the efficiencies of the polarizer and the analyzer, the reflectivities R_0 and R_F cancel out in the quantities

$$\sigma_0 = \frac{I_{off,off} - I_{off,on}}{I_{off,off} + I_{off,on}}, \ \sigma_F = \frac{I_{on,on} - I_{on,off}}{I_{on,on} + I_{on,off}}. \quad (10)$$

Thus, knowing the product P_0P_A and the four intensities, one can find both f_1 and f_2.

The product can be found by using the equation [7]

$$P_0P = [(1+\sigma^{-1})f - 1]^{-1}, \quad (11)$$

which holds for a flipper with an efficiency f and an analyzer with an efficiency P; P_0 is the initial beam polarization; $\sigma = (J_{off} - J_{on})/(J_{off} + J_{on})$, where J_{off} and J_{on} are the intensities measured in the absence of the sample with the flipper switched off and on. Formula (11) used with the data [4] on measurements in the absence of the sample yields an estimation $P_0P_A = 0.98$ ($B = 3$ mT).

Using the integral Bragg peak intensities for reflection from the non-magnetic ML ($I_{off,off}$, $I_{off,on}$) and the mirror flipper ($I_{on,off}$, $I_{on,on}$), one can infer from Eqs. (8) and (9) that $f_1 = 0.975 \pm 0.005$ and $f_2 = 0.995 \pm 0.005$ for neutrons with the wavelength 0.189 nm ($B = 3$ mT). It means that the spins of about 2.5% neutrons within the Bragg peak are not flipped by the mirror flipper. In other words, the NSF portion of the Bragg-reflected neutrons is about 2.5%.

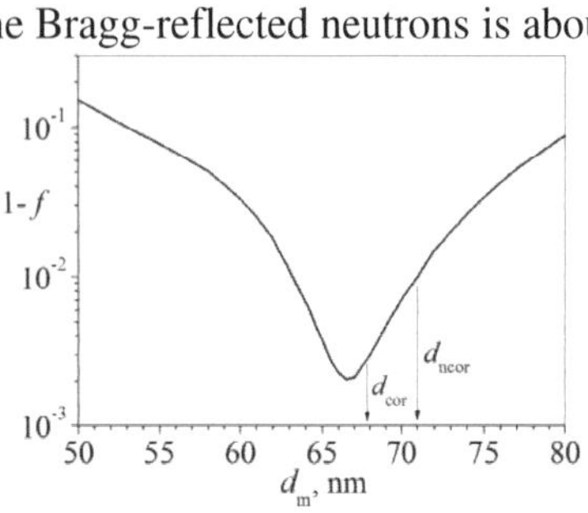

Figure 5. Calculated NSF portions in the Bragg peak integral reflectivity for the spin flipper as a function of the CoFe layer thickness d_m. The arrows point to the thicknesses d_{ncor} and d_{cor} obtained with Eq. (5) and Eq. (12), respectively.

The NSF portion in the Bragg peak integral reflectivity (Fig. 5) is calculated for the mirror flipper as a function of the magnetic layer thickness d_m, the other parameters being as designed; the nuclear and magnetic potentials for the CoFe layer are as found in Ref. [5]; the thickness of a TiO_2 oxide formed on top of the Ti layer is 3 nm [6]. Here and later the NSF portions are calculated with the exact numerical method (generalized matrix method [8]). The NSF integral (solid curve in Fig. 5) is minimum (0.21%) at a thickness $d_m = 66.5$ nm, which is noticeably less than the designed thickness 71.1 nm, obtained according to (5). Therefore, Eq. (5) should be revised. Account of the spin-dependent refraction in the magnetic layer (Fig. 6) gives a corrected thickness for the precession by an angle π at the Bragg maximum q_B:

$$d_{cor} = \pi\left(\sqrt{q_B^2 - q_-^2} - \sqrt{q_B^2 - q_+^2}\right)^{-1}. \quad (12)$$

The correction improves the estimation: $d_{cor} = 67.8$ nm for $q_B = 0.5$ nm^{-1} with the same parameters of the CoFe layer. The thicknesses not corrected (d_{ncor}) and corrected (d_{cor}) for refraction are shown in Fig. 5 with the arrows. One can see that the thickness d_{cor} corresponds to the NSF integral equal to 0.3%. The thickness d_{ncor} corresponds to 1% for the NSF integral. However, q_B was found to differ from 0.5 nm^{-1}, its shift to 0.48 nm^{-1} increases the NSF integral up to 2.5 %, matching the experimental value. So one may hope that the efficiency will approach the theoretical maximum 99.8%, when the thickness is corrected.

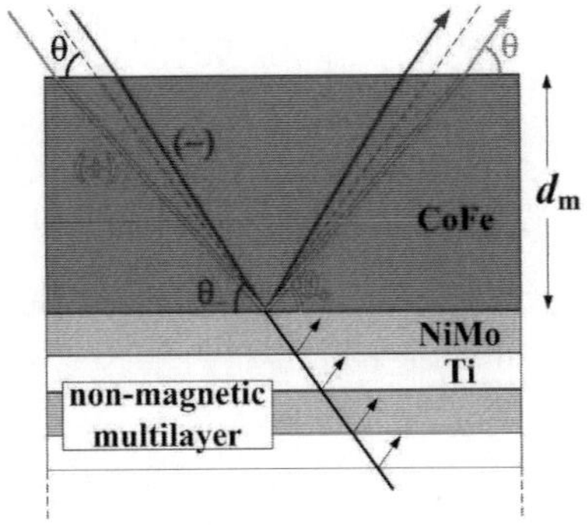

Figure 6. Taking account of the spin-dependent refraction in the magnetic layer gives a better estimation for the optimum thickness d_m. Non-refracted rays are represented with dashed lines. To avoid inessential complications, the spin-independent refraction in the non-magnetic ML is not shown. The protective Ti layer on top of the structure is not shown for the same reason.

Unlike flippers based on transmission through thin Fe foils [9], multilayer-backed flippers can combine monochromatization of a polarized beam with flipping spins of the monochromatized neutrons. However, if the polarized beam is already monochromatic, it may lead to intensity loss. To avoid this, supermirror-backed flippers with reflectivities close to 1 in a wide q-range can be used. The NSF portion of the reflected neutrons is calculated as a function of q in Fig. 7. The theoretical efficiency exceeds 99% in a q-range about 10%. The working q-range exceeds that of the multilayer-backed flipper. Due to the reflectivities close to 1 in the entire working range and due to increased angular width and wavelength spread, the gain in intensity will be up to 4 times. In addition, there is no need to precisely adjust the magnetic layer thickness with the ML spacing.

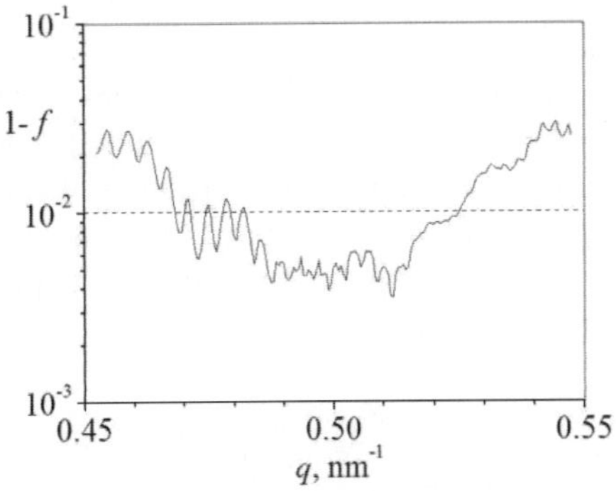

Figure 7. The calculated NSF portion of neutrons reflected from a non-magnetic NiMo/Ti (m = 3) supermirror backing a CoFe (66.5 nm) layer is represented as a function of q. Smaller oscillations due to the interference of neutron waves reflected at numerous interfaces in the supermirror are smoothed by averaging over $\Delta q = 0.001$ nm^{-1}.

4. Performance of mirror spin turners

The test of a neutron mirror flipper verified the feasibility of producing not only mirror flippers, but also mirror spin turners. Performance of a mirror spin turner is shown in Fig. 8. Here the z-axis is chosen along the incident beam polarization vector $\mathbf{P}_0$ and the y-axis belonging to the ($\mathbf{P}_0$, $\mathbf{B}_{in}$) plane, the x-axis is perpendicular to this plane (Fig. 8). Represent the vector $\mathbf{P}_0 = P_0\,(0, 0, 1)$ as the sum of the component $\mathbf{P}_{\|}$ parallel to the precession axis and the precessing component $\mathbf{P}_{\perp}$:

$$\mathbf{P}_0 = \mathbf{P}_{\|} + \mathbf{P}_{\perp},$$

$$\mathbf{P}_{\|} = P_0\cos\chi\,(0, \sin\chi, \cos\chi) = P_0\,(0, \sin\chi\cos\chi, \cos^2\chi), \quad (13)$$

$$\mathbf{P}_{\perp} = \mathbf{P}_0 - \mathbf{P}_B = P_0\,(0, -\sin\chi\cos\chi, \sin^2\chi), \quad (14)$$

where χ is the angle between $\mathbf{P}_0$ and $\mathbf{B}_{in}$.

Assume that the polarization vector $\mathbf{P}_r$ of reflected neutrons is a result of Larmor precession of $\mathbf{P}_0$ about $\mathbf{B}_{in}$. The precession plane is perpendicular to $\mathbf{B}_{in}$ lying in the (y,z) plane. Due to finite divergence and wavelength spread, precession angles vary about a designed value φ. Therefore, $\mathbf{P}_r$ results from averaging polarization vectors weighted by respective intensities over different precession angles. Note that the polarization vector $\mathbf{P}$ obtained by precession by φ ($P = P_0$) can be written as a sum of $\mathbf{P}_{\|}$ and a vector in the precession plane, decomposable into components along $\mathbf{P}_{\perp}$ and along the orth $\mathbf{e}_x$ perpendicular to $\mathbf{P}_{\perp}$ (Fig. 8):

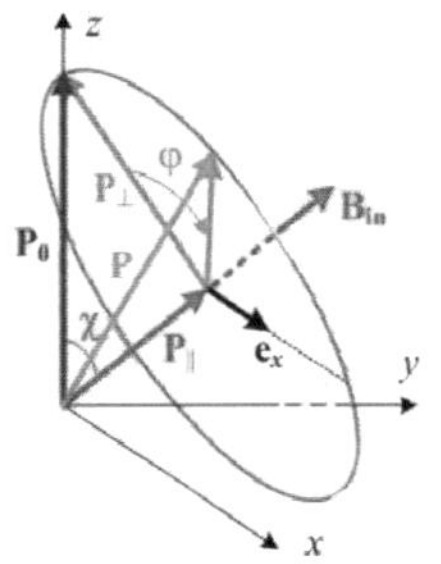

Figure 8. Performance of a spin turner (designations are given in the text).

$$\mathbf{P} = \mathbf{P}_\parallel + P_\perp \sin\varphi\, \mathbf{e}_x + \mathbf{P}_\perp \cos\varphi = P_0\,(\sin\chi\sin\varphi,\ \sin\chi\cos\chi(1 - \cos\varphi),\ \cos^2\chi + \cos\varphi\sin^2\chi), \quad (15)$$

where $P_\perp = P_0\sin\chi$ is the precessing component length.

Eq. (15) can be used to design spin turners. Particularly, the condition $P_z = \cos^2\chi + \cos\varphi\sin^2\chi = -1$ is equivalent to $\cos\varphi = 1 - 2/\sin^2\chi$, satisfied only with $\chi = \pm\pi/2$, $\varphi = \pm\pi$. It was realized in the mirror flipper. The condition $P_x = \sin\chi\sin\varphi = \pm 1$ for a $\pi/2$ (zx) spin-turner is satisfied with $\chi = \pm\pi/2$, $\varphi = \mp\pi/2$ and the condition $P_y = \sin\chi\cos\chi(1 - \cos\varphi) = \sin(2\chi)(1 - \cos\varphi)/2 = \pm 1$ for a $\pi/2$ (zy) spin-turner is satisfied with $\chi = \pm\pi/4$, $\varphi = \pm\pi$. Thus, one can turn spins from the initial z-direction to $-z$, $\pm x$- and $\pm y$-directions (Fig. 9).

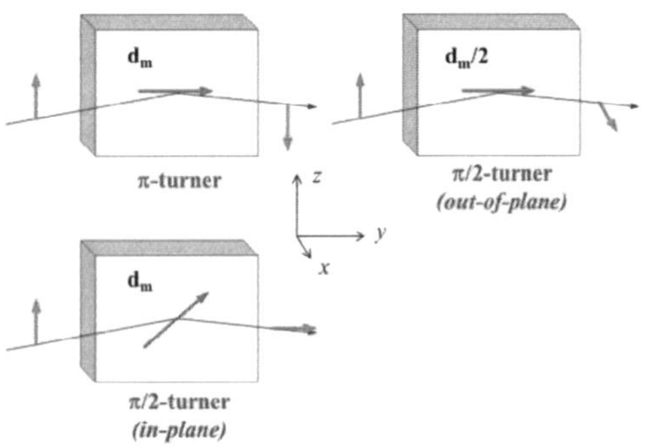

Figure 9. Rotating the mirror flipper about its surface normal by π/4, we obtain a π/2 spin turner leaving spins in the mirror plane. With magnetic layers thinner by half we obtain a π/2 spin turner that may turn initially in-plane spins to the direction perpendicular to the mirror surface.

For a deviation $\Delta\varphi$ from a designed precession angle φ the resultant polarization vector can be found by replacing φ with $\varphi + \Delta\varphi$ in (15). Its projection onto the designed direction $\mathbf{P}/P$ is

$$p_{\Delta\varphi} = \mathbf{P}_{\Delta\varphi}\mathbf{P}/P = P_0\,[\cos^2(\chi) + \sin^2\chi\cos^2\Delta\varphi]. \quad (16)$$

The spin turner efficiency is defined for an initially polarized beam as the portion of neutrons reflected with the spin parallel to a given direction **P**:

$$\varepsilon(\Delta\varphi) = (1 + p_{\Delta\varphi}/P_0)/2 = 1 - \sin^2\chi\,\sin^2(\Delta\varphi/2). \quad (17)$$

Since the Larmor precession angle for a mirror spin turner is inversely proportional to q, one finds

$$\Delta\varphi = -\varphi\Delta q/q, \quad (18)$$

where $\Delta q = q - q_\varphi$ and q_φ is related to the designed precession angle φ.

In our experiment with the mirror flipper: $\chi = \pi/2$, $\varphi = \pi$, hence $\varepsilon(\Delta\varphi) = \cos^2(\Delta\varphi/2)$, which is consistent with the probability of the spin projection onto **P**. The NSF portions

$$1 - f = \sin^2(\Delta\varphi/2) = \sin^2[\pi(q - q_\varphi)/2q], \quad (19)$$

calculated with $q_\varphi = q_B = 0.5$ nm^{-1} for the Bragg peak range of q, are represented in Fig. 10 by the light blue curve. The neutron reflectivity R (pink curve) and the NSF portion (blue curve) for the multilayer-backed flipper with a CoFe (66.5 nm) layer and other parameters as designed are also represented in Fig. 10. Superposition of the waves in the opposite spin states, each resulting from interplay of numerous waves reflected from numerous interfaces in the multilayer-backed flipper leads to distortion of the Larmor precession behavior (cf. blue and light blue curves). The difference between the two curves is only due to the difference in the spin-up and spin-down neutron potentials of the CoFe layer, resulting in different reflection amplitudes (modules and phases) for the opposite spin components. The difference between the amplitudes is smaller at greater q, so the performance of a mirror flipper tuned to a greater q_B will be better.

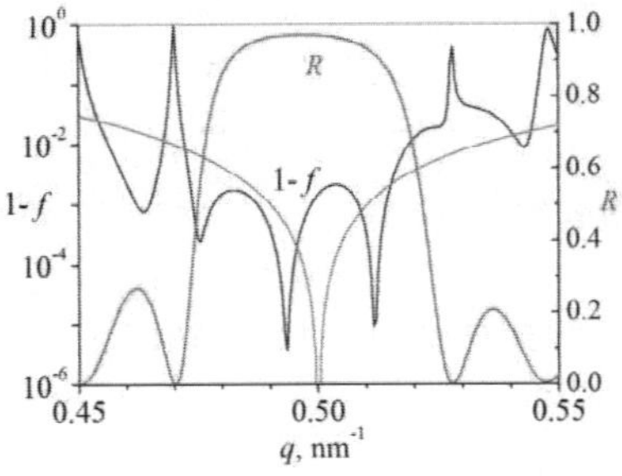

Figure 10. The neutron reflectivity R (pink curve) and the NSF portion 1-f calculated as functions of q with Eq. (21) for a purely Larmor precession in the CoFe (66.5 nm) layer (light blue curve) and with exact numerical approach for the multilayer-backed flipper (blue curve).

5. Hyperpolarizer for a reflectometer

Mirror flippers open new possibilities, one of which is a beam hyperpolarization [1], when the separation of neutrons with the opposite spins is followed by the flipping of the 'wrong' spins. Consider a scheme of such a hyperpolarizer for a neutron reflectometer with a monochromatic beam (Fig. 11). Here a polarizing filter reflects only spin-up neutrons, transmitting spin-down neutrons which are then reflected at a different glancing angle by a mirror flipper. The two formed beams converge at the sample, the two specularly reflected beams are resolved by PSD. Thus, the q-range of interest can be covered in about 2 times quicker. The intensities of beams 1 and 2 can be written as sums of spin-up and spin-down neutron intensities:

$$\begin{pmatrix} I_1^+ \\ I_1^- \end{pmatrix} = \begin{pmatrix} R_p^+ & 0 \\ 0 & R_p^- \end{pmatrix} \begin{pmatrix} I_0/2 \\ I_0/2 \end{pmatrix} = \frac{I_0}{2} \begin{pmatrix} R_p^+ \\ R_p^- \end{pmatrix}, \tag{20}$$

$$\begin{pmatrix} I_2^+ \\ I_2^- \end{pmatrix} = \begin{pmatrix} (1-f)R_F & fR_F \\ fR_F & (1-f)R_F \end{pmatrix} \begin{pmatrix} (1-R_p^+)T_s & 0 \\ 0 & (1-R_p^-)T_s \end{pmatrix} \begin{pmatrix} I_0/2 \\ I_0/2 \end{pmatrix}. \tag{21}$$

where I_0 is the intensity of the initial non-polarized beam, $R_p^{\pm}$ and R_F are the reflectivities of the polarizing filter and the mirror flipper with an efficiency f, T_s is the transmittance of the polarizing filter substrate. Respectively, total intensities and the polarization of beams 1 and 2 are

$$I_1 = I_0(R_p^+ + R_p^-)/2, \quad P_1 = P_p, \tag{22}$$

$$I_2 = I_0 R_F T_s (2 - R_p^+ - R_p^-)/2, \quad P_2 = P_p \frac{2f-1}{2/(R_p^+ + R_p^-) - 1}, \tag{23}$$

where $P_p = (R_p^+ - R_p^-)/(R_p^+ + R_p^-)$ is the efficiency of the polarizing filter.

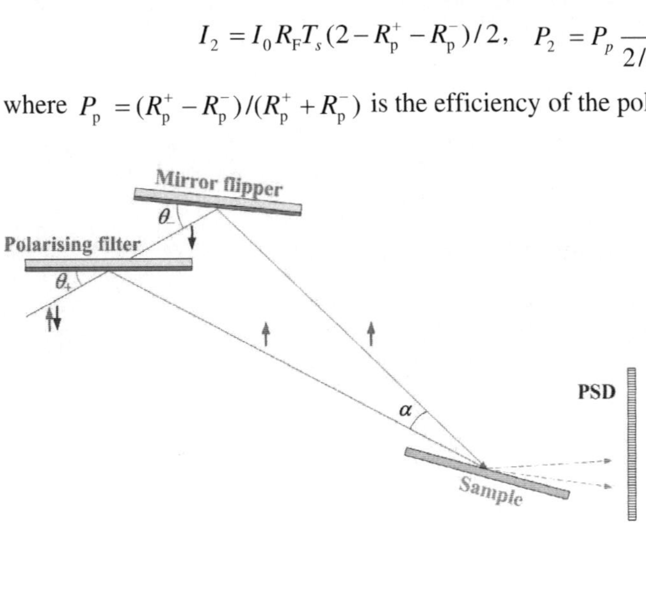

Figure 11. Scheme of a hyperpolarizer for a monochromatic beam reflectometer. The angle α between the beams formed by the polarizing filter and the mirror flipper is supposed to noticeably exceed their divergences.

The gain in the intensity obtained with such a hyperpolarizer is

$$G_h = (I_1 + I_2)/I_1 = 1 + R_F T_s [2/(R_p^+ + R_p^-) - 1]. \tag{24}$$

To achieve a good polarization P_2, the factor in square brackets should be close to 1. Therefore, the intensity gain G_h is mainly defined by T_s and R_F. Note that G_h is independent of the mirror flipper efficiency f. On the other hand, P_2 is proportional to P_p and $(2f–1)$. R^+_p

For very moderate parameters $R_p^+ = 0.9$, $R_p^- = 0.01$, $R_F = 0.9$, $T_s = 0.8$, $f = 0.975$, the performance of the hyperpolarizer is $P_1 = 0.978$, $P_2 = 0.776$, $G_h = 1.862$. For advanced parameters $R_p^+ = 0.99$, $R_p^- = 0.005$, $R_F = 0.95$, $T_s = 0.9$, $f = 0.99$ the performance is $P_1 = 0.99$, $P_2 = 0.96$, $G_h = 1.864$. To improve P_2, the polarizing filter substrate can be coated on both sides. Then for the two sets of the parameters, P_2 is 0.854 and 0.975, respectively. In order to form two highly polarized beams, R_p^+ should be as close as possible to 1. Therefore, total reflection of spin-up neutrons from a sufficiently thick magnetic layer is worth considering.

Of interest is also the option of using a non-magnetic mirror instead of the mirror flipper. Then two beams of opposite polarization will be formed. Again total measurement time can be reduced up to 2 times. In this case formulas (20-24) are also applicable with $f = 0$. Both options (mirror flipper and non-magnetic mirror) can be built in one device.

6. Conclusion

Experiments with a neutron mirror flipper verify the feasibility of producing not only mirror flippers, but also mirror spin turners. Such spin turners can be directly used only with narrow and well-collimated beams, as used in neutron reflectometry. Further developments of multichannel designs and supermirror-based spin turners may appreciably lift the restrictions. Moreover, they can be made broad-band, which is of special interest for applications at impulse reactors and spallation sources. Combining NSO elements, one can build compact devices, including those for 3D-polarization and 3D-analysis, as well as spin manipulators and hyperpolarizers.

Acknowledgements

The work was supported by the Federal target program of Ministry of Education and Science of Russian Federation (project No. RFMEFI61614X0004).

References

[1] Pleshanov N K 2014 *J. Phys.: Conf. Ser.* **528** 012023
[2] Pleshanov N K 2015 *J. Surf. Invest.: X-ray, Synchrotron Neutron Tech.* **9** 24
[3] Williams W G 1988 *Polarized neutrons* (Oxford: Clarendon Press)
[4] Pleshanov N K, Syromyatnikov V G 2016 *Nucl. Instrum. Methods* A **837** 40
[5] Pleshanov N K, Peskov B G, Schebetov A F, Syromyatnikov V G, Chen B, Huang C Q, Li X X 2007 *Physica* B **397** 62
[6] Matveev V A, Pleshanov N K, Gerashchenko O V, Bairamukov V Yu 2014 *J. Surf. Invest.: X-ray, Synchrotron Neutron Tech.* **8** 991
[7] Pleshanov N K: accepted for publication in *Nucl. Instrum. Methods* A
[8] Pleshanov N K 1996 *Z. Phys.* B **100** 423
[9] Kraan W H, Plomp J, Krouglov T V, Bouwman W G, Rekveldt M Th 2003 *Physica* B **335** 247

IOP Conf. Series: Journal of Physics: Conf. Series **862** (2017) 012022 doi:10.1088/1742-6596/862/1/012022

An ultra-compact in-situ ^{3}He polarizer for high Q-range SANS: on-beam test and future upgrades

Zahir Salhi, Earl Babcock, Kendal Bingöl, Denis Starostin, Patrick Pistel, Nils Lumma, Aurel Radulescu, Alexander Ioffe

[1] Julich Centre for Neutron Science (JCNS) at the Heinz Maier-Leibnitz Zentrum (MLZ), Lichtenbergstaße. 1, 85747 Garching Germany
[2] Forschungszentrum Julich GmbH ZEA-1, Leo-Brandt-Straße, 52425 Julich Germany

E-mail: z.salhi@fz-juelich.de

Abstract. We present an ultra-compact ^{3}He polarizer to be used as a polarization analyzer for separation of incoherent background for soft matter research. The ^{3}He will be polarized in-situ within a very compact, 18 cm long, magnetic cavity that can be placed between the sample and the detector tank of small-angle scattering diffractometer KWS2 at MLZ [1]. This allows for the angular coverage of 38°, that corresponds to the maximal momentum transfer Q_{max}=0.85 Å^{-1} for neutrons with wavelength λ=4.5Å. The Q_{max} can be extended to 1.28 Å^{-1} following the recent upgrade of KWS2 for neutron wavelength λ=3Å. The optical pumping will be done by the diode laser array bar frequency narrowed by the ultra-compact volume Bragg grating. The full system will be readily hand transportable and thus useful as an instrument add-on.

1. Introduction.

Incoherent background can create an intrinsic problem for standard small angle neutron scattering measurements. Biological samples contain hydrogen which is a strong incoherent scatterer thus creating an intrinsic source of background that makes determination of the coherent scattering parameters difficult in special situations. This can especially be true for the Q-range from around 0.1-0.5 Å^{-1} where improper knowledge of the background level can lead to ambiguity in determination of the samples structure parameters. Polarization analysis is a way of removing this ambiguity by allowing one to distinguish the coherent from incoherent scattering, even when the coherent scattering is only a small fraction of the total scattered intensity [2]. ^{3}He spin filters are ideal for accomplishing this task because they permit the analysis of large area and large divergence scattered neutron beams.

This work describes a design for a new ultra compact SEOP [3-4] polarized ^{3}He neutron spin filter that can be used for this application. The magnetic system has been completed and tested. The mechanical components for online polarization, such as a compact laser system, the oven and the results of in-situ test will be described in this document.
Our system is based on a 17 cm long mu-metal solenoid. Fully capped mu-metal solenoids can be used to create very uniform magnetic fields. However, for neutron applications there must be holes in the caps to allow access for incident and scattered neutron beams. These holes, especially the larger one required for the scattered beam, undoubtedly create field gradients that significantly reduce the lifetime of the polarized ^{3}He in the cavity. To decrease the effect of these gradients we have used an approach that combines partial shielding and the compensation of the holes in the end caps with additional coils (figure1).
A 1x1 cm^2 beam is assumed to be incident on the sample, which would be a typical soft matter sample located in the brown holder directly before the right end cap. The cell is placed within 2cm of the 1.5x1.5 cm^2 hole in this end cap. After passing through the continuously polarized ^{3}He cell of 6cm diameter and 5cm length a square cross section beam can pass the partially shielded opening in the end cap on the left of the drawing.
Expansion of the laser beam to D=6 cm will be done with optic inside the magnetic cavity and the expanded beam will be reflected longitudinally along the neutron beam via a gold coated Si-wafer as shown in Figure 1. The collimation optics will include a telescope utilizing a large diameter 45 degree parabolic metal mirror as the last element which will be located inside the magnetic cavity.

The heating system use static air heated by electric cartridge heaters mounted in aluminum heat sinks. These stainless steel electric cartridge heaters (220 V CSS-series from Omega Engineering) now used for heating have not been found to decrease the expected ^{3}He polarization, or room temperature T_1 lifetime. There is indeed some magnetic components to these heaters, namely the nickel plated lead wires, however in our installation, the heater is placed inside the before mentioned aluminum heat-sinks, about 10 cm away from the cell, with the wires exiting away from the cell position. The aluminum heat sink is grounded to limit pickup of electrical noise by the NMR FID system [5-7].

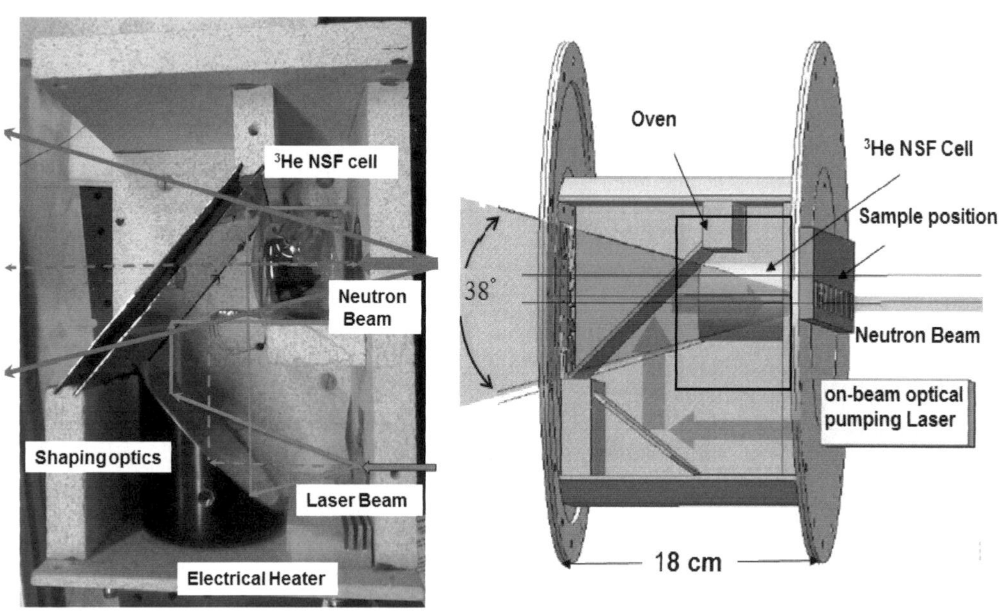

Figure 1. A diagram of the KWS2 magnetic system (right), oven and optics (left). Cut-away drawing of the magnetic cavity showing the optical access (red arrows) and the neutron access, violet.

2. Test of the magnetic cavity

The magnetic cavity has been constructed and tested for the polarized ^3He magnetic lifetime. All of the field coils, the main solenoid with 1 turn/mm and the end-hole compensating coils have been optimized with respect to the number of turns to operate in series from a single power supply. Initial experimental optimization was performed with the 3-axis hall probe mounted to a rail to make 1-D longitudinal field maps of the magnetic cavity. The current of the compensating coils was varied to minimize the magnetic field gradients at the desired cell position of 2-8 cm from the inside of the front of the cavity. Then the number of turns in the compensating coils was adjusted for the series operation. As the result, the final number of turns did not vary by more than one or two turns from calculations.

After this optimization a 5 cm diameter 5 cm long SEOP cell with 0.5 bar ^3He pressure and 930 hour lifetime was polarized externally in our lab and placed in the magnetic cavity aiming the measurements of the ^3He lifetime. The cell was placed at the distance of 2 cm from the front 1.5x1.5 cm^2 hole in the magnetic cavity (right side in Figure 1). Figure 2 shows an NMR free induction decay measurement of the ^3He lifetime [5-6].

After subtracting of the contribution for the intrinsic lifetime of the cell which includes ^3He dipole-dipole relaxation [8] and the wall relaxation of this cell [9], we obtain the measured ^3He magnetic lifetime of 1450 hours for 1 bar pressure or the averaged over the cell equivalent magnetic field gradient of about 2.5x10^{-4} cm^{-1}. Such magnetic performance proves that this non-conventional, ultra short solenoid cavity with rather large openings for scattered beams can provide the required, if not excellent, performance.

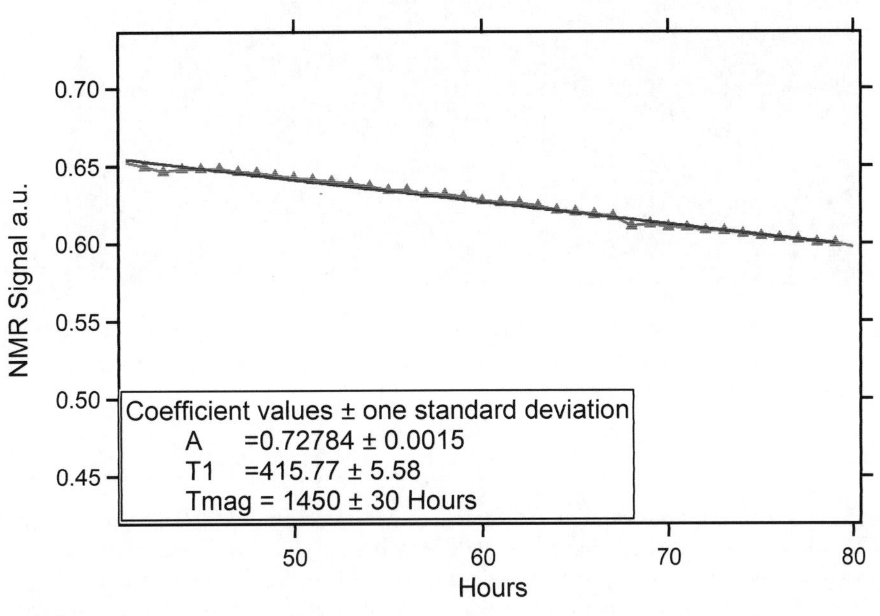

Figure 2. Measured ^3He relaxation time of 415 hours for 0.5 bar ^3He cell placed 2 cm from the entrance aperture of the magnetic cavity. Subtraction of the cells self-relaxation time yields a cell averaged equivalent gradient of 2.5x10^{-4}cm^{-1}.

3. Laser systems upgrade

Our laser systems have also been upgraded. Formerly diode array bars (DAB) with external cavities using Littrow-type configuration were used [10] as narrow-band optical pumping light has been shown to provide the best SEOP performance. However, it was found later that a new compact method using chirped volume Bragg gratings, CVBG, provides a higher absolute ^{3}He polarization [11]. Additionally, CVBG narrowed lasers give higher laser intensity output for laser diodes of the same power, and contain much fewer optical elements making the cavity more compact, stable and robust for long term operations required for in-situ ^{3}He polarizer applications. A photo of the new laser cavity is shown in figure 3, where the laser mounted inside a newly designed compact box for mounting the laser and cavity optics is shown, as well as external connections to the power cables and cooling water. This box also provides environmental protection and increased laser security. Two such lasers are used to pump the large 0.85 liter volume of the ^{3}He NSF cell used on reflectometer MARIA [12], however for the application in question, where a 0.17 litre cell is used one laser is sufficient.

Figure 3. A picture the laser-box of the current volume Bragg grating narrowed laser configuration. This optical setup fit in a custom made enclose box to enclose and mount all the optics and provide connections to power cable and cooling water

4. On-beam test

After several tests and off line optimization, the compact cavity was installed at SANS instrument KWS2 at MLZ for absolute measurements of the ^{3}He polarization using the neutron transmission technique [13]. It was placed directly after the neutron sample position and next to the window of the detector tank in order to keep the maximum detector coverage as would be done in a typical polarizer SANS measurement (figure 4).

As a first step, the system was installed near the KWS2 instrument two days before the on-beam test. The cell [14] with known opacity (^{3}He number density-length-cross section product [He]/σ) was polarized to saturation and then placed at the sample position as described above. The ^{3}He polarization was monitored using an NMR system.

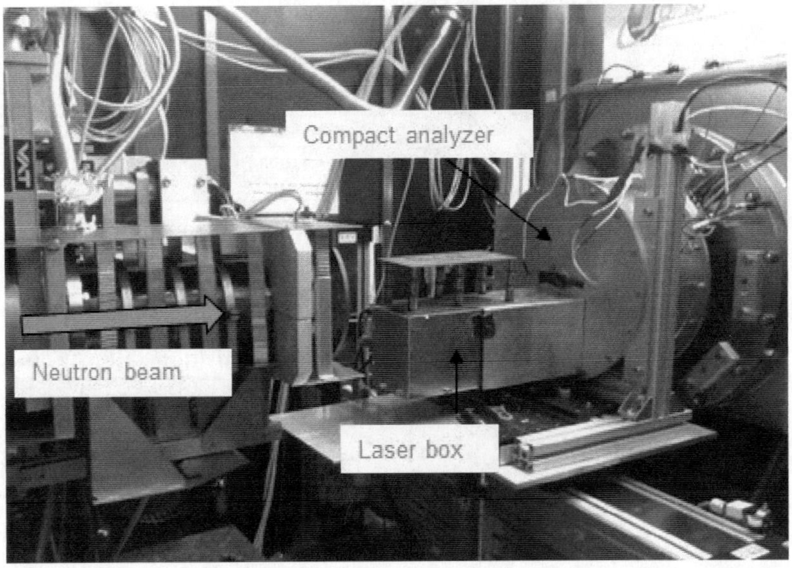

Figure 4. The ultra-compact cavity installed at the sample environment of KWS2 (SANS instrument)

The neutron transmission of the unpolarized incident beam through the ^{3}He cell was monitored as a function of time. The relative neutron transmission, i.e. the ratio of transmission T of an unpolarized neutron beam through a polarized ^{3}He cell to the transmission T_0 through the unpolarized cell is:

$$T/T_0 = \cosh([\mathrm{He}]l\sigma P_{\mathrm{He}}\,\lambda). \quad (1)$$

Here [He] is the ^{3}He number density, σ - the spin dependent ^{3}He neutron absorption cross section, λ - the neutron wavelength, l - the cell length and P_{He} - the ^{3}He polarization. Thus, by knowing the ^{3}He number "density-length-cross section" product, [He]$l\sigma$, and the neutron wavelength, the measurements of the unpolarized neutron transmission is a straightforward method to obtain the absolute ^{3}He polarization.

The product [He]$l\sigma$ of a ^{3}He cell can be determined from the unpolarized neutron transmission of an unpolarized ^{3}He cell as a function of neutron wavelength. The unpolarized neutron transmission of an unpolarized ^{3}He cell is:

$$T_0(\lambda) = T_{\mathrm{E}}\, exp([\mathrm{He}]l\sigma\,\lambda). \quad (2)$$

Indeed, by measuring the transmission as a function of neutron wavelength and fitting the function 2, one obtains the value of the exponential which is the number density-length-cross section product, where T_{E} is the empty cell transmission.

The results of the on-beam test are shown in figure 5. From this data one can see that the maximal obtained ^{3}He polarization is 56%. Though this polarization is insufficient for a high neutron performance of the SANS analyser, however improvements and further optimization should allow us to approach the level of performance achieved in other in-situ polarizers made by our group [12]. These upgrades are discussed in the next section.

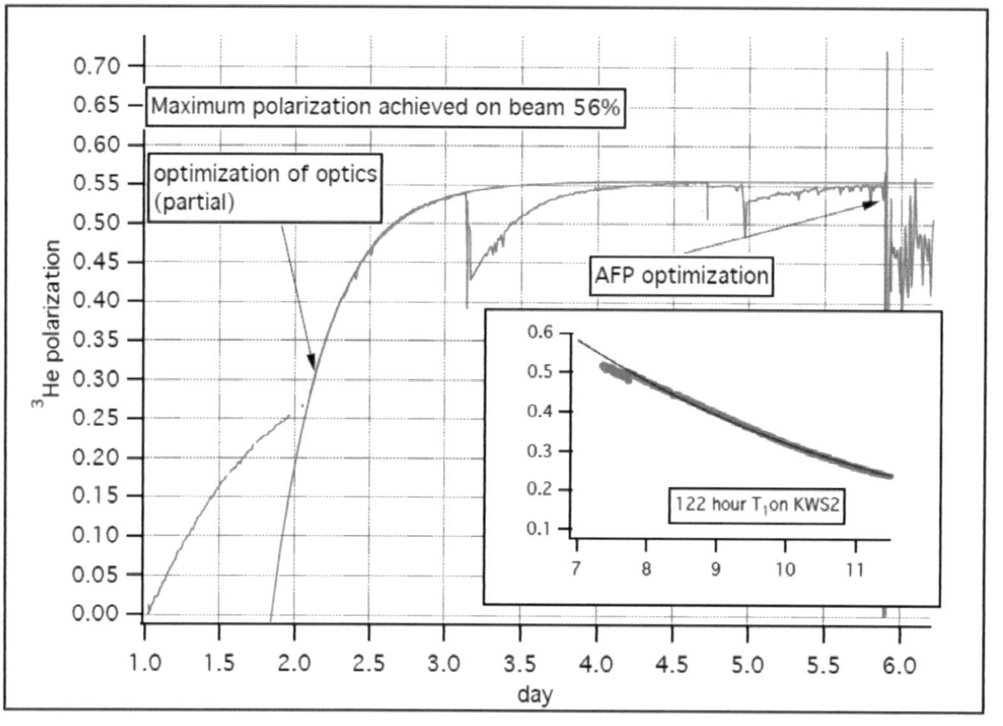

Figure 5. The ^{3}He polarization as function of time obtained during on-beam test at KWS2, as insert, the measured on beam relaxation time for the polarized cell. Maximum polarization achieved on beam 56% with 7% neutron transmission.

The in-situ ^{3}He relaxation time for this experiment was measured at the end of the test by turning off the lasers and the cell heating while monitoring the decay of the polarization using NMR system., and the fit gave T_1=122hours which is sufficiently long compared to the spin-exchange rate and doesn't significantly affect the maximum achievable polarization.

5. Future upgrades

During our on beam test we faced two problems. First, a difficult optics alignment because of the limited access to the laser and expansion-collimation optics inside the compact and closed cavity. In figure 5 one can see how the improvement of the polarization build-up time was achieved by just a partial optimization of the optics alignment of the laser beam. Second, μ-metal plates and the sample position was getting hot and reaching a temperature of nearly 100°C because of the direct contact with the oven inside the cavity. Here one should keep in mind that the sample position, which is placed within 3 cm from the front of the ^{3}He NSF cell, must not be significantly heated to avoid negative influence on the performance of the sample environment and, in turn, samples contained within it because of soft matter and biological samples are typically temperature sensitive.

To solve these problems the system is undergoing further testing, optimization and mechanical refinement before eventual user operations. The optimization and refinement include appropriate cooling of the system and the sample (to avoid denaturation of proteins placed 3cm from the oven), laser collimation and optics adapted to a such small oven (better illumination of ^{3}He cell with narrowed laser), mechanical design and fixation of the system to allow an easy use (automatically placed on and out of the beam when needed). The new proposed design is shown in figure 6. The system consists of aluminum box with plates fixed ½ cm from each side of the magnetic cavity to not reduce the angular coverage of the detector, the lateral sides were used to fix two cooling units to cool efficiently the system through ½ cm gap between the μ-metal and the aluminum plates. All cables connections are fixed via feed through on the lateral side of the box to facilitate an easy handling.

IOP Conf. Series: Journal of Physics: Conf. Series **862** (2017) 012022 doi:10.1088/1742-6596/862/1/012022

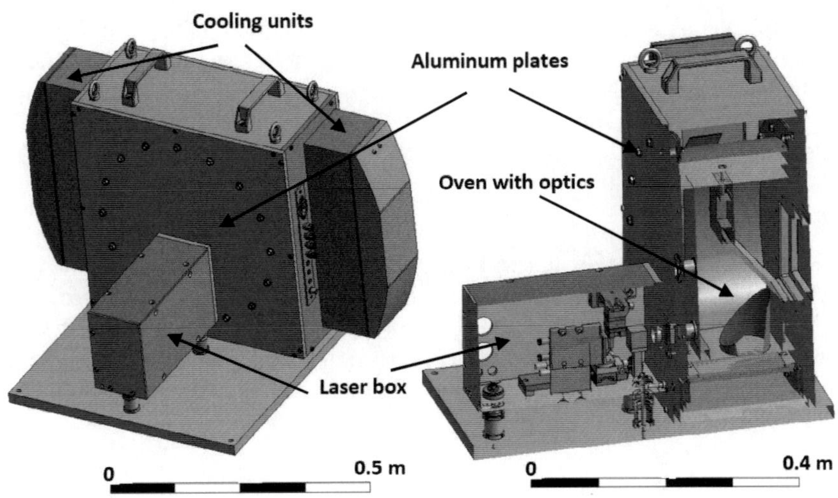

Figure 6. A new proposed design of the ultra-compact cavity including cooling unit, laser box, laser collimation and optics adapted to a such small oven

In order to optimize the cooling power needed to keep the temperature at the cell position of about 200°C needed for the optical pumping, and less than 30°C at the sample position, we have performed several FEM calculations of the heat transfer inside the cavity using ANSYS software [15]. The first calculations of the heat transfer was performed using the old design of the compact cavity without cooling units, to compare the simulation results with results obtained during the on-beam experimental test. The result of simulations (figure 7) are in good agreement with temperatures measured using PT100 temperature sensor at the surface of the μ-metal plates and inside the oven at the cell position. One can see that the temperature at sample position reach 132°C.

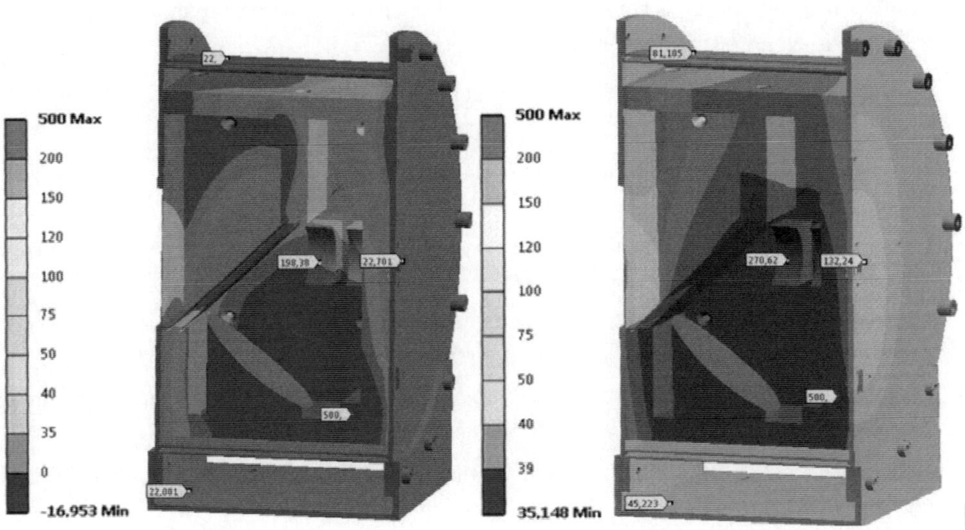

Figure 7. Results of the heat transfer calculations using ANSYS software. Right: without cooling unit, using the old cavity design; temperature at the sample position is 132°C. Left: for the new proposed design with cooling units; temperature at the sample position is 23°C.

The second series of calculations were performed using the new design taking into account the gap of 1mm between the oven and the μ-metal plates. One can see from figure 7 that with the new system we can obtain optimal parameters (200°C at the cell position and less than 30°C at sample position) with cooling power of 250 W and the same heating power as in the old design. As shown in figure1 the sample is located in the brown holder directly before the right end cap outside the cavity and can be additionally cooled to just above the freezing temperature for biological samples in order to prevent them from degrading.

The system is now under construction and will use two cooling units, each with 150 W cooling power that will allow for a better optimization of the alkali-metal vapor density. An improved mechanical fixation of the optics and the laser box will also allow for the better alignment of the optic and the illumination of the cell. A small window at the back side of the oven will allow us to control the optics inside the oven and check the complete illumination of the ^3He cell.

6. Conclusions

We have designed, prototyped, and tested an ultra-compact SEOP ^3He polarizer to be used as the in-situ analyzer for SANS diffractometer KWS2. The magnetic cavity design assures the relative field gradients less than 10^{-4} cm^{-1} and a large detector coverage of ±19 deg. Due to the difficulty of laser alignment (optic inside the closed oven) and the high temperature at the sample position and mu-metal plates (impossibility to increase further the temperature to optimize the alkali metal vapor density) the first on-beam test gave maximal polarization of 56%. Based on simulation results, the system will undergo upgrades and further tests to improve the maximum reachable on-beam polarization. Once the critical mass of components is available we will proceed with neutron sample measurements and data collection and analysis techniques.

References

[1] A. Radulescu et al., J. Vis. Exp. 118, e54639, 2016

[2] E. Babcock, Z. Salhi, M-S. Appavou, A. Feoktystov, V. Pipich, A. Radulescu, V. Ossovyi, S. Staringer, A.Ioffe; Polarization analysis with He3 spin filters for separating coherent from incoherent scattering in soft matter studies, Physics Procedia 42 (2013) 154 – 162

[3] T.Walker and W. Happer, Spin-exchange optical pumping of noble-gas nuclei. Rev Mod Phys, vol. 69, no. 2, pp. 629–42, 1997.

[4] F. D. Colegrove, L. D. Schearer, and G. K.Walters, "Polarisation of helium-3 gas by optical pumping," Physical Review, vol. 132, pp. 2561–2572, 1963.

[5] G. D. Cates, D. J. White, T. Chien, S. R. Schaefer, and W. Happer. Spin relaxation due to inhomogeneous static and oscillating magnetic _elds. Phys. Rev. A, 38, 1988.

[6] R. Jacob, J. Teter, B. Saam, W. Chen, and T. Gentile, "Low-field orientation dependence of 3He relaxation in spin-exchange cells," Physical Review A,vol. 69, pp. 021401–1–4, 2004.

[7] G. D. Cates, S. R. Schaefer, and W. Happer. Relaxation of spins due to field inhomogeneitiesin gaseous samples at low magnetic _elds and low pressure. Phys. Rev. A, 37,1988.

[8] N. R. Newbury, A. S. Barton, G. D. Cates, W. Happer, and H. Middleton. Gaseous 3He-3He magnetic dipolar spin relaxation. Phys. Rev. A, 48(6), 1993. doi:10.1103

[9] J. Schmiedeskamp, W. Heil, E. Otten, R. Kremer, A. Simon, and J. Zimmer,"Paramagnetic relaxation of spin polarized 3He at bare glass surfaces," The European Physical Journal D, 2006

[10] E. Babcock, B. Chann, I Nelson and T Walker 2005Applied Optics 44 3098–3104

[11] W. Chen, T. Gentile, Q. Ye, T Walker and E Babcock 2014 J. Appl. Phys. 116 014903

IOP Conf. Series: Journal of Physics: Conf. Series **862** (2017) 012022 doi:10.1088/1742-6596/862/1/012022

[12] E. Babcock, S. Mattauch, A. Ioffe; High level of 3He polarization maintained in an on-beam 3He spinfilter using SEOP; Nuclear Instruments and Methods Volume 625, Issue 1, 1 January 2011, Pages 43–46

[13] G.L Green, A.K Thompson, M .S. Dewey , A method for accurate determination of the polarization of a neutron bean using He3 spin filter NIM A 356(1995) 177-180

[14] Z. Salhi, E Babcock, P Pistel and A Ioffe, 3He Neutron Spin Filter cell development program at JCNS Journal of Physics Conference Series 07/2014; 528(1):012015.

[15] ANSYS® Academic Research, http://www.ansys.com/, heat transfer simulation, ANSYS, Inc.

IOP Conf. Series: Journal of Physics: Conf. Series **862** (2017) 012023 doi:10.1088/1742-6596/862/1/012023

Data processing workflow for time of flight polarized neutrons inelastic measurements

Andrei T Savici[1], Igor A Zaliznyak[2], V Ovidiu Garlea[1] and Barry Winn[1]

[1]NScD, Oak Ridge National Laboratory, Oak Ridge, TN., USA
[2]CMPMSD, Brookhaven National Laboratory, Upton, NY, USA

E-mail: saviciat@ornl.gov

Abstract. We discuss the data processing workflow for polarized neutron scattering measurements performed at HYSPEC [1] spectrometer at the Spallation Neutron Source, Oak Ridge National Laboratory. The effects of the focusing Heusler crystal polarizer and the wide-angle supermirror transmission polarization analyzer are added to the data processing flow of the non-polarized case. The implementation is done using the Mantid [2] software package.

1. Introduction

The polarized neutron scattering has proved to be an invaluable tool for distinguishing between nuclear and magnetic cross-sections. Recent advances in instrumentation allow scattering measurements using large area, position-sensitive detectors. Such experiments are routine for unpolarized neutrons. However, polarization analysis on wide angle detectors pose some challenges, which we are discussing in this paper.

Traditionally, polarized inelastic neutron investigations of the sample dynamical properties were carried out using triple axis spectrometers [3]. A schematic representation of such an instrument is shown in Figure 1. A monochromatic neutron beam is produced when neutrons from the source (moderator) are Bragg-reflected by the monochromator crystal. To produce a polarized neutron beam, one can use a monochromator made of Heusler crystals, magnetized so that they only reflect one neutron spin state, or insert a polarized ^{3}He cell, which only transmits one spin state along the monochromator to sample flight path. The incident beam polarization can be rotated using an adiabatic guide field, and inverted with respect to this direction using a Mezei flipper. From the scattered beam, an analyzer crystal selects only neutrons with a certain final energy. The post-sample instrument components can select a particular neutron polarization, using a setup similar to the one described for the incident beam. The analysis workflow for polarized experiments is very similar to unpolarized experiments. In the case

[1] This manuscript has been authored by UT-Battelle, LLC under Contract No. DE-AC05-00OR22725 with the U.S. Department of Energy. The United States Government retains and the publisher, by accepting the article for publication, acknowledges that the United States Government retains a non-exclusive, paid-up, irrevocable, worldwide license to publish or reproduce the published form of this manuscript, or allow others to do so, for United States Government purposes. The Department of Energy will provide public access to these results of federally sponsored research in accordance with the DOE Public Access Plan (http://energy.gov/downloads/doe-public-access-plan).

IOP Conf. Series: Journal of Physics: Conf. Series **862** (2017) 012023 doi:10.1088/1742-6596/862/1/012023

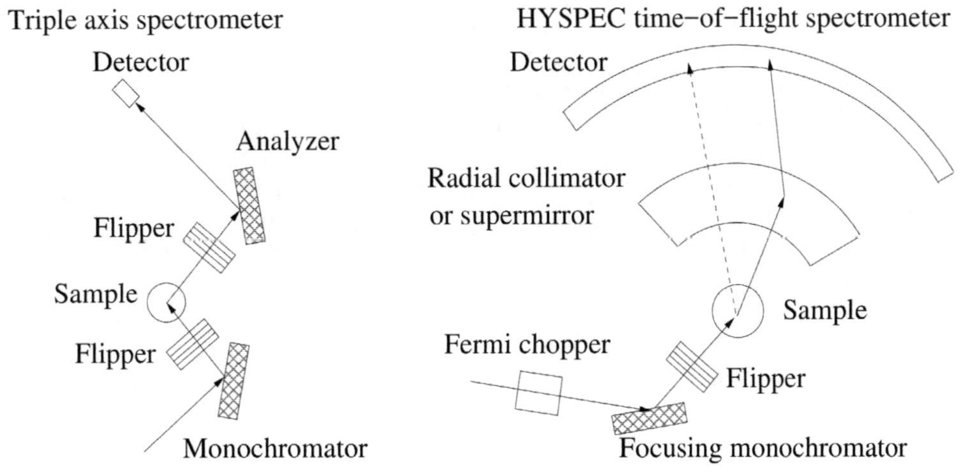

Figure 1. Comparison of the polarized neutron scattering setup for a triple axis spectrometer and a time-of-flight direct geometry spectrometer. For the HYSPEC spectrometer, a radial collimator, or a supermirror analyzer can be inserted in the scattered neutron path between the sample and the detector. Neutrons are not deflected in the radial collimator (dashed path), but their trajectory is changed for the supermirror analyzer case (solid path).

that Heusler crystals are used, the scattering angle and reflectivity are different from the case of non-polarized setup, but, with a single detector, calibrating the polarization efficiency is straightforward.

The HYSPEC spectometer [1] at the Spallation Neutron Source, Oak Ridge National Laboratory has recently started measurements in the fully polarized mode. The incident neutron beam is monochromated by a short-blade ("rotating collimator") Fermi chopper. The beam is then reflected by a vertically focusing crystal array, thus increasing the flux on the sample at the expense of vertical resolution. For unpolarized experiments, the focusing array made of pyrolitic graphite (PG) is used. For polarized measurements, a Heusler crystals array is used. In unpolarized or half-polarized measurements, a radial collimator, which defines the scattering volume viewed by detectors is placed immediately after the sample, at a fixed position (not oscillating), well in front of the detector array. For fully polarized experiments, the radial collimator is replaced with the similarly sized, 60°-wide-angle supermirror array, which selects a single neutron polarization. The scattered neutrons that passed through a collimator, or a supermirror, are detected by the two-dimensional (2D) position sensitive array of detectors (160 vertical tubes, 1.2 m long, 128 pixels each) covering ≈ 0.27 sR ($w \times h = 60° \times 15°$). A schematics of the HYSPEC setup is shown in Figure 1. Details of the geometry of the instrument can be found in a complementary paper in the same conference proceedings [4]. Similarly to the other SNS instruments, the data on HYSPEC is collected in the event mode, where information about each detected scattered neutron is individually stored.

The principal advantage of the time-of-flight spectrometers for the inelastic neutron scattering measurements is in the capability of surveying large volumes of the reciprocal space at a time. In the current polarized setup this requires properly accounting for the transmission probability through the instrument, which is heavily dependent on the neutron's path and energy.

2. Characterization of the polarizing components

In a polarized neutron scattering experiment, it is important to maximize both the polarization efficiency and the transmission through the instrument, which both need to be assessed experimentally. In the HYSPEC setup, characterizations of different components of the instrument were performed using an isotropic incoherent scatterer, made of TiZr alloy. In order to characterize the dependence of the transmission on the horizontal scattering angle, one can group all pixels in a vertical tube, to get better statistics. The grouping occurs after conversion of data from time-of-flight to energy transfer, so the different secondary flight paths are taken into account. The effect of this procedure on the vertical momentum transfer is comparable in magnitude with the effect of the focussing monochromator (the size of Bragg peaks on the detector array is about one third of the tube length in vertical direction).

Measurements of the reflectivity of the Heusler monochromator (not shown in this paper) exhibit the same intensity pattern in the detector as for the pyrolytic graphite monochromator (no tube to tube variation). The reflectivity ratio is slightly incident energy dependent, and independent of the energy resolution of the incident beam (Fermi chopper frequency).

For the comparative evaluation of the transmission through the supermirror array and the radial collimator, we measured the same TiZr alloy sample in both cases using the PG monochromator (higher flux). As seen in Figure 2 (top left panel), the scattering intensity from

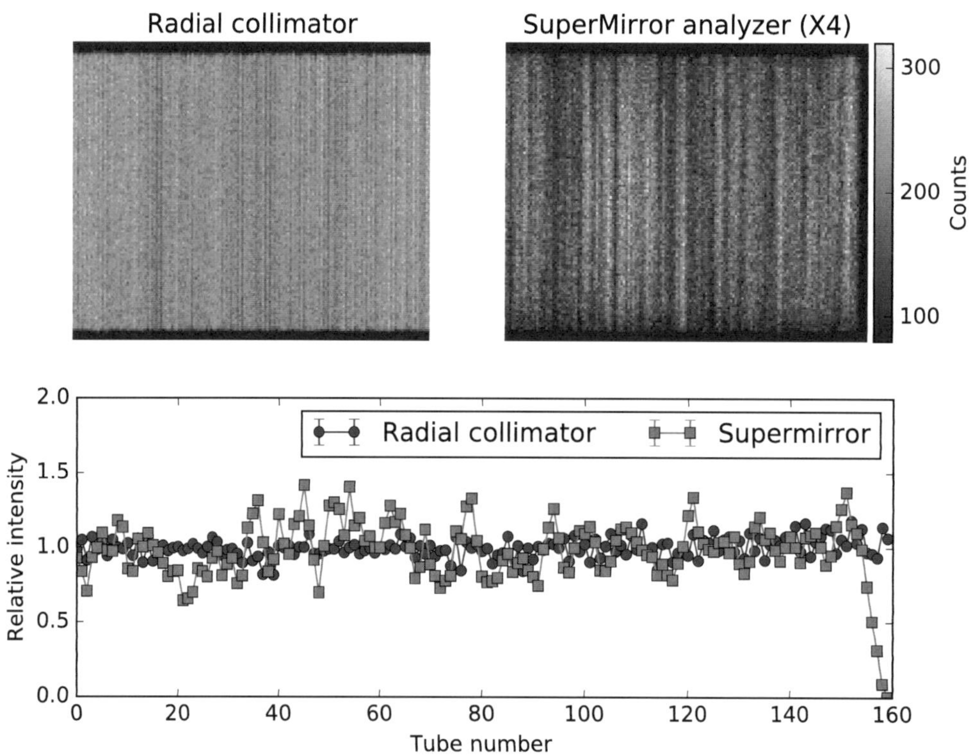

Figure 2. Position dependent transmission of the radial collimator and the supermirror analyzer, as seen in the corresponding images of the HYSPEC detectors, measured with 3.8 meV incident energy. The radial collimator has a uniform transmission (top left). The supermirror analyzer intensity (top right - multiplied by 4 to be on the same scale) shows angle-dependent inhomogeneity. The relative intensity variation of the elastic line integrated in energy and over the tube length is shown at the bottom.

such an incoherent source measured with the radial collimator after sample is relatively flat. In the case of the supermirror polarizer, the transmission is markedly inhomogeneous (top right panel). Notably, the last few detector tubes on the right side are not illuminated at all. The reason for this is the deflection of the beam by the supermirror analyzer. While the geometry of the analyzer-detector channels on HYSPEC is similar to that on the D7 instrument at ILL [5], their angular positions are referenced by the scattering angle at the entrance to the analyzer channel (neutronic positions). These are different from the corresponding angles at the exit of the analyzer, while the angles of the detector pixels, which are about 4 m away from the analyzer, are parameterized by their physical positions and correspond to the angles at the analyzer entrance, as in the case of the unpolarized neutrons. To exclude colormap bias when looking at the angular dependent intensity pattern, we integrated the elastic scattering in each tube. The relative intensity of each tube, with respect to the average for the configuration, is shown in the bottom panel of Figure 2. For the supermirror analyzer only the first 155 tubes were considered in the calculation of the average. We found that the intensity pattern on the detector for the supermirror array transmission is sample position dependent. This is not surprising since the small angular acceptance of a single analyzer channel, combined with the short sample to analyzer distance, yield a small horizontal acceptance interval for each individual channel. To account for this effect, at the end of each experiment we measure the incoherent scattering from a Vanadium reference placed at the sample position. The dimensions of the Vanadium reference are compatible with the acceptance width of the analyzer system (6.3 mm diameter).

For the transmission measurements, we masked (in software) the top and the bottom 8 pixels of the tubes and the tubes shadowed by the beam deflection in the analyzer. The neutron count integrated over thus defined PSD area for the elastic scattering at different energies is shown in Figure 3. The measured TOF data was transformed to the energy transfer, and a Gaussian fit of the elastic line was used for evaluating the intensity on top of a linear background. The total count in the peak was corrected for the energy dependent detector efficiency and normalized to the same proton charge. Thus obtained intensities in the left panel of Figure 3 contain the energy dependent incident flux. In order to obtain the transmission through the supermirror array, we divide the corresponding measured intensity with the one for the radial collimator case, and multiply by 2 to account for the relative fraction of neutrons with one spin polarization: both polarization are transmitted by the radial collimator, but only one by the supermirror array. For the ideal polarizer, one would expect a transmission $T = 1$. The resulting dependence of the polarizer transmission on the neutron energy (E_F) is shown in Figure 3. We found that we can empirically describe this dependence as an exponential, $T \sim e^{-E_F/E_0}$.

Since the transmission through the supermirror at each given incident energy is position-(angle-)dependent (as seen in Figure 2), and the overall average transmission varies with the energy, we checked if the energy dependence of the transmission is the same for each tube, or not. The results are shown in Figure 4. The transmission for each tube is shown on the left side. Each curve has been fitted with the same exponential dependence as the overall transmission,

$$T = T_0 e^{-E_F/E_0} \,. \tag{1}$$

The scatter plot of the T_0 and E_0 parameters against each other is shown in the right panel of Fig. 4. We found that the points cluster around the values obtained by fitting the intensity integrated over all detectors. Future instrument upgrades, which will allow fine-tuning the supermirror alignment could perhaps narrow the spread of the E_0 parameter, reducing the regular tube-dependent variation of the transmission at a given energy. Finally, the present procedure only accounts for the energy dependence of the neutron beam deflection in the supermirror by virtue of the energy-dependent transmission calibration. While this introduces a systematic error in the energy-wave vector calculation, this error is small, while implementing in Mantid [2] a procedure which would account for both the varying transmission and the varying

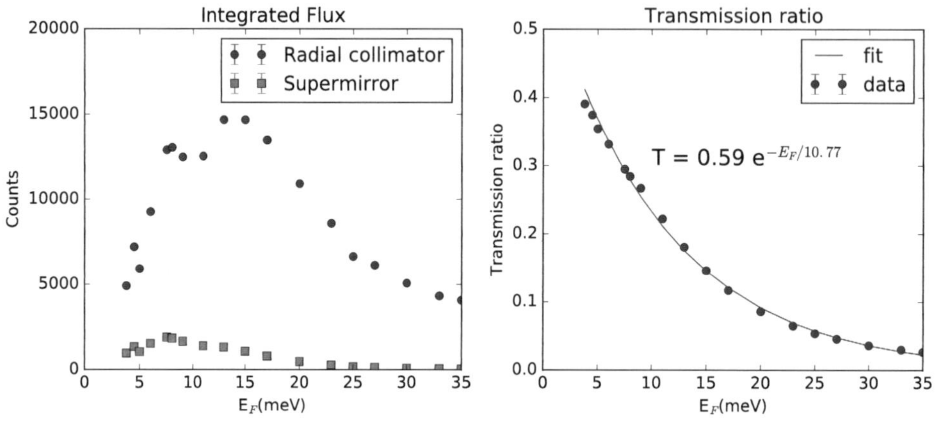

Figure 3. Energy dependent flux at the detector measured with radial collimator or supermirror after the sample. (left) absolute values (right) transmission through the supermirror measured as a ratio of the fluxes through the supermirror and the radial collimator, multiplied by 2.

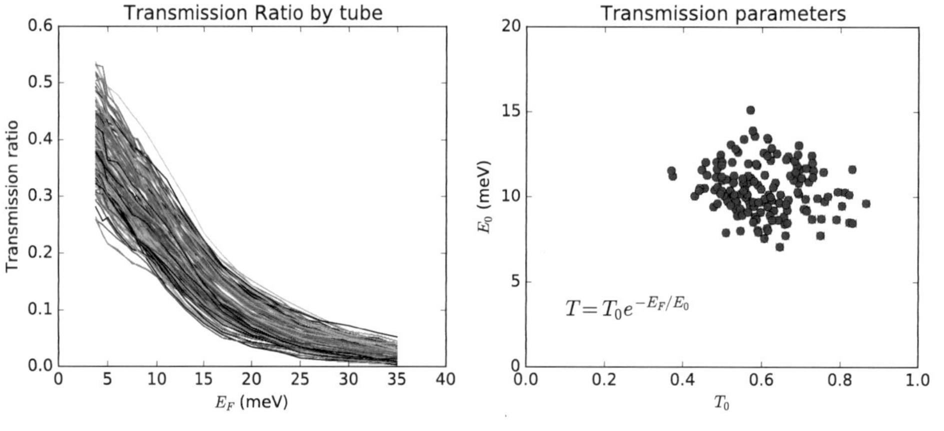

Figure 4. Energy dependent transmission for detector tubes. (left) Raw data showing an exponential dependence (right) Fit parameters (see text for details)

beam deflection is challenging. We nevertheless plan such work for the future.

The last step in characterization of the supermirror array is a measurement of the polarization efficiency. Results are shown in Figure 5. The Heusler alloy monochromator has been used to polarize the incident beam. For the isotopic incoherent scattering, like in the case of the TiZr alloy used for this measurement, we expect to have intensity only in the non spin flip channel. The flipping ratio is then defined as the ratio of scattered intensities in the non spin flip and the spin flip channels, or with the flipper ON and the flipper OFF, correspondingly, in the case of the HYSPEC setup,

$$F = \frac{I^{++}}{I^{+-}} = \frac{I^{ON}}{I^{OFF}} \ . \tag{2}$$

The analysis was performed on each individual tube by fitting the corresponding elastic line

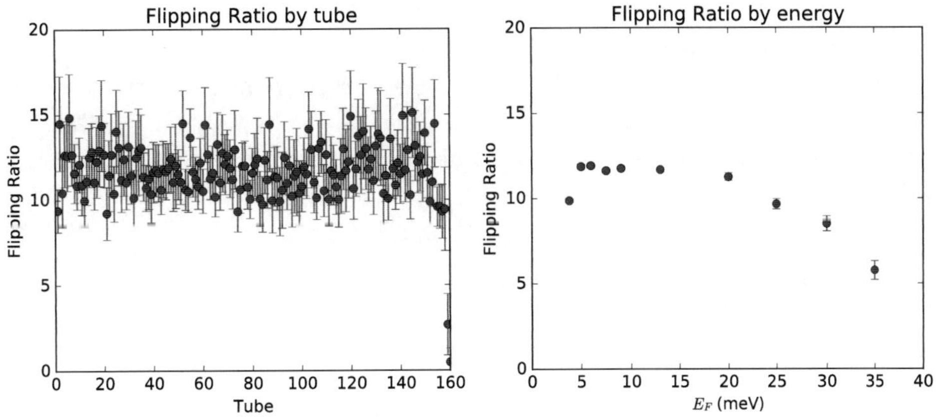

Figure 5. Measurement of system-wide flipping ratio. (left) At constant energy (13 meV) the flipping ratio is independent of the detector tube. (right) Energy dependence integrated over the whole instrument

intensities to Gaussian profiles on top of a linear background. We observe no angular dependence of the flipping ratio in the detectors that are not shadowed by beam deflection in the supermirror analyzer. This is in contrast to the result for the angle-dependent transmission seen in Figures 2 and 4. The right panel of Figure 5 presents the energy dependence of the flipping ratio obtained by adding counts from all detectors. It is relatively constant up to about 20 meV, with a gradual decrease at higher energies. The reason for the lower flipping ratio measured at incident energy of 3.8 meV remains unclear.

3. Data reduction workflow

Data reduction is the process of transforming the time-of-flight event data into quantities that have some physical meaning in the sample parameter space, independent of the measurement details. For the case of HYSPEC, the original data is stored as a collection of time of flights when neutron events were collected. The final outcome of the reduction, the scattering cross section, or the cumulative dynamical structure factor (i. e. the sum of the structure factors of all scatterers in the neutron beam), is stored as a histogram in terms of momentum and energy transfer. The workflow that we describe in this paper is implemented using the Mantid [2] software suite. It tries to follow, as much as possible, the workflow for unpolarized experiments. If not explicitly marked as polarized only steps, individual items in the workflow are common to both polarized and unpolarized experiments.

(1) Pre-process the incoherent scattering data from a Vanadium, or a TiZr alloy sample: the angle- and energy-dependent transmission and the flipping ratio are stored for later use. This step is not performed for unpolarized neutron case, where the transmission is uniform.
(2) Load data.
(3) Correct the detector positions, accounting for the deflection of the scattered beam passing through the supermirror analyzer. The same deflection, corresponding to neutrons scattered elastically, is used for all scattered neutrons. This step is not needed for the unpolarized, or half-polarized (polarized incident beam, unpolarized scattered beam) measurements.
(4) Correct the time of flight for the (incident-energy-dependent) moderator emission time.
(5) Evaluate and subtract the time-independent background.

(6) Convert events from time-of-flight to energy transfer.
(7) Correct for the energy-dependent efficiency of the detectors.
(8) Optionally, convert from the scattering cross-section to the dynamical structure factor.
(9) Convert events from detector positions and energy transfer coordinates to momentum and energy transfer coordinates. Event data at this point is not on a regular grid.
(10) Correct for spin flip efficiency [5] (polarized experiments only).
(11) Histogram data onto a regular grid in reciprocal space.
(12) Combine data from multiple measurements and/or multiple detectors, correctly accounting for the statistical weights
(13) Correct for the energy dependent polarizer transmission.

If needed, the data can be histogrammed in energy prior to step 9 in order to obtain/save a histogram in energy transfer for each detector (typically, as a SPE/NXSPE file). Several programs [6, 7] use this format to perform the next steps for the unpolarized measurement.

Flipping ratio might be dependent on the sample orientation (like in the case of superconducting samples), so for statistical reasons we must perform this step before histogramming. The energy dependent polarizer transmission could in principle be performed at the same step, but some users would prefer the option of looking at the data without this correction.

4. Conclusions

This paper describes the challenges of data processing for polarized neutron experiments at the HYSPEC spectrometer, and the algorithms that we have developed to meet these challenges. Different hardware elements of the polarized setup were characterized, and the effects were considered in the data reduction workflow. Data output of the proposed workflow will be separated into different components (nuclear, magnetic, etc.) [8, 9]. Several tasks remain to be performed, such as accounting for the energy-dependent beam deflection by the supermirror analyzer, or understanding its effect on the resolution of the instrument.

This work was supported in part by the Paul Scherrer Institut by loaning the supermirror analyzer to Oak Ridge National Laboratory.

References

[1] Winn B, Filges U, Garlea V O, Graves-Brook M, Hagen M, Jiang C, Kenzelmann M, Passell L, Shapiro S M, Tong X and Zaliznyak I 2015 *EPJ Web of Conferences* **83** 03017 URL `http://dx.doi.org/10.1051/epjconf/20158303017`

[2] Arnold O, Bilheux J, Borreguero J, Buts A, Campbell S, Chapon L, Doucet M, Draper N, Leal R F, Gigg M, Lynch V, Markvardsen A, Mikkelson D, Mikkelson R, Miller R, Palmen K, Parker P, Passos G, Perring T, Peterson P, Ren S, Reuter M, Savici A, Taylor J, Taylor R, Tolchenov R, Zhou W and Zikovsky J 2014 *Nuclear Instruments and Methods in Physics Research Section A: Accelerators, Spectrometers, Detectors and Associated Equipment* **764** 156 – 166 ISSN 0168-9002 URL `http://dx.doi.org/10.1016/j.nima.2014.07.029`

[3] Moon R M, Riste T and Koehler W C 1969 *Phys. Rev.* **181**(2) 920–931 URL `http://link.aps.org/doi/10.1103/PhysRev.181.920`

[4] Zaliznyak I A, Savici A T, Garlea V O, Winn B, Filges U, Schneeloch J, Tranquada J M, Gu G, Wang A and Petrovic C 2017 *EPJ Web of Conferences* Unpublished, this PNCMI conference proceedings URL `https://arxiv.org/pdf/1610.06018v3.pdf`

IOP Conf. Series: Journal of Physics: Conf. Series **862** (2017) 012023 doi:10.1088/1742-6596/862/1/012023

[5] Stewart J R, Deen P P, Andersen K H, Schober H, Barthélémy J F, Hillier J M, Murani A P, Hayes T and Lindenau B 2009 *Journal of Applied Crystallography* **42** 69–84 URL http://dx.doi.org/10.1107/S0021889808039162

[6] Azuah R, Kneller L, Qiu Y, Tregenna-Piggott P, Brown C, Copley J and Dimeo R 2009 *J. Res. Natl. Inst. Stand. Technol.* **114** 341

[7] Ewings R A, Buts A, Le M D, van Duijn J, Bustinduy I and Perring T G 2016 *Nuclear Instruments and Methods in Physics Research Section A: Accelerators, Spectrometers, Detectors and Associated Equipment* **834** 132 – 142 ISSN 0168-9002 URL http://www.sciencedirect.com/science/article/pii/S016890021630777X

[8] Schärpf O and Capellmann H 1993 *physica status solidi (a)* **135** 359–379 ISSN 1521-396X URL http://dx.doi.org/10.1002/pssa.2211350204

[9] Ehlers G, Stewart J R, Wildes A R, Deen P P and Andersen K H 2013 *Review of Scientific Instruments* **84** 093901 URL http://scitation.aip.org/content/aip/journal/rsi/84/9/10.1063/1.4819739

IOP Conf. Series: Journal of Physics: Conf. Series **862** (2017) 012024 doi:10.1088/1742-6596/862/1/012024

Neutron depolarization measurements of magnetite in chiton teeth

M Seifert[1,2], M Schulz[1,2], G Benka[2], C Pfleiderer[2], S Gilder[3]

[1] Heinz Maier-Leibnitz Zentrum (MLZ), Technical University of Munich, D-85748 Garching, Germany

[2] Physics Department, Technical University of Munich, D-85748 Garching, Germany

[3] Department of Earth and Environmental Sciences, Ludwig-Maximilians-Universität München, D-80539 Munich, Germany

E-mail: marc.seifert@frm2.tum.de

Abstract. Magnetite constitutes one of the most abundant magnetic minerals in the Earth's crust. In the single domain state, magnetite often carries the magnetic remanence in rocks due to its stable and strong magnetic remanence. Hence it is of keen interest to paleomagnetists who study the ancient magnetic field preserved in the rock record. The extremely small size range and vulnerability to oxidation of single domain magnetite makes synthetization and preservation virtually impossible. Consequently, most experimental work on magnetite under pressure is carried out on multidomain magnetite. The radula of the marine mollusc chiton (Polyplacophora) is one of the few natural sources of single domain magnetite. We have performed a comparative study on samples of chiton radula in a vibrating sample magnetometer (VSM) and with the newly evolving neutron depolarization imaging (NDI) technique. Despite a constant offset between the VSM and NDI data in the coercivity we find a good agreement between the two techniques.

1. Introduction

The spontaneous magnetization of magnetite (Fe_3O_4) is produced within an inverse spinel structure where superexchange coupling occurs between iron and oxygen ions in two magnetic sublattices. Magnetite's magnetic properties are governed by interatomic distances and angles in each sublattice, which in turn dictate the magnetocrystalline, magnetostrictive, and magnetostatic energies, all of which are anisotropic and vary with temperature, pressure and external magnetic fields [1, 2]. Magnetite constitutes one of the most abundant magnetic minerals in the Earth's crust. In the single domain state, magnetite often carries the magnetic remanence in rocks due to its stable and strong magnetic remanence, hence it is of keen interest to paleomagnetists who study the ancient magnetic field preserved the rock record.

Although the influence of temperature on magnetic remanence has long been a subject of research, much less is known about how pressure modifies the magnetic properties of magnetite. Understanding how and why magnetite's remanent intensity and magnetization direction change under pressure, and upon decompression from high pressure, are important to properly interpret magnetic field anomalies, the role of shock in meteorite impact craters and the magnetic field strength in the solar system preserved in meteorites.

Although single domain magnetite carries a stable remanence over solar-system time scales, most experimental work under pressure is carried out on multidomain magnetite because single domain magnetite's extremely small size range (ca. 30 nm to a few 100 nm depending on its shape) and vulnerability to oxidation makes it virtually impossible to synthesize and preserve. However, a few animals synthesize single domain magnetite, one being the marine mollusc, chiton (Polyplacophora) whose radula consists of a conveyor-belt like array of teeth that consist of three distinct mineral zones: a core of apatite (like what humans have for teeth) covered by a thin band of lepidocrocite, then a thicker layer of magnetite [3, 4]. The magnetite is precipitated in a way that restricts individual grain sizes to be within the single domain range [5].

Gilder et al. [6] measured the acquisition of isothermal remanent magnetization, direct field demagnetization, and alternating field demagnetization of multi-domain (MD) and single domain (SD) magnetite under hydrostatic pressures to 6 GPa. The single domain material came from chiton teeth. They found significant differences in the pressure behavior of the magnetizations of MD and SD magnetite calling for further investigations.

Another important aspect of the magnetic properties of magnetite concerns the first-order Verwey transition, which is a metal-to-insulator transition due to a slight rhombohedral distortion of the lattice. Carporzen and Gilder [7] studied the effects of stress on the Verwey transition for stoichiometric, synthetic multidomain magnetite and for natural multidomain magnetite having both relatively low and high degrees of oxidation. Low temperature measurements of the magnetic moments were carried out after pressure release. Their results unambiguously showed an increase in the Verwey transition temperature with increasing pressure that ranged from 1 K/GPa for stoichiometric magnetite to 3 K/GPa for highly oxidized magnetite for pressures up to about 5 GPa. These findings could make the Verwey transition suitable for use as a geobarometer in cases where high pressures ($>$ 1 GPa) are involved, such as in meteorite impact craters. Until now, no one has carried out similar experiments on SD magnetite.

In order to further explore the magnetic properties of magnetite from chiton radula, we carried out a comparative study between "classic" magnetometry using a vibrating sample magnetometer (VSM) with the newly developed neutron depolarization imaging (NDI) technique at the instrument ANTARES at the Heinz Maier-Leibnitz Zentrum, Garching. Imaging with polarized neutrons has proven to be sensitive to magnetic field distributions on the macroscopic [8, 9, 10, 11] and the microscopic [12, 13, 14] scale. In this publication we show the potential of this new technique which is intrinsically sensitive to different microscopic parameters than standard magnetometry for applications to paleomagnetic materials.

2. Methods and Experimental setup

The neutron depolarization imaging (NDI) technique is based on the combination of a neutron imaging beam line using a position sensitive detector with a neutron polarization analysis setup. It enables the spatially resolved measurement of the influence of a sample on the neutron polarization. A typical NDI setup is sketched in Fig. 1. The neutron beam first passes a transmission polarizer in which magnetic neutron supermirrors reflect neutrons with spin direction up out of the beam such that the beam is polarized in down direction. A Mezei spin flipper allows to flip the spin direction from down to up if it is turned on. Guide fields with magnetic fields of the order of B=1 mT have been installed to prevent depolarization of the beam due to the earth's magnetic field and stray fields from electrical components. A Helmholtz magnet enables to apply fields up to 300 mT at the sample position. Both positive and negative fields can be applied due to the installation of horizontal guide fields which allow for an adiabatic coupling of the neutron spin to the field of the magnet. After the magnet an analyzer is located which works the same way as the polarizer. The subsequent 2d neutron detector, which is not shown in the figure, uses a 100 μm thick LiF/ZnS scintillator to convert the neutron beam into visible light. A CCD camera with a pixel size of 6.5 μm and an objective set to 1:3.7

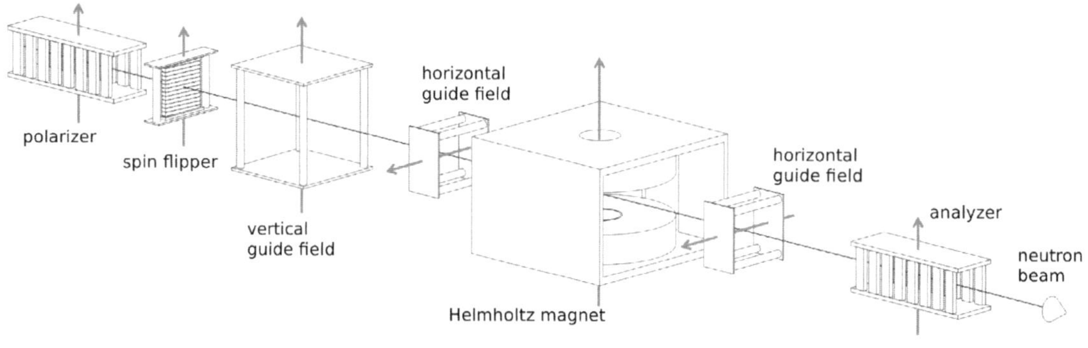

Figure 1. Sketch of a neutron depolarization imaging (NDI) setup. The neutron beam first passes a polarizer which polarizes the beam. A subsequent spin flipper allows the rotation of the polarization by π. Vertical and horizontal guide fields prevent a depolarization of the neutron beam due to the earth's magnetic field. The sample is placed in a Helmholtz magnet. The installation of horizontal guide fields allows to apply positive and negative fields at the sample position. An analyzer which is identical to the polarizer is placed right before the 2d detector.

magnification detects the light which is proportional to the incoming neutron intensity.

A 18 mm pinhole close to the neutron source was used to adjust the beam divergence. The detector position from the pinhole is 9.0 m which gives a L/D of 500. The distance from the sample to the detector is approximately 0.8 m. As the polarizers operate only with cold neutrons a neutron velocity selector was used to set the neutron wavelength to $\lambda = 4.2$ Å with a resolution of $\Delta\lambda/\lambda = 10\%$.

The polarization P of the neutron beam at the detector is determined by

$$P = P_0 \cdot \frac{I_\downarrow - I_\uparrow/f}{I_\downarrow + I_\uparrow/f}. \tag{1}$$

P_0 is the instrument polarization, a normalization factor which accounts for the polarization efficiencies of the polarizers and possible depolarization along the flight path of the neutron. I is the intensity at the detector where the indices depict the polarization direction after the spin flipper, i.e. whether the spin flipper is turned off ($\downarrow$) or on ($\uparrow$). As the flipping efficiency f of the spin flipper for the given setup is very close to 1 [17] it can be omitted. This evaluation is done pixelwise and therefore gives the spatial distribution of the neutron beam polarization in the field of view of the detector. The measured polarization was around $P_0 = 0.80$ and not completely uniform over the field of view due to the beam divergence and the used polarizers. For each data point 5 spin-up and 5 spin-down images were taken with an exposure time of 45 s each to achieve good statistics. Artifacts in the images caused by gamma radiation (white spots) were removed by median filtering the stack of images.

When a neutron moves through a magnetic field not parallel to the spin direction of the neutron, the spin precesses around the magnetic field B with the Larmor frequency $\omega_L = -\gamma B$ where γ is the gyromagnetic ratio of the neutron $\gamma = 183$ MHz/T. This effect is used in the neutron depolarization imaging method for the detection of ferromagnetism [15]. Ferromagnetic materials form ferromagnetic domains to decrease the energy of stray fields. On its path through such a domain configuration a neutron sees in each domain a magnetic field pointing in a different direction. Due to Larmor precession the neutron acquires a certain Larmor phase after passing the ferromagnetic sample. Each neutron has a slightly different path through the sample, sees a different domain configuration and therefore collects a different Larmor phase.

IOP Conf. Series: Journal of Physics: Conf. Series **862** (2017) 012024 doi:10.1088/1742-6596/862/1/012024

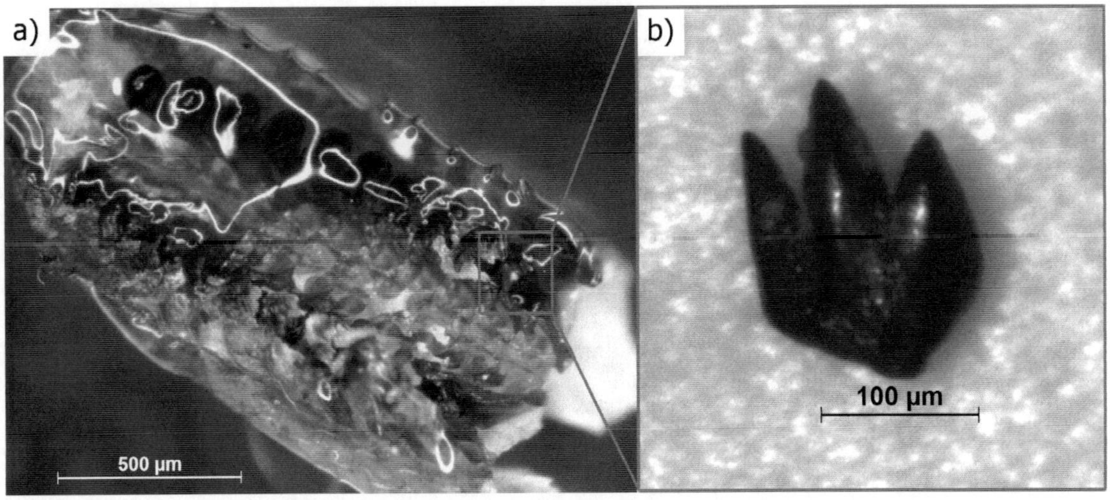

Figure 2. (a) Microscopic image of the chiton radula part that was investigated using neutron depolarization imaging. The black chiton teeth are arranged in two rows. (b) shows a magnification of one chiton tooth from the radula (indicated by the red frames)

Since the polarization is the average over the neutron spins at the detector in a certain area, a ferromagnetic sample leads to a decrease of neutron polarization. According to Ref. [16] the polarization for the case of small spin rotations per magnetic domain is given by

$$P = P_0 \exp\left(-\frac{1}{3}\gamma^2 B'^2 d \frac{\delta}{v^2}\right). \tag{2}$$

B' is the average magnetic field per domain, d is the sample thickness in beam direction, δ is the average domain size and v is the neutron velocity. As v and d are known we can obtain the product $B'^2\delta$ from the polarization data. When determining B' with magnetometer measurements the average domain size can be accessed.

Radulas were extracted from freshly deceased specimens of Acanthochitona fascicularis. We took a sample from two different radula (sample 1 and 2) and measured their hysteresis loops using a LakeShore MicroMag 3900 at 300 K, 140 K and 80 K, the latter two being above and below the Verwey transition. In order to measure the Verwey transition temperature, we cooled the radula in a low (Earth's) field from 300 K to 80 K, briefly applied a 1 T field, and then warmed the sample to 140 K while measuring its magnetic remanence every 1 K.

The sample investigated using NDI is shown in Fig. 2a and is depicted as "NDI sample" in the following. It is the anterior part of a chiton radula which was stored in ethanol to prevent oxidation of the teeth's magnetite. A magnification of a single chiton tooth is shown in Fig. 2b. The volume of the radula part is 1.7 mm × 0.8 mm × 1.0 mm. The chiton radula was mounted on an aluminium sample holder in a cryostat. Additionally, a neutron absorbing cadmium mask with a 2 mm circular hole was mounted before the sample. Magnetization hysteresis loops of the NDI sample were also acquired at 80 and 300 K using a Quantum Design Physical Property Measurement System (PPMS).

3. Results

Fig. 3a shows VSM data of sample 1 from a hysteresis loop at T=80 K. The applied field was ramped from +0.5 T to -0.5 T and then back to +0.5 T. To determine the coercivities B_{c1} and

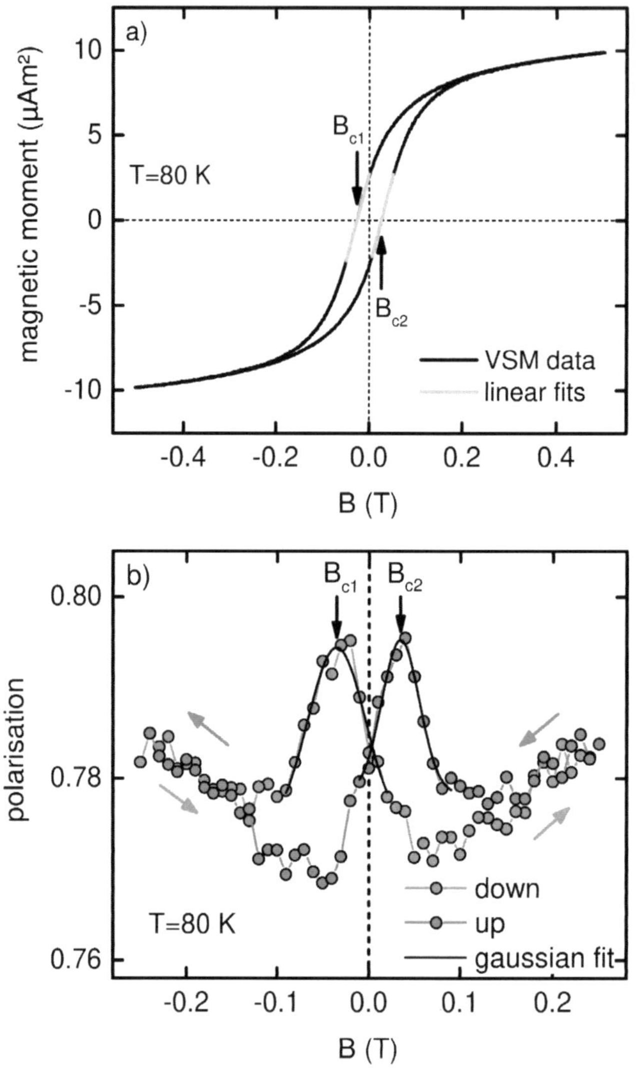

Figure 3. Comparison of VSM and neutron depolarization imaging data. (a) shows the magnetic moment of a whole chiton radula as a function of the applied magnetic field B at a temperature of 80 K measured using a VSM. A hysteresis loop from +0.5 T to −0.5 T to +0.5 T was recorded. Lines (green) were fit to the VSM data close to the zero crossing of the x-axis. The intersection of these lines with the zero magnetic moment line gives the coercivities $B_{c1/2}$. (b) Neutron polarization as a function of the magnetic field B at T=80 K is shown in (b). The applied fields range from +0.25 T to −0.25 T. The coercivities $B_{c1/2}$ were evaluated by fitting a Gaussian to the peaks close to B=0 (indicated as black lines).

IOP Conf. Series: Journal of Physics: Conf. Series **862** (2017) 012024 doi:10.1088/1742-6596/862/1/012024

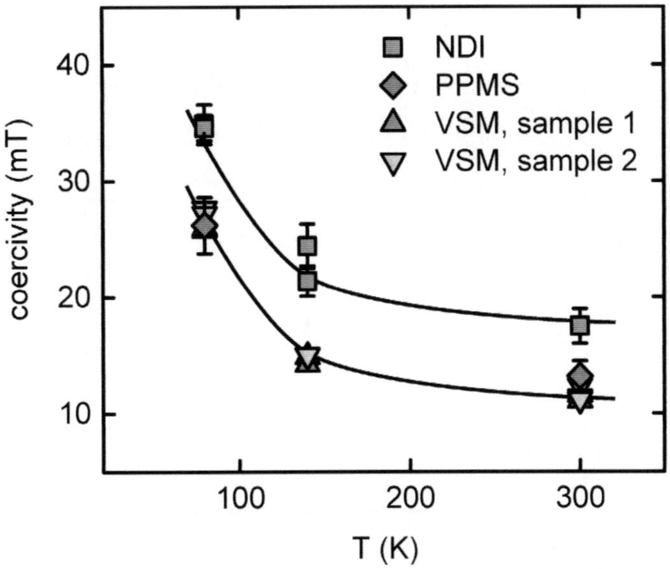

Figure 4. Coercivities obtained from hysteresis loops as a function of temperature. Two whole chiton radula were measured using a VSM (sample 1 and 2). The corresponding data is indicated as orange and yellow triangles. Error bars for these data points are not shown since they are smaller than the symbols. Furthermore, coercivities obtained from NDI measurements are shown as green squares. A significant offset is visible. The black lines are a guide to the eye and are the same except for an offset. For a comparison, the coercivities of the sample from the NDI measurements were also determined using a PPMS (depicted as lilac rhombs).

B_{c2}, lines were fitted to the data close to the points where the magnetic moment is zero. The intersections of the lines with the x-axis gives the coercivities.

Fig. 3b shows the neutron polarization of the NDI sample as a function of the applied field. Since the chiton radula is too small to be resolved in detail at the detector the polarization was averaged over the entire pinhole area, therefore, no spatial information is available. As the size of the sample was much smaller than the pinhole, only a small fraction of the beam is depolarized which results in small changes in polarization of only a few per cent.

The field was ramped from +0.25 T to -0.25 T and then back to +0.25 T. The up and down measurements are indicated in orange and green, respectively. Both data sets show peaks which are shifted from B=0 due to hysteresis. In zero field the polarization is comparably high. Small applied fields increase the average domain size in field direction at the expense of antiparallel domains and therefore decrease the polarization. Higher fields turn more and more domains into the direction of the applied field. This leads to smaller acquired Larmor phases which weakens the depolarization effect. Therefore, the polarization increases for higher fields again. The coercivities were determined as the centers of two Gaussians which were fitted to both peaks in the polarization signal.

Such magnetic field loops were measured using the VSM, the PPMS and the NDI method at the temperatures 80, 140 and 300 K. The coercivities from these measurements determined by the methods described above are shown in Fig. 4 as a function of temperature. All data sets show a decrease of the coercivity with increasing temperature. The VSM measurements are depicted as orange (sample 1) and yellow (sample 2) triangles. No error bars are shown for these data points as they are smaller than the symbols. The black line connecting the VSM data points is a guide to the eye. The corresponding coercivities obtained from NDI data are plotted

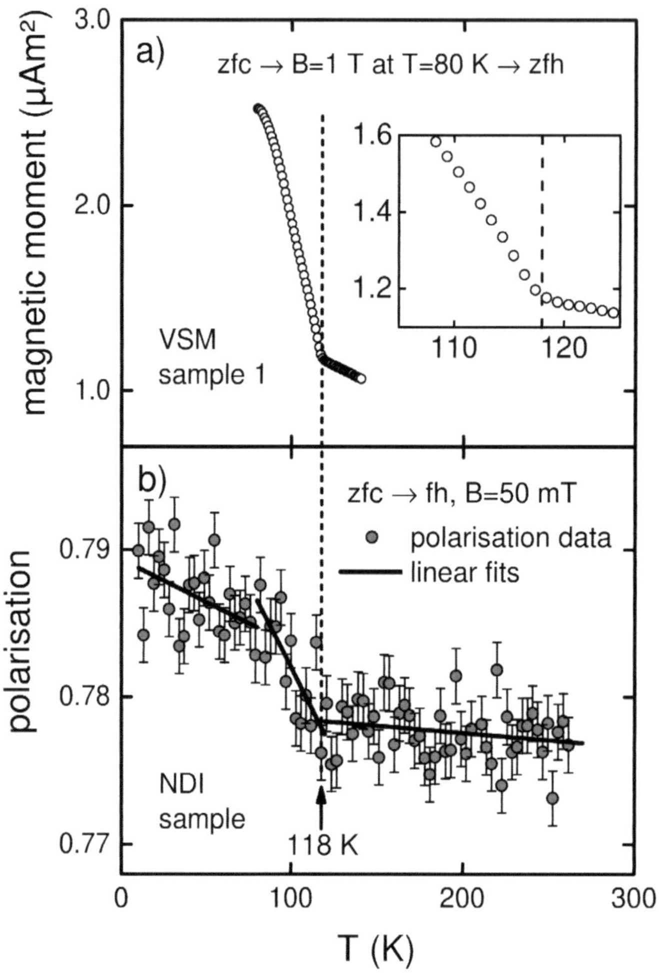

Figure 5. (a) shows VSM data from sample 1 as a function of temperature. The measurements were taken after applying a field of 1 T at 80 K during heating in zero field (zfh). The sharp kink at 118 K is caused by the Verwey transition. In (b) the neutron polarization is shown as a function of temperature. The sample was cooled in zero field (zfc), then a field of 50 mT was applied. The measurement was performed during heating with applied field (fh). Three lines (black) were fit to the data points in the temperature ranges < 80 K, 80-120 K and > 120 K. A kink is visible at approximately 118 K.

as green squares. The error bars were obtained from the error values of the corresponding fits. Again, a black line is shown as a guide to the eye. The NDI data shows the same behavior as the VSM data except for an offset of about 7 mT. For a better comparison the radula used for the NDI measurements was also investigated using a PPMS at 80 and 300 K. The corresponding data points are depicted as lilac rhombs.

Furthermore, temperature sweeps using the VSM and NDI are shown in Fig. 5a and b, respectively. For the VSM measurement sample 1 was cooled down in zero field. At T=80 K a field of 1 T was applied to saturate the sample. The data was then acquired during heating in zero field up to 140 K. For the NDI measurement (see Fig. 5b) the sample was cooled down in zero field and measured during heating with a magnetic field of B=50 mT. The NDI data are noisier than the magnetic field sweep from Fig. 3 as only 3 up and down images per data point with 30 s exposure time were taken. The errors were determined from the standard deviation of the data points in the near-constant region above 120 K. Three lines were fitted to the polarization data in the temperature regions <80 K, 80-120 K and >120 K. A kink in the signal is visible approximately at the same temperature as in the VSM measurement.

4. Discussion

The coercivities obtained from conventional magnetometers (VSM and PPMS) and NDI measurements shown in Fig. 4 indicate good qualitative agreement except for an offset of about 7 mT. Although biological samples as the investigated chiton radula are susceptible to natural variation, this can be discarded as a explanation for the offset. The coercivities obtained by PPMS measurements of the NDI sample confirm the SD magnetite coercivities of sample 1 and 2. A possible influence of remanence due to the iron yoke of the used magnet is smaller than 1 mT. However, NDI is sensitive to other microscopic parameters compared to the VSM and PPMS methods. One possible explanation could be a magnetic screening of domains within a MD grain proposed by Xu and Merrill [18].

In both the VSM and NDI measurements shown in Fig. 5a kink at approximately 118 K is visible. This abrupt change in the slope of both signals is caused by the Verwey transition, a reconfiguration of the atomic lattice which also affects the magnetic behavior.

5. Conclusion and outlook

In conclusion, the NDI method reproduces the magnetometer measurements except for an offset of the obtained coercivities. As NDI measurements are not directly depending on the magnetization but also on microscopic parameters as the average domain size δ, additional or complementary results compared to the magnetometer measurements are expected. The origin of the offset may be explained by macroscopically screened domains and is subject of further investigations.

Since the chiton teeth are a comparably small sample for NDI we plan to use focussing neutron guides to achieve higher counting rates and therefore higher statistics. First measurements with the neutron guides report a gain in neutron intensity of a factor of 20. It is also planned to increase the density of magnetite in the sample volume by removing the biological material of the radula which should result in a larger depolarization signal. Of particular interest is also the pressure dependence of the Verwey transition. A new pressure cell has been developed for our NDI setup enabling to apply pressures up to 20 GPa.

6. Acknowledgments

The authors wish to thank the staff at MLZ for their help in conducting the experiments. Financial support from the German Science Foundation (DFG) in the research unit "Quantum phase transitions" (FOR960), and the Augsburg-Munich transregional collaborative research network "From electronic correlations to functionality" (TRR80) is gratefully acknowledged.

References

[1] Hodych J 1977 *Can. J. Earth Sci.* **14** 2047-61
[2] Dunlop D J and Özdemir Ö 1997 *Rock Magnetism* (Cambridge: Cambridge Univ. Press)
[3] Wealthall R J, BrookerL R, Macey D J and Griffin B J 2005 *J. Morphology* **265** 165-71
[4] Shaw J A, Macey D J, Brooker L R, Stockdale E J, Saunders M and Clode P L 2009 *J. Morphology* **270** 588-600
[5] Kirschvink J and Lowenstam H A 1979 *Earth Planet. Sci. Lett.* **44** 193-204
[6] Gilder S, Le Goff M, Chervin J C and Peyronneau J 2004 *Geophys. Res. Lett.* **31** L10612

[7] Carporzen L and Gilder S A 2010 *J. Geophys. Res.* **115** B05103
[8] Kardjilov N, Manke I, Strobl M, Hilger, Treimer W, Meissner M, Krist T and Banhart J 2008 *Nat. Phys.* **4** 399-403
[9] Piegsa F, van den Brandt B, Hautle P, Kohlbrecher J and Konter J 2009 *Phys. Rev. Lett.* **102** 145501
[10] Treimer W, Ebrahimi O and Karakas N 2012 *Appl. Phys. Lett.* **101** 162603
[11] Treimer W, Ebrahimi O, Karakas N and Prozorov R 2012 *Phys. Rev. B* **85** 184522
[12] Schulz M 2010 *Radiography with polarized neutrons* Ph.D. thesis, Technical University of Munich
[13] Schulz M, Neubauer A, Masalovich S, Mühlbauer M, Calzada E, Schillinger B, Pfleiderer C and Böni P 2010 *J. Phys.: Conf. Ser.* **211** 012025
[14] Schulz M, Schmakat P, Franz C, Neubauer A, Calzada E, Schillinger B, Böni P and Pfleiderer C 2011 *Physica B: Cond. Matter* **406** 2412-14
[15] Halpern O and Holstein T 1942 *Phys. Rev.* **59** 960-81
[16] Mitsuda S, Yoshizawa H and Endoh Y 1992 *Phys. Rev. B* **45** 9788
[17] Schmakat P 2016 *Neutron Depolarisation Measurements of Ferromagnetic Quantum Phase Transitions, Wavelength-Frame Multiplication Chopper System for the Imaging Instrument ODIN at the ESS* http://mediatum.ub.tum.de/node?id=1278491
[18] Xu S and Merrill R T 1990 *J. Geomag. Geoelectr.* **42** 637-52

IOP Conf. Series: Journal of Physics: Conf. Series **862** (2017) 012025 doi:10.1088/1742-6596/862/1/012025

Polarization analysis for magnetic field imaging at RADEN in J-PARC/MLF

Takenao Shinohara[1,*], Kosuke Hiroi[1], Yuhua Su[1], Tetsuya Kai[1], Takeshi Nakatani[1], Kenichi Oikawa[1], Mariko Segawa[1], Hirotoshi Hayashida[2], Joseph D. Parker[2], Yoshihiro Matsumoto[2], Shuoyuan Zhang[2], Yoshiaki Kiyanagi[3]

[1]J-PARC Center, Japan Atomic Energy Agency, Tokai, Ibaraki 319-1195, Japan
[2]Comprehensive Research Organization for Science and Society, Tokai, Ibaraki 319-1106, Japan
[3]Nagoya University, Chikusa-ku, Nagoya, Aichi 464-8601, Japan

E-mail: takenao.shinohara@j-parc.jp

Abstract. Polarized neutron imaging is an attractive method for visualizing magnetic fields in a bulk object or in free space. In this technique polarization of neutrons transmitted through a sample is analyzed position by position to produce an image of the polarization distribution. In particular, the combination of three-dimensional spin analysis and the use of a pulsed neutron beam is very effective for the quantitative evaluation of both field strength and direction by means of the analysis of the wavelength dependent polarization vector. Recently a new imaging instrument "RADEN" has been constructed at the beam line of BL22 of the Materials and Life Science Experimental Facility (MLF) at J-PARC, which is dedicated to energy-resolved neutron imaging experiments. We have designed a polarization analysis apparatus for magnetic field imaging at the RADEN instrument and have evaluated its performance.

1. Introduction

Neutron imaging is one of the fundamental techniques used to visualize the inner structure of objects nondestructively by using the neutron's unique characteristics such as high transmittance through massive objects and high sensitivity to light elements. As a neutron has a magnetic moment, it can also interact directly with magnetic fields. Hence the use of polarized neutrons enables us to visualize the distribution of magnetic fields by means of a change in neutron polarization [1, 2]. On the other hand, owing to improvements in the imaging devices and an increase of neutron beam intensity, a new neutron imaging technique, namely energy-resolved neutron imaging, has been developed. In this technique, analysis of the energy dependent neutron transmission has been performed position by position to produce quantitative distribution maps of physical quantities, for example, crystallographic structure information such as phase, texture, and strain, by Bragg-edge imaging [3], and elemental composition and temperature by Resonance absorption imaging [4, 5]. The neutron energy analysis technique using pulsed neutrons has also been applied to polarized neutron imaging, and we have developed a new magnetic field imaging technique, with which the magnetic field strength and direction can be treated quantitatively by analyzing the wavelength dependence of neutron polarization using the polarized pulsed neutron beam and the time-of-flight (TOF) method [6, 7].

In the MLF of J-PARC, a new instrument named "RADEN" has been constructed at the beam line BL22 [8, 9]. This is the first instrument dedicated to pulsed neutron imaging experiments. The main purpose of RADEN is to conduct energy-resolved neutron imaging experiments efficiently by fully utilizing the nature of the neutron beam produced by the short pulsed neutron source i.e., fine energy or wavelength resolution by means of the TOF method and simultaneous availability of a wide energy range neutron beam from meV to keV. In addition, it has a role as the state-of-the-art neutron imaging facility in Japan, where it is possible to conduct experiments with sufficient neutron flux, selectable beam collimation and field of view, and monochromatic/white beam imaging capabilities. Because magnetic field imaging using polarized pulsed neutrons is regarded as one of the important applications of RADEN, a neutron polarization analysis apparatus has been prepared from the first stage of instrument operation. In this paper we report the design of the polarization analysis apparatus for polarized pulsed neutron imaging and its performance evaluated by commissioning studies using the neutron beam.

2. Design of the polarization analysis apparatus for magnetic field imaging

2.1. Magnetic field imaging using polarized pulsed neutrons

First, we explain the basis of the magnetic field imaging technique using polarized pulsed neutrons. As a neutron has a spin angular moment, it experiences torque in a magnetic field, causing the neutrons to precess around the magnetic field. The resulting precession angle ϕ of a neutron spin can be expressed by the following equation,

$$\phi = \gamma B t = \frac{\gamma m \lambda}{h} \int_{\mathrm{path}} B \, \mathrm{d}s \tag{1}$$

where γ is the gyromagnetic ratio of the neutron, t is the time for a neutron to traverse a magnetic field B, λ is the neutron wavelength, m is the neutron mass and h is Planck's constant. As is shown by equation (1), the precession angle ϕ depends on both the neutron wavelength and the accumulated magnetic field strength along the neutron beam path. Hence, when the absolute value of ϕ is measured, the path integral of the magnetic field strength can be evaluated using a monochromatic neutron beam with a given neutron wavelength. However, what we can measure is not ϕ but the neutron spin polarization P, which is defined as $P=(n_+-n_-)/(n_++n_-)$ where n_+ and n_- are the number of neutrons with spin-up and spin-down states. According to the relation between ϕ and P that is discussed in reference [10], P undergoes a sinusoidal change against ϕ and wraps when ϕ increases above 2π. On the other hand, because ϕ is proportional to λ, P also shows a periodic oscillatory behavior against λ. Since the frequency of the neutron wavelength dependent P is equal to $\gamma m/h\int_{\mathrm{path}} B \, \mathrm{d}s$ from equation (1), we can evaluate the accumulated field strength without measuring ϕ. Thus performing polarization analysis position by position using a polarized pulsed neutron beam, whose wavelength can be easily determined by means of TOF analysis, it is possible to map the projection of the magnetic field strength distribution quantitatively for uniformly oriented fields.

Moreover in general a magnetic field is treated as a vector and hence the neutron polarization P is also expanded to a vector quantity $\boldsymbol{P}$. The change in $\boldsymbol{P}$ after transmission through a magnetic field region $\boldsymbol{B}$ can be written as

$$P = D(n,t)P_0. \tag{2}$$

Here $D(\boldsymbol{n},t)$ is a rotation matrix depending on a unit vector along the magnetic field direction $\boldsymbol{n}=\boldsymbol{B}/|\boldsymbol{B}|$ and t [8], and $\boldsymbol{P}_0$ is the initial polarization. Since the diagonal and off-diagonal terms of $D(\boldsymbol{n},t)$ can be expressed using vector elements of $\boldsymbol{n}$, namely (n_x, n_y, n_z), as in the following equations,

$$D_{ii} = 1 - (1 - cos\,\phi) \cdot (1 - {n_i}^2) \tag{3}$$

$$D_{ij} = 1 - (1 - cos\,\phi) n_i n_j - n_k \, sin\,\phi \;,\; i \neq j \neq k \tag{4}$$

where $i, j, k = x, y, z$, it is obvious that the magnetic field direction can be derived from the analysis of the wavelength dependence of these terms when the field direction does not change along the beam trajectory. To obtain every matrix element it is necessary to control the incoming neutron spin direction in thee-dimensions and to analyze the spin direction of the outgoing neutron after traversing the magnetic field region in three-dimensions as well. This means that the three-dimensional neutron polarization analysis similar to spherical polarimetry [11] is required. While introducing a tomographic approach is strictly necessary to fully reconstruct the tree-dimensional magnetic field distribution [12], such a technique enables us to produce maps quantitatively visualizing the distributions of both the magnetic field strength and direction. A pilot experimental study for the quantification of a magnetic field by means of 3D polarization analysis with pulsed neutrons has demonstrated that the magnetic field strength and direction in a solenoid coil was successfully deduced within an accuracy of less than 3% [13]. Building on this previous work, we have constructed an improved polarization analysis system for the RADEN beam line.

2.2. *Design of the polarization analysis apparatus*

For the neutron polarization analysis we have designed an apparatus as shown in Figure 1. Here the x, y, and z directions were taken along the horizontal, vertical, and neutron beam directions, respectively. The apparatus consists of three main components, i.e., neutron polarization devices, spin-controlling devices, and guide fields. We describe the details of each component below.

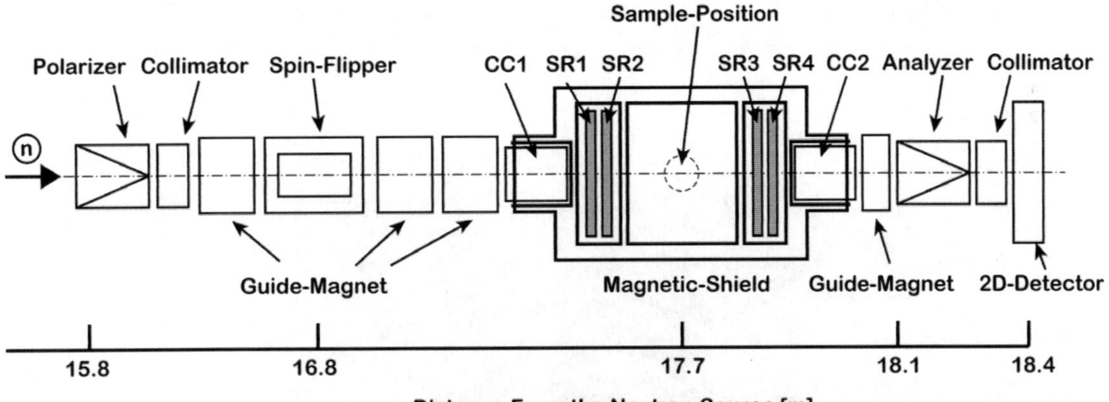

Figure 1. Schematic illustration of the polarization analysis apparatus for RADEN. The neutron beam comes from the left side of this illustration. SR1 to SR4 refer to spin rotators, and CC1 and CC2 are coupling coils.

2.2.1. Polarization devices. The polarizer and analyser are the most important devices for the polarization analysis. For magnetic field imaging, a large beam size and a short distance between the sample and the detector, which affects the accessible smallest spatial resolution, are desired. In addition, the capability to use a wide wavelength range neutron beam is required. Because the dynamic range of the magnetic field strength that can be evaluated by means of polarization analysis depends on the range of frequencies that can be analyzed from the wavelength dependent neutron polarization, a broader available wavelength range is better. Based on these requirements, we have designed a magnetic super mirror polarizer, and the same device has been used as an analyzer. The drawing of the polarizer (analyzer) and its photograph are shown in Figure 2. By means of a multi-V-shape arrangement of magnetic mirrors, it covers a cross section of 60 mm x 60 mm with a length shorter than 180 mm. The coating material is Fe/Si, and the substrate material is single crystal Si wafer with a thickness of 0.3 mm. The m value is 4.5. The tilt angle of each mirror was set at ±12.0 mrad so as to achieve the critical neutron wavelength of 1.5 Å. This critical wavelength was determined so as to cover the peak of the neutron flux spectrum provided by the J-PARC/MLF's liquid hydrogen moderator, which is located at a neutron wavelength around 2.5 Å, to allow efficient measurements using polarized neutrons at the MLF. The magnetic mirrors are aligned vertically and spin-up neutrons are reflected in the horizontal plane.

A magnetic housing, which consists of NdFeB permanent magnets and pure iron plates producing a vertical magnetic field of more than 60 mT, was used to apply a magnetic field on magnetic super mirrors to magnetize them. This housing was designed not to produce strong stray fields around it, since devices used in our system, described later, are sensitive to the environmental fields and respond in an undesirable way. The polarizer and the analyzer were used in combination with collimators consisting of stacked Gd coated Si wafers, which were placed behind the polarizer and the analyzer to remove the reflected neutrons. The acceptable horizontal beam divergence was designed to be ±19 mrad by taking the natural beam divergence, which was determined by the size of aperture located at about 8 m upstream from the collimators, into account.

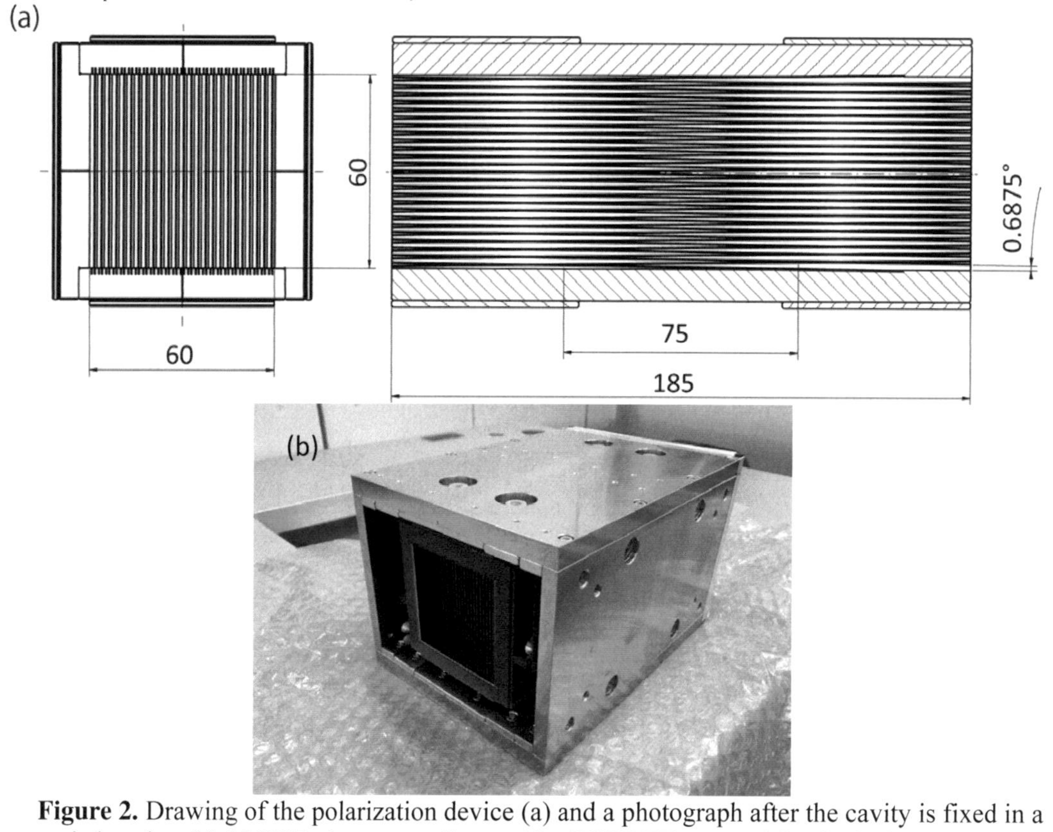

Figure 2. Drawing of the polarization device (a) and a photograph after the cavity is fixed in a magnetic housing (b). N-BK7 plates manufactured by SCHOTT are used for the body of the cavity.

2.2.2. Spin controlling devices. To control the neutron spin direction, we installed a spin flipper and a 3D spin rotation system with magnetic shielding. An adiabatic fast passage (AFP) type flipper [14] was adopted as a spin flipper because it works effectively for a broad wavelength range of neutrons. In this device a static field is generated along the y direction by a pair of pure iron plates that are magnetized by electromagnets. The strength of this field decreases along the z direction to produce a spatial magnetic field gradient. A solenoid coil is placed between the magnetized iron plates to provide an RF field along the z axis orthogonal to the static field by applying an alternating current of around 100 kHz (Figure 3). The static field is tuned by changing the applied current to the electromagnets so that a resonance occurs around the middle of the spin flipper.

The 3D spin rotation system consists of two pairs of spin rotators (SR1 to SR4 in Figure 1) and a magnetic shielding box. This system looks like the MuPAD device developed for spherical neutron polarimetry [15]. A spin rotator is a planar precession coil wound by 1 mm-ϕ Al wire and is used to turn a neutron spin by $\pi/2$ (Figure 4). The outer field of the coil is absorbed by a permalloy yoke and a surrounding permalloy shield. Then two coils with permalloy shields are arranged orthogonally and again placed together in another permalloy shield. Finally two pairs of spin rotators are mounted inside an outer permalloy shield. The first pair of spin rotators is located upstream of sample to turn the incident neutron spin direction, and the second pair is located downstream of the sample to turn the transmitted neutron spin. The reason why we have constructed such a strict magnetic shielding is that the spin rotators are very sensitive to the environmental field and it is important to ensure a zero field environment for spherical polarimetry by eliminating magnetic fields not only from outside but also from the precession coils themselves. To apply this system to the pulsed neutron beam, a dumping current proportional to $1/t$ is applied to the spin rotators synchronized with the generation time of pulsed neutrons. Then we can get a constant neutron spin rotation angle of $\pi/2$ for a wide neutron wavelength range.

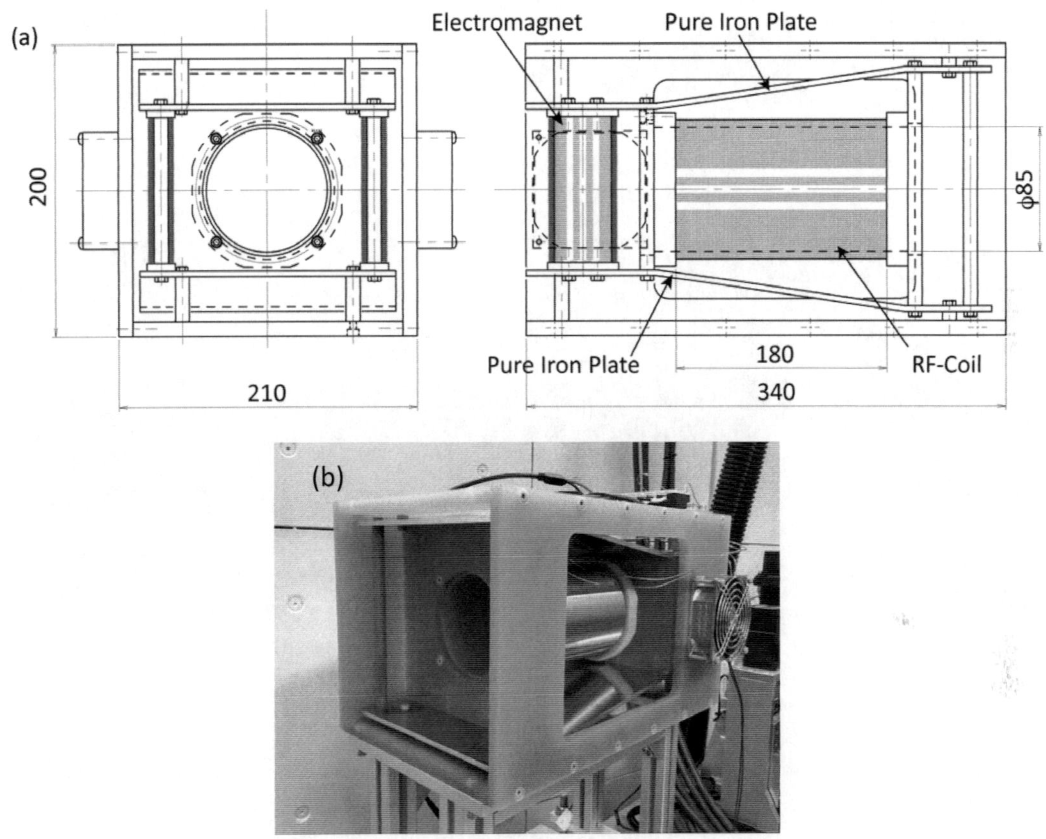

Figure 3. Drawing of the AFP spin flipper (a) and its photograph (b).

In addition, since the polarized neutron beam has to enter the zero field area inside the magnetic shield, it is necessary to avoid possible depolarization by a sudden decrease of the guide magnetic field and stray fields around the apertures of the beam pass. In this system, two coupling coils (CC1 and CC2 in Figure 1) made by Al-wire are placed at the entry and exit of the outer magnetic shield to preserve the neutron spin along the y direction (Figure 5). The magnetic field of the coupling coil is connected to the outer guide field by splitting the Al-wire at one side. The opposite side of the coupling coil, i.e., that facing the inside of the magnetic shield, is closed to create a screen of Al-wire that acts as a boundary for the magnetic field, like a current sheet. The neutrons come into the field of the coupling coil from the outer magnetic field adiabatically and then go through the field of the coupling coil to zero field non-adiabatically at the entrance of the outer magnetic shield. At the exit of the magnetic shield, this process occurs in reverse. Similar to the precession coil of the spin rotator, the coupling coil was surrounded by a permalloy yoke and a shield to absorb stray fields.

Figure 4. Photographs of the spin rotator. Left side is a photograph of the precession coils, and right is that of the assembled spin rotators inside a permalloy magnetic shield. The surface of the Al-wire was coated by an aluminum oxide layer for insulation.

Figure 5. Photographs of coupling coils. The left photograph shows the coils' open side facing outside of the magnetic shield box and right displays the screen side facing inside. The coils are tightly surrounded by a permalloy yoke.

2.2.3. Guide fields. The guide field plays an important role to preserve neutron spin direction and to avoid possible depolarization. We placed several magnetic field guides to fill the gap between devices as shown in Figure 1. A magnetic field guide is composed of two opposing pure iron plates and Al-Ni-Co permanent magnet pillars placed between plates. The distance between the iron plates is 120 mm and their width is 120 mm. The size of the cross section of this guide was determined to make a uniform magnetic field in the central 60 mm x 60 mm area, which corresponds to the maximum beam size, by means of magnetic field calculation. The field strength is about 6 mT in the middle of the guide.

2.3. Magnetic field simulations

Magnetic field simulation has been performed to confirm the magnetic field connection along the neutron beam passage and the uniformity of the field in the vertical plane perpendicular to the beam propagation. A commercial magnetic field calculation software ELF/MAGIC, which has the capability to perform electromagnetic field analysis using the boundary element method, was used. The calculation model is shown in Figure 6(a). In this calculation we omitted the RF coil of the spin flipper, spin rotators and non-magnetic components since they don't affect the characteristics of the static magnetic field in the system related to transportation of the polarized neutron beam. The magnetic field simulation results, shown in Figure 6(b), clearly indicated that the magnetic field is smoothly connected along the beam direction except inside the magnetic shield box and that the difference along the vertical direction is small. On the other hand, to guarantee that the neutron spin follows the guide magnetic field, the adiabaticity condition indicating the relation between the Larmor frequency ω_L and the frequency of rotation of the magnetic field ω_B, which is expressed as equation (5), should be fulfilled.

$$\Gamma = \frac{\omega_L}{\omega_B} \gg 1, \quad \omega_B = \frac{ds}{dt}\left|\frac{\partial n}{\partial s}\right|, \tag{5}$$

where $\boldsymbol{n}$ is the magnetic field direction and s is the neutron beam path. The ratio Γ is called the adiabaticity parameter. We calculated Γ from the magnetic field simulation results along the entire neutron beam path and found that it exceeded 10 over the whole system except inside the magnetic shield. This Γ value is large enough to effectively hold the neutron spin direction along the guide field. According to the results of these simulation studies, we concluded that the designed polarization analysis apparatus possesses enough capability to preserve the spin direction of the polarized neutron beam with a large beam size homogeneously over the whole system.

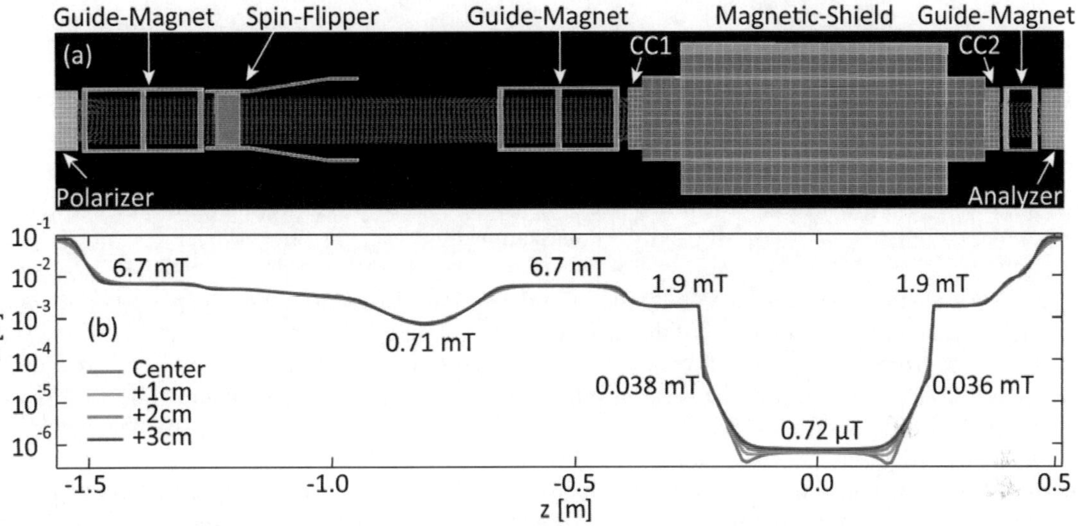

Figure 6. Schematic drawing of magnetic field calculation model (a) and the obtained magnetic field distributions along the beam direction at several beam height (b).

3. Polarization analysis performance

Performance of the polarization analysis apparatus has been studied at the RADEN instrument using the nGEM detector [16] as a two-dimensional neutron detector. The spatial resolution was 0.8 mm, and the active area was 100 mm x 100 mm. Distance between the source and the detector was 19 m. The available wavelength range was defined using a double disk chopper installed in the up-stream beam line shield. In addition, high energy neutrons and prompt gamma-rays generated at the time of neutron production were eliminated using a T_0 chopper [9]. The beam was shaped to the size of 45 mm x 45 mm by a B_4C slit located in front of the polarizer. The beam power at that time was 200 kW.

First, we measured the polarization distribution without turning on the spin rotators. Here the polarization P was replaced by $P=(I_{off}-I_{on})/(I_{off}+I_{on})$ where I_{off} and I_{on} are the measured neutron intensity with the spin flipper switched off and that with the spin flipper on, respectively [17]. Since the efficiency of our spin flipper was very close to 1, as confirmed from the experimental results of the flipping ratio and the polarization data provided by the polarizer manufacturer, the measured P can properly describe the performance of our apparatus. Figure 7(a) is an image of the two-dimensional polarization distribution obtained at the wavelength of 3.5 Å and clearly demonstrates a flat distribution in the xy plane. We evaluated the standard deviation of the polarization along the x

direction to be 2.0% and that along the y direction to be 0.3% within the beam area. The small deviation along the y direction is evidence that vertically homogeneous guide magnetic fields are smoothly connected along the beam path. On the other hand, the larger deviation along the x direction can be attributed to the difference in performance of individual magnetic mirrors. The wavelength dependence of the polarization is shown in Figure 7(b). The polarization rises from around 1Å and quickly increases up to 80% at a wavelength of about 2 Å. Then it exceeds 90% for wavelengths longer than 3 Å. This behavior agrees with the original design, and this system is deduced to possess adequate polarization analysis performance.

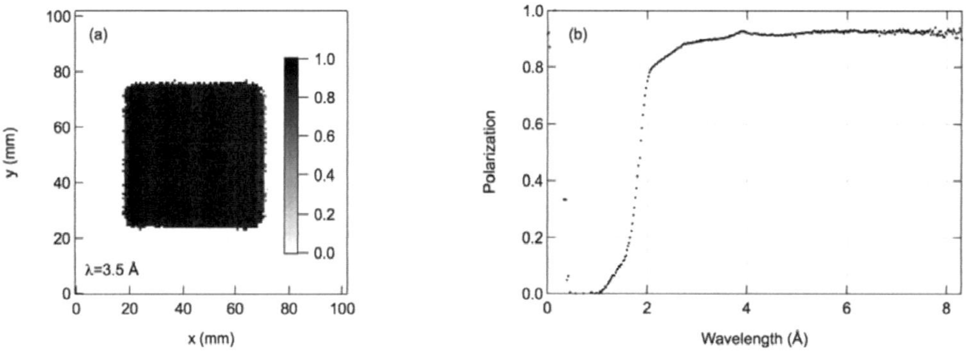

Figure 7. (a) Polarization distribution map (λ= 3.5 Å) and (b) wavelength dependence of neutron polarization.

Next, we operated the spin rotators to measure the spatial distribution of each spin rotation matrix element. The neutron spin direction of the incident beam was aligned with the y direction at first. Then it was rotated from the y to the z direction by switching on SR1, and from the y to the x direction by SR1 and SR2. The z element of the neutron spin polarization vector of the transmitted beam was then analyzed by switching on SR4, and the x element by SR3 and SR4. With SR3 and SR4 switched off, the y element of the polarization vector could be analyzed. In this way, every element of the spin rotation matrix of equation (2) was measured by switching on and off the appropriate combination spin rotators. The obtained images for all spin rotation matrix elements and their wavelength dependences are shown in Figure 8, 9 and Figure 10. From these results, it was found that the diagonal terms demonstrated similar behavior as the original one shown in Figure 7(b), and that the off-diagonal terms were almost constant with a value close to 0. Accordingly, the spin rotators were clearly verified to rotate the neutron spin direction into the proper direction and to cause no significant deterioration of the polarization. However when looking in detail at the profiles along the horizontal axis (x direction) and the vertical axis (y direction) for the off-diagonal terms shown in Figure 9 and their wavelength dependences shown in Figure 10, we can recognize that a small but nonzero polarization of less than 10% remains. Since this small deviation of the polarization from 0 is attributed to an insufficient tuning of the spin rotators, a precise adjustment of the applied current to the spin precession coil is needed to improve the accuracy of the spin rotation angle. Moreover, the uniformity along the vertical direction was slightly worse for the off-diagonal terms as compared to the diagonal terms. This spatial variation originates from a small magnetic field difference between the central region and top and bottom parts of the coupling coils. To improve the uniformity of the polarization distribution, a more careful connection of the outer guide field and the coupling coil field is required by placing additional guides or by overlapping both fields more deeply.

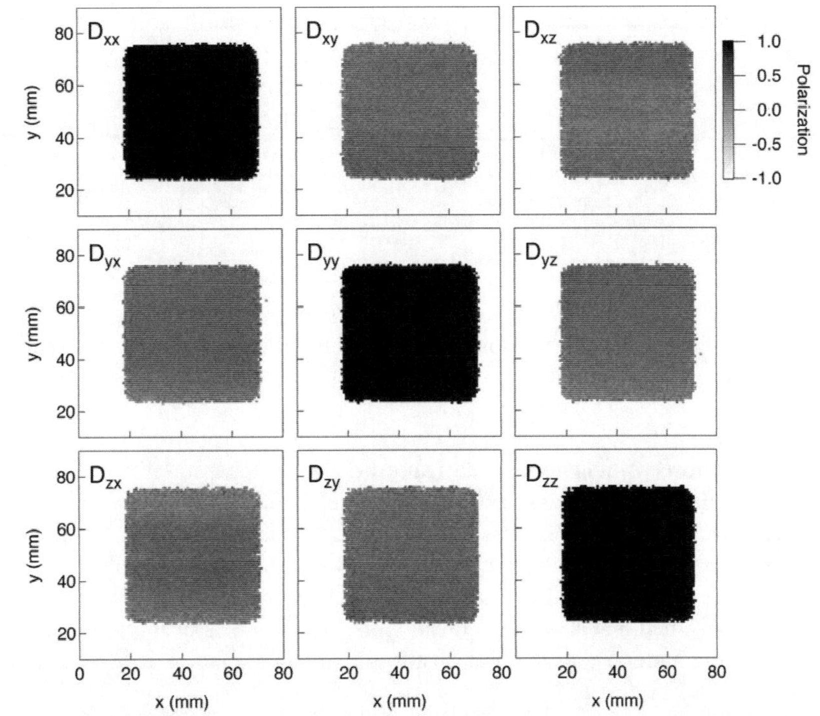

Figure 8. Distribution maps of spin rotation matrix elements (λ= 3.5 Å).

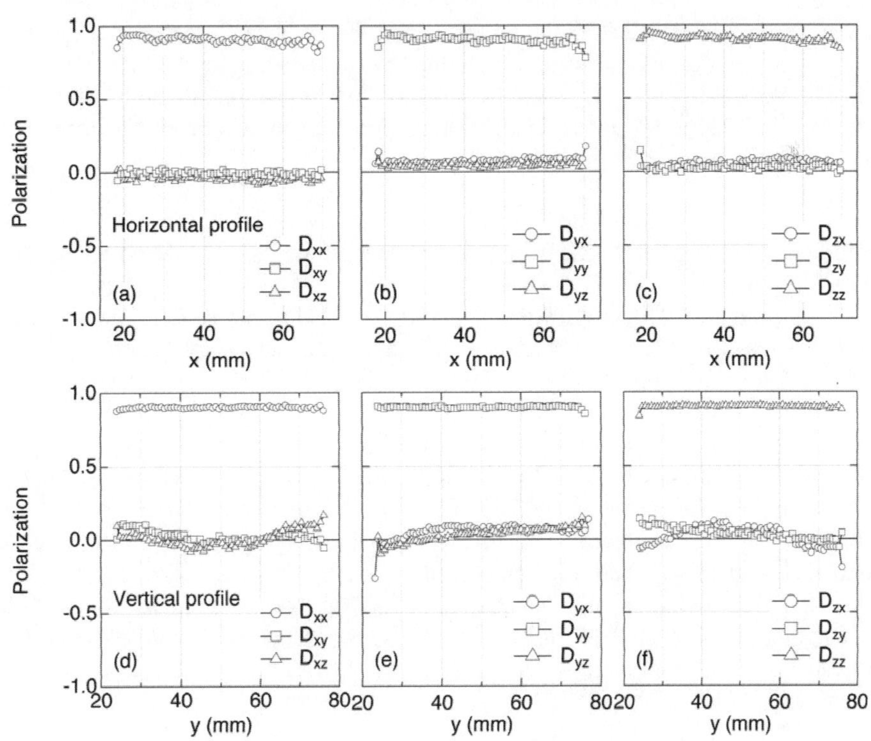

Figure 9. Polarization profiles along the horizontal ((a) to (c)) and vertical axes ((d) to (f)) for the spin rotation matrix elements (λ= 3.5 Å).

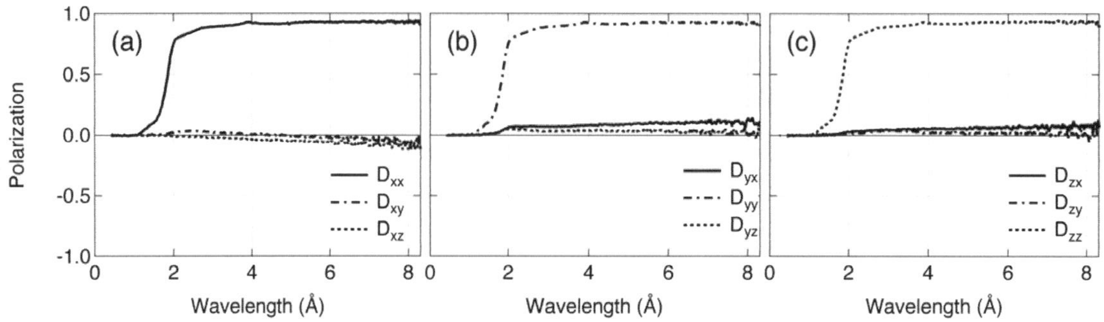

Figure 10. Wavelength dependences of spin rotation matrix elements. The spin direction of the incident beam is parallel to (a) *x* direction, (b) *y* direction, and (c) *z* direction.

4. Conclusion

We have constructed a three-dimensional polarization analysis apparatus for magnetic field imaging experiments using a polarized pulsed neutron beam at the RADEN instrument at J-PARC/MLF. We decided the arrangement of the devices based on the magnetic field calculation and on the evaluation of the adiabaticity parameter along the neutron beam trajectory. From the test experiment, it was confirmed that a good polarization over 90% over a wide wavelength range with a flat spatial distribution could be obtained and that the neutron spin direction was properly controlled by the spin rotators. Therefore, we conclude that our polarization analysis apparatus can be effectively used for quantitative magnetic field imaging.

Acknowledgement

This work was partially supported by the Grant-in-Aid for Science Research (S) 23226018 from the Japan Society for the Promotion of Science and by the Photon and Quantum Basic Research Coordinated Development Program from the Ministry of Education, Culture, Sports, Science and Technology, Japan. The performance evaluation experiments have been performed under a user program (Proposal No. 2015I0022).

References

[1] Kardjilov N, Manke I, Strobl M, Hilger A, Treimer W, Meissner M, Krist T and Banhart J 2008 *Nat. Phys.* **4** 399
[2] Kardjilov N, Manke I, Hilger A, Strobl M, Banhart J, 2011 *Materials Today* **14** 248256
[3] Sato H, Takada O, Iwase K, Kamiyama T, Kiyanagi Y, 2010 *J. Phys.: Conf. Ser.* **251** 012070
[4] Kaneko K, Kamiyama T, Kiyanagi Y, Sakuma T, Ikeda S 1999 *J. Phys. Chem. Solids* **60** 1499
[5] Kai T, Maekawa F, Oshita H, Sato H, Shinohara T, Ooi M, Harada M, Uno S, Otomo T, Kamiyama T, Kiyanagi Y 2013 *Phys. Proc.* **43** 111
[6] Shinohara T, Sakai K, Ohi M, Kai T, Harada M, Oikawa K, Maekawa F, Suzuki J, Oku T, Takata S, Aizawa K, Arai M and Kiyanagi Y 2011 *Nucl. Instr. and Meth. A* **651** 121
[7] Strobl M, Kardjilov N, Hilger A, Jericha E, Badurek G, Manke I, 2009 *Phys. B* **404** 26112614
[8] Shinohara T, Kai T 2015 *Neutron News* **26** 11
[9] Shinohara T, Kai T, Oikawa K, Segawa M, Harada M, Nakatani T, Ooi M, Aizawa K, Sato H, Kamiyama T, Yokota H, Sera T, Mochiki K, Kiyanagi Y 2016 *J. Phys: Conf. Series* **746** 012007
[10] Rekveldt M T 1941 *Z. Phys.* **259** 391
[11] Tasset F 1989 *Physica B* **156–157** 627
[12] Hochhold M, Leeb H, Badurek G 1996 *J. Magn. Mang. Mater.* **157/158** 575
[13] Wada N, Shinohara T, Sato H, Hasemi H, Kamiyama T and Kiyanagi Y 2015 *Phys. Proc.* **69** 427
[14] Weinfurter H and Badurek G 1989 *Nucl. Instr. and Meth. A* **275** 233

IOP Conf. Series: Journal of Physics: Conf. Series **862** (2017) 012025 doi:10.1088/1742-6596/862/1/012025

[15] Janoschek M, Klimko S, Gähler R, Roessli B, and Böni P 2007 *Phys. B* **397** 125
[16] Shoji M, Uno S, Uchida T, Murakami T, Sekimoto M, Koike T, Nakano E, Yamamoto K, Hoshi Y, Nagashima F, 2012 *J. Inst.* **7** C05003
[17] Mitsuda S, Endoh Y 1985 *J. Phys. Soc. Jpn.* **54** 1570

IOP Conf. Series: Journal of Physics: Conf. Series **862** (2017) 012026 doi:10.1088/1742-6596/862/1/012026

On the analysis of time-of-flight spin-echo modulated dark-field imaging data

Morten Sales,[a;b] Jeroen Plomp,[c] Wim G. Bouwman,[c] Anton S. Tremsin,[d] Klaus Habicht[e] and Markus Strobl[a;f]

[a]Nano-Science Center, Niels Bohr Institute, University of Copenhagen, DK-2100 Copenhagen, Denmark,
[b]Department of Physics, Technical University of Denmark, DK-2800, Kgs. Lyngby, Denmark,
[c]Faculty of Applied Sciences, Delft University of Technology, 2629 JB Delft, The Netherlands,
[d]University of California at Berkeley, Space Sciences Laboratory, Berkeley 94720, California, USA,
[e]Helmholtz-Zentrum Berlin fuer Materialen und Energie GmbH, D-14109 Berlin, Germany, and
[f]European Spallation Source ESS-AB, Science Division, SE-22100 Lund, Sweden.

E-mail: markus.strobl@esss.se

Abstract. Spin-Echo Modulated Small Angle Neutron Scattering with spatial resolution, i.e. quantitative Spin-Echo Dark Field Imaging, is an emerging technique coupling neutron imaging with spatially resolved quantitative small angle scattering information. However, the currently achieved relatively large modulation periods of the order of millimeters are superimposed to the images of the samples. So far this required an independent reduction and analyses of the image and scattering information encoded in the measured data and is involving extensive curve fitting routines. Apart from requiring a priori decisions potentially limiting the information content that is extractable also a straightforward judgment of the data quality and information content is hindered. In contrast we propose a significantly simplified routine directly applied to the measured data, which does not only allow an immediate first assessment of data quality and delaying decisions on potentially information content limiting further reduction steps to a later and better informed state, but also, as results suggest, generally better analyses. In addition the method enables to drop the spatial resolution detector requirement for non-spatially resolved Spin-Echo Modulated Small Angle Neutron Scattering.

1. Introduction

While Spin-Echo Small Angle Neutron Scattering (SESANS) [1] is an established small angle neutron scattering technique (SANS) extending the range of SANS to very small (V-SANS) and ultra small (USANS) angles, Spin-Echo Modulated Small Angle Neutron Scattering (SEMSANS) has been introduced only recently [2,3] based on spatial beam modulation induced by spin-echo as proposed and proven earlier [4-6]. SEMSANS can be seen as a variation of SESANS, using less precession regions and providing the option to place the sample behind the spin-manipulation region, hence in a field free region where also magnetic samples can be measured. In contrast to SESANS, not the final beam polarisation, but a spatial beam modulation is measured and analysed, the damping of which provides the same information content as a SESANS polarisation measurement (2,3,7,8). While for SEMSANS a spatially resolved detection is required, it offers on the other hand the potential for spatially resolved SANS studies in an imaging configuration, because the sample can be placed close to the detector and in both techniques a reasonable beam divergence can be used, allowing for a useful field of view for imaging in SEMSANS [8]. In fact, with the spatially modulated intensity on the detector, SEMSANS resembles the highly successful method to measure dark-field images (DFI) with grating interferometers in neutron imaging instruments [9]. It has been shown recently that both methods are equivalent to SESANS in their capability to measure quantitative small angle scattering [7]. However, the higher fexibility of the magnetic field set-up of SEMSANS in contrast to a grating set-up, which is optimised for a single wavelength, makes SEM DFI much more efficient for quantitative SANS studies with spatial resolution and in particular in a Time-of-Flight (ToF) mode also able to take advantage of modern pulsed spallation sources.

2. SEMSANS

SEMSANS uses inclined field surfaces to map neutron small angle scattering information into the amplitude dampening of a spatially modulated beam [2-6,8]. The final spin rotation angle of polarised neutrons at the detector depends only on the lateral position with respect to the direction of the inclination of the field surfaces when the focusing condition [4]

$$B_1L_1=B_2L_2 \qquad \text{eq. 1}$$

is fulfilled, where B1 and B2 and L1 and L2 are the magnetic fields in the triangular precession regions and their respective distances from the detector. Due to the fact that the final precession angle at the detector is independent of the trajectory and origin in the source, this results in a linearly changing final accumulated spin angle across the beam at the detector. Hence, together with a polarisation analyser downstream the field regions, the result is a spatially modulated intensity across the beam at the detector position with a period given by [5]:

$$\zeta = \frac{\pi \tan\theta_0}{c\lambda(B_2 - B_1)}, \qquad \text{eq. 2}$$

where c = 4.632x1014 $T^{-1}m^{-2}$, λ the neutron wavelength and θ_0 the inclination angle of the field surface to the optical axis.

Detecting such modulation in a SEMSANS experiment requires the ability of the detection system to resolve the spatial modulation, which is in one direction only. So far this has been achieved by a combination of slits with a scanning mode, [3,10,11], by gratings with corresponding periods (comparable to neutron grating interferometers) [3,10] or with detectors with 2-dimensional spatial resolution [2,8,12]. Modulation periods of reported measurements to date were of the order of millimeters, which can hence still be resolved by neutron imaging detectors [13].

The correlation length or simply the size-parameter probed with such set-up when applied to small-angle scattering can similar to SESANS [1] be given as the Spin-Echo length,

$$\delta^{SE} = \frac{c\lambda^2 L_S(B_2 - B_1)}{\pi \tan\theta_0} = \frac{\lambda L_S}{\zeta}, \qquad \text{eq. 3}$$

with L_S being the sample to detector distance.

If a small angle scattering sample is inserted in the beam, it will cause a dampening of the amplitude of the spatial modulation as the scattering function will redistribute intensities between the minima and maxima, which can be described as a convolution of the modulation function with the scattering function [9]. The visibility of the modulated beam is defined as

$V=(I_{max}-I_{min})/(I_{max}+I_{min})$ eq. 4

where $(I_{max}+I_{min})$ provides the transmission of the sample when normalized with the empty beam measurement $((I_{0max}+I_{0min})$. Hence the normalized visibility V/V_0 is a transmission corrected entity where V (V_0) is the visibility of the spatially modulated signal, I (I_0), with (without) sample in the beam.

The normalised visibility (V/V_0) can then be modeled by a SESANS correlation function, G, according to [7]:

$$\frac{V(\delta^{SE})}{V_0} = \exp\left\{\Sigma t\left[G(\delta^{SE}) - 1\right]\right\}$$ eq. 5

like described analogue for SESANS in Ref. [1,14,15] and for grating interferometry in [7]. Here t is the sample thickness and Σ is the total scattering cross section of the sample, which e.g. for a two-phase system can be given as $\Sigma=\lambda^2(\Delta\rho)^2\phi(1-\phi)\xi$ [14], with $\Delta\rho$ being the scattering length density contrast, ϕ the volume fraction of one phase, and ξ the correlation length perpendicular to the neutron beam in the probed scattering direction, which is across the modulation.

A scan of δ^{SE}, the correlation length parameter probed (eq. 3), is required for quantitative SANS measurements. This can be achieved according to eq. 3 by either scanning wavelength, sample distance or values of the modulation period, e.g. through the magnetic fields B_1 and B_2, which however have to further fulfill the focusing condition (eq. 1) for the modulation. In order to model and fit data directly with the real space correlation function G, they have to be normalized by sample thickness and for ToF experiments utilizing an extended wavelength band by wavelength square, such that the data is of the form

$$\frac{\ln\left[\frac{V(\delta^{SE})}{V_0}\right]}{\lambda^2 t} = \Sigma\left[G\left(\delta^{SE}\right) - 1\right]$$ eq. 6

However, the actual extraction of the modulation parameters and in particular the off-set providing the attenuation contrast and the visibility providing the SANS information is the main challenge for data reduction and analyses. A spin-echo modulated beam does not provide a constant modulation for an extended cross-section of the beam as the echo condition is only perfectly fulfilled in a particular lateral position and can be framed by a decaying curve as an envelope for the maxima and minima of the spatial modulation like it is well known in neutron spin-echo applications. Additionally, data are affected by statistics. Because in SANS experiments samples can be considered homogeneous with regards to the beam cross section there are several possibilities to extract a single V/V_0 value, however. The most consuming analyses method is to model and fit the modulation curves fully, while the available maxima and minima can also be averaged in order to calculate a single averaged visibility value V and V_0, respectively, which is sufficient for such kind of measurement. This is technically also done in case of using an analyser grating, which can be done when the modulation is kept constant in a ToF measurement [12] or in a monochromatic mode [3] and which by a step scan provides a single modulation curve with a non-spatially resolved detector.

IOP Conf. Series: Journal of Physics: Conf. Series **862** (2017) 012026 doi:10.1088/1742-6596/862/1/012026

3. SEM-DFI

However, the SEMSANS set-up bears the big advantage as compared to SESANS, that it can be applied combined with imaging in a DFI approach [8] in analogy to that with Talbot Lau grating interferometers [9]. This is the case because here the sample can be placed close to the detector with 2-dimensional spatial resolution, which allows spatial correlation, i.e. image resolution, and an extended beam can be applied where the modulation can be analysed locally, due to the large beam divergence enabled by the technique. Hence, a combined application of SEMSANS and imaging requires an imaging detector with significant 2D spatial resolution. In analogy to grating interferometry the SE modulation can then be utilized in a neutron dark-field imaging mode providing an attenuation contrast image and a spatially resolved SAS analyses of the sample, simultaneously probing two different length scale regimes [8,9].

Because in grating interferometry the modulation periods are fixed by the grating geometries and the set-up needs to be optimized for a particular wavelength, it is much less flexible to probe a broad range of correlation lengths than such SEM-DFI approach [16]. However, modulations achieved with grating set-ups are in the range of micrometers and not directly resolved by the detector but by a scan of an anaylser grating within every pixel. Therefore the spatial resolution is decoupled from the modulation resolution and the correlation lengths probed are of the order of around a micrometer.

In the SEM imaging approach however, periods are of the order of millimeters so far and superposed to the real space image of the sample with entangled spatial resolutions of both the modulation, i.e. SANS signal, and the real space attenuation data. Currently achieved SE modulations are probing size ranges of about 20-200nm, but the data reduction requires significantly different strategies. In principle for the attenuation contrast the full spatial resolution at the detector can be utilized, which is defined by the collimation conditions given in terms of the L/D ration, the sample to detector distance d and indeed the spatial resolution of the imaging detector [13]. L denotes the distance of the source, in the case of a pinhole set-up the pinhole, to the sample while D denotes its dimension, and together with d they define the image blur $b=d(L/D)^{-1}$. However, in current data reduction and analyses the extraction of the attenuation contrast also relies on a full period fitting as will be outlined below. The SANS signal in contrast is smeared over ranges of at least the largest modulation period that provides significant SANS signal in terms of modulation damping. This to date limits the spatial resolution of the quantitative SANS information to the order of mm with this technique. This can only be improved by improving the magnetic field technology and achieving shorter correlation lengths δ_{SE} on smaller periods ζ.

The spatial resolution sought in the SANS signal is the only difference between SEMSANS and the corresponding SEM-DFI approach. While for SEMSANS the sample has to be assumed homogeneous, the opposite is the case for SEM-DFI with respect to attenuation, density, structure parameters and sample thickness. Therefore the possible spatial dependencies have to be taken into account in such applications and the visibility V, which could in SEMSANS be averaged over the beam becomes a function of the spatial dimensions across the beam V(x,y), where it is assumed that x constitutes the modulation direction. This in particular complicates the data reduction to pure wavelength dependent attenuation images and the relative visibility images at each $\delta_{SE}(\lambda)$. Up to date a complex and consuming fitting procedure was applied to the images fitting a sine function on approximately one period and assigning the corresponding offset and visibility values for the attenuation and DFI contrast to the central pixel of the fitted range, while shifting this range pixel by pixel across the x-direction of the image, repeating for every row of y. The required binning of images has to be decided beforehand in order to minimize the fitting routine resources, as fitting accumulates to approximately $1x10^6$ sine fits for a standard 1024x1024 pixel image.

4. Novel Data reduction and analyses

Here we introduce a novel approach to this data reduction, which does not only simplify the data extraction significantly, but also enables regaining full spatial resolution for the attenuation in SEM-DFI and has the potential to overcome the requirement for a spatially resolved detection for straightforward SEMSANS applications. This approach takes advantage of the polarized neutrons required for a spin-echo technique, which implies that two separate measurements with opposite initial spins (I↑ and I↓) and hence modulations shifted by a phase difference of π, can be recorded. Consequently the data can first be reduced to an attenuation corrected modulation, which has the potential to provide all required parameters in every single pixel without the requirement for extensive fitting procedures. Defining

$$M(x) = \frac{I\uparrow(x) - I\downarrow(x)}{I\uparrow(x) + I\downarrow(x)} = \frac{2A(x)}{2I} = \frac{(A_{max} - A_{min})sin(\frac{2\pi}{\zeta}x)}{2I} \qquad \text{eq. 7}$$

where x is the lateral direction of modulation. (Note that the dimension of y is ignored here as all operations can be expanded to y without any implications on the procedure.) The transmitted modulation averaged intensity I, i.e. the offset parameter required for the attenuation analyses, has the value

$$2I = I\uparrow(x) + I\downarrow(x) = (I\uparrow_0(x) + I\downarrow_0)e^{-\mu t} = 2I_0e^{-\mu t} \qquad \text{eq. 8}$$

which implies that the position independent attenuation $e^{-\mu t}$ in the case of SEMSANS can be extracted straightforwardly without fitting. I_0 is the modulation averaged incident intensity. While this applies to SEMSANS where the sample has to be considered homogeneous and hence the attenuation constant over the field of view, the attenuation can vary even over a single period in the case of SEM-DFI applications. Note that potential variations of the incident intensity do not play a role as they vanish in the normalization of I/I_0 when extracting the attenuation factor.

In contrast to earlier approaches measurements with both initial spin states carry all relevant modulation parameter information for every position x as can be seen in the schematic of figure 1. The only exception are positions where I↑=I↓ and the close proximity of such positions due to statistics, which will be discussed further down.

Otherwise the normalization

$$\frac{I\uparrow(x)+I\downarrow(x)}{I\uparrow_0(x)+I\downarrow_0(x)} = e^{-\mu(x)t(x)} \qquad \text{eq. 9}$$

straightforwardly provides the local attenuation $e^{-\mu(x)t(x)}$ and the normalization

$$\frac{M(x)}{M_0(x)} = \frac{V(x)}{V_0(x)} = \frac{(A_{max}-A_{min})I_0(x)}{(A_{0max}-A_{0min})I(x)} \qquad \text{eq. 10}$$

which however does not overcome the spatial resolution restriction described earlier for the DFI data. In contrast for the attenuation data the conventional geometric resolution calculation holds to a good approximation despite the redistribution of intensity through small angle scattering, because overall the intensity not scattered or scattered within the spatial resolution limit can be assumed dominant in most cases, as is in conventional imaging.

IOP Conf. Series: Journal of Physics: Conf. Series **862** (2017) 012026 doi:10.1088/1742-6596/862/1/012026

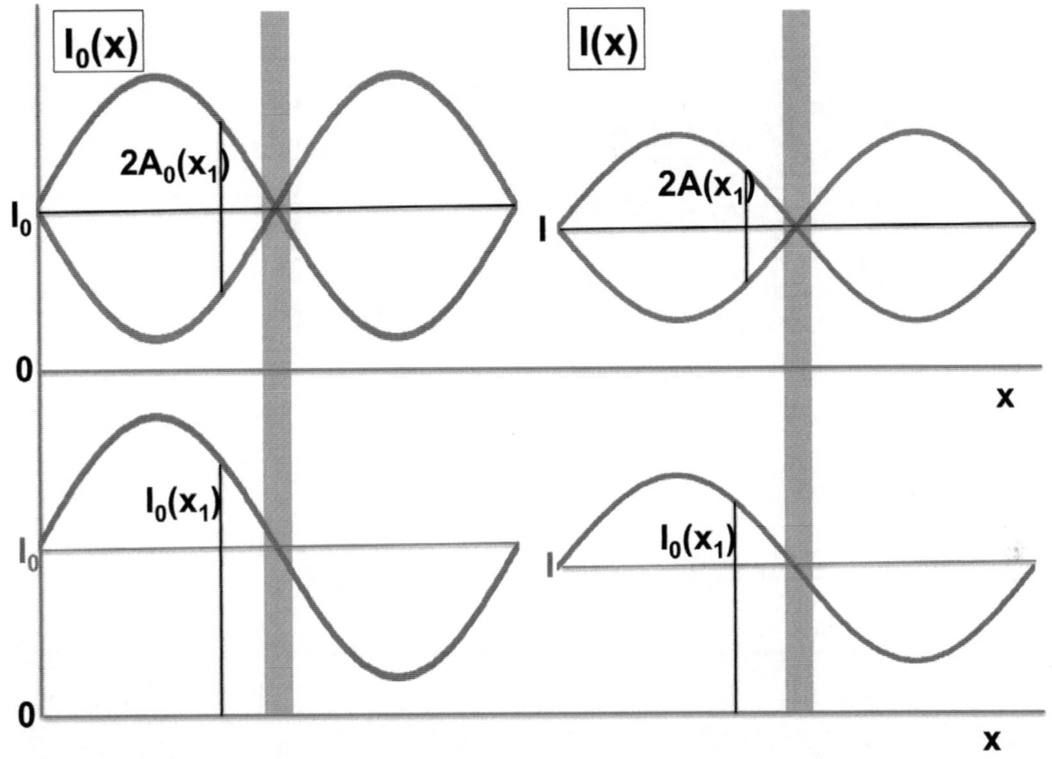

Figure 1. Schematic comparison of information content per pixel utilizing information of spin-up/spin-down measurements (top) versus a conventional single curve approach (bottom). The left side displays schematics of open beam and the right side of a sample measurement, in every case for one period and constant average intensities over such. It becomes obvious that for the top image all information on relative amplitude, i.e. visibility and offset can be extracted for every single point along x independently, while in the bottom example a whole period or a large fraction of it is required to extract such parameterization.

All these operations can be performed straightforwardly as image operations, i.e. by adding, subtracting and dividing the 2D image data. This enables immediate assessment of data and flexible binning at any time in order to improve statistics and assessment. The "blind areas" around $I\uparrow=I\downarrow$ do not affect SEMSANS measurements without spatial resolution and in DFI imaging mode when done in ToF with changing periods they change with wavelength and hence lead to minor missing sampling in every specific position of an image, certainly beyond spatial resolution for the SANS signal (compare figure 2). Hence they are not an obstacle for data analyses at all.

However, an absolute precondition in order to enable such direct reduction is the phase stability. That means the set-up must enable sufficient stability of the modulation phase all through the measurement of the sample and empty beam.

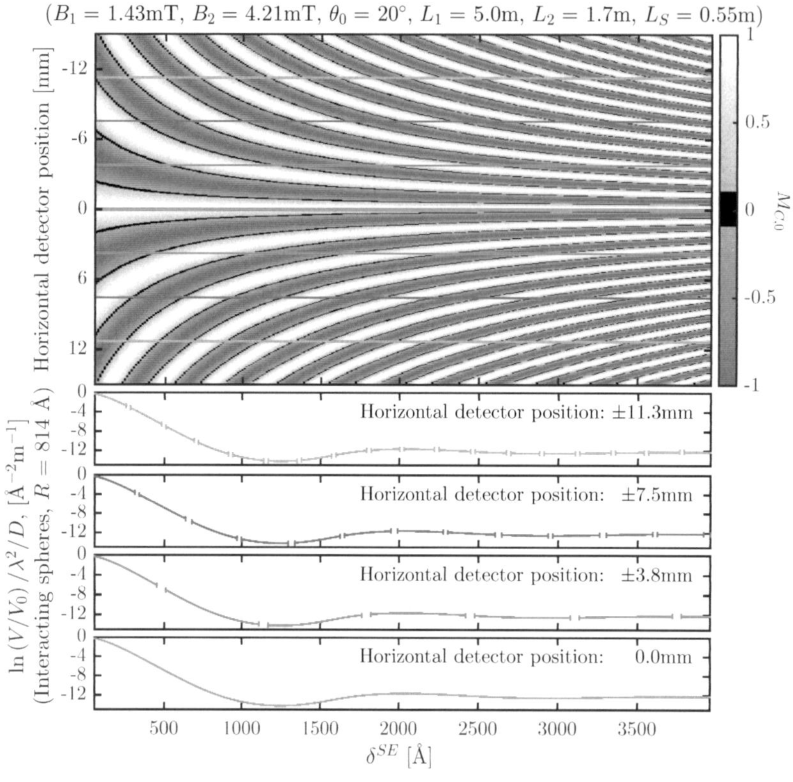

Figure 2. Calculations of scattering curve "dead spots" at different detector positions for a sample of hard spheres in high concentration investigated by SEMSANS using the novel pixel-by-pixel analysis. Each curve corresponds to the signal in a 55 × 55 μm^2 pixel and it can be seen that at the echo position a scattering signal can be extracted for the full Spin-Echo range, whereas at positions away from echo losses of signal occur when M_0 gets close to zero.

5. Measurement and analyses example

Measurements were performed at the Reactor Institute Delft at TU Delft on the SEMSANS setup described in [8], at a thermal source, where the pulsed beam was created using two co-rotating choppers in optically blind mode pro- viding a wavelength resolution of $\delta\lambda/\lambda \sim 5\%$ [17]. The triangular field coils were placed at L_1 = 5.0 m and L_2 = 1.7 m from the detector, with fields of B_1 = 1.43 mT and B_2 = 4.21 mT and inclinations angles of θ_0 = 20°. Our sample setup, which consisted of a combination of three different sample types, was placed at L_S = 0.55 m upstream of the detector. The supermirror multi-channel polariser was oriented so that initial polarisation was vertical whereafter two adiabatic π/2-rotators were used to choose between spin up or down setting and to rotate the neutron spin into the horizontal plane. Two vertical precession fields were used between polariser and analyser to control the neutron spin and before the analyser a second adiabatic π/2-rotator pair was used to rotate the neutron spins back to the vertical plane. The supermirror multi-channel analyser was oriented in this plane, such that effects from the structure of the analyser (and polariser) would be perpendicular to the modulation and therefore having no influence on the observed modulation. The Spin-Echo range covered was from about 10 nm to 175 nm. Since our proof-of-principle measurements were performed at a low-flux instrument, we examined three different regions of interest (ROI) covering areas with iron powder, an empty beam area, and dispersed polystyrene colloids in D2O respectively. Each ROI corresponds roughly to a detector area of approximately 50 mm^2 with a detector-pixel size of 55 ×

55 µm^2 [18,19]. Exposure time for every single measurement with spin-up and spin down and with or without samples in the beam was of the order of 3.5 hours, making the total measurement time approximately 14 hours.

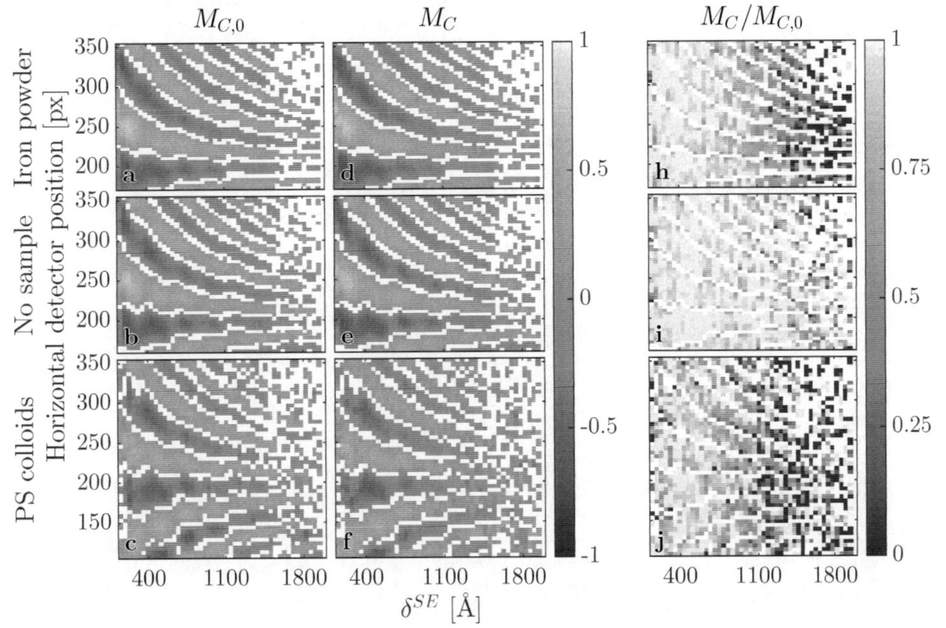

Figure 3. Images a-f show data from the ROIs which has been averaged along the vertical direction on the detector, where the colorscale corresponds to the attenuation corrected modulation and white areas are without information due to either "dead spots" or low counting statistics. The rows corresponds to the three different sample areas, and images a-c (d-f) are without (with) the sample in the beam. Images h-j show the normalised visibility (eq. 10).

To improve the quality of the obtained images for each Spin-Echo length the images were first filtered using an inverse scale space filter [20] to reduce noise and were averaged over the vertical direction of the ROI, i.e. perpendicular to the modulation direction, which was horizontal. Obviously at an instrument with sufficiently high flux and quality of components the vertical resolution can be as good as the collimation and detector resolution allow. Figure 3 shows the modulation as a function of Spin-Echo length and horizontal detector position. The normalised modulation, $M(x,y)/M_0(x,y)$, was calculated and is shown in figure 3 h-j, where white stripes indicate the positions in time and space with missing information due to the "blind areas" for the DFI signal analyses. As it can be seen in figure 3 h-j (the signal in h and j much faster drops into low visibility regions indicated by green and blue color), and even better in the curves of figure 4, the normalised visibility and the modulation expectedly decrease with increasing Spin-Echo length when there is a small angle scattering sample present. It can also be seen that for long Spin-Echo lengths (long wavelengths) there is a drop in statistics due to the reduced flux of the neutron source at these wavelengths. However, having a higher flux and/or a longer measuring time (possibly in combination with a cold instead of thermal neutron source) will improve statistics and the accessible range of correlations lengths in this regime.

Applying the conventional fitting data analyses for each Spin-Echo length (time frame) and ROI the modulated data in the horizontal detector direction was fitted with a sinusoidal function thereby obtaining the $V(\delta_{SE})$ and V_0 information [8] displayed as the "sine fitting" curves in figure 4, where they are compared to curves obtained through our novel approach of dividing the modulation functions achieved by the spin-up and spin-down measurements directly. Reference curves for the three sample types (iron powder, no sample, and polystyrene colloids) are shown as well. Concerning the "blind areas" (figure 2), $V_0 < 0.1$ was set as the threshold for determining their location in order to leave them out of the analyses to avoid corresponding errors due to division by zero. There is a high level of agreement between the curves from the two different reduction methods for data analysis, though at long Spin-Echo lengths the novel approach seems to be more robust with respect to noisier data and shows better precision and better correspondence with the reference curves.

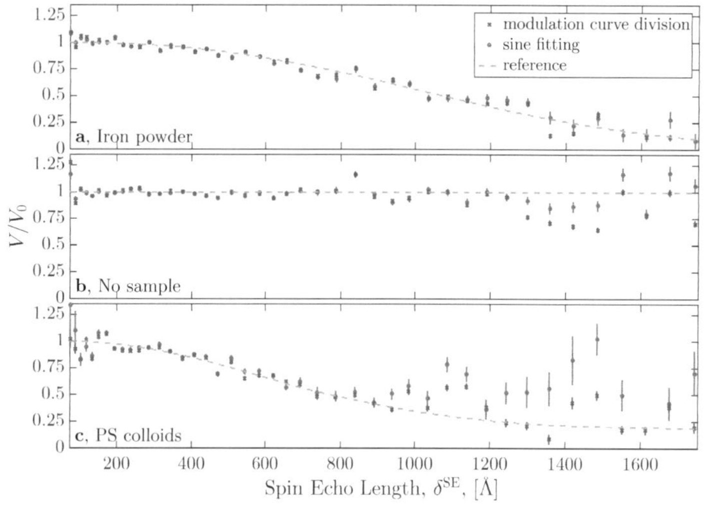

Figure 4. Comparison of relative visibilities dependent on the Spin-Echo length obtained through sine fitting and modulation curve division. The reference curves in a and c are based upon fits of higher resolution SESANS measurements performed individually on the same samples at RID. The model used for the reference curve in a is of a random two-phase media (Andersson et al., 2008) and the model used for the reference curve in c is for hard spheres [21]. For the reference curve in b, V/V_0 is equal to 1 for all Spin-Echo lengths as there is no small angle scattering sample present.

6. Discussion and conclusion

The results clearly demonstrate the feasibility of the approach and its advantages with respect to data reduction effort. It is obvious that the novel strategy enables faster assessment and feedback to measurement and data recording, reduces the resources and increases the flexibility and efficiency for data treatment. It is also demonstrated that the appearance of small blind areas do not affect the data negatively and with respect to the real resolution conditions not at all. Apart from enormous simplification of data treatment in particular for imaging applications the approach would also allow to eliminate the requirement for a high spatial resolution detector or a slit or grating scan for SEMSANS applications. The use of a grating with a duty cycle of about 0.5 and set to the maximum of the spin-up modulation is sufficient for all analyses as all parameters can be extracted for a single sampling position over the modulation. In a ToF approach where the period varies with wavelength in a basic set-up the constant period approach introduced by Sales et al., [12] is suited to overcome that issue. This will in future when higher fields can be realized also enable measurements at very small modulation periods beyond detector resolution limits. In addition such local approach is more robust with respect to altering attenuation over significantly broad periods as well as towards the decay of the modulation amplitude off the echo position. What is, however, maybe even more important overall, is that this simplified data reduction is significantly more user friendly.

References

[1] Rekveldt M T 1996 *Nucl. Instr. Meth. B* **114** (3-4), 366-370. http://dx.doi.org/10.1016/0168-583X(96)00213-3
[2] Strobl M, Tremsin A S, Hilger A, Wieder F, Kardjilov N, Manke I, Bouwman W G and Plomp J 2012 *J. Appl. Phys.* **112** (1), 014503. http://dx.doi.org/10.1063/1.4730775
[3] Strobl M, Wieder F, Duif C, Hilger A, Kardjilov N, Manke I and Bouwman W G 2012 *Phys. B Condens. Matter* **407** (21), 4132. http://dx.doi.org/10.1016/j.physb.2012.06.036
[4] Gaehler R 2007 *Phys. B* **397** (1-2), 1 200. http://dx.doi.org/10.1016/j.physb.2007.02.034
[5] Bouwman W G, Duif C P and Gaehler R 2009 *Phys. B Condens. Matter* **404** (17), 2585-2589 http://dx.doi.org/10.1016/j.physb.2009.06.052
[6] Bouwman W G, Duif C P, Plomp J, Wiedenmann A and Gaehler R 2011 *Phys. B Condens. Matter* 406(12), 2357-2360. http://dx.doi.org/10.1016/j.physb.2010.11.069
[7] Strobl M 2014 *Sci. Rep.* **4**, 7243. http://dx.doi.org/10.1038/srep07243
[8] Strobl M, Sales M, Plomp J, Bouwman W G, Tremsin A S, Kaestner A, Pappas C and Habicht K 2015 *Sci. Rep.* **5**:16576 | DOI: 10.1038/srep16576
[9] Strobl M, Gruenzweig C, Hilger A, Manke I, Kardjilov N, David C and Pfeiffer F 2008 *Phys. Rev. Lett.* **101** (12), 123902. http://dx.doi.org/10.1103/PhysRevLett.101.123902
[10] Sales, M., Plomp, J., Habicht, K. & Strobl, M. (2015). J. Appl. Crystallogr. **48**(1), 92-96. http://dx.doi.org/10.1107/S1600576714025916
[11] Li F, Parnell S R, Bai H, Yang W, Hamilton W A, Maranville B B,Ashkar R, Baxter D V, Cremer J T and Pynn R 2016 *J. Appl. Cryst.* **49**, 55-63. http://dx.doi.org/10.1107/S1600576715021573
[12] Sales M, Plomp J, Habicht K, Tremsin A, Bouwman W G, Strobl M 2016 *Rev. Sci. Instrum.* **87**, 063907. http://dx.doi.org/10.1063/1.4954727
[13] Strobl M, Manke I, Kardjilov N, Hilger A, Dawson M and Banhart J 2009 *J. Phys. D: Appl. Phys.* **42** 243001 (21pp). doi:10.1088/0022-3727/42/24/243001
[14] Andersson R, van Heijkamp L F, de Schepper I M and Bouwman W G 2008 *J. Appl. Crystallogr.* **41** (5), 868-885.URL: http://dx.doi.org/10.1107/S0021889808026770
[15] Krouglov T, de Schepper I M, Bouwman W G and Rekveldt M T 2003 *J. Appl. Crystallogr.* **36** (1), 117-124. URL: http://dx.doi.org/10.1107/S0021889802020368
[16] Strobl M, Betz B, Harti R P, Hilger A, Kardjilov N, Manke I, Gruenzweig C 2016 *J. Appl. Cryst.* **49** doi:10.1107/S1600576716002922.
[17] van Well A 1992 *Phys. B* **180-181**, 959-961. http://dx.doi.org/10.1016/0921-4526(92)90521-S
[18] Tremsin A, Vallerga J, McPhate J and Siegmund O 2015 Nucl. Instr. Meth. A **787**, 20-25. http://dx.doi.org/10.1016/j.nima.2014.10.047
[19] Tremsin A S 2012 Neutron News **23** (4), 35-38. http://dx.doi.org/10.1080/10448632.2012.725341
[20] Kaestner A, Lehmann E and Stampanoni M 2008 *Adv. Water Resour.* **31** (9), 1174-1187. http://dx.doi.org/10.1016/j.advwatres.2008.01.022
[21] Kinning D J and Thomas E L 1984 *Macromolecules* **17** (9), 1712-1718. http://dx.doi.org/10.1021/ma00139a013

IOP Conf. Series: Journal of Physics: Conf. Series **862** (2017) 012027 doi:10.1088/1742-6596/862/1/012027

Magnetic properties and spin structure of MnO single crystal and powder

X Sun[1], E Feng[2], Y Su[2], K Nemkovski[2], O Petracic[1], T Brückel[1]
[1] Jülich Centre for Neutron Science JCNS and Peter Grünberg Institut PGI, JARA-FIT, Forschungszentrum Jülich GmbH, Leo-Brandt Str. 52425 Jülich, Germany.
[2] Jülich Centre for Neutron Science JCNS at Heinz Maier-Leibnitz Zentrum MLZ, Forschungszentrum Jülich GmbH, Lichtenberg Str. 1, 85748 Garching, Germany.

Email: x.sun@fz-juelich.de

Abstract. Zero field cooled (ZFC)/Field Cooled (FC) magnetization curves of a bulk MnO single crystal show a peculiar peak at low temperatures (~ 40 K) similar to the low temperature peak observed in MnO nanoparticles. In order to investigate the origin of this peak, the spin structure of a MnO single crystal has been studied and compared with a single phase powder sample using magnetometry and polarized neutron scattering. Both magnetometry and polarized neutron diffraction results confirm the antiferromagnetic (AF) phase transition at the Néel temperature T_N of 118 K, in both powder and single crystal form. However, the low temperature peak in the ZFC/FC magnetization curves is not observed in single phase MnO powder. To better understand the observed behavior, ac susceptibility measurements have been employed. We conclude that the clear peak in the magnetic signal from the single crystal originates from a small amount of ferrimagnetic (FiM) Mn_2O_3 or Mn_3O_4 impurities, which is grown at the interfaces between MnO crystal twins.

1. Introduction

As one of the first studied materials using neutron scattering, bulk MnO is a textbook AF with a rocksalt crystal structure [1]. A paramagnetic (PM)-to-AF phase transition at the Néel temperature of 118 K is usually observed in bulk MnO [1,2]. However, in recent studies on MnO nanoparticles the expected feature at the Néel temperature is absent in ZFC/FC magnetization curves, but instead a low temperature peak at ~25 K is found [3–6]. This phenomenon in particle systems was explained by either superparamagnetism, superspin glass behavior, diluted AF states, finite size or surface effects [7–11]. However, we observe a similar peak at ~ 40 K in the ZFC/FC curves of a MnO single crystal.

One can speculate that this feature may either originate from a nanosize effect as in the nanoparticles or intrinsically from the frustrated superexchange interactions in MnO. One should note that the AF spin state of MnO is still not well understood despite numerous investigations [1,12–17]. The magnetic moments of Mn^{2+} are reported to align parallel within its (111) planes, while magnetic moments between neighboring planes arrange antiparallel [1,16]. However, the spin direction is reported to be perpendicular [14,16,17] to the [111] direction in contrast to common text book knowledge [18]. Moreover, a similar low temperature anomaly has already been observed in powder [19] and polycrystalline MnO [15]. In order to shed light on this low temperature peak in the magnetometry, we studied the magnetic behavior on a MnO single crystal using dc magnetometry, ac susceptometry and polarized neutron scattering. The results are compared with annealed MnO powder.

For the MnO powder a specific annealing procedure (see below) guaranteed single phase MnO with no Mn_2O_3 and Mn_3O_4 impurities.

2. Experimental

The commercially available MnO single crystal was obtained from Surface Net GmbH with dimensions 5 x 5 x 2 mm^3. From a pre-characterization using a Laue diffraction camera we observe two to three crystal twins. Further structural characterizations using X-ray diffractometers confirm the fcc crystal structure of MnO at room temperature. Moreover, no reflection from other phases is observed. The MnO powder (Alfa Aesar GmbH, 99.99%) was heated at 920°C for 150 hours in vacuum to remove nano-sized particles and potentially existing impurity phases Mn_3O_4 and Mn_2O_3. Subsequently, the powder was heated to 1000°C for several hours to avoid oxidation after cooling down to room temperature [14]. Magnetometry and ac susceptibility measurements have been performed using a SQUID (Superconducting Quantum Interference Device) magnetometer and a PPMS (Physical Property Measurement System) from Quantum Design. The ac excitation amplitude was 0.08 mT. Polarized neutron scattering was performed at the Diffuse Neutron Scattering (DNS) instrument at the Heinz Maier-Leibnitz Zentrum (MLZ) [20,21] with λ = 3.3 Å for powder and 4.2 Å for the single crystal.

3. Results and discussions

ZFC/FC magnetization curves of the MnO single crystal are shown in Fig. 1. Besides the expected AF-to-PM phase transition at T_N = 118 K, an additional peak at T_P (peak temperature) $\approx$ 40 K is observed. A similar peak was observed previously in both MnO nanoparticles [3–6] and powder [19] and was explained by superparamagnetism or spin-glass behavior. However, the ZFC/FC curves on the MnO single crystal in Fig. 1 rather show a FM-like behavior, i.e. a steep drop above T_p and an overall shape of a typical FM order parameter. One should note that X-ray diffraction and X-ray fluorescence analysis on this sample does not show any impurity.

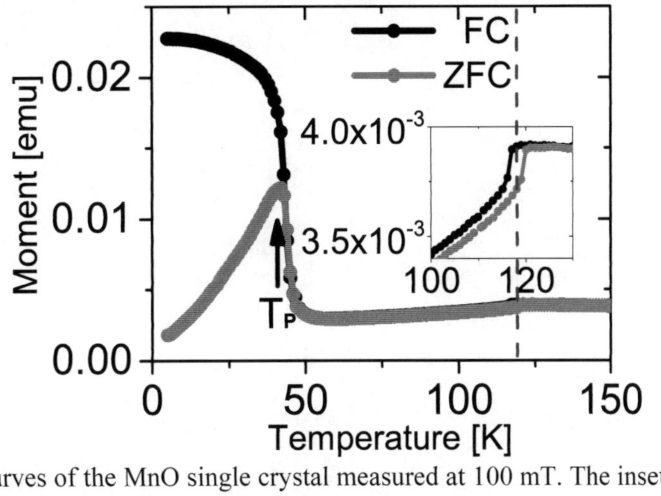

Figure 1: ZFC/FC curves of the MnO single crystal measured at 100 mT. The inset shows the enlarged view near the Néel temperature of MnO at 118 K.

In order to study the origin of this low temperature peak at 40 K in the MnO single crystal, polarized neutron diffraction is performed (Fig. 2). The orientation of the single crystal is fixed on the sample holder so that the $(1\bar{1}0)$ axis is perpendicular to the scattering plane. According to ref. [22,23], only magnetic contributions with spin incoherent background are measured in the spin-flip (sf) channel and the full nuclear coherent contribution with a little bit of incoherent background are measured in the non-spin-flip (nsf) channel, when the polarization ($\vec{P}$) of the neutrons is parallel to the scattering vector ($\vec{Q}$). After corrections of the background, flipping ratio and detector efficiency,

Fig. 2(a) shows the magnetic contribution measured at 4 K. Fig. 2(b) shows the temperature dependence of the magnetic $\left(\frac{\bar{1}}{2}\,\frac{\bar{1}}{2}\,\frac{\bar{1}}{2}\right)$ Bragg peak. The AF order parameter of MnO vanishes as expected at $T_N = 118$ K. However, no magnetic phase transition at $T_P \approx 40$ K is found. The temperature dependence of the magnetic $\left(\frac{\bar{1}}{2}\,\frac{\bar{1}}{2}\,\frac{\bar{3}}{2}\right)$ Bragg peak shows a similar behavior like the magnetic $\left(\frac{\bar{1}}{2}\,\frac{\bar{1}}{2}\,\frac{\bar{1}}{2}\right)$ Bragg peak (data not shown). However, at the position of the $(\bar{1}\ \bar{1}\ \bar{1})$ nuclear peak of MnO ($Q = 2.5$ Å^{-1}) a small feature is observed at about 40 K before background, flipping ratio and detector efficiency corrections (Fig. 2(c)). This $(\bar{1}\ \bar{1}\ \bar{1})$ nuclear peak is located near the expected Q-value of the (2 0 2) and (2 1 1) Bragg peaks of ferrimagnetic (FiM) Mn_3O_4. Taking the broadening of the Bragg peak into account, this might be due to the FiM-to-PM transition of Mn_3O_4 (Curie temperature ≈ 42 K) [24,25]. One should note that for a single crystal a higher oxidation state (Mn_2O_3) is not expected. The drop observed at 118 K in Fig. 2(c) is very likely due to the imperfect separation from the nuclear component. Another explanation of this drop is the extinction effect because of the structural distortion of MnO below the Néel temperature.

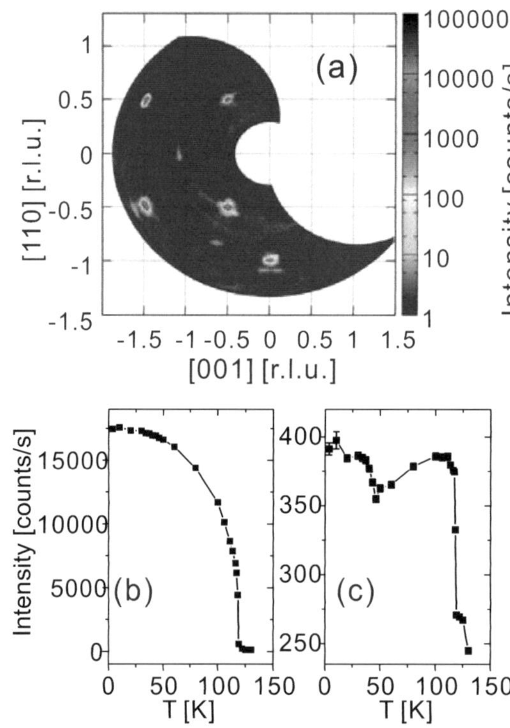

Figure 2: Polarized neutron diffraction of (a) the MnO single crystal (sf channel) measured at 4 K. (b) The intensity of the magnetic $\left(\frac{\bar{1}}{2}\,\frac{\bar{1}}{2}\,\frac{\bar{1}}{2}\right)$ Bragg peak of MnO single crystal as function of temperature. (c) The intensity of the sf-channel at the position of $(\bar{1}\ \bar{1}\ \bar{1})$ Bragg peak of the MnO single crystal as function of temperature.

Further measurements have been performed on the MnO single crystal to shed light on the behavior of the low temperature peak in the ZFC/FC measurements. AC-susceptibility results show two peaks in the $\chi'(T)$ curve (Fig. 3). Both peaks, the one at 40 K and the second at 118 K have no frequency dependence, which is expected for FM/FiM or AF systems [2]. The peak at 118 K indicates the phase transition of bulk MnO from the AF to the PM state. The peak at 40 K very likely marks the PM to FiM transition of Mn_3O_4 since the temperature value matches the Curie temperature of 42 K.

However, this feature at 40 K is not observed away from the $(\bar{1}\,\bar{1}\,\bar{1})$ peak of MnO. Therefore it is possibly due to intergrowth of Mn_3O_4 inside MnO at the interfaces between MnO crystal twins.

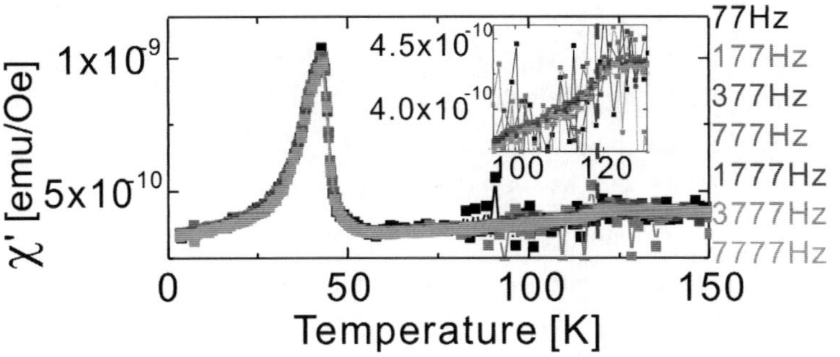

Figure 3: AC-susceptibility measurements of MnO single crystal. The inset is an enlarged view near the Néel temperature of MnO.

From the above measurements we conclude that the MnO single crystal contains Mn_3O_4 impurities. We assume that the imperfect crystal has grain boundaries and/or dislocations at which oxidation toward Mn_3O_4 occurred. These impurities are invisible in X-ray scattering due to a negligible amount of the Mn_3O_4. Interestingly, magnetometry is able to reveal these impurities. This is due to the fact that a FM or FiM subsystem has a much larger magnetic moment compared to a perfectly ordered AF phase, particularly at relatively small applied magnetic fields as in our case. The amount of Mn_3O_4 impurities is estimated to be about 0.034 mg assuming the 40 K peak in the ZFC curve originates completely from Mn_3O_4. This is about 0.3% of the sample mass.

To further investigate whether the low temperature peak originates from a higher oxidation state or from intrinsical magnetic behavior of MnO alone, a single phase MnO powder sample (after the annealing procedure) was compared with the single crystal results.

Fig. 4(a) shows the ZFC/FC magnetization curves obtained on this MnO powder. As expected only a feature at the Néel temperature of MnO (118 K) is observed. Fig. 4(b) displays the AC-susceptibility measurements of MnO powder. As in the ZFC/FC curves, only one peak indicating the Néel temperature of MnO is found. In addition, this peak is frequency independent as much as one can estimate from the data. The relatively large noise is due to the extremely small signal as expected from a perfect AF at zero applied dc field and hence perfectly compensated AF sublattices.

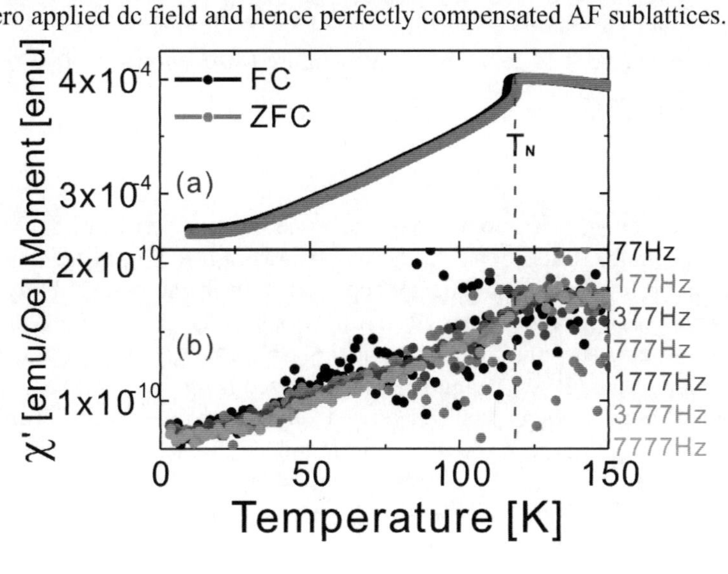

Figure 4: (a) ZFC/FC magnetization curves of MnO powder measured at 100 mT. (b) AC-susceptibility data, χ' on MnO powder performed at various frequencies. The Néel temperature T_N of MnO is marked by the dashed line.

Polarized neutron diffraction was also performed on MnO powder (Fig. 5). Using xyz-polarization analysis [22,23], we obtained separated scattering intensities which are shown in Fig. 5(a). The magnetic $\left(\frac{1}{2}\,\frac{1}{2}\,\frac{1}{2}\right)$ and $\left(\frac{3}{2}\,\frac{1}{2}\,\frac{1}{2}\right)$ Bragg peaks can be found at Q = 1.25 $Å^{-1}$ and Q = 2.3 $Å^{-1}$, respectively. The nuclear (111) peak is located at Q = 2.5 $Å^{-1}$. The AF-to-PM phase transition of MnO is found at $T_N \approx 120$ K from the magnetic $\left(\frac{1}{2}\,\frac{1}{2}\,\frac{1}{2}\right)$ (Fig. 3(b)) Bragg peak. In addition, no feature near 40 K is observed in both magnetic Bragg peaks which agrees with the single crystal results.

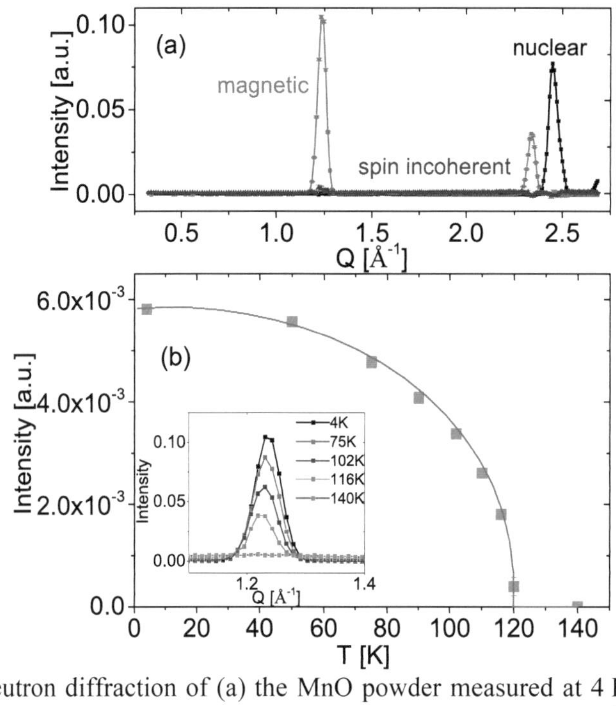

Figure 5: Polarized neutron diffraction of (a) the MnO powder measured at 4 K. (b) The intensity of the magnetic $\left(\frac{1}{2}\,\frac{1}{2}\,\frac{1}{2}\right)$ Bragg peak as function of temperature (inset shows the $\left(\frac{1}{2}\,\frac{1}{2}\,\frac{1}{2}\right)$ peak at various temperatures). The red line is a guide to the eye.

4. Conclusion

Polarized neutron scattering, magnetometry and ac susceptometry results of the MnO single crystal show an anomaly at about 40 K additional to the well-known AF to PM phase transition of MnO at 120 K. This feature at 40 K shows no frequency dependence in the ac susceptibility results, which is different from the spin-glass behavior reported in a previous study [19]. We conclude that the MnO single crystal contains ferrimagnetic Mn_3O_4 impurities. Interestingly, the Mn_3O_4 impurity is only observed at the position of the $(\bar{1}\,\bar{1}\,\bar{1})$ Bragg peak in polarized neutron diffraction. This indicates that the higher oxidation phase Mn_3O_4 crystallizes very likely at the interfaces between crystal twins. These impurities cannot be observed by X-ray scattering alone due to a negligible amount of Mn_3O_4. Magnetometry and susceptometry can reveal these impurities since a ferro- or ferrimagnetic subsystem has a much larger magnetic moment compared to a perfectly ordered antiferromagnetic subsystem. Moreover, we exclude that the low temperature peak in magnetometry is due to an intrinsic magnetic behavior of MnO since the single phase MnO powder does not show this feature but only regular antiferromagnetic behavior.

Acknowledgments
We would like to thank Hailey Williamson, Pankaj Thakuria, Thomas Müller, Paul Hering, Karen Friese and Jörg Perßon for their help with the structural characterizations of the single crystal and the annealing procedure of the powder sample. We would also like to thank Berthold Schmitz for the technical supports with the magnetometry measurements. We thank ZEA-3: Analytik in Forschungszentrum Jülich for the X-ray fluorescence analysis of the MnO single crystal.

References

[1] Shull C G, Strauser W A and Wollan E O 1951 Neutron Diffraction by Paramagnetic and Antiferromagnetic Substances *Phys. Rev.* **83** 333–45

[2] McGuire T R and Happel R J 1959 The magnetic susceptibility of an MnO single crystal *J. Phys. Radium* **20** 424–6

[3] Chatterji T, Su Y, Iles G N, Lee Y-C, Khandhar A P and Krishnan K M 2010 Antiferromagnetic spin correlations in MnO nanoparticles *J. Magn. Magn. Mater.* **322** 3333–6

[4] Wang C H, Baker S N, Lumsden M D, Nagler S E, Heller W T, Baker G A, Deen P D, Cranswick L M D, Su Y and Christianson A D 2011 Antiferromagnetic order in MnO spherical nanoparticles *Phys. Rev. B* **83** 214418

[5] Lee Y-C, Pakhomov A B and Krishnan K M 2010 Size-driven magnetic transitions in monodisperse MnO nanocrystals. *J. Appl. Phys.* **107** 9E124

[6] López-Ortega A, Tobia D, Winkler E, Golosovsky I V, Salazar-Alvarez G, Estradé S, Estrader M, Sort J, González M A, Suriñach S, Arbiol J, Peiró F, Zysler R D, Baró M D and Nogués J 2010 Size-dependent passivation shell and magnetic properties in antiferromagnetic/ferrimagnetic core/shell MnO nanoparticles *J. Am. Chem. Soc.* **132** 9398–407

[7] Bedanta S, Barman A, Kleemann W, Petracic O and Seki T 2013 Magnetic Nanoparticles: A Subject for Both Fundamental Research and Applications *J. Nanomater.* **2013** 1–22

[8] Batlle X and Labarta A 2002 Finite-size effects in fine particles: magnetic and transport properties *J. Phys. D. Appl. Phys.* **35** R15–R42

[9] Petracic O, Chen X, Bedanta S, Kleemann W, Sahoo S, Cardoso S and Freitas P P 2006 Collective states of interacting ferromagnetic nanoparticles *J. Magn. Magn. Mater.* **300** 192–7

[10] Benitez M J, Petracic O, Tüysüz H, Schüth F and Zabel H 2011 Fingerprinting the magnetic behavior of antiferromagnetic nanostructures using remanent magnetization curves *Phys. Rev. B* **83** 134424

[11] Ebbing A, Hellwig O, Agudo L, Eggeler G and Petracic O 2011 Tuning the magnetic properties of Co nanoparticles by Pt capping *Phys. Rev. B* **84** 012405

[12] Li Y-Y 1955 Magnetic Moment Arrangements and Magnetocrystalline Deformations in Antiferromagnetic Compounds *Phys. Rev.* **100** 627–31

[13] Roth W L 1958 Magnetic structures of MnO, FeO, CoO, and NiO *Phys. Rev.* **110** 1333

[14] Shaked H, Faber, Jr. J and Hitterman R L 1988 Low-temperature magnetic structure of MnO: A high-resolution neutron-diffraction study *Phys. Rev. B* **38** 11901–3

[15] Woodfield B F, Shapiro J L, Stevens R, Boerio-Goates J and Wilson M L 1999 Critical phenomena at the antiferromagnetic transition in MnO *Phys. Rev. B* **60** 7335–40

[16] Goodwin A L, Tucker M G, Dove M T and Keen D A 2006 Magnetic Structure of MnO at 10 K from Total Neutron Scattering Data *Phys. Rev. Lett.* **96** 047209

[17] Frandsen B A and Billinge S J L 2015 Magnetic structure determination from the magnetic pair distribution function (mPDF): ground state of MnO *Acta Crystallogr. Sect. A, Found. Adv.* **71** 325–34

[18] Blundell S 2001 *Magnetism in Condensed Matter* (Oxford: Oxford University Press)

[19] Hauser J J and Waszczak J V 1984 Spin-glass transition in MnO *Phys. Rev. B* **30** 5167–71

[20] Schweika W and Böni P 2001 The instrument DNS: polarization analysis for diffuse neutron scattering *Phys. B Condens. Matter* **297** 155–9

[21] Su Y, Nemkovskiy K and Demirdiş S 2015 DNS: Diffuse scattering neutron time-of-flight spectrometer *J. large-scale Res. Facil.* **A27** 1–4

[22] Schärpf O and Capellmann H 1993 The XYZ-Difference Method with Polarized Neutrons and the Separation of Coherent, Spin Incoherent, and Magnetic Scattering Cross Sections in a Multidetector *Phys. stat. sol.* **135** 359–79

[23] Stewart J R, Deen P P, Andersen K H, Schober H, Barthélémy J-F, Hillier J M, Murani A P, Hayes T and Lindenau B 2009 Disordered materials studied using neutron polarization analysis on the multi-detector D7 *J. Appl. Crystallogr.* **42** 69–84

[24] Winkler E, Sinnecker J P, Novak M A and Zysler R D 2011 Dynamic study of the internal magnetic order of Mn3O4 nanoparticles *J. Nanoparticle Res.* **13** 5653–9

[25] Boucher B, Buhl R and Perrin M 1971 Magnetic Structure of Mn3O4 by Neutron Diffraction *J. Appl. Phys.* **42** 1615–7

IOP Conf. Series: Journal of Physics: Conf. Series **862** (2017) 012027 doi:10.1088/1742-6596/862/1/012027

Supporting information

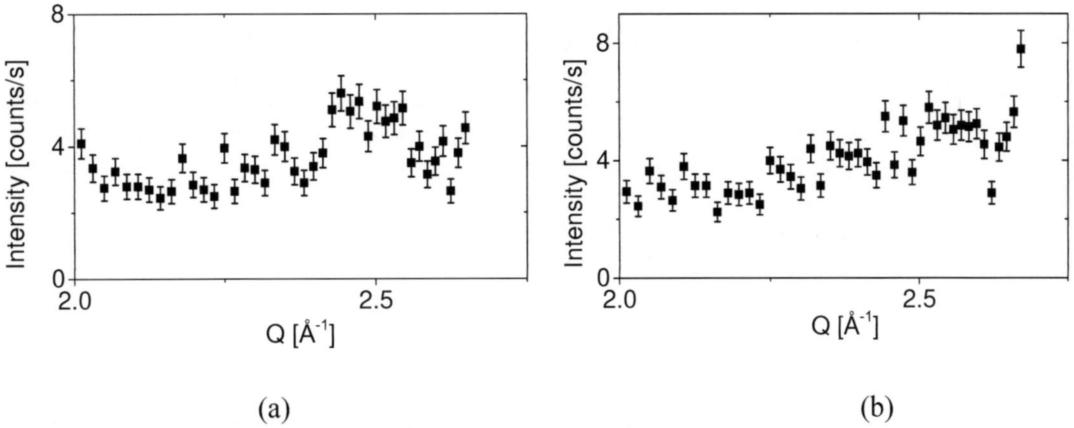

Fig. 1: Q-dependence measured at the angle of (a) magnetic $(\frac{1}{2}\frac{1}{2}\frac{1}{2})$ peak (71°) and (b) 86° at 4 K. No obvious peak can be observed within the error of the background at Q = 2.5 Å^{-1}.

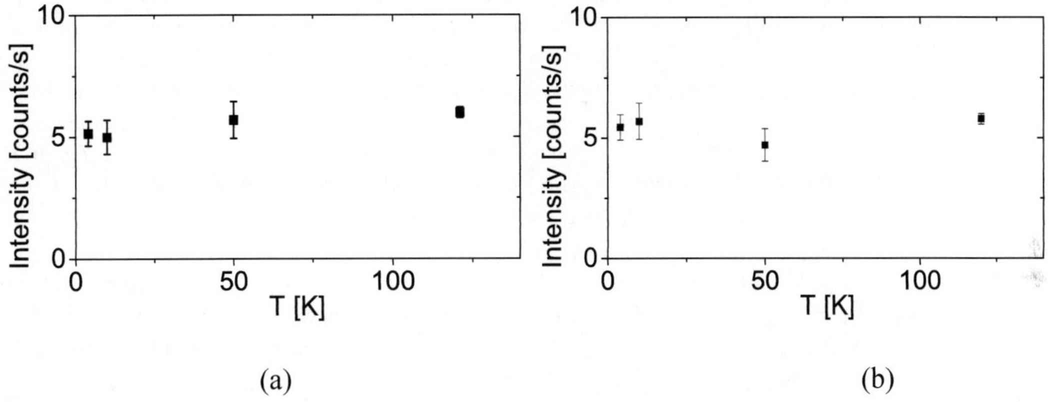

Fig. 2: Temperature dependence of the intensity at Q = 2.5 Å^{-1} away from the $(\bar{1}\bar{1}\bar{1})$ nuclear peak of MnO at (a) 71° and (b) 86°.

New compact neutron supermirror transmission polarizer

V G Syromyatnikov [1,2], V M Pusenkov [1]

[1] Petersburg Nuclear Physics Institute, National Research Center "Kurchatov Institute", Gatchina, St. Petersburg, 188300, Russia

[2] Department of Physics, St. Petersburg State University, Ulianovskaya 1, Petrodvorets, St.Petersburg, 198504, Russia

E-mail: syromyatnikov_vg@pnpi.nrcki.ru

Abstract. A new compact neutron supermirror transmission polarizer is suggested. The polarizer consists of a set of plates transparent to neutrons placed in the magnet gap. There are no air gaps between the plates. Polarizing supermirror coating without absorbing underlayer is deposited on the polished surfaces of the plates. Magnetic and nonmagnetic layers of the supermirror coating as well as the material of the plates have nearly equal neutron-optical potentials for spin-down neutrons. There is a considerable difference between neutron-optical potentials of layers in the supermirror structure for spin-up neutrons. As a result, spin-up neutrons reflect from the supermirror coating and deviate from their initial trajectories whereas spin-down neutrons do not practically reflect from the coating and, consequently, do not deviate from their initial trajectories. Thus, spin-down neutrons dominate near the axis of distribution of intensity on the angle for the beam transmitted through this polarizer, i.e., the beam is substantially polarized. Application is discussed of this polarizer in a research facility for small angle scattering of monochromatic neutrons with wavelengths $\lambda = 4.5 \div 20$ Å. The polarizing cross section of the beam of this facility is 30x30 mm^2. Calculated parameters are presented of a polarizer on silicon plates with supermirror CoFe/TiZr (m = 2) coating. The suggested polarizer is compared with solid state bender, S-bender and widely known transmission neutron polarizer V-cavity in the same spectral range. Two polarizers are used to cover the wavelength range $\lambda = 4.5 \div 20$ Å: the first one whose length is 50 мм covers the range $\lambda = 4.5 \div 10$ Å and the second one whose length is 21.2 мм covers the range $\lambda = 10 \div 20$ Å. The length of each of these polarizers is more than 30 times smaller than that of V-cavity! On the other hand, basic parameters of the proposed polarizer, polarization of the beam falling on the sample P and transmission coefficient T^- of the main spin component, exceed those of V-cavity. T^- = 0.8 – 0.9 for both polarizers and for each wavelength range. Polarization P is very high. P is better than -0.99 for wavelength range $\lambda = 12.5 \div 20$ Å at the beam divergence of 24 mrad.

1. Introduction

Beam former is used in Small-Angle Neutron Scattering (SANS) facilities for monochromatization and collimation of neutron beams. In order SANS facilities be suitable for investigation of magnetic properties of materials, the beam former should also be able to polarize the beam.

Such a beam former-polarizer is known which is based on spatial spin resonance of neutron beam [1, 2]. It is used in some SANS facilities (see e.g., [3, 4]). However, this beam former has significant drawbacks: bulkiness and complexity in operation (because it contains four elements: polarizer, magnetic resonator, spin-flipper, and analyzer). Monochromatic beam produced by this polarizer has a significant non-monochromatic background caused by imperfect polarization by polarizer and analyzer. Besides, it is difficult to switch from polarized to unpolarized neutrons in SANS facilities with such a beam former.

A beam former is also used in SANS facilities for polarization of neutron beams which contains bender - multichannel polarizing supermirror neutron guide described in [5]. Neutron beam deviates from its initial trajectory when passing through the polarizing bender. Due to this circumstance, much time is needed for rebuilding of the SANS facility when switching from polarized to unpolarized neutrons. The beam former presented in Fig. 1 has become the most widely spread one in SANS facilities in recent 15 years. A neutron beam having a broad spectral distribution falls on a sample after successive passing through the velocity selector (1), the transmission polarizer (2), and the collimation system (3).

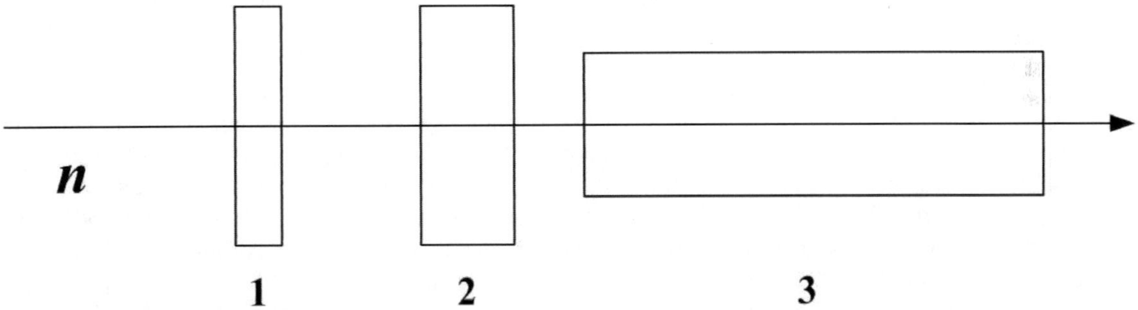

Figure 1. Sketch of beam former for SANS facility. 1 - velocity selector, 2 - transmission polarizer, 3 - collimation system.

This beam former allows dealing with both polarized and unpolarized neutrons so that the trajectory of the beam does not change upon switching between these two regimes. This circumstance is in the origin of the wide spreading of the beam former of this type.

Variants of use are described in [6] of the V-cavity, S-benders, and elliptic neutron guides as transmission polarizers.

The transmission polarizer is frequently done in the form of V-cavity. The V-cavity of one of the first types is described in [7] which is used in the range of wavelengths $\lambda = 4.5 \div 20$ Å. Basic characteristics of this polarizer are the following: the coefficient of transmission of neutrons with the main spin component of the beam decreases gradually from 0.7 for 6 Å to 0.2 for 18 Å; the polarizing efficiency is not worth than 0.93 for wavelengths in the range 4 Å < λ < 10 Å and it decreases gradually to 0.8 upon wavelength increasing to $\lambda = 18$ Å. The beam axis does not change when passing through the polarizer. The polarizer V-cavity consists of a straight neutron guide and two equal long polarizing supermirrors on silicon substrates placed inside this neutron guide. Walls of the neutron guide are covered by natural Ni (m = 1). Each mirror contains a set of rectangular polished silicon plates which are lined up and which edges are pressed to each other. A polarizing supermirror *CoFe/Si* coating (m = 2) covers the plates. These long polarizing supermirrors are crossed with each other. The axis of the beam (of the neutron guide) forms with each of them a small angle θ = 8.33 mrad. The angle between supermirrors is equal to 2θ. Angle θ is defined as $\theta = \alpha_c \cdot \lambda_{\min}$, where α_c is a critical angle of the supermirror and $\lambda_{\min}$ = 4.8 Å is the shortest wavelength of the neutron spectrum. The V-cavity is placed in a saturating magnetic field. The set of plates is cased. Neutrons having (+) spin component (i.e., neutrons with spin component oriented parallel to the guiding magnetic field) are reflected from one of the plates of the V-cavity by an angle smaller than the critical one. Then, they are reflected from walls of the neutron guide with $m = 1$ by an angle greater than the critical one. As a result, the divergence increases of the neutron beam with (+) spin component that leads to their absorption either after falling on the wall by an angle greater than the critical one, or after reflection from the silicon mirror, or in the collimation system. Thus, the number of neutrons with (+) spin component is much smaller in the beam passed through the V-cavity than that with (-) spin component (i.e., the passed neutron beam has a negative polarization). The extensive length of V-cavity $L = 1800$ mm is defined as $L = d/2\theta$, where $d = 30$ mm is the neutron beam width used in the facility and $\theta = 8.33$ mrad. Both the straight neutron guide whose walls are covered by natural nickel ($m = 1$) and the magnetic system for the V-cavity have the same length L. Taken together, they form a heavy, bulky and lengthy unit which is difficult to handle. For example, it is necessary to remove the entire bulky polarizer with magnetic system in order to transfer to nonmagnetic system investigation. Besides, it is difficult to make the polarizer because it is difficult to line up silicon plates on such a large length of 1800 mm and to fix them. Moreover, such a polarizer is placed inside expensive and lengthy (~ 1800 mm) magnetic system.

Supermirror neutron polarizers are known on substrates which are transparent for neutrons [8-13] including transmission polarizers. A supermirror transmission solid state bender on silicon substrates is discussed in [12]. The bender consists of a stack of thin silicon wafers, which thickness is 0.16 mm, coated with polarizing FeCo/Si supermirrors ($m = 2.3$). Being inserted in an unpolarized neutron beam, this device works as a spin splitter by transmitting the spin down component without any change in its flight path and reflecting the spin up component. The bender cross section is 20*100 mm, where the captured beam width is 20 mm and its height is 100 mm. The bender length is 54 mm. The bender was tested using a monochromatic beam having the wavelength of 4.72 Å together with the solid state collimator. Rocking curves were measured for the neutron intensity transmitted through both the bender and the collimator. Triangular transmission with a maximum of 54% of the spin down component and a mean flip ratio above 100 were reached over an angular range with a FWHM of 0.37 degree. A spin down neutrons can be extracted with the angle width of 1 degree with the mean flip ratio of the order of 120 ($P = 0.983$). The transmission coefficient of such a beam through the bender was found to be of the order of 0.67.

A supermirror solid state *S*-bender is presented in [13]. This polarizer does not deflect the axis of the beam passing through it as well. An 80 mm long polarizing *S*-bender with a large cross-section of 30*100 mm has been simulated, built and tested. The results of the experiment show for the wavelength of 4.4 Å a homogenous polarization of the order of $P = 0.987$ across the bender cross-section and a high transmission 68 % of the spin up component of the neutron beam. The angle divergence of the beam at the end of this polarizer was 1.85 degrees. The *S*-bender length is 80 mm. The bender design has the advantage of delivering the polarized beam in the direction of the incoming beam.

The aim of the present work is to work out a neutron transmission supermirror polarizer of a new type for a SANS facility which provides a high polarizing efficiency and the transmitting coefficient of the main spin component of the beam. The new polarizer must be substantially more compact, cheaper, and easy to handle and use.

2. Design and principle of operation of the suggested polarizer

The polarizer design (side view) is shown in Fig. 2. A set of neutron transparent plates P is sandwiched between two broken polished metal surfaces 1 and 2 (the punch and the matrix) so that each plate of thickness d forms a broken asymmetric neutron guide channel with supermirror walls. This channel is shown in Fig. 3. As is shown in the figure, the channel height (plate thickness) is d, the front part of the channel (plate) has length L_1 and it makes with the beam axis angle θ_1, and the corresponding parameters the output part of the channel are L_2 and θ_2, respectively. The width of the channel along the axis perpendicular to the plane of the figure is equal to L_3 (not shown in the figure) is determined by the width of the incident beam.

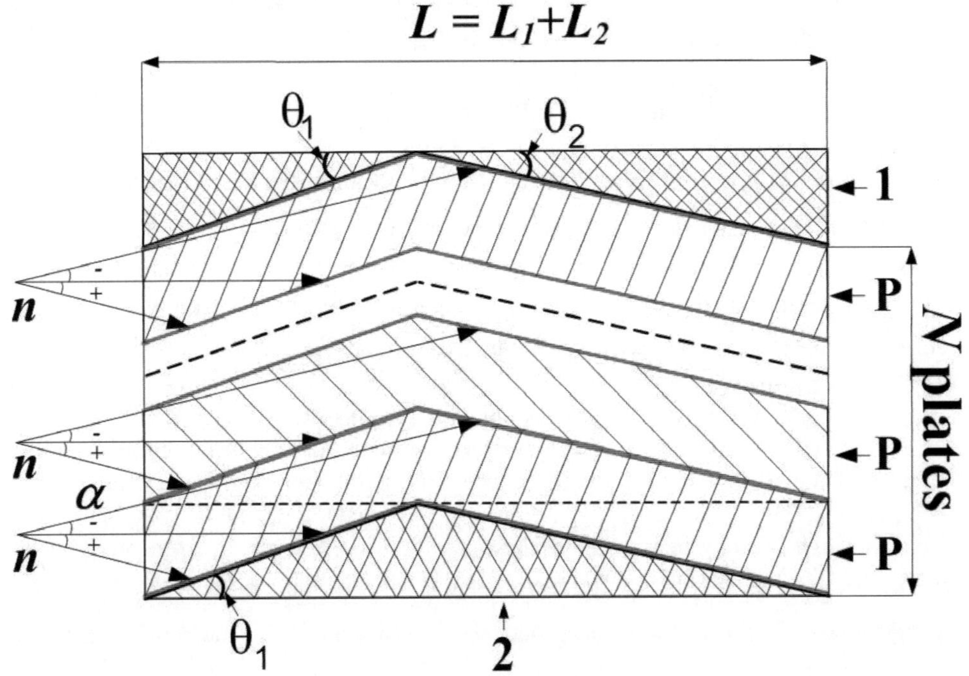

Figure 2. General view of the proposed compact neutron transmission supermirror polarizer.

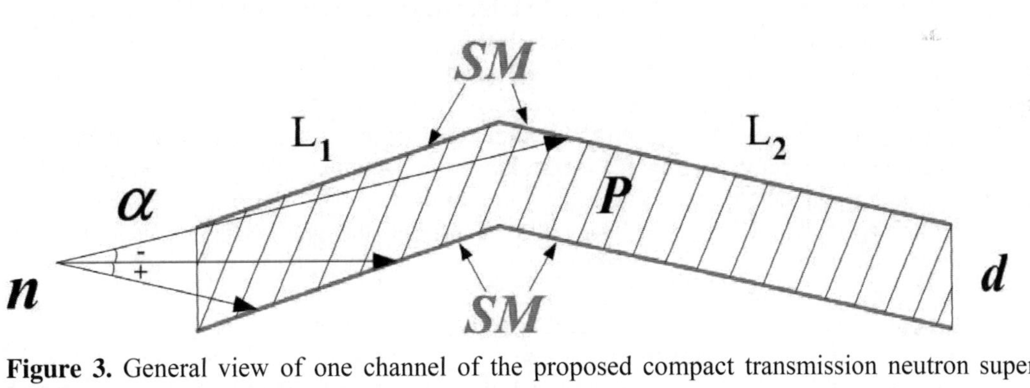

Figure 3. General view of one channel of the proposed compact transmission neutron supermirror polarizer.

A divergent neutron beam ***n*** comes to the entrance of each channel (Figs. 2 and 3). Entrance angle $\boldsymbol{\alpha}$ of neutrons is counted off from horizontal line. The sign of $\boldsymbol{\alpha}$ is defined as it is shown in Figs. 2 and 3. A polarizing supermirror coating <SM> covers all polished surfaces of width ***d***. For neutrons with (-) spin component of the beam, the neutron-optical potentials are close to each other of the supermirror coating layers and the material of the plate so that the critical angle α_c^- is close to zero for the boundary "material-supermirror". Slight exceeding is allowed of the plate potential over potentials of layers of the supermirror. For neutrons with (+) spin component of the beam, the neutron-optical potentials of its layers differ significantly from each other so that the corresponding critical angle α_c^+ is great for the same boundary. Polarizer plates are in the field of the magnetic system of the polarizer.

This field lies in supermirror plane and perpendicular to the plane of the figure (Fig. 3). The applied field saturates the magnetic layers of supermirrors deposited on the plates. The number of broken asymmetric channels (plates) and their width L_3 are defined by the required beam cross section used in the facility which is equal to $h \times L_3$, where h is the beam height. *N* such channels are used which are pressed to each other without air spaces, where *N* is defined as $h = N \cdot d$.

Channel parameters are given by the following relationships:

$\theta_1 = \lambda \cdot \alpha_c^+$, where λ is the smallest wavelength of neutrons in the beam formed by velocity selector; α_c^+ is the critical angle of the supermirror for (+) spin component of the beam; γ is required divergence of the beam coming to the sample through the polarizer and the collimation system; $\theta_2 = \theta_1 - \gamma$; d is plates thickness made from a material which absorbs neutrons weakly (e.g., silicon, quartz, sapphire etc.); $L_1 = d / \theta_1$; $L_2 = d / \theta_2$; L_1 *и* L_2 are plates lengths.

The device operates as follows.

The neutron flux coming from the neutron guide and having wide angular and spectral distributions enters the entrance of the velocity selector, where a monochromatic beam is formed with wavelength λ and with a width $\Delta\lambda / \lambda$ defined by the selector. Then, the resulting beam comes to the entrance of the proposed polarizer. Trajectories of neutrons with (-) spin component do not practically change when coming through the polarizer because these neutrons do not reflect from boundaries "material-supermirror" (as soon as $\alpha_c^- \sim 0$). In contrast, neutrons with (+) spin component do reflect from boundaries "material-supermirror" (as soon as $\alpha_c^+ > 0$) and their trajectories change considerably.

Consequently, the number of neutrons with (-) spin component is much greater than that of neutrons with (+) spin component at the exit of the polarizer near the axis of the angle distribution of intensity. Then, the transmitted beam has negative polarization.

Neutron fluxes with (+) and (-) spin components coming from the polarizer go through the collimation system and come to the sample. Divergences of these fluxes are determined by divergence of the collimation system. The reduction is more pronounced of intensity of (+) spin component than that of (-) spin component because the divergence of flux of (+) spin component is enhanced much greatly upon coming through the polarizer. The polarization of the beam falling on the sample depends on divergence of the collimator. Then, the choice of this divergence depends on the physical problem to be solved using the facility.

3. The new compact neutron supermirror transmission polarizer on silicon

As an example of the suggested transmission polarizer, we consider the polarizer in which silicon is chosen as a material that absorbs neutrons weakly. This polarizer is planned to use in beam former for SANS-2 facility at reactor PIK. The structure *CoFe/TiZr* ($m = 2$) [14] without absorbed sublayer *TiZrGd* is used as the polarizing supermirror covering on silicon. For neutrons with (-) spin component, potentials are equal to zero of layers *TiZr* and fully magnetized layers *CoFe*, whereas the silicon has small positive potential. Thus, a small jump of potential is on the boundary between silicon and supermirror for neutrons with (-) spin component reflection from which is insignificant for such neutrons. Price is very small of polished silicon plates of width 0.3 mm and the supermirror coating *CoFe/TiZr* ($m = 2$). The beam cross section for SANS-2 is $30 \times 30\ mm^2$, i.e. $h = L_3 = 30\ mm$. At *d=0.3* mm, the number of plates (channels) *N* in polarizer is equal to 100.

Neutron flux leaving the neutron guide with nickel coating ($m = 1$) comes to the entrance of the velocity selector. Monochromatic beam with wavelength λ and $\Delta\lambda / \lambda = 0.1$ made by the velocity selector comes to polarizer.

IOP Conf. Series: Journal of Physics: Conf. Series **862** (2017) 012028 doi:10.1088/1742-6596/862/1/012028

Let us compare the suggested former with *V*-cavity for $\lambda = 4.5 \div 20$ Å. Two one-type compact polarizers of the suggested construction (for $\lambda = 4.5 \div 10$ Å and $\lambda = 10 \div 20$ Å) are needed to cover this range of wavelengths.

<u>The geometrical parameters of polarizers:</u>

1. <u>The first polarizer:</u> $\lambda = 4.5 \div 10$ Å, $d = 0.3\ mm$ - thickness of silicon plates, supermirror coating - *CoFe/TiZr* ($m - 1.81$) without sublayer *TiZrGd*, $\theta_1 = 15\ mrad$, $\theta_2 - 10\ mrad$, $L_1 = 20\ mm$, $L_2 = 30\ mm$, $\alpha_c = 3.1\ mrad/$Å (for reflection from silicon), $\gamma = 5\ mrad$.

Length of the first polarizer $L = L_1 + L_2 = 20 + 30 = 50\ mm$ that is 36 times smaller than the length of V-cavity!

2. <u>The second polarizer:</u> $\lambda = 10 \div 20$ Å, $d = 0.3\ mm$, supermirror coating - *CoFe/TiZr* ($m = 1.81$) without sublayer *TiZrGd*, $\theta_1 = 31\ mrad$, $\theta_2 = 26\ mrad$, $L_1 = 9.7\ mm$, $L_2 = 11.5\ mm$, $\alpha_c = 3.1\ mrad/$Å (for reflection from silicon), $\gamma = 5\ mrad$.

Length of the second polarizer $L = L_1 + L_2 = 9.7 + 11.5 = 21.2\ mm$ that is 85 times smaller than the length of V-cavity!

The first or the second polarizer will be placed in the beam depending on the neutron wavelength λ to be used in the facility which is defined by the velocity selector.

As it follows from calculations, the both polarizers have non-symmetric form: $L_2 > L_1$. Let us consider a symmetric system. In this case, if a neutron reflects twice from walls of the first part of the polarizer, it would reflect twice by the same angle from walls of the second part of the polarizer so that the outlet angle (with respect to the beam axis) would be equal to the inlet one. In addition, the transmission of spin-up neutrons would increase considerably for angles α near the beam axis. Hence, the polarization would decrease for this angle range. Implementation of the asymmetry in the polarizer allows eliminating this effect.

Dependence is shown in Fig. 4 for (+) spin component of the beam of the parameter λ/θ (θ is the glancing angle) of the modeling reflection coefficient from supermirror polarizing coating *CoFe/TiZr* ($m = 1.81$) evaporated on polished silicon substrates. This dependence has been used in calculations.

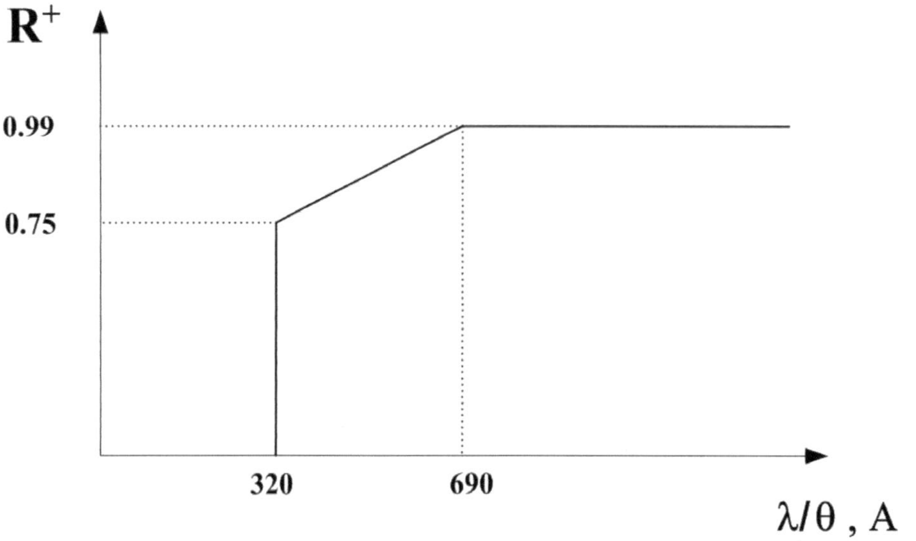

Figure 4. Modeling reflection coefficient of spin-up neutrons from supermirror polarizing coating *CoFe/TiZr* ($m = 1.81$) evaporated on polished silicon substrates for reflection from silicon in dependence of parameter λ / θ.

Parameters of polarizing supermirror CoFe/TiZr have been used in calculations. Upon reflection from such supermirror from vacuum, the cut of the curve reflectivity corresponds to $\lambda / \theta = 310$ Å whereas for the area of the total reflection λ/θ = 600 Å. Upon reflection from such supermirror from silicon, a shift of these cuts arises due to the refraction in silicon. As a result, one obtains 320 и 690 Å (Fig. 4) for the corresponding values because the critical angle in silicon α_c = 0.812 mrad/ Å or $(\lambda/\theta)_c$ = 1232 Å.

Calculations have been carried out for each polarizer of dependence on the incoming angle α of intensities at the polarizer exit of (+) I^+ and (-) I^- spin components of monochromatic beam for some neutron wavelengths. By the reason discussed above, for I^-, the beam divergence does not change upon coming through the polarizer. The beam intensity of this spin component has been just slightly reduced due to neutron absorption in silicon. Then, I^- has been large. Calculation of angle distribution of I^+ at the polarizer exit has been carried out mainly using software based on so-called "short section approximation" method which is described in [15]. This method allows to trace the trajectory of each neutron coming through practically any neutron-optical device. As a result, I^+ and I^- can be calculated with high accuracy at the exit of this device.

Calculation of I^+ is substantially more complicated than that of I^- because, apart from absorption of the beam in the silicon, a reflection (including multiple reflection) should be taken into account from supermirror walls of polarizer channels. Besides, a transmission through these walls without reflection should be taken into account of a small part of neutrons which takes place because the reflection coefficient is smaller than unity at $\lambda / \theta = 320 \div 690$ Å (see Fig. 4).

Results of calculations are shown in Fig. 5 for I^+ and I^- at $\lambda = 5.5$ Å for the first polarizer dependence on α, of intensity I at the entrance of the polarizer and polarization

$P = \dfrac{I^+ - I^-}{I^+ + I^-}$ of the transmitted beam. Neutron absorption in silicon has been taken into account. As expected, P turns out to be negative and the beam divergence of (+) spin component is increased due to the reflection from supermirror walls of the polarizer. The latter circumstance is in the origin of complicated structure of angle distribution of I^+ which has three separated peaks. As it is seen from the figure, the area $\alpha \approx -6 \div 16\ mrad$ has small amount of neutrons of (+) spin component. In contrast, the width of the angle distribution of (-) spin component does not practically change upon coming through the polarizer. As a consequence, the polarization of the transmitted beam in the angle range $\alpha \approx -6 \div 10\ mrad$ is extremely high being close to -1. Results similar to those presented in Fig. 5 have been obtained also for wavelengths $\lambda = 4.5, 6.5, 7.5, 8.5, 9.5$ Å from the working range of the first polarizer.

Results are shown in Fig. 6 of the corresponding calculations for the second polarizer at $\lambda = 12.5$ Å. It is seen that all graphics are similar to those presented in Fig. 5. Results similar to those presented in Fig. 6 have been obtained also for wavelengths $\lambda = 10.5, 11.5, 14.0, 16.0, 18.0, 20.0$ Å from the working range of the second polarizer.

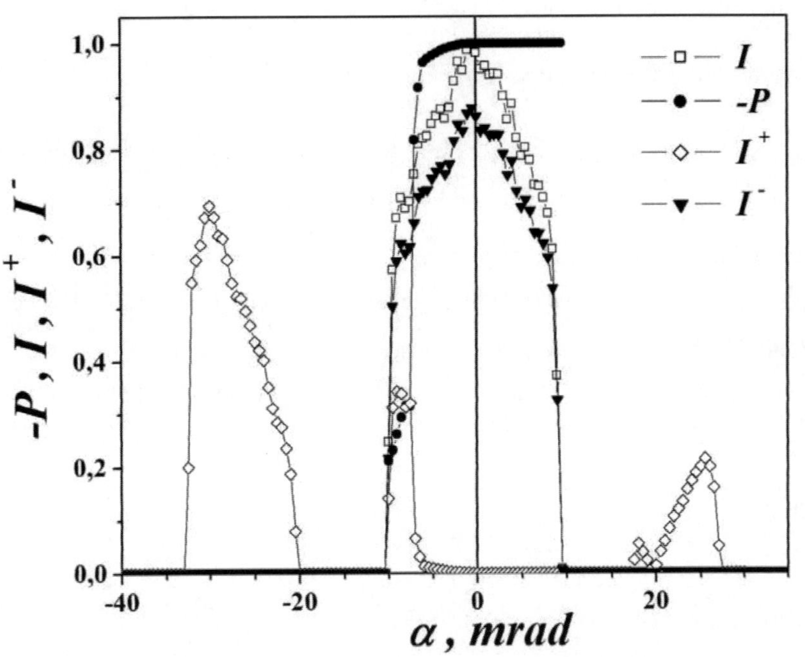

Figure 5. Calculated dependencies on angle α (see Figs. 2 and 3) of normalized intensities I^+ and I^- for, respectively, (+) and (-) spin components of monochromatic neutron beam with $\lambda = 5.5$ Å transmitted through the first polarizer. Dependencies on α are also shown of intensity at the entrance of the polarizer I and polarization P of the transmitted beam.

IOP Conf. Series: Journal of Physics: Conf. Series **862** (2017) 012028 doi:10.1088/1742-6596/862/1/012028

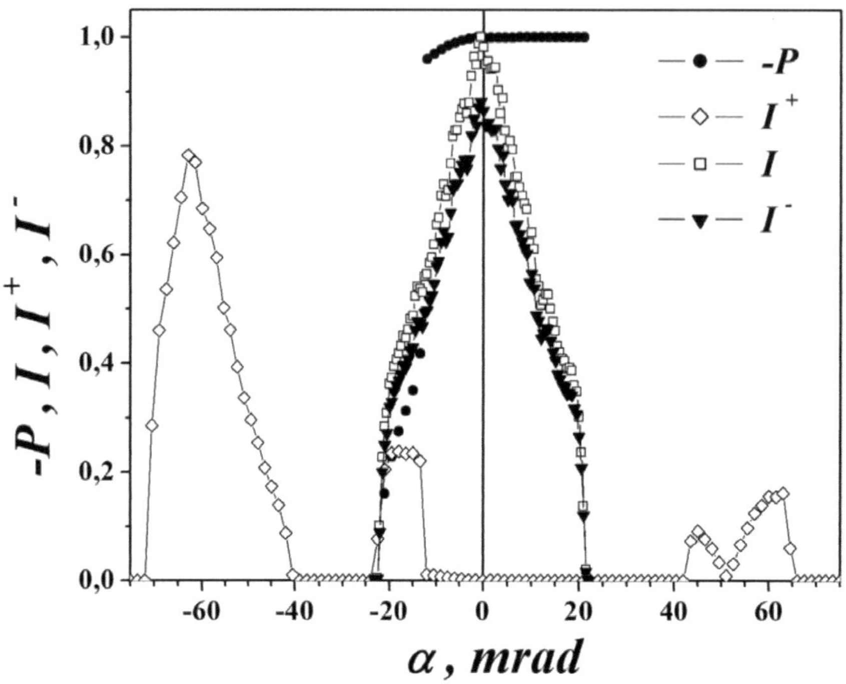

Figure 6. Same as in Fig. 5, but for the second polarizer and for $\lambda = 12.5$ Å.

Calculated wavelength dependencies are shown in Figs. 7 and 8 of polarization P and transmission coefficient T^- (in which the absorption in silicon is taken into account) of neutrons of (-) spin component for the first and the second polarizers, respectively. As it is seen from the Fig. 7, T^- (curve 4) is large and it decreases from 0.9 to 0.8 upon wavelength decreasing from $\lambda = 4.5$ Å to $\lambda = 9.5$ Å. Curves 1, 2 and 3 for P are shown in Fig. 7 are obtained for angular divergences of the beam at the exit of polarizer of 8 mrad ($\pm$ 4 mrad), 10 mrad ($\pm$ 5 mrad) and 12 mrad ($\pm$ 6 mrad), respectively. As it is seen from these dependencies, the polarization is better than -0.99 in the whole spectral range $\lambda = 4.5 \div 10$ Å and for $\lambda = 5.5 \div 10$ Å for the divergence of 8 mrad and 12 mrad, respectively.

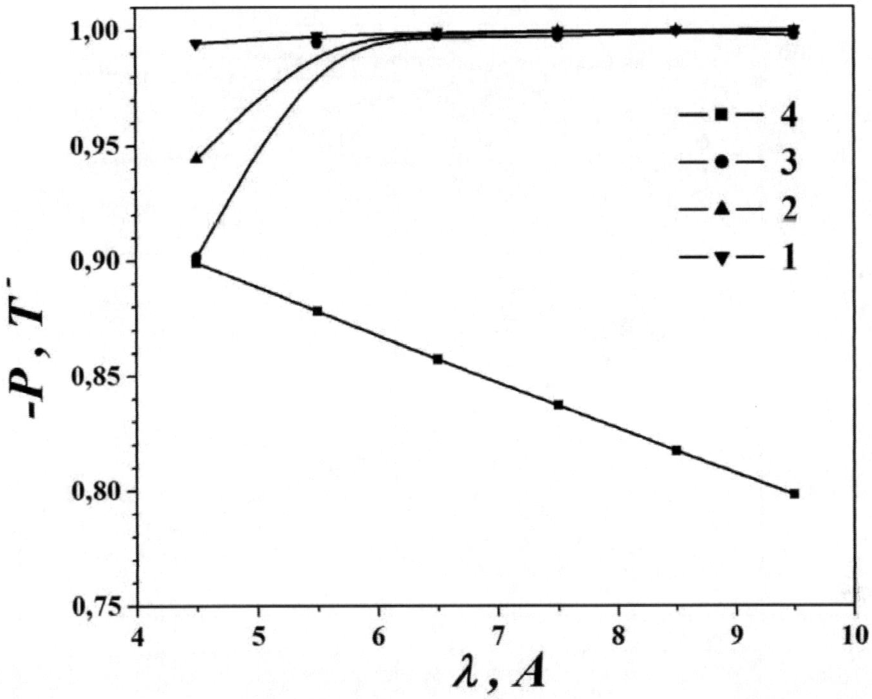

Figure 7. Calculated wavelength dependencies of polarization P (curves 1-3) and transmission coefficient T^{-} (curve 4) (in which the absorption in silicon is taken into account) of neutrons of (-) spin component (curve 4) for the beam transmitted through the first polarizer. Curves 1, 2 and 3 are obtained for angular divergences of the beam at the exit of polarizer of 8 mrad ($\pm$ 4 mrad), 10 mrad ($\pm$ 5 mrad) and 12 mrad ($\pm$ 6 mrad), respectively.

IOP Conf. Series: Journal of Physics: Conf. Series **862** (2017) 012028 doi:10.1088/1742-6596/862/1/012028

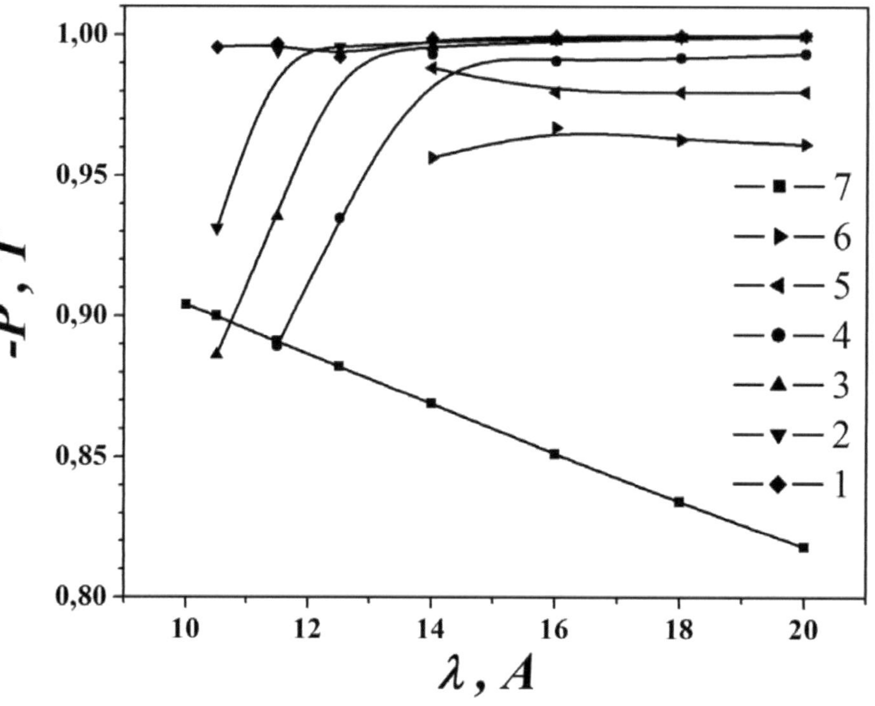

Figure 8. Calculated wavelength dependencies of polarization P (curves 1-6) and transmission coefficient T^- (in which the absorption in silicon is taken into account) of neutrons of (-) spin component (curve 7) for the beam transmitted through the second polarizer. Curves 1, 2, 3, 4, 5, and 6 are obtained for angle divergences of the beam at the exit of polarizer of 12 mrad ($\pm$ 6 mrad), 18 mrad ($\pm$ 9 mrad), 24 mrad ($\pm$ 12 mrad), 30 mrad ($\pm$ 15 mrad), 33 mrad ($\pm$ 16.5 mrad), and 36 mrad ($\pm$ 18 mrad), respectively.

As it is seen from Fig. 8, the transmission coefficient T^- (curve 7) is also high in the second polarizer. It decreases from 0.905 to 0.82 upon wavelength increasing from $\lambda = 10.5$ Å to $\lambda = 20$ Å. Six curves for P are shown in Fig. 8 found for α =12 mrad ($\pm$ 6 mrad), 18 mrad ($\pm$ 9 mrad), 24 mrad ($\pm$ 12 mrad), 30 mrad ($\pm$ 15 mrad), 33 mrad ($\pm$ 16.5 mrad), and 36 mrad ($\pm$ 18 mrad). As it is seen from these dependencies, the polarization is better than -0.99 in the whole spectral range $\lambda = 10.5 \div 20$ Å, at $\lambda = 11.5 \div 20$ Å, at $\lambda = 12.5 \div 20$ Å, and at $\lambda = 14 \div 20$ Å for the divergence of 12 mrad, 18 mrad, 24 mrad, 30 mrad, respectively. The absolute value of polarization of the transmitted beam decreases (remaining, however, quite high) upon further increasing of the angle divergence. For example, $P = -0.98 \div -0.99$ and $P = -0.95 \div -0.97$ for the range $\lambda = 14 \div 20$ Å and angle divergences of 33 mrad and 36 mrad, respectively. For solution of many physical problems in which the small polarization decreasing at $\lambda = 14 \div 20$ Å is inessential, it is possible to substantially increase the intensity of the beam incident on sample by tuning the collimator before the sample on the divergence of 33 - 36 mrad.

4. Discussion and Conclusions

As it follows from the above consideration, basic parameters (P and T^-) of the suggested compact transmission neutron supermirror polarizer (kink polarizer) are high. T^- = 0.8 – 0.9 for both polarizers and for each wavelength range. Polarization P is very high. P is better than -0.99 for both polarizers for wavelength range $\lambda = 4.5 \div 10$ Å at the beam divergence of 8 mrad and for wavelength range $\lambda = 12.5 \div 20$ Å at the beam divergence of 24 mrad. That means that the basic parameters (P and T^-) of the suggested kink polarizer exceed those of widely used transmission neutron supermirror polarizer V-cavity whereas its length (including its magnetic system) is more than 30 times smaller! Besides, its fabrication and operation in experiment are substantially simpler.

Let us compare the suggested polarizer (kink polarizer) with the above mentioned solid state bender [12]. For the accurate comparison, polarizing coating with m = 2.3 should be used instead of that with m = 2 and silicon plates of the width 0.16 mm instead of 0.3 mm. We obtain after a calculation using the above formulas that the length of the first and the second polarizer is 20.7 mm and 8.8 mm, respectively. Then, the first polarizer is 2.6 and 3.9 times as small as the bender [12] and the solid state S-bender [13], correspondingly! For the wavelength of 4.5 Å, the transmission coefficient T^- increases from 0.9 to 0.96. The beam divergence with the polarization P > 0.99 remains 8 mrad.

Thus, the beam polarization at the exit of the suggested polarizer is not worse than that in the above considered solid state benders [12, 13]. At the same time, the divergence of the beam with large polarization is more than two times smaller that leads to the luminosity reduction. On the other hand, the length of the suggested polarizer is 2.6 times as small as that of the considered polarizer that increases the transmission.

The polarization is comparable in all the considered facilities, the transmission is higher for the kink polarizer but its angle divergence is several times smaller than that in the solid state benders. At the same time, the length of the suggested polarizer is 2.6 and 3.9 times smaller than those of considered benders.

When using such benders [12, 13] for neutrons with wavelengths greater than 10 Å the transmission coefficient for them becomes even smaller compared to kink polarizer. In order to reduce these losses can be reduced length benders, but them be bent with a large beam cross-section becomes difficult. There is no such problem in the kink polarizer.

It should be noted also that the main parameters of the suggested kink polarizer meet the requirements for the transmission polarizer in SANS-2 facility at reactor PIK (National Research center "Kurchatov Institute", Russia).

5. Acknowledgements

The work was supported by the Federal target program of Ministry of Education and Science of Russian Federation (project No. RFMEFI61614X0004).

References

[1] Drabkin G M 1962 *ZhETF* **43** 1107 (in russian)
[2] Agamalyan M M, Drabkin G M, Sbitnev V I 1988 *Physics Reports* **168** (5) 265

[3] Runov V V, Il'in D S, Runova M K, Radzhabov A K 2012 *Letters to ZhETF* **95** (9) 530 (in russian)
[4] Lebedev V M, Lebedev V T, Orlov S P, Margolin B Z, Morozov A M 2014 *Fizika Tverdogo Tela* **56** (1) 160 (in russian)
[5] Schaerpf O 1989 *Physica B* **156-157** 631
[6] Boeni P, Munzer W, Ostermann A 2009 *Physica B* **404** 2620
[7] Keller N, Krist Th, Danzig A, Keiderling U, Mezei F, Wiedenmann A 2000 *Nuclear Instruments and Methods in Physics Research A* **451** 474
[8] Syromyatnikov V G, Schebetov A F, Soroko Z N 1993 *Nuclear Instruments and Methods in Physics Research A* **324** 401
[9] Syromyatnikov V G, Schebetov A F, Soroko Z N 1994 *Physica B* **198** 224
[10] Syromyatnikov V G, Alexeev V L 2001 *Physica B* **297**/1-4 148
[11] Krist Th 2004 *Nuclear Instruments and Methods in Physics Research A* **529** 50
[12] Krist Th, Peters J, Shimizu H M, Suzuki J, Oku T 2005 *Physica B* **356** 197
[13] Krist Th, Rucker F, Brandl G, Georgii R 2013 *Nuclear Instruments and Methods in Physics Research A* **698** 94
[14] Pleshanov N K, Kolyvanova N G, Metelev S V, Peskov B G, Pusenkov V M, Syromyatnikov V G, Ul'yanov V A, Schebetov A F 2005 *Physica B* **369** 234
[15] Pusenkov V M, Schebetov A, Gibcus H P M, Gommers R M, Labohm F, de Haan V O, van Well A A 2002 *Nuclear Instruments and Methods in Physics Research A* **492** 105

IOP Conf. Series: Journal of Physics: Conf. Series **862** (2017) 012029 doi:10.1088/1742-6596/862/1/012029

Neutron Larmor diffraction with double and single precession arm

A.A. van Well and M.T. Rekveldt

Department of Radiation Science and Technology, Faculty Applied Sciences, Delft University of Technology, 2629JB Delft, The Netherlands

a.a.vanwell@tudelft.nl

Abstract

A review is given of double and single arm Larmor diffraction. With the former a lattice-spacing resolution down to 10^{-6} can be obtained. The latter is a good high-resolution alternative if the sample or sample environment disturbs the magnetic field, e.g. ferromagnetic samples or applied magnetic fields. By choosing the tilt angle of the precession fields the optimum resolution can be set at a scattering angle at choice. The resolution for both single-crystal and polycrystalline samples is discussed in depth and is compared with conventional neutron-diffraction techniques.

1. Introduction

One of the areas of research where high-resolution diffraction is crucial is crystal-structure determination of low-symmetry and/or many-component systems. Then, the overlap of Bragg peaks at higher wave-vector transfers Q, corresponding to the smaller lattice parameters d is the limiting factor is the structure refinement. Another class of diffraction studies only investigates a limited number of Bragg peaks. Here the interest lies in the precise determination of a limited number of lattice parameters and/or the distribution thereof. Examples for accurate determination of lattice parameter can be found in the field of phase transitions with very small changes in the lattice parameters [1], weak magneto-elastic effect [2] or engineering materials such as different types and/or heat treatments of inconel [3]. Precise determination of the lattice-parameter distribution, or shape of the Bragg peak, gives unique information in material science. It can be used to discriminate between crystallite-size distribution and micro-strain effects [4], or to study dislocations in composite materials [5].

The latter class of studies is the subject of this paper. The use of Larmor precession of polarized neutrons offers a tool to label the energy transfer or the scattering angle with very high precision. The first application was the spin-echo technique to perform high resolution quasi-elastic neutron scattering introduced by Mezei [6]. The Larmor coding of scattering angle was proposed by Pynn [7], and later applied in small-angle scattering [8] and reflectometry [9]. The application in diffraction, with lattice-parameter resolution of $\Delta d/d \approx 10^{-6}$, the so-called double-arm Larmor diffraction was is introduced in [10]. This technique cannot be applied for magnetic samples. Then, an alternative is the one-arm Larmor diffraction technique [11] with a resolution in between the double-arm Larmor diffraction and the conventional, monochromatic or time-of-flight, diffractometers. The subject of this paper is the comparison of these four techniques. In Section 2 the performance of the monochromatic and time-of-flight diffractometer is discussed. In Section 3 the double-arm Larmor diffraction method

IOP Conf. Series: Journal of Physics: Conf. Series **862** (2017) 012029 doi:10.1088/1742-6596/862/1/012029

is introduced. In Section 4 the single-arm version is treated, with emphasis on the high-resolution focusing condition, and the difference in resolution for mosaic crystalline and poly-crystalline samples. Section 5 discusses the comparison of the different techniques, and the potential application of both Larmor methods.

2. Conventional diffractometers

The vast majority of the present neutron diffractometers can be divided in two types: the monochromatic diffractometer, generally located at continuous neutron sources and the time-of-flight diffractometers, most often used at pulsed sources.

2.1. Monochromatic diffractometers

This mean components of this type of diffractometers are the monochromator characterized by the lattice spacing d_M and mosaic spread β, set at a Bragg angle θ_M, , selecting the wavelength λ, the sample position, and an area of detectors covering a range of scattering angles 2θ. The beam divergences between source and monochromator, monochromator and sample, sample and detector are characterized by α_1, α_2, and α_3, respectively. The angular resolution, full width half maximum, $\Delta(2\theta)$ of the diffractometer is given by

$$\Delta(2\theta) = \sqrt{U \tan^2\theta + V \tan\theta + W} \, , \quad (1)$$

where U, V and W are functions of α_1, α_2, α_3, β and θ_M [12,13]. The minimum value of $\Delta(2\theta)$ is $\Delta(2\theta)_{\min} = \sqrt{W - V^2/4U}$ for $\tan\theta = -V/2U \approx \tan\theta_M$. The lattice-spacing resolution is through the Bragg relation given by

$$\frac{\Delta d}{d} = \cot\theta\Delta\theta \, . \quad (2)$$

In Fig. 1(a) the angular resolution for a number of state-of-the-art monochromatic diffractometers is displayed. The corresponding resolution in lattice spacing is shown in Fig. 1(b). Note that the best resolution $\Delta d/d \approx 10^{-3}$ is only reached in a limited d-range.

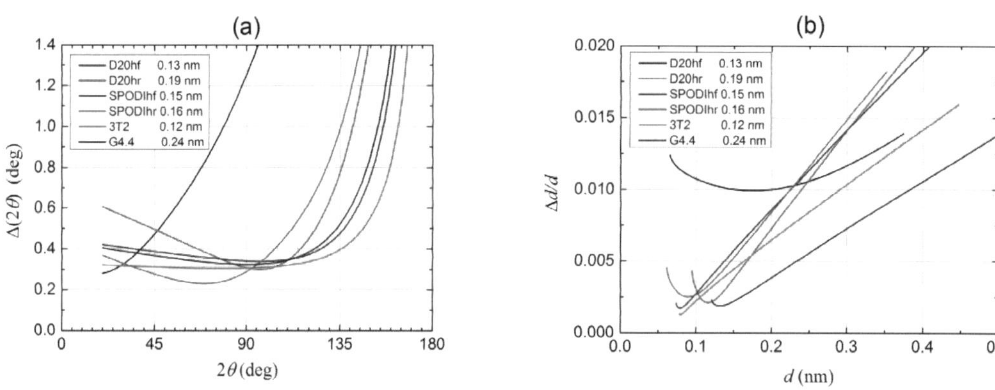

Figure 1. (a) Angular resolution as a function of scattering angle of some monochromatic diffractometers with (b) corresponding resolution in lattice parameter. D20hf and D20hr are a high-flux, Cu (200), and high-resolution, Ge (711) setting of D20, ILL, Grenoble [14], SPODIhf and SPODIhr are a high-flux and high-resolution setting of SPODI, FRM2, Munich [15], 3T3 and G4.4 are diffractometers at LLB [16], Saclay.

2.2. *Time-of-flight diffractometers*

This type of diffractometers is used at pulsed sources. Essentially the instrument consists of a primary flight path from source to sample, a secondary flight path from sample to detector and an array of detectors covering a large range of scattering angles $2\theta_B$. The resolution is given by [17]

$$\left(\frac{\Delta d}{d}\right)^2 = \left(\frac{\Delta t}{t}\right)^2_{\text{mod}} + \left(\frac{\Delta L}{L}\right)^2 + \frac{\cot^2\theta_B}{4}\left(\Delta\theta^2 + \Delta\phi^2\right), \tag{3}$$

where Δt is the pulse width of the source, t the time of flight from the source to the detector, L the distance from source to detector, with DL its uncertainty, $\Delta\theta$ and $\Delta\phi$ the beam divergences of the incoming and scattered beam, respectively. In Fig. 2 the resolution of the high resolution diffractometer HRPD [18] and the engineering materials diffractometer Engin-x [19], both at ISIS, UK, are shown. In contrast with the monochromatic instruments a large range of lattice parameters is covered at one scattering angle as a result of polychromatic nature of the technique.

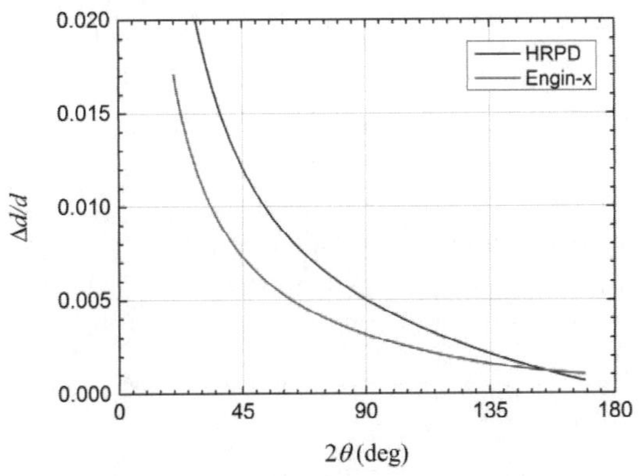

Figure 2. Lattice-spacing resolution of the time-of-flight diffractometer HRPD, [18] and Engin-x [19], ISIS, UK.
HRPD: $\sqrt{\Delta\theta^2 + \Delta\phi^2}$ = 10 mrad, $\Delta t/t = 5\cdot 10^{-4}$ (the best time-resolution value, i.e. for λ = 0.1 nm). Engin-x $\sqrt{\Delta\theta^2 + \Delta\phi^2}$ = 6 mrad, $\Delta t/t = 1\cdot 10^{-3}$ (valid for λ = 0.2 nm).

3. Double-arm Larmor diffraction

The double-arm Larmor diffraction setup, for the first time described by Rekveldt *et al.* [10], is schematically shown in Fig. 3. The main parts are two well-defined Larmor precession regions with the same, tuneable magnetic field B, one placed in the incoming polarized neutron beam and one in the scattered beam. Before entering the detector that is placed at a scattering angle $2\theta_0$ the polarization of the beam is analyzed by a second polarizer. Here we only describe the geometry with parallel magnetic fields. The anti-parallel, or spin-echo, geometry is described in the Appendix. We first consider the sample to be a *single crystal* with lattice planes parallel to the faces of the precession fields, *i.e.* α = 0 in Fig.3. Then $\mathbf{Q} = \mathbf{G}_{hkl}$. The precession angle in the first and second arm is given by

$$\varphi_1 = \omega_L \frac{L}{v_{1\perp}} = \frac{\omega_L m L}{\hbar}\frac{1}{k_{1\perp}} = 2\pi c B L \frac{1}{k_{1\perp}} \quad \text{and} \quad \varphi_2 = \omega_L \frac{L}{v_{2\perp}} = 2\pi c B L \frac{1}{k_{2\perp}}. \tag{4}$$

Here, the Larmor frequency $\omega_L = \gamma B$ with the neutron's gyromagnetic ratio $\gamma = 1.382\cdot 10^8$ s^{-1}T^{-1}, $c = \gamma m/h = 4.632\cdot 10^{14}$T^{-1}m^{-2}, L is the length of the magnetic field perpendicular to the lattice planes, $k_{1\perp} = k_{2\perp} = k\sin\theta_0 = G_{hkl}/2$, and k is the neutron's wave number. The total precession is given by

$$\varphi_t = \varphi_1 + \varphi_2 = 2\pi cBL\left(\frac{1}{k_{1\perp}} + \frac{1}{k_{2\perp}}\right) = \frac{8\pi cBL}{G_{hkl}} = 4cBLd_{hkl}\,, \tag{5}$$

independent of the beam divergence $\Delta\theta$ or the neutron's wavelength λ. Note, that $\Delta\theta$ and λ are directly related through Bragg's law.

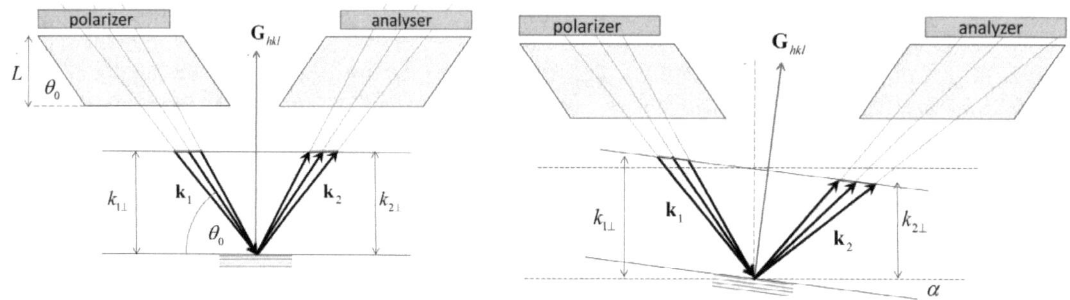

Figure 3. Schematic representation of a double-arm Larmor diffractometer. Left: single-crystal sample with lattice planes under investigation parallel to the faces of the precession regions. Right: misaligned single crystal.

If the crystal is misaligned with an angle α then $k_{1\perp}$ will be larger and $k_{2\perp}$ smaller, see Fig.3, resulting in cancellation of the effect of α in first order. This is shown mathematically as follows.

$$k_{1\perp} = k\sin(\theta_B + \alpha) \text{ and } k_{2\perp} = k\sin(\theta_B - \alpha). \tag{6}$$

Using the Bragg relation $k = G_{hkl}/(2\sin\theta_B) = \pi/(d_{hkl}\sin\theta_B)$ and the Taylor expansion $1/\sin(\theta_B+\alpha) = 1/\sin\theta_B\left(1-\cot\theta_B\alpha + 0.5(1+2\cot^2\theta_B)\alpha^2 + O(\alpha^3)\right)$ Eq.(5) reads to second order of α

$$\varphi_t = \varphi_1 + \varphi_2 = 4cBLd_{hkl}\left(1+\frac{1}{2}(1+2\cot^2\theta_B)\alpha^2\right). \tag{7}$$

The resolution in determining the lattice parameter d for this setup is given by

$$\left(\frac{\Delta d}{d}\right)^2 = \left(\frac{\Delta BL}{BL}\right)^2 + \frac{1}{4}(1+2\cot^2\theta_B)^2\alpha^4. \tag{8}$$

If we now consider a *polycrystalline material*, or powder, all orientations of the lattice planes are present and the role of α in Eq.(8) is taken over by the beam divergence, resulting in a distribution of α with spread

$$\alpha^2 = \Delta\theta^2 + \Delta\phi^2\,, \tag{9}$$

where $\Delta\theta$ and $\Delta\phi$ are the divergences of the incoming and scattered beam, respectively. The intensity measured at the detector is given by $I(BL) = I_0(1+P(BL))$ with I_0 the shim intensity and, if we only consider one Bragg peak,

$$P(BL) = \langle \cos\varphi_t \rangle = G(BL)\cos\left(4c\langle BL\rangle\langle d_{hkl}\rangle\right). \tag{10}$$

The period of the cosine determines the average lattice spacing and the damping envelope is given by $G(B) = \int f(\varphi - \varphi_0)\cos(\varphi)d\varphi$, where the distribution $f(\varphi)$ - see Eq.(7) - is determined by the distribution of the lattice spacing d_{hkl}, the field-integral aberrations, and angular uncertainties. In d_{hkl}–space these contributions are convolutions. This results in multiplying the Fourier transforms of these contributions in *BL*-space, leading to

$$G(BL) = G_{BL}(BL)G_\alpha(BL)G_d(BL). \tag{11}$$

G_d is the Fourier transform of the lattice-spacing distribution and G_{BL} and G_α describe the instrumental resolution. G_{BL} is determined by magnetic field *B* and its length *L* and can be measured by using a perfect single crystal as a calibration sample.

If more lattice parameters are measured simultaneously, The resulting intensity will be a sum of damped cosines with different periods and different damping. In Fig. 4 an example is shown of a perfect measurement, i.e. no beam divergence, no aberrations, i.e. $G_{BL} = G_\alpha = 1$ of a sample with two lattice parameters d = 0.2 and 0.5 nm, both normally distributed with a 0.5% standard deviation, corresponding with full width at half maximum of 1.9%. To the 1000 calculated intensities I, with spacing $\Delta BL = 1.36\cdot10^{-6}$ Tm, 10% statistics was added (corresponding with a shim intensity of 100 counts). Note, that the contribution of the 0.5 nm d-spacing is damped faster because of its larger width Δd. The resulting Fourier transform is shown in Fig.4. It shows that in this case 1000 measuring points with very modest statistics remarkably good lattice-parameter distribution functions can be reproduced. Note, that the oscillation of the background noise is an artefact of the Fourier transform.

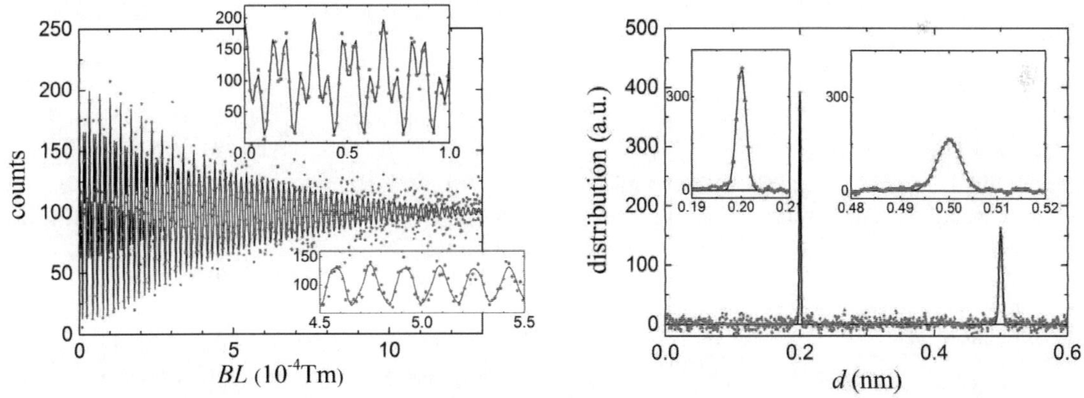

Figure 4. Simulation of a Larmor-diffraction experiment with perfect instrumental resolution. The model system has two lattice parameters *d*, each with a Gaussian distribution with 0.5% standard deviation (right, black line). Left: Fourier transform (black line) with 10% statistics added (red dots), representing experimental data. Note, that at the smaller *BL* values a beating representing both *d* values is present (inset top), while for larger *BL* meanly the contribution of *d* = 0.2 nm is present (inset bottom). The Fourier transform of the noisy data is displayed in the right panel (red dots).

4. Single-arm Larmor diffraction

4.1. Diffraction from perfect single crystal

If the sample is magnetic or has spin-flip scattering the double-arm setup may cause problems. Then the single-arm geometry, first described in [11], is a good alternative. In that case the analyzer is placed between the precession field and the sample, see Fig. 5. Whereas in the double-arm geometry the change of $k_{1\perp}$ as a result of the misalignment α is in first order compensated by the change of $k_{2\perp}$, see Fig. 3, this is not the case in the single-arm geometry, what will lead to a more modest resolution, as will be derived in the following. We will consider the single crystal and polycrystalline material separately.

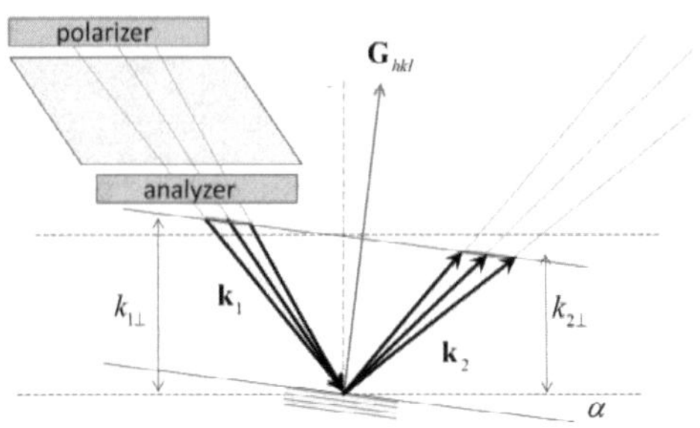

Figure 5. Schematic representation of a single-arm Larmor diffractometer with the Larmor precession device in the incoming beam. The angle α indicates the misalignment of the lattice planes with respect to the face of the precession region.

For a perfect single crystal, i.e. no mosaic spread, misaligned with angle α with respect to the face of the precession field, as indicated in Fig. 6, the Larmor precession phase is given by

$$\varphi_t = \varphi_1 = 2\pi cBL\frac{1}{k_{1\perp}} = 2cBLd_{hkl}\left(1-\cot\theta_B\alpha + O\left(\alpha^2\right)\right). \tag{12}$$

This leads to a resolution that depends linearly on the spread in α:

$$\left(\frac{\Delta d}{d}\right)^2 = \left(\frac{\Delta BL}{BL}\right)^2 + \cot^2\theta_B\Delta\alpha^2 \tag{13}$$

4.2. Powder diffraction

Whereas for the single-crystal case the small parameter is the misalignment α, for a polycrystalline sample the small parameters are the divergences of the incoming and the scattered beam. We define all angles with respect to the face of the precession field, see Fig.6.

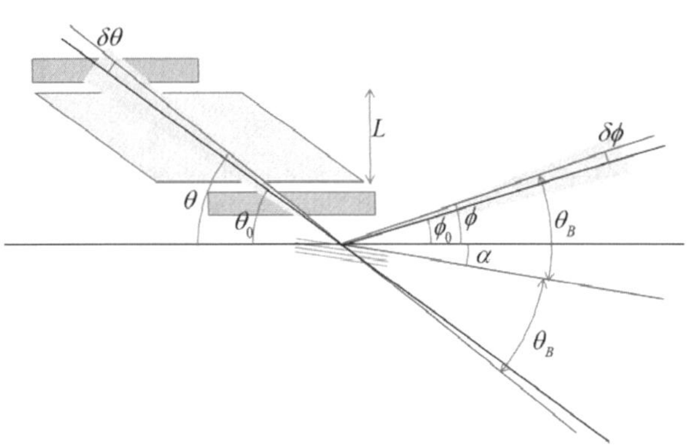

Figure 6. Schematic representation of a single-arm Larmor diffractometer with Larmor precession device in the incoming beam for the description of a powder sample. The nominal incoming and scattered beam (characterized by θ_0 and ϕ_0) are indicated by the black lines. A possible beam with deviations and are shown as red lines (characterized by θ and ϕ).

The nominal incoming angle is θ_0. We follow the neutron path entering at incoming angle

$$\theta = \theta_0 + \delta\theta \tag{14}$$

and scattered at angle ϕ, determined by the nominal angle ϕ_0 and deviation $\delta\phi$

$$\phi = \phi_0 + \delta\phi . \tag{15}$$

The Bragg angle is given by

$$2\theta_B = \theta + \phi . \tag{16}$$

The relation between the actual Bragg angle θ_B and the nominal Bragg angle $\theta_{B0} = (\theta_0 + \phi_0)/2$ is given by $\theta_B = \theta_{B,0} + \delta\theta/2 + \delta\phi/2$. Note that the scattering-plane orientation $\alpha = \theta_B - \phi$ is not necessarily a small parameter. Using $k_{1\perp} = k\sin\theta$ and $k = G_{hkl}/(2\sin\theta_B) = \pi/(d_{hkl}\sin\theta_B)$, a Taylor expansion of $\varphi_1 = 2cBLd_{hkl}\sin\theta_B/\sin\theta = 2cBLd_{hkl}\sin(\theta_{B,0} + \delta\theta/2 + \delta\phi/2)/\sin(\theta_0 + \delta\theta)$ with small parameters $\delta\theta$ and $\delta\phi$ yields, in first order,

$$\varphi_1 = 2cBLd_{hkl}\frac{\sin\theta_{B,0}}{\sin\theta_0}\left[1 + \left(\cot\theta_{B,0} - 2\cot\theta_0\right)\frac{\delta\theta}{2} + \cot\theta_{B,0}\frac{\delta\phi}{2}\right]. \tag{17}$$

If we assume $\delta\theta$ and $\delta\phi$ to be uncorrelated, which is the case for a polycrystalline sample, having distributions with spread $\Delta\theta$ and $\Delta\phi$, then this leads to the resolution

$$\left(\frac{\Delta d}{d}\right)^2 = \left(\frac{\Delta BL}{BL}\right)^2 + \frac{1}{4}\left(\cot\theta_{B,0} - 2\cot\theta_0\right)^2\Delta\theta^2 + \frac{1}{4}\cot^2\theta_{B,0}\Delta\phi^2 \tag{18}$$

The contribution of the divergence of the incoming neutron beam to the resolution cancels in first order, if the focussing condition

$$\cot\theta_{B,0} = 2\cot\theta_0 \tag{19}$$

is met, see Fig.7. It shows that by tuning the tilt angle the minimum resolution, that is of the order $\Delta d/d \approx 10^{-3}$ can be chosen to be at a scattering angle of choice.

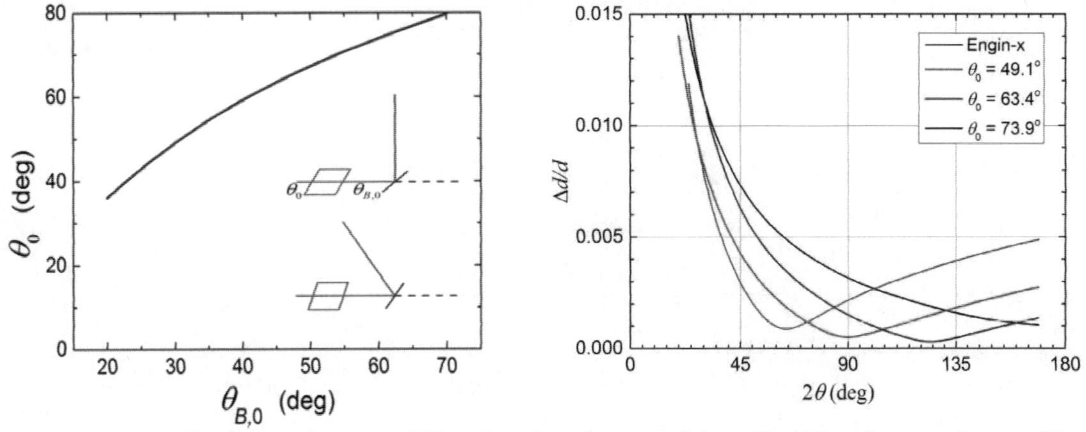

Figure 7. Left: Single-arm Larmor diffraction focusing condition, Eq.(19), for a polycrystalline sample. Right: resulting resolution for three different focusing conditions, corresponding $2\theta_B = 60^{\circ}$, 90° and 120°. Here the divergences of the incoming beam, $\Delta\theta$, and scattered beam, $\Delta\phi$, were taken 5

and 2 mrad, respectively. As a comparison the resolution of Engin-x is shown, with the same divergences.

4.3. Mosaic single crystal

For a powder the divergences $\delta\theta$ and $\delta\phi$ are uncorrelated, i.e. for any combination of divergences of incoming and scattered beam reflection planes are available. In the case of a mosaic crystal these parameters are correlated: assuming $\Delta\phi < \Delta\theta$, then for every value of $\delta\phi$ a range of $\delta\theta$ is available satisfying $\delta\theta = -\delta\phi + 2\delta\alpha$ for all $\delta\alpha$ within the mosaic spread $\Delta\alpha$. Then, for $\Delta\alpha < \Delta\theta$, Eq.(17) should be replaced by

$$\begin{aligned}\varphi_1 &= 2cBLd\frac{\sin\theta_{B,0}}{\sin\theta_0}\left[1+\left(\cot\theta_{B,0}-2\cot\theta_0\right)\frac{-\delta\phi+2\delta\alpha}{2}+\cot\theta_{B,0}\frac{\delta\phi}{2}\right] \\ &= 2cBLd\frac{\sin\theta_{B,0}}{\sin\theta_0}\left[1+\left(\cot\theta_{B,0}-2\cot\theta_0\right)\delta\alpha+\left(\cot\theta_0-\cot\theta_{B,0}\right)\delta\phi\right],\end{aligned} \tag{20}$$

and then the resolution is given by

$$\left(\frac{\Delta d}{d}\right)^2 = \left(\frac{\Delta BL}{BL}\right)^2 + \left(\cot\theta_{B0}-2\cot\theta_0\right)^2\Delta\alpha^2 + \left(\cot\theta_{B0}-\cot\theta_0\right)^2\Delta\phi^2. \tag{21}$$

In the double-arm geometry the mosaic spread only enters in second order, and is independent of the divergences of incoming and scattered beam, see Eq.(8). However, in the single-arm geometry the mosaic spread, in combination with the divergence $\Delta\phi$ of the scattered beam, will determine the resolution. Some examples are given in Fig. 8. For a perfect single crystal the minimum resolution is reached for scattering angle $2\theta_B = 2\theta_0$, conform Eq.(13). For increasing mosaic spread the minimum shifts to smaller scattering angles approaching the focusing value for powders, Eq.(19), if the mosaic spread approaches the divergence $\Delta\phi$ of the incoming beam.

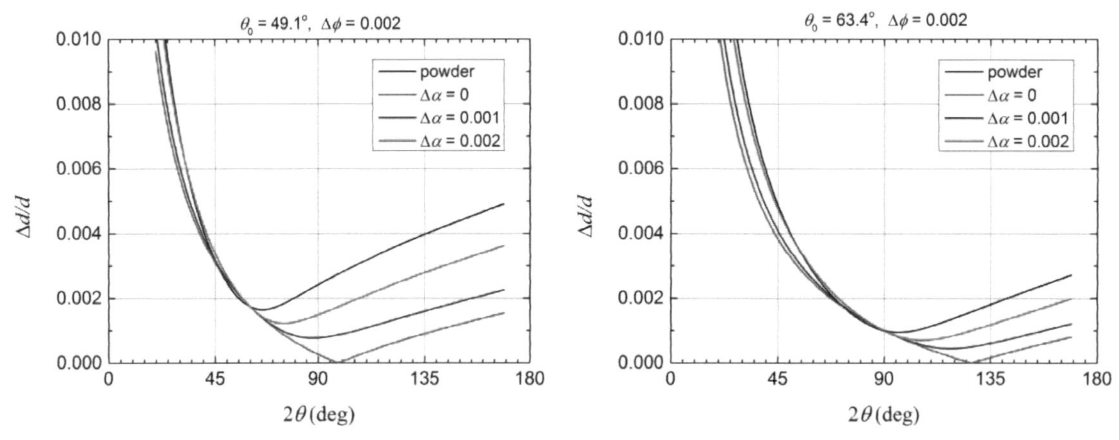

Figure 8. Resolution of a mosaic single crystal sample for three values of the mosaic spread $\Delta\alpha$ and two values of the precession-field tilt angle θ_0. The divergence of the scattered beam is taken $\Delta\phi = 2$ mrad. As comparison, the resolution for a polycrystalline sample (with $\Delta\theta = 5$ mrad and $\Delta\phi = 2$ mrad) is shown (black line), also displayed in Fig. 7.

IOP Conf. Series: Journal of Physics: Conf. Series **862** (2017) 012029 doi:10.1088/1742-6596/862/1/012029

5. Discussion

Double-arm Larmor diffraction yields a resolution of $\Delta d/d \approx 10^{-6}$ that is orders of magnitude better than conventional diffractometers. For magnetic samples the single-arm geometry is a good alternative that may reach a resolution down to $\Delta d/d \approx 10^{-5}$ for single crystals and better than 10^{-3} for powder samples around the focussing condition, that can be set at any scattering at choice by changing the tilt angle of the precession region. For polycrystalline samples, the contribution of the divergence of the incoming beam is cancelled in first order at this focussing condition. This may be used to separate overlapping diffraction lines around this scattering angle while the data at the other scattering angles are used to increase the statistics. If the Larmor modulation is performed in the scattered beam instead of the incoming beam, a similar focussing condition will cancel the contribution of the divergence of the scattered beam.

One of the advantages of Larmor diffraction compared with monochromatic diffraction is the considerable gain in intensity because the former makes use of the total wavelength spectrum and accepts much larger beam divergences. Due to technical restrictions, the double-arm configuration has in general a limited range of scattering angles, while in the single-arm version a large range can be used. The argument that is often used against Larmor techniques that long measuring times are needed because one has to perform a large B scan is not valid. The important quantity is the total number of neutrons acquired. This is shown in Fig. 5 where the individual points have very poor statistics and the extracted information about the lattice-space distribution is of high quality. However, as a result of the Fourier technique the information in the tails of the distribution is less accurate.

A crucial part of Larmor diffraction is the accurate dimensioning of the precession fields. In practice this is established by using two radio-frequency neutron flippers per precession region. These have been implemented in the triple-axes spectrometers TRISP (FRM II, Munich) [20], FLEXX (HZB, Berlin) [21] and ZETA (ILL, Grenoble) [22]. These instruments are operated in a single wavelength mode, ideally suited for the investigation of a single Bragg peak. A general disadvantage of the Fourier technique is the fact that the contribution of low-intensity Bragg peaks is difficult to detect in the measured data, that are mainly determined by the high-intensity peaks. For this reason Larmor diffraction is especially suited for determining the intensity and shape of a limited number of Bragg peaks, e.g. for strain scanning. Usually this technique is used at a scattering angle of 90 degrees. Fig.7. shows that at that angle Larmor diffraction is considerably better than Engin-x, ISIS. The Fourier technique is also applied by making use of a Fourier chopper, examples being FSS at the former GKSS, Geesthacht [24] and FSD, IBR-2, Dubna [25]. The advantage of the Larmor diffraction is that there are no rotating parts and that a many orders in Fourier space can be covered electronically.

The comparison of applying the technique at a steady-state (reactor) and a pulsed (spallation) source is not straight forward. If only one Bragg peak is investigated the reactor sources are better compared to the present spallation sources. Because the sample acts as a monochromator the neutron intensity will be considerably higher in the former. If the Larmor diffractometer makes use of the complete spectrum it is in principle possible to measure more Bragg peaks simultaneously, as shown in ref [11]. However, in that case it will be difficult to measure low-intensity peaks. This problem can be circumvented at a pulsed source. Then the different time-of-flight channels will only contain a few Bragg peaks, even at relaxed time-of-flight resolution. The ESS, Lund, will be the ideal source for the Larmor diffraction technique. It combines a high average neutron flux, comparable with ILL, with a sufficient time-of-flight resolution. Having a broad-wavelength spectrum makes it possible to measure many Bragg peaks at one scattering angle in focussing condition. Possible application may be found in multiple-phase systems where precise d spacing and distribution for all phases can be determined; for separating two overlapping Bragg peaks; in engineering materials determining Q-dependent line broadening to discriminate grain-size and micro-strain contributions; the information from more lattice planes will give information about anisotropic strain.

References

[1] C. Pfleiderer, P. Böni, T. Keller, U.K. Rößler, A. Rosch, Science **316** (2007) 1871
[2] N. Martin, L.P. Regnault, S. Klimko, J. Phys: Conf. Series **340** (2012) 012012
[3] J. Repper, T. Keller, M. Hofmann, C. Krempaszky, W. Petry, E. Werner, Acta Materialia **58** (2010) 3459
[4] I. Lucks, P. Lamparter, E. J. Mittemeijer, J. Appl. Cryst. **37** (2004) 300
[5] T.C. Bor, H.H.M. Cleveringa, R. Delhez, E. van der Giessen, Mater. Sci. Engin. A**309–310** (2001) 505
[6] F. Mezei, Zeitschrift für Physik A **255** (1972) 146
[7] R. Pynn, in *Neutron Spin Echo*, ed. F. Mezei, Springer-Verlag, Berlin, 1980
[8] T. Keller, R. Gähler, H. Kunze, R. Golub, Neutron News **6** (1995); M.T. Rekveldt, Nucl. Instr. Meth. Phys. Res. B **114** (1996) 366
[9] M.T. Rekveldt, Physica B **234-236** (1997) 1135
[10] M.T. Rekveldt, W.H. Kraan, J. Neutron Res. **8** (1999); M.T. Rekveldt, T. Keller, R. Golub, Europhys. Lett. **54** (2001) 342
[11] M.T. Rekveldt, J. Plomp, A.A. van Well, J. Appl. Cryst. **47** (2014) 436
[12] G. Caglioti, A. Paoletti, F.P. Ricci, Nucl. Instr. **3** (1958) 223
[13] A.W. Hewat, Nucl. Instr. Meth. **127** (1975) 361
[14] T.C. Hansen et al., Meas. Sci. Technol. **19** (2008) 034001
[15] M. Hoelzel, A. Senyshyn, N. Juenke, H. Boysen, W. Schmahl, H. Fuess, Nucl. Instr. Meth. A **667** (2012) 32
[16] http://www-llb.cea.fr/en/Web/hpr_web/HPRWEB5.php
[17] C.G. Windsor, *Pulsed Neutron Scattering*, (1981) London: Taylor and Francis.
[18] R.M. Ibberson, Nucl. Instr. Meth. A **600** (2009) 47
[19] J. R. Santisteban, M.R. Daymond, J.A. Jamesb, L. Edwards, J. Appl. Cryst. **39** (2006) 812
[20] T. Keller, B. Keimer, J. Large-scale Res. Fac. A **37** (2015) 1
[21] F. Groitl, T. Keller, D.L. Quintero-Castro, K. Habicht, Rev. Sci. Instr. **86** (2015) 025110
[22] N. Martin, L.P. Regnault, S. Klimko, J.E. Lorenzo, R. Gähler, Physica B **406** (2011) 2333
[23] H. Pöyry, P. Hiismäki, Virjo, Nucl. Instr. Meth. **126** (1975) 421
[24] J. Schröder, V.A. Kudryashev, J.M. Keuter, H.G. Priesmeyer, J. Larsen, A. Tiitta, J. Neutron Res. **4** (1994) 129; V.A. Kudryashev. H.G. Priesmaeyer, J.M. Keuter, J. Schröder, R. Wagner, Nucl. Instr. Meth. Phys. Res. B **101** (1995) 484
[25] A.M. Balagurov, G.D. Bokuchava, E.S. Kuzmin, A.V. Tamonov, V.V. Zhuk, Z. Kristallogr. Suppl. **23** (2006) 217; A.M. Balagurov, I.A. Bobrilov, G.D. Bokuchava, V.V. Zhuravlev, V.G. Simkin, Phys. Part. Nucl. **46** (2015) 249

IOP Conf. Series: Journal of Physics: Conf. Series **862** (2017) 012029 doi:10.1088/1742-6596/862/1/012029

Appendix: Double-arm spin-echo geometry

In the spin-echo geometry the direction of the magnetic field in the second arm is opposite to that of the first arm. The total precession angle for a perfect single crystal with misalignment angle α is

$$\varphi_t = \varphi_1 - \varphi_2 = 4cBLd_{hkl}\left(-\cot\theta_B\alpha + O\left(\alpha^3\right)\right). \tag{A1}$$

and

$$P(BL) = G(BL) = \int f(\varphi)\cos(\varphi)d\varphi \tag{1}$$

where the distribution $f(\varphi)$ is determined by both the spreads in lattice spacing d_{hkl} and the misalignment parameter α.

$$\left(\frac{\Delta\varphi}{\varphi}\right)^2 = \left(\frac{\Delta BL}{BL}\right)^2 + \left(\frac{\Delta d_{hkl}}{d_{hkl}}\right)^2 + (\Delta\alpha)^2 \tag{A2}$$

For a *single crystal* sample the variation is lattice parameter may be neglected and $P(BL)$ directly yields the mosaic spread. For a *polycrystalline* samples $P(BL)$ yields the distribution of the lattice parameter, i.e. the Fourier transform of the Bragg peak. In ref [7] is stated that in this setup the peak shape can be determined with ‘extremely fine’ resolution. However, we show that for a polycrystalline sample the peak is broadened by the angular resolution of the setup. Here, the same arguments hold as for the parallel-field geometry.

IOP Conf. Series: Journal of Physics: Conf. Series **862** (2017) 012030 doi:10.1088/1742-6596/862/1/012030

Polarized neutron scattering on HYSPEC: the HYbrid SPECtrometer at SNS

Igor A Zaliznyak[1], Andrei T. Savici[2], V. Ovidiu Garlea[2], Barry Winn[2], Uwe Filges[3], John Schneeloch[1], John M. Tranquada[1], Genda Gu[1], Aifeng Wang[1], Cedomir Petrovic[1]

[1]CMPMSD, Brookhaven National Laboratory, Upton, NY 11973 USA
[2]NScD, Oak Ridge National Laboratory, Oak Ridge, TN 37831 USA
[3]Paul Scherrer Institut, Laboratory for Scientific Developments and novel Materials, Villigen PSI, Switzerland

E-mail: `zaliznyak@bnl.gov`

Abstract. We describe some of the first polarized neutron scattering measurements performed at HYSPEC [1–4] spectrometer at the Spallation Neutron Source, Oak Ridge National Laboratory. We discuss details of the instrument setup and the experimental procedures in the mode with full polarization analysis. Examples of polarized neutron diffraction and polarized inelastic neutron data obtained on single crystal samples are presented.

1. Introduction

Neutron polarization analysis (NPA) [5–8] provides an unprecedented range of capabilities, both for developing new neutron scattering techniques [9–11], and for elastic and inelastic neutron scattering studies of novel magnetic states and excitations. Polarized neutron measurements allow one to identify the directions of magnetic moments and their fluctuations, distinguish between the structural and magnetic scattering features, and separate the magnetic and the non-magnetic background. There are a number of recent examples, which show how NPA enables disentangling the nature of complex magnetic states [11–13], understanding the nature of magnetic excitations in quantum magnets [14–18], cuprate [19, 20] and heavy fermion [21] superconductors, and probing novel types of magnetic dynamics, such as chirality fluctuations [11, 22–24]. Recent discoveries of unusual temperature-induced dynamical magnetism [25, 26] and unexpected, "forbidden" structural modes in the iron pnictides and chalcogenides [27, 28] emphasize the special significance of polarization analysis for strongly-correlated electron systems and unconventional superconductors. Only via NPA can these features be uniquely identified; an example of such a polarized neutron measurement performed on HYSPEC will be presented below.

Traditionally, polarized inelastic neutron experiments were carried out using triple-axis spectrometers (TAS) [5, 10, 29, 30], with narrow and nearly plane-parallel, collimated beams of small angular divergence. While significantly limiting the throughput, the corresponding instrument geometries are well suited for employing diverse neutron beam polarization methods, such as Bragg reflection from a Heusler crystal, or total reflection from a supermirror, or transmission through a reasonably small vessel filled with polarized ^{3}He. However, much of the recent progress in neutron spectroscopic techniques has been associated with the development of time-of-flight (TOF) direct geometry chopper spectrometers (DGCS) at spallation sources, which feature large detector arrays covering up to $\approx 2-3$ sr solid angle [31, 32]. Such wide-open geometry is in conflict with the requirements of most traditional beam polarization and NPA methods, which is further exacerbated by the usage of broad neutron energy bands, including thermal and high-energy epi-thermal neutrons.

One approach to meeting the requirements of wide-angle, broad-band scattered beam polarization analysis relies on developing wide-angle ^{3}He cells, which can be placed between the sample and the detectors. Such a method can provide high polarization efficiency and good transmission in a broad energy range, but at the same time has several significant deficiencies. Firstly, the volume of a wide-angle ^{3}He cell typically limits significantly the diameter of sample environments that can be used with it. Secondly, the neutron beam polarization is determined by the ^{3}He polarization, which typically is time-dependent and has to be periodically regenerated. For data acquired in the event mode, which is now standard on modern instruments at spallation sources, this can be adequately treated in the data analysis software, although the disadvantage of the labor-intensive experimental procedure associated with maintaining the ^{3}He polarization

Figure 1: HYSPEC instrument configured for polarized beam measurements (a) and the beamline components involved in the setup, (b)-(f). The incident beam is polarized by the vertically focussing magnetized Heusler crystal array, (b), which by the means of an automated vertical translation stage is interchangeable with the co-axially mounted pyrolytic graphite (PG) monochromator, (c), used for the un-polarized measurements. The coils mounted at the sample table provide adiabatic guide field of up to ~ 50 Gauss at the sample, (d), whose orientation can be freely tuned. The radial collimator, (e), used for the un-polarized measurements is replaced by the identically sized supermirror analyzer, (f).

remains. This problem can be alleviated via an in-situ ^{3}He polarization strategy using the spin-exchange optical pumping (SEOP) [33, 34]. However, this approach is technically challenging and entails significant constraints on instrument movement and the setup geometry. A more significant problem stems from the extreme sensitivity of the ^{3}He system to magnetic fields, which restricts the usage of magnetic fields in the sample environment and typically limits the wide-angle NPA to only a single (vertical) guide field direction, thus markedly reducing its potential. Although recent work on magnetic holding field environments [35, 36] for ^{3}He indicates that this limitation can be overcome, this is achieved at the cost of limiting the opening angle of the holding field coils, and therefore limiting the angular acceptance.

An alternative approach, which is largely free of these limitations, relies on developing a wide-angle supermirror polarizer (SP) system [1, 37–40]. Such systems are low-maintenance, easy to operate, and are weakly sensitive to magnetic fields, thus allowing the XYZ-polarization analysis [41, 42]. However, they have a strongly energy-dependent transmission, which is very low in the thermal to epithermal neutron energy range, thus limiting significantly the neutron energy bandwidth amenable to NPA.

The HYSPEC instrument [1–3] at the Spallation Neutron Source (SNS), Oak Ridge National Laboratory, was designed with the goal of implementing wide-angle polarization analysis. HYSPEC's design accommodates both the ^{3}He polarization analyzer (PA) option, which is still at the development stage [33, 34], and the supermirror PA, which was delivered to SNS in July of 2015 and is now in service. The supermirror polarization analyzer has been designed and built by the Paul-Scherrer Institute (PSI), where the associated unique supermirror manufacturing and assemblage technology was also developed. HYSPEC operation in fully polarized mode became available since the end of 2015, and here we report on some of the first polarized neutron experiments on HYSPEC with full NPA.

2. General overview of HYSPEC design and specifications

HYSPEC is a high-intensity, medium-resolution thermal neutron TOF spectrometer located at SNS beam line 14B viewing a cryogenic coupled-hydrogen moderator [3, 4]. The neutron beam is delivered by a $\approx$ 38.7 m long, 4 cm wide by 15 cm tall neutron guide (the initial $\approx$ 4 m section expands the beam height from 13.2 cm to 15 cm) with $m = 3(2)$ outer(inner) supermirror coating. The $\approx$ 24 m long central section of the guide is curved ($R = 2558.7$ m), and offsets the guide exit by $\approx$ 16 cm, well beyond the line of sight of the source. The resulting guide cutoff is at $E_i \approx 120$ meV, with a factor $\approx$ 2 beam attenuation at $E_i = 90$ meV.

The thermal neutron beam delivered by the guide is Bragg-reflected by a vertically focussing crystal array located $\approx$ 39 m from the source and 1.8 m from the sample. Focussing the 15 cm tall beam increases neutron flux on a typical 2 cm tall sample by a factor 3 to 6, at the expense of the vertical wave vector resolution. The latter is determined by the angular view of the 15 cm tall beam incident on the crystal array, from the sample position, which is $\approx 4.8^\circ$, or about 1/3 of the detector height. The beam is cleaned of contamination by a set of disk choppers and a T_0 chopper similar to those on ARCS and SEQUOIA [31], and is monochromatized by a short-blade ("rotating collimator") Fermi chopper. The latter is located 1.81 m upstream of the vertically focussing crystal array, which is placed in a rotating drum shield and is commonly called the "monochromator" because its design is similar to that of a focussing crystal monochromator of a triple axis spectrometer [29]. The Fermi chopper rotation frequency can be set between 30 Hz and 420 Hz, in 30 Hz increments, which yields the elastic energy resolution full width at half maximum (FWHM) from $\approx$ 10% down to $\approx$ 2.3% of E_i. The latter is roughly equivalent to halving the resolution FWHM of the triple axis spectrometer at the same E_i.

The sample stage and the detector vessel on HYSPEC move in registry with the rotating drum shield when the incident neutron energy is changed. The sample-monochromator optics is similar in design to a that of a triple-axis spectrometer. The cantilevered open air sample

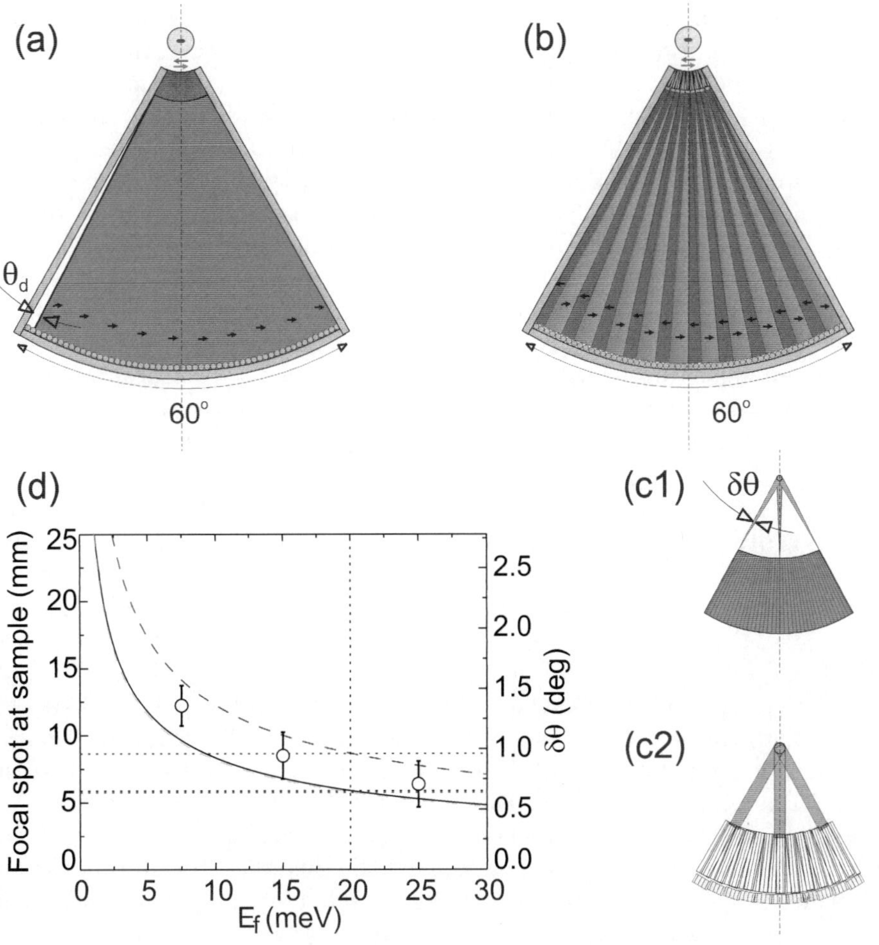

Figure 2: Schematics of the scattered neutron polarization analysis on HYSPEC: (a) as currently operational, with a radial supermirror polarizer that transmits the deflected neutrons with one polarization and absorbs neutrons with the opposite polarization, and (b) the initially considered multi-splitter-polarizer setup [1, 30], where both the deflected and the transmitted neutron beams with the opposite polarizations are separated on the detector and measured simultaneously. The sample scattering volume accepted by the radial supermirror polarizer, (a), is determined by the supermirror critical angle and the distance to the focal spot at the sample position, (c1), while in the initially considered multi-polarizer setup, (b), the accepted sample size is determined by the width ($\sim$ 20 mm) and the collimation ($\sim 20'$) of each polarizer channel, (c2). The radial polarizer effectively acts as a radial collimator, where the angular acceptance is determined by the supermirror critical angle and is energy-dependent, and so is the accepted sample size, (d). The solid line in (d) shows the effective Gaussian FWHM angular acceptance (right scale) and the sample size (left scale) corresponding to $4.5\theta_c^{Ni}$, assuming $m = 3$ on the outer and $m = 1.5$ on the inner surface of each supermirror channel [39]. Originally it was planned to coat the inner surface with $m = 3$ and the outer surface with $m = 1.5$ but to avoid undefined bend of the glass substrates (coming from different stress applied by the different number of multilayers) both coatings were made with m=3. The upper broken line shows the square-window acceptance for $m = 4.5$ and the lower broken line for $m = 3$. The horizontal dotted lines illustrate typical values for $E_i = 20$ meV (vertical dotted line). The symbols show the FWHM of the beam profile at the sample position determined experimentally by scanning the translation of an incoherent scatterer (vertical plastic rod of 5mm diameter) in the beam.

stage, which provides sample rotation, translation, and tilts is attached with a rigid arm to the monochromator drum shield and moves together with the drum when the drum rotates. The collimators, slits, flippers, and other auxiliary primary beam line components are mounted on the optic stage attached to the drum. The entrance to the scattered neutron flight path in the Ar-filled detector vessel is defined by a radial collimator (in an unpolarized mode), at about 50 cm from the sample axis. This permits the use of a variety of sample environments.

The scattered neutron energy is determined by the time-of-flight between the sample and the detectors. The detector bank is a cylindrical array of 20 8-packs of 1.2 m long, 1″ diameter ^{3}He linear position sensitive (128-pixel) detector tubes vertically mounted on the outside of the back wall of the detector vessel. Detectors are positioned coaxially with the sample rotation axis, at a nominal 4.5 m distance from it. The HYSPEC detector bank covers 60° horizontal by 15° vertical, for a nominal 0.25 sr solid angle (the actual physical coverage by detector pixels is 0.226 sr [31]). The detector vessel is attached with a rigid arm to the sample stage, employs an air pad system to enable motion, and can independently rotate around the sample axis to cover scattering angles from 0° up to 135° [3, 4].

3. HYSPEC setup and the measurement procedure in the polarized beam mode

The HYSPEC setup used for polarized beam measurements is shown in Figure 1(a). A vertically focusing (horizontally flat) array of magnetically saturated Heusler (Cu_2MnAl) crystals mounted in a magnetic yoke, Fig. 1(b), polarizes and vertically focuses the reflected neutron beam onto the sample. It is positioned in the drum shield in place of a pyrolytic graphite (PG) array, Fig. 1(c), using a vertical translation stage. The (initially horizontal) polarization of the neutron beam reflected by the horizontally magnetized Heusler array is aligned to the vertical direction by a static magnetic guide field provided by permanent magnets installed around the incident neutrons beam path, inside the drum shield. A sample, Fig. 1(d), is mounted inside a variable-temperature environment, which is placed on a standard 2-axes goniometer inside a water-cooled coil system capable of providing an adiabatic magnetic guide field of up to $\sim$ 50 Gauss. The direction of the guide field can be freely adjusted as required by experiment using the instrument control software. In particular, this allows a XYZ polarization analysis [41–43]. With coils, the sample tilts are limited to $\pm 2.5°$, or $\approx \pm 5°$, depending on the sample environment. Nevertheless, coils could can accommodate a range of sample environments, including 5 K to 600 K close-cycle refrigerators, 1.5 K to 325 K pumped-^{4}He "orange" cryostats, and a dilution refrigerator.

For the un-polarized measurements, an $\approx$ 4.3° radial collimator (RC, 0.67° channel spacing) is placed at the entrance of the detector vessel, $\approx$ 53 cm after the sample, Fig. 1(e). For the polarized beam measurements, RC is replaced by the similarly positioned, magnetically saturated, yoked 60°-wide-angle multi-channel radial supermirror-polarizer (RSP) array, Fig. 1(f) [38–40]. The calibration measurements performed using an incoherent scatterer (plastic, Vanadium, or TiZr alloy) show that while the transmission of the radial collimator is sufficiently flat as a function of scattering angle, the transmission of the RSP is rather inhomogeneous, with variations of up to $\pm \approx 40\%$. Although with the suitable calibration and correction procedure [44] this does not present operational problems, an improved alignment procedure is presently discussed, which has the goal to overcome this behavior. The RSP transmission is also strongly energy dependent, decreasing from $\approx 40\%$ at 5 meV to about 10% at 20 meV [44]. Because polarization selection is achieved by reflection(s) from the supermirror, the RSP deflects the transmitted polarized beam by an angle, θ_d, which is determined by the curvature of the supermirror channel, Fig. 2(a). The latter has been optimized to maximize the transmission while still avoiding the direct view, and thus keeping the high polarization efficiency [39, 40]. The curvature of the supermirror channel was calculated with the formula $R = L^2/8d$, where L is the length of the analyzer (171 mm) and d is the channel opening at the exit side (0.566 mm) plus the substrate thickness of 0.2 mm. Finally, the radius was adjusted

IOP Conf. Series: Journal of Physics: Conf. Series **862** (2017) 012030 doi:10.1088/1742-6596/862/1/012030

to 460 cm. Accounting for the scattered beam offset on the detector and the scattering angle dependence of the RSP transmission is carried out in MANTID [45] data preprocessing, by adjusting the detector efficiency for the corresponding events. The (optional) accounting for the energy-dependent angle-averaged RSP transmission, $T \approx T_0 e^{-E_F/E_0}$ ($T_0 \approx 0.59$, $E_0 \approx 10.8$), is performed at a later stage, on the histogrammed data. A detailed description of the polarized data analysis procedures is given in the accompanying paper [44].

The size of the sample scattering volume visible to detectors is determined by the size of the focal spot of either RC, or RSP. The latter is defined by the effective collimation, $\delta\theta$, and the distance (≈ 53 cm) between the sample and the RC, or RSP entrance, respectively, Fig. 2(c1). While in the case of a rather coarse RC the size of the focal spot is well matched to the nominal 4 cm wide (maximum) illuminated sample size as determined by the beam (guide) width, in the case of the RSP the focal spot size is much smaller and is also energy-dependent, Fig. 2(d). This is because the effective collimation of the narrow channels of the RSP is controlled by the critical angle of the supermirror coating, which is small and depends on the neutron wavelength. The resulting reduced accepted sample scattering volume imposes considerable penalty in intensity for larger samples, and also requires careful alignment of the sample on the rotation axis, and co-alignment of the sample axis with that of the RSP. A misalignment of the sample with respect to the rotation axis could move it in/out of the RSP focal view as a function of the rotation angle resulting in a spurious intensity variation, a potential source of a systematic error. This problem is easily solved when the analyzer has an adjustable rotation as foreseen in the design.

Currently, the polarizer alignment mechanism, which is an important feature envisioned in the design, is not yet installed. With the default, fixed RSP alignment used at present there is an offset of ≈ 5 mm ($0.5°$) between the RSP focal spot and the sample rotation axis on HYSPEC. This improves the RSP polarization efficiency reducing the probability for transmission of the undesired neutron polarization by $\approx 20\%$ (the polarization efficiency increases from $\approx 94\%$ to $\approx 95\%$) at the cost of a similar, $\approx 25\%$ intensity loss for $E_i \approx 6$ meV neutrons [40]. During the design phase of the RSP is was decided to concentrate on a high polarization mode at low E_i. Now we know, this offset is disadvantageous on HYSPEC, as it leads to a progressively larger loss of flux from the sample at higher E_i, and also introduces an ill-controlled systematic error associated with the rotation of a single crystal sample. The commissioning of the polarizer alignment mechanism, which would allow centering the RSP focal spot at the sample is planned for the near future.

The sample size limitations imposed by the RSP potentially could be alleviated in the beam-splitter [30] multi-analyzer setup, Fig. 2(b),(c2), initially considered for HYSPEC [1]. However, the potential gain in sample size visible to detectors, which is determined by the width (2 cm) of a single channel, would be outweighted by the loss from the $20'$ collimation required in such a setup. There would also be significant disadvantages associated with the poorly controlled polarization mixing on detector, leading to a detector- and energy-dependent flipping ratio (FR), and a complicated measurement and calibration procedure. The follow-up detailed comparative studies and simulations of the different polarization analysis options [38, 39] have favored the radial supermirror polarizer implemented on HYSPEC [3].

4. Polarization analysis

4.1. Spin-flip and non-spin-flip intensity measurement

The HYSPEC polarizing supermirror array can be magnetized either up, or down using the dedicated electromagnet charging station. The default mode of operation is to have the RSP magnetized so that it transmits neutron polarization opposite to the polarization of the incident beam, which is reflected by the Heusler crystal array (magnetized horizontally) and adiabatically rotated to the vertical direction by a suitably coupled permanent magnetic guide field. The measurement with the Mezei spin-flipper in front of sample turned off then yields a spin-flip

(SF) scattering intensity and the measurement with the flipper on yields a non-spin-flip (NSF) signal. With this choice, the flipper efficiency (which is very close to 1, [3]) shows up in the NSF intensity and does not affect the SF intensity.

The finite efficiencies of the polarizing elements lead to a finite probability, p, for the non-spin-flip sample scattering intensity, I_{NSF}, to be detected with the flipper off and contribute to the measured spin-flip intensity, $I_{SF}^{exp} = pI_{NSF} + (1-p)I_{SF}$, and vice versa, for the I_{SF} to contribute to the measured $I_{NSF}^{exp} = (1-p)I_{NSF} + pI_{SF}$. This probability is quantified by the experimentally measured flipping ratio, FR, which is determined as the ratio of the scattered intensities measured in the non spin flip, I_{NSF}^{exp}, and the spin flip, I_{SF}^{exp}, channels, or with the flipper ON and the flipper OFF, respectively, for a non-magnetic nuclear Bragg scattering, I_{nuc}. If we denote the instrument flipping ratio, $F = 1/p$, as an inverse probability, p, for a neutron scattered by sample in the NSF channel to be measured in the SF channel, and vice versa, then $I_{SF}^{exp} = pI_{NSF} = \frac{1}{F}I_{nuc}$, and $I_{NSF}^{exp} = (1-p)I_{NSF} = \frac{F-1}{F}I_{nuc}$,

$$FR = \frac{I_{NSF}^{exp}}{I_{SF}^{exp}} = \frac{1}{p} - 1 = F - 1. \tag{1}$$

The corrected sample scattering intensities, I_{SF} and I_{NSF}, are then obtained from the measured ones in a conventional way, by applying this measured flipping ratio correction,

$$\begin{aligned} I_{SF} &= \frac{1-p}{1-2p}I_{SF}^{exp} - \frac{p}{1-2p}I_{NSF}^{exp} = \frac{FR}{FR-1}I_{SF}^{exp} - \frac{1}{FR-1}I_{NSF}^{exp}, \\ I_{NSF} &= \frac{1-p}{1-2p}I_{NSF}^{exp} - \frac{p}{1-2p}I_{SF}^{exp} = \frac{FR}{FR-1}I_{NSF}^{exp} - \frac{1}{FR-1}I_{SF}^{exp}. \end{aligned} \tag{2}$$

4.2. The configuration of the guide field and XYZ polarization analysis

The neutron polarization at the sample position, $\mathbf{P}$, can be adiabatically guided to point in any predetermined direction with respect to the instrument coordinate frame using the guide field provided by the coil system, Fig. 1(a), [3]. In a conventional triple-axis setup with narrow, collimated beams, this uniquely determines the orientation of the polarization, $\mathbf{P}$, with respect to the wave vector transfer, $\mathbf{Q} = \mathbf{k}_i - \mathbf{k}_f$ ($\mathbf{k}_i$ and $\mathbf{k}_f$ are the incident and the scattered neutron wave vectors). In a wide-angle TOF setup, on the other hand, the detector bank registers neutrons with scattered wave vectors, $\mathbf{k}_f$, filling a large volume of phase space and having a wide range of directions and magnitudes. As a result, the direction of $\mathbf{Q}$ varies significantly as a function of the energy transfer and the position of the detector pixel, which defines the direction of the scattered neutron wave vector, $\mathbf{k}_f$. This direction is conveniently parameterized by two angles, the scattering angle in the horizontal plane, $S2$, between $\mathbf{k}_i$ and the horizontal projection of $\mathbf{k}_f$, and the vertical angle, η_v, of $\mathbf{k}_f$ with respect to the horizontal plane. S2 is determined by the position of a (vertical) detector tube in the detector bank and of the detector bank itself, while η_v is determined by the pixel position on the detector tube.

From wave vector conservation, it is easy to obtain the expression for the angle, $\theta_{\mathbf{Q}}$, between $\mathbf{Q}$ and $\mathbf{k}_i$ (the incident beam direction),

$$\sin\theta_{\mathbf{Q}} = -\frac{k_f}{Q}\sin(2\theta_s), \tag{3}$$

where $2\theta_s$ is the sample scattering angle, i.e. the angle between $\mathbf{k}_i$ and $\mathbf{k}_f$ in the scattering plane. In general, the scattering plane, which is defined by these two wave vectors is not horizontal (for out-of-plane detector pixels), but is at an angle α to the horizontal plane, where

$$\sin\alpha = \frac{\sin\eta_v}{\sin(2\theta_s)}, \quad \cos\alpha = \frac{\sin(S2)}{\sin(2\theta_s)}\cos\eta_v = \frac{\tan(S2)}{\tan(2\theta_s)}. \tag{4}$$

The scattering angle is given by,

$$\cos(2\theta_s) = \cos(S2)\cos\eta_v, \tag{5}$$

so $2\theta_s > S2$ for $S2 < 90°$. For small scattering angles, the scattering plane can be strongly tilted even if η_v is small, e.g. for small $\eta_v \approx S2$, $\alpha \approx 45°$. The angle with respect to the horizontal plane for the wave vector transfer, $\eta_{\mathbf{Q}}$, is given by a relation similar to Eq. (3),

$$\sin\eta_{\mathbf{Q}} = -\frac{k_f}{Q}\sin\eta_v. \tag{6}$$

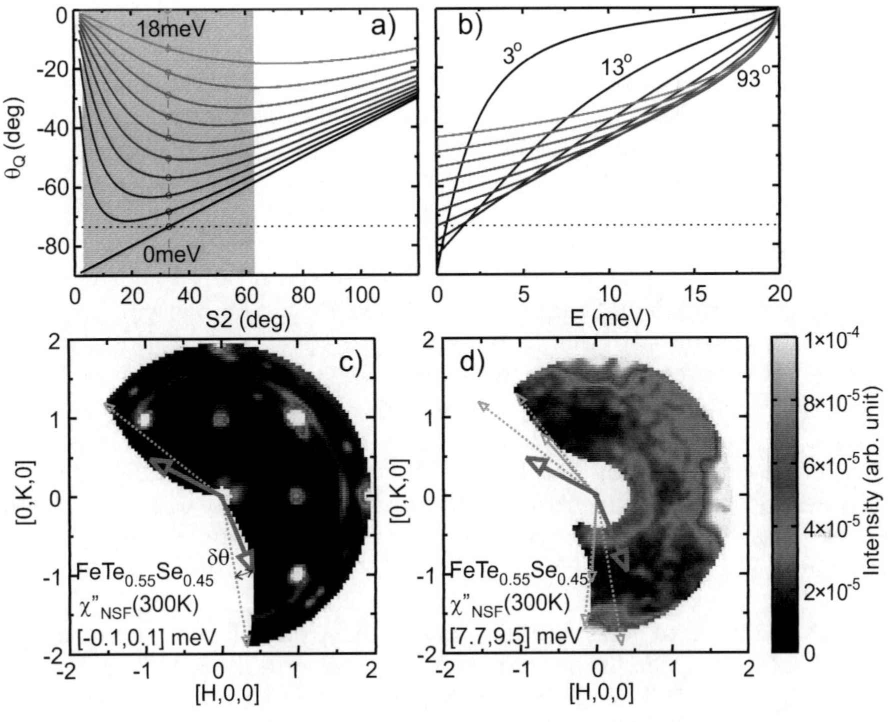

Figure 3: Dependence of the angle between the incident beam direction, $\boldsymbol{k}_i$, and the wave vector transfer, $\boldsymbol{Q}$, on the scattering angle in the horizontal plane, $S2$, for different energy transfers, (a), and on the energy transfer for different scattering angles, (b), for $E_i = 20$ meV. Curves in (a) correspond to energy transfers from 0 meV to 18 meV, with the step of 2 meV, from bottom to top. The shaded area shows the 60° range of scattering angles accepted by the analyzer-detector system; the vertical dotted line and symbols correspond to the detector center, positioned at $S2 = 33°$. Curves in (b) correspond to scattering angles from 3° to 93°, with the step of 10° (dark to light). In measurements with the horizontal guide field (HF) presented here, the neutron polarization, $\boldsymbol{P}$, which is determined by the guide field at the sample, is set parallel to the wave vector transfer at zero energy transfer for the scattering angle at the center of the detector, dotted horizontal line in (a), (b). This is illustrated by the thick solid grey arrows in (c), where the slice of the measured elastic intensity is shown (two arrows correspond to the limits of the sample rotation range). The dotted grey arrows show the respective elastic wave vector transfers at the detector edges. The same arrows are shown on the inelastic data slice in (d) together with the thin solid and dotted arrows, which illustrate the (rotated) wave vector transfer, $\boldsymbol{Q}$, at the same detector scattering angles for the energy transfer $E = 8.6(9)$ meV.

Even though detector vertical acceptance on HYSPEC is relatively small, $\eta_v < 7.5°$, for small wave vector transfers, $Q < k_f$, $\cos(2\theta_s) > \frac{k_i}{2k_f}$, $\eta_{\mathbf{Q}}$ can be large. For neutron energy-loss scattering, $k_i \geq k_f$, $\left(\frac{Q}{k_f}\right)^2 = \left(\frac{k_i}{k_f} - 1\right)^2 + 4\frac{k_i}{k_f}\sin^2\theta_s \geq 4\sin^2\theta_s$, and an upper bound for $\eta_{\mathbf{Q}}$ is given by,

$$\sin\eta_{\mathbf{Q}} \leq \frac{\sin\eta_v}{2\sin\theta_s}. \tag{7}$$

Here the equality corresponds to elastic scattering, where $\eta_{\mathbf{Q}}$ is the largest. For the top and the bottom detector pixels on HYSPEC, $\eta_v \approx 7.5°$, this gives (in the elastic channel) $\eta_{\mathbf{Q}}$ of $\approx 68°$, $\approx 23°$, $\approx 13°$ and $\approx 7°$ for the horizontal scattering angle, $S2$, of $3°$, $18°$, $33°$ and $63°$, respectively. The increase of $\eta_{\mathbf{Q}}$ with η_v is most pronounced at low angles, $S2 \leq 2\eta_v$. Only for $\theta_s \geq 30°$ ($S2 \geq 60°$), where $\sin\eta_{\mathbf{Q}} \leq \sin\eta_v$ holds, $\eta_{\mathbf{Q}}$ can be safely neglected if η_v is small.

What the above considerations show, is that the direction of $\mathbf{Q}$ varies strongly with respect to the plane defined by the two vectors, $\mathbf{k}_i$ and $\mathbf{P}$, whether it is horizontal ($\mathbf{P} \perp Z$), or vertical ($\mathbf{P} \| Z$). This variation is well beyond the paraxial approximation naïvely implied by small η_v, and is especially dramatic at small $S2$ angles. It presents a challenge for the polarization analysis on HYSPEC, where magnetic scattering is routinely measured down to $S2 \leq 3°$. In general, the XYZ polarization analysis for a strongly out-of-plane scattering is quite complicated and requires one to account for the variation of the angle $\eta_{\mathbf{Q}}$ as a function of detector pixel and time ($\mathbf{Q}$ and E). Methods for performing the XYZ NPA in such cases were considered in Refs. [41–43]. The advanced procedures for full NPA on HYSPEC are currently in development.

In the case of a coherent scattering in the horizontal plane, such as Bragg diffraction, the out-of-plane angular deviation of the scattered neutron wave vector is determined by the vertical resolution, and is $\eta_v \leq \pm 2.5°$ on HYSPEC. The assumption that $\mathbf{Q}$ is in the horizontal scattering plane, then typically constitutes a reasonable approximation; the deviations must be treated in the resolution correction formalism for NPA, which needs to be developed. The dependence of the angle between $\mathbf{Q}$ and the incident beam direction (Eq. 3) for scattering in the horizontal plane is illustrated in Fig. 3. It shows $\theta_{\mathbf{Q}}$, as a function of the scattering angle, $S2$, for a setup with $E_i = 20$ meV, which was used in the polarized inelastic measurements on $FeTe_{0.55}Se_{0.45}$ presented below (Fig. 5).

For the full XYZ NPA, at least 3 orientations of neutron polarization on sample have to be measured [41–43]. In a polarized neutron experiment on a TAS with narrow, collimated beams, this is usually accomplished by performing measurements using the vertical guide field (VF, $\mathbf{P} \| Z$) and two mutually perpendicular horizontal field (HF) directions: along the wave vector transfer ($\mathbf{P} \| \mathbf{Q}$) and pependicular to it ($\mathbf{P} \perp \mathbf{Q}$). In the initial polarized measurements on HYSPEC reported here, we followed a similar procedure. The guide field defining the neutron polarization $\mathbf{P}$ at the sample was either vertical (VF), or horizontal (HF) and such that condition $\mathbf{P} \| \mathbf{Q}$ was satisfied for elastic scattering at the center of the detector bank (we still call this configuration $\mathbf{P} \| \mathbf{Q}$). According to Eq. 3, in the case of elastic scattering ($k_i = k_f$, $Q = 2k_f \sin\theta_s$),

$$\theta_{\mathbf{Q}} = \theta_s - 90°, \tag{8}$$

which also follows from the scattering geometry. In our measurements, the detector bank was positioned to span horizontal scattering angles $S2$ in the $[-63°, -3°]$ range, and HF ($\mathbf{P}$) was directed at $\theta_{\mathbf{Q}} = -106.5°$ (Fig. 3, (a,b) illustrates the opposite scattering side, where $S2$ is in the $[3°, 63°]$ range and $\theta_{\mathbf{Q}} = -73.5°$). Although the deviation of $\mathbf{P}$ from the $\mathbf{Q}$ direction as a function of energy transfer (E_f) is large, it is only $\pm 15°$ across the whole detector bank for the elastic scattering, and even less than that for the angular range where Bragg reflections are observed (Fig. 4). Hence, neglecting the deviation of $\theta_{\mathbf{Q}}$ in the elastic channel is reasonably accurate (to within $\sim 10\%$). The wave vector range spanned by the detector bank and its rotation between

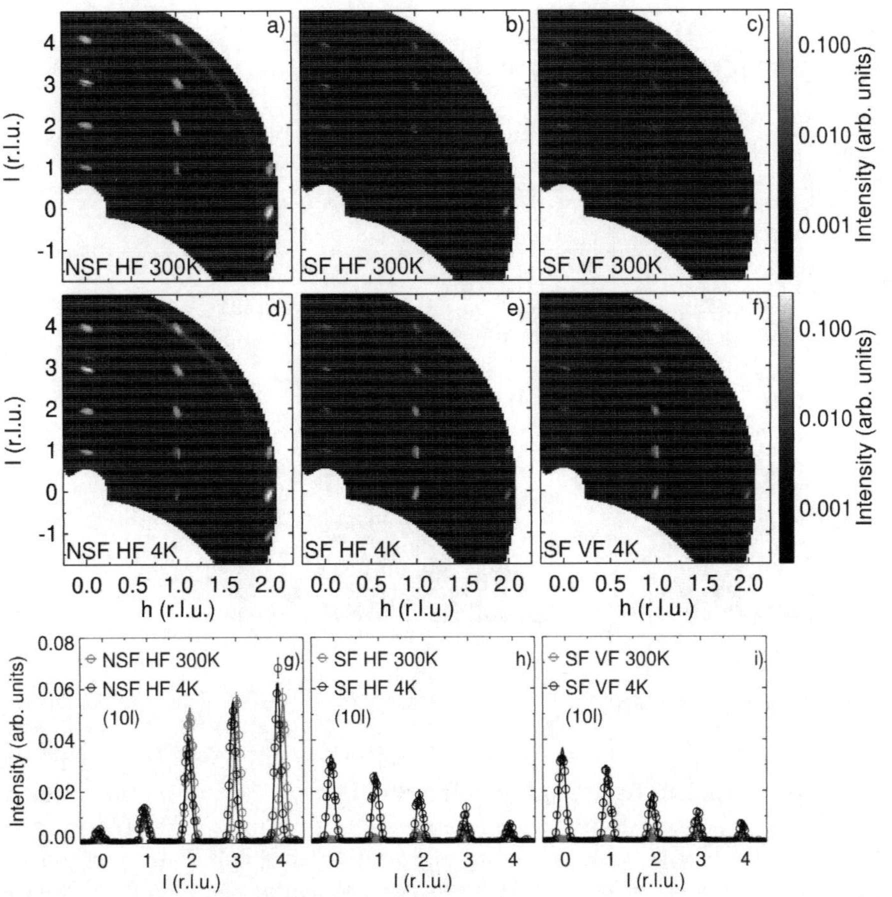

Figure 4: Polarized neutron diffraction from the $YbMnBi_2$ ($\approx$ 0.76 g) sample in the $(h, 0, l)$ plane measured on HYSPEC with E_i = 15 meV and ν_{Fermi} = 120 Hz. The data was collected by rotating the sample with a 0.5° step within the range of 130°, with about 25 s counting time at each angle setting. The measurement was repeated in the non-spin-flip (NSF, flipper on) and the spin-flip (SF, flipper off) channels with the horizontal guide field (HF) at the sample, at 300 K (a,b) and at 4 K (d,e), and in the SF channel with the vertical guide field (VF), also at 300 K (c) and 4 K (f). Apart from a small shift in position resulting from the temperature-dependent lattice parameters, there is essentially no change between 300 K and 4 K for $(0, 0, l)$ peaks, (a-f), and for $(1, 0, l)$ peaks in the NSF channel, (g), where the nuclear lattice scattering dominates. There is, however, significant intensity arising at 4 K at the $(1, 0, l)$ lattice positions, which shows up in the spin-flip channel, (h,i), manifesting the appearance of the antiferromagnetic order.

the elastic and the inelastic ($E \approx 8.6$ meV) slices of the typical HYSPEC data measured in the polarized mode with E_i = 20 meV are illustrated in Fig. 3, (c) and (d), respectively.

5. Polarized neutron diffraction in $YbMnBi_2$

Figure 4 presents the results of the polarized neutron diffraction measurements on a candidate Weyl metal $YbMnBi_2$ [46]. Measurements were performed on a 0.8 g high quality single crystal, with E_i = 15 meV neutrons and the Fermi chopper at 120 Hz. Each guide field/flipper configuration required about 100 min. counting time for 130° crystal rotation. The nuclear Bragg intensities measured in a magnetically disordered phase at 300 K indicate a flipping ratio FR = 14(1), consistent for all guide field directions and for different Bragg peaks (sample

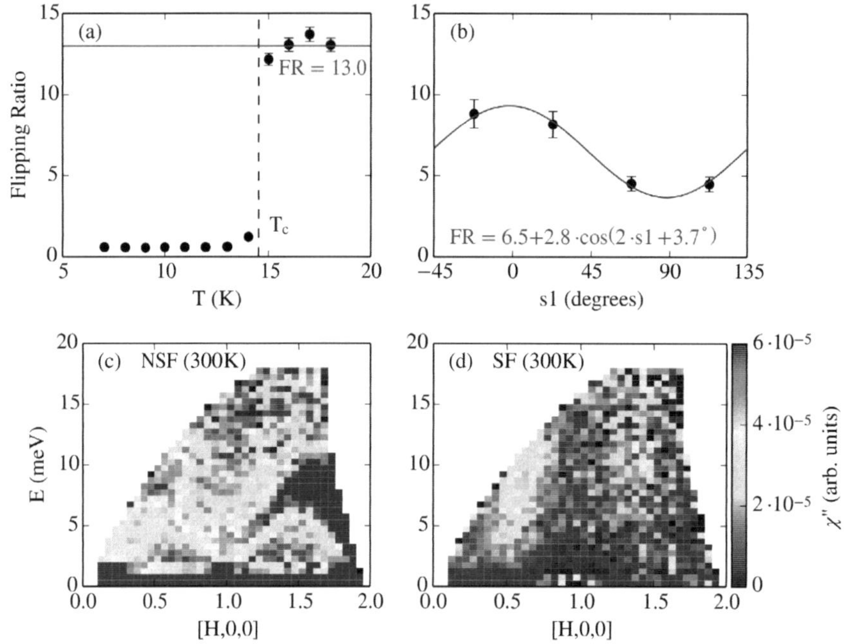

Figure 5: Data from the polarized inelastic measurement on HYSPEC (E_i = 20 meV, ν_F = 120 Hz) of a superconducting $FeTe_{0.55}Se_{0.45}$ single crystal (m = 23.4 g, mosaic $\approx 2.2°$ FWHM) in the $(h, k, 0)$ zone. (a) Bragg flipping ratio measured on warming after the sample was cooled through the superconducting transition temperature, T_c, in a horizontal guide field of $\approx$ 20 Gauss. (b) Bragg flipping ratio at T = 5 K in the same horizontal guide field ($\approx$ 20 G), as a function of sample rotation angle after the sample is cooled in zero magnetic field. The imaginary part of the dynamical susceptibility obtained from the non-spin-flip (NSF), (c), and the spin-flip (SF), (d) intensity measured at T = 300 K by correcting for energy-dependent polarizer transmission and the detailed balance factor.

rotation angles). The data shown in Fig. 4 was corrected for this flipping ratio according to Eq. 2.

$YbMnBi_2$ crystallizes in a tetragonal (P4/nmm) symmetry, with alternating layers of square nets of Bi atoms hosting Dirac-like charge carriers, and square-lattice magnetic layers of Mn^{2+}, with 2 Mn per unit cell, structurally similar to those in cuprates and Fe-based superconductors, and with Yb spacer layers. $YbMnBi_2$ orders antiferromagnetically below $\sim$ 300 K. The antiferromagnetic and crystallographic unit cells are identical, and therefore magnetic and nuclear Bragg positions coincide. However, magnetic and nuclear scattering can be separated using neutron polarization analysis at HYSPEC, as shown in Fig. 4. Measurements confirm the magnetic structure determined with unpolarized neutrons and a refined Mn magnetic moment of $4.3(1)\mu_B$, oriented along the c-axis.

6. Inelastic magnetic and non-magnetic scattering in $FeTe_{1-x}Se_x$

Magnetic excitations in the parent materials of the chalcogenide family, $Fe_{1+y}Te$, and in the superconducting versions of it obtained by partial substitution of Te by Se, or S, reveal rather unusual behavior, which does not comply with the standard spin-wave description. In particular, dynamical magnetism increases in intensity with increasing temperature [25], a rare

IOP Conf. Series: Journal of Physics: Conf. Series **862** (2017) 012030 doi:10.1088/1742-6596/862/1/012030

and unexpected phenomenon. This unusual temperature dependence of magnetic intensity suggests a non-trivial interaction between local spins, which give rise to the bulk of the spectral intensity, and the itinerant band electrons, which also contribute to it. So far, scattering patterns have been identified as magnetic based on their **Q**-dependence, but magnetic and lattice contributions can only be distinguished using polarized neutrons. The maxima in the energy dependence of the magnetic intensity occur near the top of the acoustic phonon dispersion (~ 8 meV), at the exact positions where diffuse columns of magnetic scattering meet the phonons. Very similar observations were made in superconducting $FeTe_{1-x}(S,Se)_x$ samples [47, 48], where temperature-induced dynamical magnetism was also observed.

The magnetism and the lattice in iron chalcogenides are closely coupled. A bond-order wave (BOW) transition leading to ferro-orbital order in the ground state associated with formation of zigzag Fe-Fe chains stabilizes the bicollinear antiferromagnetism [27]. This latter transition entails lowering of the lattice symmetry so that structure factor for the (1,0,0) lattice Bragg reflection, which is forbidden at high T, becomes non-zero. Surprisingly, an acoustic-phonon-like dispersion, centered at (1,0,0), is observed in the inelastic spectra in both the monoclinic P21/m and the tetragonal P4/nmm structural phases [28]. This mode is unexpected because although in the monoclinic phase the phonon mode is accompanied by a Bragg peak at $\mathbf{Q} = (1, 0, 0)$, in the tetragonal phase elastic Bragg scattering at this **Q** is forbidden by unit cell symmetry and no accompanying Bragg peak was observed despite the continued observation of an acoustic phonon mode.

Using polarized neutrons on HYSPEC enables us to ascertain which scattering is magnetic or structural in origin, and allows to understand the interaction between electronic magnetism and the lattice degrees of freedom in this family of materials. Figure 5 presents the results of the polarized neutron inelastic scattering measurements on a single crystal sample [shown in Fig. 1(d)] of an optimally doped 11 iron chalcogenide superconductor $FeTe_{0.55}Se_{0.45}$ [48]. After the sample has been cooled through the superconducting transition at $T_c \approx 14$ K in a 20 Gauss guide field, the flipping ratio deteriorated, because the neutron beam was completely depolarized by small random magnetic fields frozen inside the sample, Fig. 5(a). This problem was largely resolved by cooling the sample in a zero field environment. Both placing the cryostat in a μ-metal shield, and on the sample table and using the compensating field created by the coil system, lead to similar results shown in Fig. 5(b). After the zero-field cooling, the flipping ratio varied between ≈ 4 and ≈ 8, as a function of the sample rotation angle, $S1$. This is understandable, because the large and irregularly shaped superconducting sample still somewhat depolarizes the neutron beam, and this depolarization depends on the sample orientation with respect to the guide field. Symmetry suggests that this dependence should be 180°-periodic as a function of $S1$, $FR = FR_0 + f\cos(2S1 + \phi_0)$, Fig. 5(b).

The flipping ratios at several different $S1$ were obtained from the SF and NSF intensities of the nuclear Bragg peaks, which were measured in the same experiment as the inelastic data [see Fig. 3(c)]. Even though FR is not very high and angle-dependent, the data can be reliably corrected using Eq. (2) and the fitted angular dependence of FR. The flipping ratio for the 300 K data presented in Fig. 5(c),(d), on the other hand, was 13(1), independent of $S1$. The data shown in the figure was corrected for this flipping ratio according to Eq. 2. The measured NSF intensity [Fig. 5(c)] clearly reveals the acoustic phonon mode coming out of the forbidden $(1, 0, 0)$ Bragg peak, joining at ≈ 8 meV with some NSF intensity at small angles, which is probably of magnetic origin. The SF intensity [Fig. 5(d)] reveals columns of magnetic scattering above a gap of ≈ 5 meV, at $\mathbf{Q} = (0.5, 0, 0)$ and $(1.5, 0, 0)$.

7. Summary and conclusions

This paper describes the instrument setup, and procedures for the polarized beam measurements on the HYbrid SPECtrometer at the SNS. We also present the first results of the diffraction and inelastic measurements with polarization analysis. Despite the significant penalty in intensity imposed by the lower (compared to PG) reflectivity of Heusler crystals, and low transmission of the supermirror-array polarization analyzer, broad surveys of the scattered SF and NSF intensities are possible for incident neutron energies as high as 20 meV. For non-ferromagnetic and non-superconducting samples the flipping ratio is high, in the range of 13 to 15. However, broad polarized beam surveys, such as performed on HYSPEC, can tolerate situations where it is reduced. In an experiment with sample rotation, the flipping ratio can be measured on several nuclear Bragg peaks throughout the sample rotation range, which allows reliable FR correction using Eq. (2). While distinguishing SF and NSF scattering is straightforward, the full XYZ polarization analysis requires a complex procedure accounting for the change in $\mathbf{Q}$ orientation as a function of time of flight and detector position, which is still in development. Another, related challenge is in devising a resolution correction procedure for the XYZ NPA, which accounts for the appreciable vertical spread of the focussed beam, whose impact is especially severe at small angles.

8. Acknowledgements

Authors are grateful to M. Hagen for his leadership role in HYSPEC construction, to M. Kenzelmann for valuable discussions, and to M. Graves-Brooks for help with the measurements. This work was supported by the Office of Basic Energy Sciences (BES), Division of Materials Sciences and Engineering, U.S. Department of Energy (DOE), under Contract No. DE-SC00112704. This research used resources at the Spallation Neutron Source, a DOE Office of Science User Facility operated by the Oak Ridge National Laboratory. This work was supported by the Paul Scherrer Institut by providing the supermirror analyzer as a temporary loan to Oak Ridge National Laboratory. This manuscript has been co-authored by UT-Battelle, LLC under Contract No. DE-AC05-00OR22725 with the U.S. Department of Energy. The United States Government retains and the publisher, by accepting the article for publication, acknowledges that the United States Government retains a non-exclusive, paid-up, irrevocable, worldwide license to publish or reproduce the published form of this manuscript, or allow others to do so, for United States Government purposes. The Department of Energy will provide public access to these results of federally sponsored research in accordance with the DOE Public Access Plan (http://energy.gov/downloads/doe-public-access-plan).

References

[1] Zaliznyak I, Ghosh V, Shapiro S and Passell L 2005 *Physica B: Condensed Matter* **356** 150 – 155 ISSN 0921-4526

[2] Shapiro S, Zaliznyak I, Passell L, Ghosh V, Leonhardt W and Hagen M 2006 *Physica B: Condensed Matter* **385–386, Part 2** 1107 – 1109 ISSN 0921-4526

[3] Winn B, Filges U, Garlea V O, Graves-Brook M, Hagen M, Jiang C, Kenzelmann M, Passell L, Shapiro S M, Tong X and Zaliznyak I 2015 *EPJ Web of Conferences* **83** 03017

[4] Details of HYSPEC technical specifications, design and configuration can be found on http://neutrons.phy.bnl.gov/HYSPEC/memos_technical_reports.shtm. More information about the instrument can be found on the HYSPEC IDT website at BNL, http://neutrons.phy.bnl.gov/HYSPEC/, or on the SNS HYSPEC page at ORNL, http://neutrons.ornl.gov/hyspec/.

[5] Moon R M, Riste T and Koehler W C 1969 *Phys. Rev.* **181**(2) 920–931

[6] Maleev S V 1961 *Zh. Éksp. Teor. Fiz **40**, 1224 – 1227 (1961); Soviet Phys. – JETP* **13** 860 – 862

[7] Izyumov Y A and Maleev S V 1962 *Zh. Éksp. Teor. Fiz **41**, 1644 – 1648 (1962); Soviet Phys. – JETP* **14** 1168 – 1171

[8] Blume M 1963 *Phys. Rev.* **130**(5) 1670–1676

[9] Pappas C, Ehlers G and Mezei F 2006 *Neutron Scattering from Magnetic Materials* ed Chatterji T (Amsterdam: Elsevier Science) pp 521 – 542 ISBN 978-0-444-51050-1

[10] Regnault L 2006 *Neutron Scattering from Magnetic Materials* ed Chatterji T (Amsterdam: Elsevier Science) pp 363 – 395 ISBN 978-0-444-51050-1

[11] Maleev S V 2002 *Physics-Uspekhi* **45** 569

[12] Schweizer J 2006 *Neutron Scattering from Magnetic Materials* ed Chatterji T (Amsterdam: Elsevier Science) pp 153 – 213 ISBN 978-0-444-51050-1

[13] Brown P 2001 *Physica B: Condensed Matter* **297** 198 – 203 ISSN 0921-4526 proceeding of the Third International Workshop on Polarised Neutrons

[14] Raymond S, Yokoo T, Zheludev A, Nagler S E, Wildes A and Akimitsu J 1999 *Phys. Rev. Lett.* **82**(11) 2382–2385

[15] Lake B, Tennant D A and Nagler S E 2000 *Phys. Rev. Lett.* **85**(4) 832–835

[16] Regnault L, Rönnow H, Boullier C, Lorenzo J and Marin C 2004 *Physica B: Condensed Matter* **345** 111 – 118 ISSN 0921-4526 proceedings of the Conference on Polarised Neutron and Synchrotron X-rays for Magnetism

[17] Lake B, Tennant D A and Nagler S E 2005 *Phys. Rev. B* **71**(13) 134412

[18] Lake B, Tennant D A, Frost C D and Nagler S E 2005 *Nature Materials* **4** 329

[19] Fong H F, Bourges P, Sidis Y, Regnault L P, Bossy J, Ivanov A, Milius D L, Aksay I A and Keimer B 2000 *Phys. Rev. B* **61**(21) 14773–14786

[20] Headings N S, Hayden S M, Kulda J, Babu N H and Cardwell D A 2011 *Phys. Rev. B* **84**(10) 104513

[21] Raymond S and Lapertot G 2015 *Phys. Rev. Lett.* **115**(3) 037001

[22] Plakhty V P, Maleyev S V, Kulda J, Wosnitza J, Visser D and Moskvin E 1999 *EPL (Europhysics Letters)* **48** 215

[23] Plakhty V P, Schweika W, Brückel T, Kulda J, Gavrilov S V, Regnault L P and Visser D 2001 *Phys. Rev. B* **64**(10) 100402

[24] Lorenzo J E, Boullier C, Regnault L P, Ammerahl U and Revcolevschi A 2007 *Phys. Rev. B* **75**(5) 054418

[25] Zaliznyak I A, Xu Z, Tranquada J M, Gu G, Tsvelik A M and Stone M B 2011 *Phys. Rev. Lett.* **107**(21) 216403

[26] Soh J H, Tucker G S, Pratt D K, Abernathy D L, Stone M B, Ran S, Bud'ko S L, Canfield P C, Kreyssig A, McQueeney R J and Goldman A I 2013 *Phys. Rev. Lett.* **111**(22) 227002

[27] Fobes D, Zaliznyak I A, Xu Z, Zhong R, Gu G, Tranquada J M, Harriger L, Singh D, Garlea V O, Lumsden M and Winn B 2014 *Phys. Rev. Lett.* **112**(18) 187202

[28] Fobes D M, Zaliznyak I A, Tranquada J M, Xu Z, Gu G, He X G, Ku W, Zhao Y, Matsuda M, Garlea V O and Winn B 2016 *Phys. Rev. B* **94**(12) 121103

[29] Shirane G, Shapiro S M and Tranquada J M 2002 *Neutron Scattering with a Triple-Axis Spectrometer* (Cambridge University Press) ISBN 9780511534881 cambridge Books Online

[30] Zaliznyak I A and Lee S H 2005 *Modern Techniques for Characterizing Magnetic Materials* ed Zhu Y ISBN 978-1-4020-8007-4

[31] Stone M B, Niedziela J L, Abernathy D L, DeBeer-Schmitt L, Ehlers G, Garlea O, Granroth G E, Graves-Brook M, Kolesnikov A I, Podlesnyak A and Winn B 2014 *Review of Scientific Instruments* **85** 045113

[32] Ewings R, Buts A, Le M, van Duijn J, Bustinduy I and Perring T 2016 *Nuclear Instruments and Methods in Physics Research Section A: Accelerators, Spectrometers, Detectors and Associated Equipment* **834** 132 – 142 ISSN 0168-9002

[33] Jiang C Y, Tong X, Brown D R, Culbertson H, Graves-Brook M K, Hagen M E, Kadron B, Lee W T, Robertson J L and Winn B 2013 *Review of Scientific Instruments* **84** 065108

[34] Jiang C Y, Tong X, Brown D R, Chi S, Christianson A D, Kadron B J, Robertson J L and Winn B L 2014 *Review of Scientific Instruments* **85** 075112

[35] Babcock E, Salhi Z and Ioffe A 2013 *Journal of the Physical Society of Japan* **82** SA030

[36] Babcock E, Salhi Z, Theisselmann T, Starostin D, Schmeissner J, Feoktystov A, Mattauch S, Pistel P, Radulescu A and Ioffe A 2016 *Journal of Physics: Conference Series* **711** 012008

[37] Stewart J R, Deen P P, Andersen K H, Schober H, Barthélémy J F, Hillier J M, Murani A P, Hayes T and Lindenau B 2009 *Journal of Applied Crystallography* **42** 69–84

[38] Filges U, Allenspach P, Hagen M and Zaliznyak I 2006 Presentation at the workshop on polarized inelastic neutron scattering (pins) at bnl (unpublished) URL `http://neutrons.phy.bnl.gov/HYSPEC/PINS_IDT_2006.shtm`

[39] Filges U, Allenspach P and Hagen M 2006 Simulations for the hyspec polarization options (unpublished) URL `http://lns00.psi.ch/mcworkshop/papers/bender_psi_filges.pdf`

[40] Filges U and Panzner T 2012 Characterization of the hyspec supermirror analyzer, experimental report (unpublished)

[41] Schärpf O and Capellmann H 1993 *physica status solidi (a)* **135** 359–379 ISSN 1521-396X

[42] Ehlers, G, Stewart, JR, Deen, PP and Andersen, KH 2015 *EPJ Web of Conferences* **83** 03004

[43] Ehlers G, Stewart J R, Wildes A R, Deen P P and Andersen K H 2013 *Review of Scientific Instruments* **84** 093901

[44] Savici A T, Zaliznyak I A, Garlea O V and Winn B 2017 *EPJ Web of Conferences* (PNCMI proceedings, this issue, unpublished)

[45] Arnold O, Bilheux J, Borreguero J, Buts A, Campbell S, Chapon L, Doucet M, Draper N, Leal R F, Gigg M, Lynch V, Markvardsen A, Mikkelson D, Mikkelson R, Miller R, Palmen K, Parker P, Passos G, Perring T, Peterson P, Ren S, Reuter M, Savici A, Taylor J, Taylor R, Tolchenov R, Zhou W and Zikovsky J 2014 *Nuclear Instruments and Methods in Physics Research Section A: Accelerators, Spectrometers, Detectors and Associated Equipment* **764** 156 – 166 ISSN 0168-9002

[46] Wang A, Zaliznyak I, Ren W, Wu L, Graf D, Garlea V O, Warren J B, Bozin E, Zhu Y and Petrovic C 2016 *ArXiv e-prints* (*Preprint* `1604.01009`)

[47] Zaliznyak I, Savici A T, Lumsden M, Tsvelik A, Hu R and Petrovic C 2015 *Proc. Natl. Acad. Sci. USA.* **112** 10316–10320 URL `http://www.pnas.org/content/112/33/10316.abstract`

[48] Xu Z, Schneeloch J A, Wen J, Božin E S, Granroth G E, Winn B L, Feygenson M, Birgeneau R J, Gu G, Zaliznyak I A, Tranquada J M and Xu G 2016 *Phys. Rev. B* **93**(10) 104517

IOP Conf. Series: Journal of Physics: Conf. Series **862** (2017) 012031 doi:10.1088/1742-6596/862/1/012031

New POLDI – project of reincarnation of a polarized neutron diffractometer at the reactor PIK

I Zobkalo[1,2], S Gavrilov[1], V Matveev[1], J Fenske[3]

[1] Condensed State Research Department, Petersburg Nuclear Physics Institute of National Research Centre Kurchatov Institute, Orlova Roshcha, Gatchina, 188300, Russia
[2] Physical Department, Saint-Petersburg State University, Ulyanovskaya 3, St. Petersburg, 198504, Russia
[3] Zentrum fur Material und Kustenforschung GmbH, Helmholz-Zentrum Geesthacht, Max-Planck-Straße 1, Geesthacht, 21502, Germany

Corresponding author's e-mail address: zobkalo@pnpi.spb.ru

Abstract. The project of a considerable modernization of the polarized neutron diffractometer POLDI is discussed. It assumes the adoption of POLDI to a broader range of magnetic investigations such as determination of magnetic structures, detailed investigation of complex magnetic structures, studies of magnetic domains, study of the magnetization density maps, magnetic form-factor particularities, local susceptibility, etc. The flexible construction should permit to use either spherical neutron polarimetry technique or flipping ratio technique. Different types of polarization system were analyzed. Original focusing fan-like bender is proposed as polarizer unit. Our simulations give evidence that for the wavelength range 1.3 - 3 Å and with suitable size, such a device can give much better efficiency than ^{3}He cells, which are often in use. The higher flux at the sample position of a factor of at least 3.3, with lower divergence and good polarization degree from 98% (1.3 Å) to above 94% (3 Å) makes the bender set-up favorable over the layout with a ^{3}He-cell.

1. Introduction.

The polarized neutron diffractometer POLDI was built by the Physikalisch-Technische Bundesanstalt (PTB), Braunschweig. Initially it was installed and operated at the 1 MW PTB research reactor. It was adapted to studies of ferromagnetic domain structures and ferromagnetic correlations in the range from 20 nm to several mm. POLDI was equipped with a set-up for full 3-dimensional depolarization analysis of the beam transmitted by the sample. In magnetized media, the neutron polarization changes during the passage through the sample, and this information can be obtained from the depolarization matrix **D**. This latter relates the initial polarization **Po** to the final polarization **P'** = **DPo**. In this way, the mean domain size δ and the magnetic texture γ in multi-domain samples could be extracted.

After the PTB reactor shutdown, POLDI was set up on the 5MW FRG-1 reactor at GKSS, where it continued to operate thanks to team from PTB under the guidance of Dr. Volker Wagner.

After that reactor was shut down, POLDI was delivered to PNPI (Gatchina), where it will be installed at new high-flux reactor PIK. The current plan of placement of POLDI on PIK is the 5th thermal channel in the main reactor hall (Fig.1), where it will be the first instrument on the channel, and the second will be another instrument, probably a polarized triple axes spectrometer.

Since we expect to have high intensity of polarized neutrons, it is natural to adapt POLDI to the broader range of magnetic investigations such as determination of magnetic structures, sound investigation of complex magnetic structures, studies of magnetic domains, study of magnetization density maps, magnetic form-factor particularities, local susceptibility, etc. Thorough magnetic investigations with polarized neutrons become more demanded as in recent years increasing attention is paid to magnetic materials with strong interacting order parameters. These include multiferroics which have some magnetic subsystems, compounds with colossal magnetoresistance and giant magnetocaloric effect, high temperature superconductors etc. The study of microscopic mechanisms responsible for interaction in such materials can ensure progress in the area of study and creation of new materials.

For the implementation of these possibilities, two main different working modes should be realized: spherical neutron polarimetry technique (SNP) [1,2] and flipping ratio technique (FRT) [3]. In both of these techniques, diffraction of polarized neutrons is used. In SNP the polarization matrix **P**(hkl) is measured for selected Bragg reflections. This matrix relates the change of neutron polarization in diffraction **P'**(hkl) = **P**(hkl)**P**o. In this case, the change of the polarization can take place due to a rotation of polarization and/or to the appearance of additional polarization in the diffraction process. The coefficients of the polarization matrix depend strongly on the particular type of magnetic scattering [2] and this permits to make unique investigations of complex magnetic structures. FRT in turn, uses the exceptional sensibility of polarized neutron scattering to weak magnetic responses. In this method, flipping ratios $R(hkl) = I^+(hkl)/I^-(hkl)$ for all possible Bragg reflections should be collected, where $I^+(hkl)$ is the reflection intensity with incident neutron spin "up" and $I^-(hkl)$ is the intensity with spin "down". In this way, one can obtain magnetization density maps, anisotropy of magnetic form-factors, etc. The essential difference between these two experimental methods is the completely opposite requirements for the magnetic field on the sample stage: for SNP there should be zero field on the sample position, which could be implemented with a zero-field device like CryoPAD [1], and for FRT there should be a very high magnetic field, which could be implemented with a cryomagnet. In order to satisfy these requirements the flexible construction of some stages is considered.

The important parts of the polarized diffractometer are polarizer and analyzer units. In order to choose the most effective way for realization of polarized neutron diffraction methods we performed a number of simulations in which we considered some different variants for the polarizer unit.

2. Simulations.

The numerical simulations were performed using the McStas software [4]. For the simulation of the PIK reactor source the following parameters have been used: temperature T=300 K and the estimation value of the brilliance for 5th PIK channel was given by $I_0 = 1.432 \cdot 10^{13}$ n/cm²sec·st.rad·Å, which is close the flux of thermal H2 channel of HFR ILL.

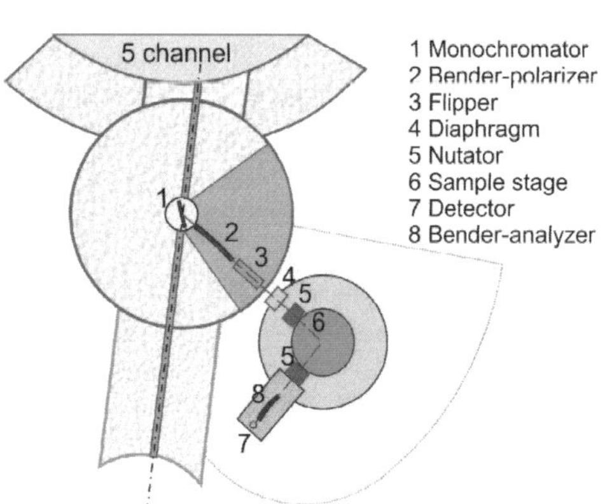

Figure 1. Placement of the diffractometer POLDI in the PIK main reactor hall on the 5th channel. Next to POLDI is supposed to be one more instrument.

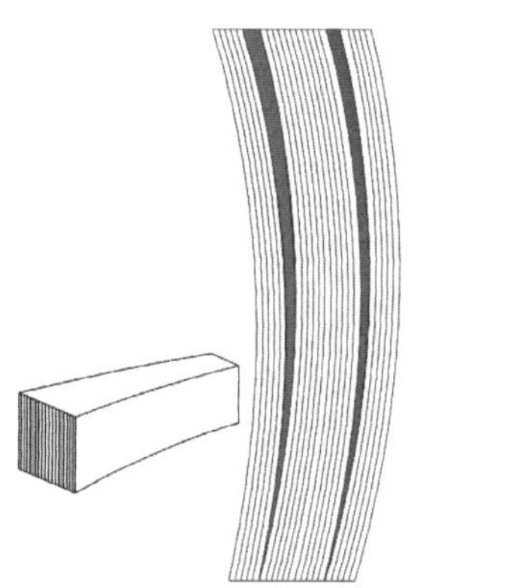

Figure 2. Three-section fan-like polarizing bender.

First, we made estimations of the use of a polarizing double focusing monochromator. Since the diffractometer POLDI will not be the only instrument installed on the channel, the most effective decision based on polarizing Heusler crystals Cu_2MnAl (111) is not suitable. As the main working wavelength λ should be ~ 1.3 Å, the take-off angle for Cu_2MnAl (d_{111}=3.362 Å) is very small and it will not be possible to place another instrument on the same beam. Therefore we considered polarizing focusing monochromator using FeCo single crystals (d = 1.771 Å). This leads to the appropriate take-off angle (~43°), though reflection efficiency is very small – about 4%. For the calculations the following parameters of the monochromator unit were chosen: 28×6 FeCo plates with 10×10 mm dimensions and mosaicity 30'; vertical focusing radius – 2202 mm, horizontal one – 16348 mm. These parameters yield a flux value Φ on the sample position equal to ~ $3.76 \cdot 10^5$ n/cm²sec.

A helium cell was chosen as polarizing unit for the next step of the simulation. In this variant we use monochromator model on the base of Cu single crystals with d_{200}=1.807 Å. The reflection efficiency of Cu (200) is ~ 50%. The number of crystals and their dimensions were taken as in the previous case. Vertical focusing radius here was 3052 mm, horizontal one – 16680 mm. The parameters of the helium cell were the following: height and radius were 100 mm, helium pressure – 2.75 bar and a polarization degree of about 70%. This configuration leads for neutrons with a wavelength of 1.3 Å to a polarization degree of about 96%. In these conditions the neutron flux on the sample position was calculated to be ~ $12.2 \cdot 10^5$ n/cm²sec, i.e. about 3 times higher, than in the case of a FeCo monochromator. The variant with FeCo polarizer was excluded from subsequent analysis. In addition, high activation of Co gives one more drawback of this monochromator.

As an alternative decision for a ^{3}He polarizer, we develop an idea of a compact focusing bender. The optimized construction of fan-like bender looks like the following. The bender consists of 3 sections as shown in Figure 2. The central section is divided into 26 channels. The width of each channel is 1.4 mm. The glass substrates have a thickness of about 0.3 mm and are coated with m=4. The outer sections consist of 9 channels and they are tilted about ±0.0167 rad with respect to the central section. The curvature of each section has a radius of 54012 mm leading to a characteristic wavelength of 1.04 Å. The length of 800 mm together with the curvature of the bender guarantees that the direct flight is blocked and that the neutrons are scattered at least once in the bender.

For the simulations, the same Cu double focused monochromator was used as for the helium cell set-up. The polarization degree for 1.3 Å has a value of > 98% and the intensity at the sample position has a value of about 4×10^6 n/cm²sec ± 1×10^5 n/cm²sec which is a factor of about 40 higher than at the FRG-1. In Figure 3 it can be seen that the polarization degree of the bender set-up is wavelength dependent. At 2.4 Å the polarization degree has a value of about 98% and drops to 94% for 3 Å. At the same time the flux decreases to ~1×10^5 n/cm²sec ± 3×10^3 n/cm²sec at 3 Å, which is a result of the Gaussian distribution of the thermal flux from the source, as can be seen in figure 4.

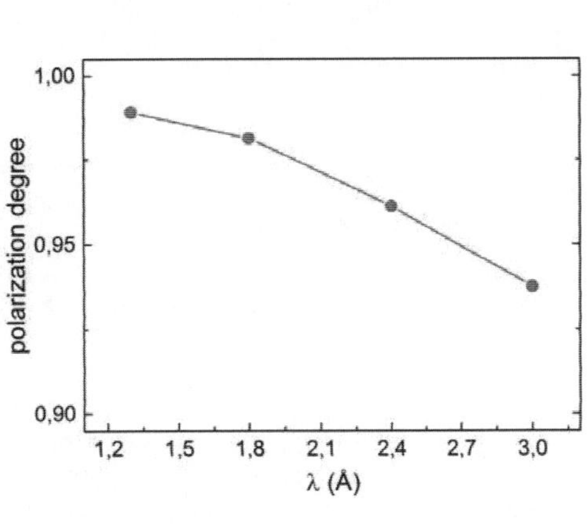

Figure 3. Polarization efficiency of a fan-like polarizing bender.

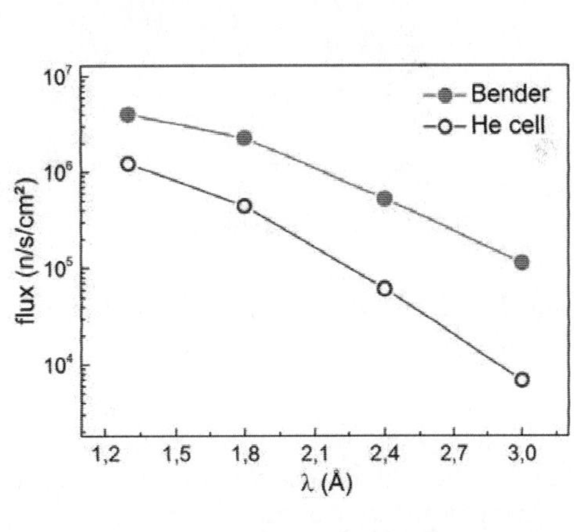

Figure 4. Neutron flux on the sample position for a fan-like polarizing bender and a helium cell.

Contrary to the bender set-up the polarization degree for the He-cell shows only a slight decrease of the polarization for higher wavelengths. For 3 Å the polarization degree decreases from 96% at 1.3 Å to a value of about 93%. On the another hand the intensity on the sample position at He cell set-up is 1.22×10^6 n/cm²sec ± 4×10^4 n/cm²sec at 1.3 Å, which is almost a factor 3.3 lower compared to the bender set up (see figure 4). For 3 Å, the flux decreases to 6.79×10^3 n/cm²sec ± 5×10^2 n/cm²sec, thus the ratio between intensities for the bender set-up and the helium cell increases considerably for longer wavelengths. The vertical divergences for both the He cell and the bender set-ups are approximately the same at 1.3 Å: ±0.65°, while at higher wavelengths it grows in different ways – up to ±1.3° at 3 Å for the He cell and to ±0.9° for the bender. Horizontal divergence for the bender set-up significantly better than for the He-cell beginning from1.3 Å, where it has a value of ±0.32° against ±0.44° (Figure 5).

3. Conclusion

The instrument layout of POLDI at the PIK reactor with a polarizing bender was compared to that of a He cell. The simulation demonstrates that bender set-up has higher flux at the sample position for a wavelength range from 1.3 Å to 3 Å and lower divergence distribution. For the bender set-up, is found that the polarization degree depends on the wavelength. For short wavelengths a polarization degree better than

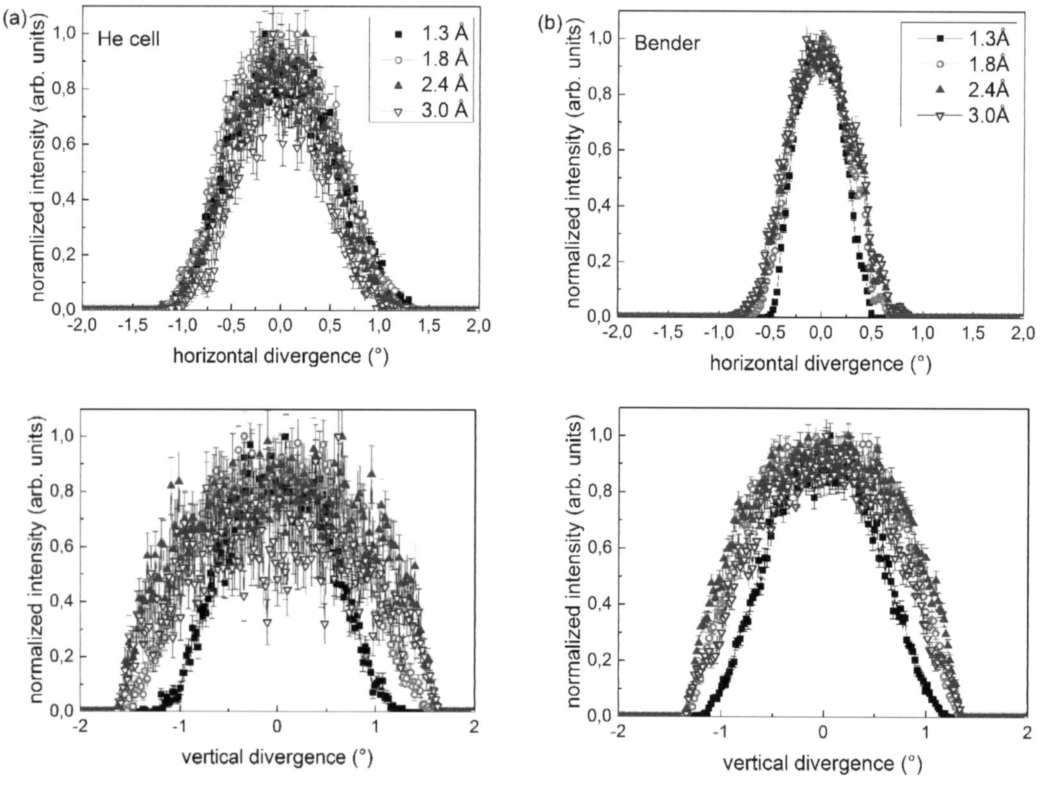

Figure 5. Horizontal and vertical divergences for different wavelengths for (a) helium cell set-up and (b) fan-like polarizing bender set-up.

98% can be reached, which drops for longer wavelengths down to about 94% for 3 Å. The polarizing He-cell on the other hand shows only a slight change of the polarization with the wavelength, here, the polarization degree decreases from 96% to only 93% for 1.3 Å and 3 Å, respectively. The higher flux at the sample position, a factor of at least 3.3, with lower divergence and still good polarization degree of above 93% makes the bender set-up favorable over the layout with the He-cell. The He-gas has to be polarized from time to time; consequently, the polarization degree is time dependent. It should be emphasized also that a bender acts as a $\lambda/2$ filter which gives one more benefit. Finally yet importantly, it can be noted that the bender set-up needs less service efforts.

Acknowledgment

The work is performed with financial support of Ministry of Science and Education of Russian Federation, the agreement on grant No. 14.616.21.0004 of September 17, 2014, a unique identifier of the agreement RFMEFI61614X0004.

References

[1] Tasset F 1989 Zero field neutron polarimetry. *Physica B* **156-157** 627
Tasset F 1994 Magnetic Structures and neutron polarimetry *JMMM* **129** 47

IOP Conf. Series: Journal of Physics: Conf. Series **862** (2017) 012031 doi:10.1088/1742-6596/862/1/012031

[2] Brown P J 2006 Spherical Neutron Polarimetry *Neutron Scattering from Magnetic Materials* ed T Chatterji (Amsterdam: Elsevier) p 215

[3] Schweizer J 2006 Polarized neutron Without Polarization analysis: The flipping Ratio Method *Neutron Scattering from Magnetic Materials* ed T Chatterji (Amsterdam: Elsevier) p 164

[4] Lefmann K and Nielsen K 1999 McStas, a General Software Package for Neutron Ray-tracing Simulations *Neutron News* **10** 20

Lefmann K and Nielsen K 2000 Monte Carlo simulations of neutron scattering instruments using McStas *Physica B* **283** 426

AUTHOR INDEX

AUTHOR INDEX